New Syllabus

# MATHEMATICS

## 6th Edition

# 1

Consultant:
Dr Yeap Ban Har

Authors:
Teh Keng Seng BSc, Dip Ed • Loh Cheng Yee 劳静仪 BSc, Dip Ed

shinglee publishers pte ltd

**SHINGLEE PUBLISHERS PTE LTD**
120 Hillview Avenue #05-06/07
Kewalram Hillview Singapore 669594
Tel: 67601388 Fax: 67623247
email: info@shinglee.com.sg
http://www.shinglee.com.sg



**First Published 1982**
**Reprinted 1983, 1984, 1985, 1986**
**Second Edition 1987**
**Reprinted 1987, 1988, 1989, 1990, 1991**
**Third Edition 1992**
**Reprinted 1992, 1993, 1994, 1995**
**Fourth Edition 1997**
**Reprinted 1997, 1999**
**Fifth Edition 2001**
**Reprinted 2002, 2003, 2004, 2005, 2006**
**Sixth Edition 2007**
**Reprinted 2007**

ISBN 978 981 237 311 3

Cover design by Dave Cheong

Printed in Singapore by KHL Printing Co Pte Ltd

# PREFACE

New Syllabus Mathematics is a series of four books. These books follow the Mathematics Syllabus for Secondary Schools, implemented from 2007 by the Ministry of Education, Singapore. The whole series covers the complete syllabus for the Singapore-Cambridge GCE 'O' Level Mathematics.

The sixth edition of New Syllabus Mathematics retains the goals and objectives of the previous edition, but has been revised to meet the needs of the users of the fifth edition and to keep materials up-to-date as well as to give students a better understanding of the contents.

All topics are comprehensively dealt with to provide students with a firm grounding in the subject. Explanations of concepts and principles are precise and written clearly and concisely with supportive illustrations and examples. Examples and exercises have been carefully graded to aid students in progressing within and beyond each level. Those exercises marked with a ✶ either require more thinking or involve more calculations.

Numerous revision exercises are provided at appropriate intervals to enable students to recapitulate what they have learnt.

Some interesting features of this series include the following:

- an interesting introduction at the beginning of each chapter complete with photographs or graphics.
- brief specific instructional objectives for each chapter.
- **Just for fun** arouses the students' interests in studying mathematics.
- **Thinking Time** encourages students to think creatively and go even deeper into the topics.
- **Exploration** provides opportunities for students to learn actively and independently.
- **For your information** provides extra information on mathematicians, mathematical history and events etc.
- **Problem Solving Tips** provide suggestions to help students in their thinking processes.
  We also introduce problem solving heuristics and strategies systemically throughout the series.
- **Your Attention** alerts students to misconceptions.

# CONTENTS


**1 Factors and Multiples** **1**
Factors and Multiples 3
Prime Numbers and Composite Numbers 4
Prime Factorisation 8
Index Notation 8
Highest Common Factor (HCF) 11
Least Common Multiple (LCM) 14
Squares and Square Roots 17
Cubes and Cube Roots 19
Mental Estimation 21
The Use of Calculators 22
Summary 25
Review Examples 1 26
Review Questions 1 27

**2 Integers** **29**
Negative Numbers 31
Integers 31
The Number Line 32
Absolute Value of an Integer 33
Addition of Integers 35
Subtraction of Integers 38
Multiplication of Integers 41
Division of Integers 44
Rules for Operating on Integers 45
Summary 48
Review Examples 2 48
Review Questions 2 49

**3 Rational Numbers** **51**
Rational Numbers 53
Ordering of Rational Numbers 54
Addition and Subtraction of Rational Numbers 55
Multiplication and Division of Rational Numbers 57
Arithmetical Operations on Rational Numbers 59
Problem Solving Involving Rational Numbers 61
Terminating and Recurring Decimals 64
Use of Calculators on Real Numbers 65
Summary 68
Review Examples 3 69
Review Questions 3 70

**4 Estimation and Approximation** **71**
Estimation and Rounding 73
Approximations in Measurements and Acccuracy 76
Rounding Off a Number to a Given Number of Decimal Places 78
Acccuracy and Significant Figures 80
Rounding a Number to a Given Number of Significant Figures 80
Rounding and Truncation Errors 83
Summary 86
Review Examples 4 87
Review Questions 4 88

**5 Fundamental Algebra** **89**
Notation in Algebra 91
Evaluation of Algebraic Expressions 94
Some Rules in Algebra 96
Use of Brackets in Simplification 98
Addition and Subtraction of Algebraic Expressions 100
Linear Algebraic Expressions with Fractional Coefficients 101
Factorisation 103
Factorisation by Grouping 105
Summary 106
Review Examples 5 107
Review Questions 5 108

**Revision Exercise I** **110**

**6 Number Sequences** **115**
Number Sequences 117

General Term in a Number Sequence 120
Problem Solving 125
Summary 133
Review Examples 6 134
Review Questions 6 135

**7 Algebraic Equations and Simple Inequalities 137**
Open Sentences 139
Simple Equations 139
Solving Simple Equations 141
Further Examples on Equations 145
Formulae 148
Construction of Formulae 151
Writing Algebraic Expressions 152
Problem Solving with Algebra 153
Inequalities 158
Properties of Inequalities 159
Equations and Inequalities 162
Summary 163
Review Examples 7 164
Review Questions 7 165

**8 Perimeter and Area of Simple Geometrical Figures 167**
Units of Area 169
Area of a Parallelogram 179
Area of a Trapezium 181
Summary 187
Review Examples 8 187
Review Questions 8 189

**9 Volume and Surface Area 191**
Concept of Volume 194
Volume of Fluids 194
Right Prisms 201
Volume of a Prism 202
Surface Area of a Prism 203
Cylinders 206
Volume of a Cylinder 207
Surface Area of a Cylinder 207
Summary 212
Review Examples 9 213
Review Questions 9 214

**Revision Exercise II 217**

**10 Ratio, Rate and Speed 223**
Ratio 225
Equivalent Ratios 225
Increase and Decrease in Ratio 228
Rate 232
Average Rate 234
Time 235
Speed and Average Speed 238
Problems Involving Speed and Average Speed 239
Summary 243
Review Examples 10 243
Review Questions 10 245

**11 Percentages 247**
Percentages, Fractions and Decimals 249
Expressing One Quantity as a Percentage of Another 250
Finding the Percentage of a Number 250
Comparing Two Quantities by Percentages 252
Percentages Greater than 100% 253
Increasing/Decreasing a Quantity by a Given Percentage 255
Discount 259
Commission 260
Value-added Tax and GST 261
Summary 263
Review Examples 11 263
Review Questions 11 265

**12 Functions and Graphs 267**
Rectangular Coordinates in Two Dimensions 269
The Rectangular or Cartesian Plane 270
Coordinates of a Point 271
The Idea of Functions 273
Ordered Pairs Satisfying a Function 274
Gradient of a Straight Line 280
Summary 286
Review Examples 12 286
Review Questions 12 288

**13 Statistics** **289**
Introduction to Numerical Data 291
Collection, Organisation and Interpretation of Data 291
Collection of Data Using a Questionnaire 293
Pictograms 294
Bar Graphs 295
Collection of Data Through Observation 300
Pie Charts 301
Collection of Data Through Interviews 304
Line Graphs 307
Frequency Tables and Histograms 310
Collection of Data by Measuring 313
Collection of Data by Using Electronic Means and the Internet 315
Grouped Frequency Table 315
Summary 320
Review Examples 13 320
Review Questions 13 322

**Revision Exercise III** **325**

**14 Basic Geometrical Concepts and Properties** **331**
Points 333
Lines, Rays, Line Segments 333
Planes 334
Intersecting Lines 334
Angles 334
The Protractor and Angle Measure 335
Different Kinds of Angles 336
Complementary Angles 338
Supplementary Angles 338
Adjacent Angles on a Line 338
Vertically Opposite Angles 339
Parallel Lines, Alternate Angles, Corresponding Angles, Interior Angles 343
Summary 350
Review Examples 14 351
Review Questions 14 353

**15 Angle Properties of Polygons** **355**
Polygons 357
Triangles 358
Angle properties of Triangles 359
Exterior and Interior Opposite Angles 359
Quadrilaterals 364
Convex Polygons 370
Sum of the Interior Angles of a Convex Polygon 371
Sum of the Exterior Angles of a Convex Polygon 372
Summary 376
Review Examples 15 377
Review Questions 15 377

**16 Geometrical Constructions** **381**
Geometrical Constructions 383
Use of Compasses 383
Bisecting an Angle 384
Bisecting a Line Segment 386
Summary 396
Review Examples 16 397
Review Questions 16 399

**Revision Exercise IV** **400**

**Answers** **406**

*In this chapter, you will learn*

- ***about prime numbers;***
- ***how to find the highest common factor (HCF) and the lowest common multiple (LCM) of two or more numbers;***
- ***how to find squares, square roots, cubes and cube roots of numbers.***

CHAPTER

1

# Factors and Multiples

## Introduction

In the picture, the alarm clock and the two watches display the same time. Suppose the alarm clock shows the correct time, while the two watches do not. If, for every minute, the watch on the left is slower by 5 seconds, while that on the right is faster by 5 seconds, how many minutes later do you think all three will display the same time again?

After learning this chapter, you will be able to solve this problem.

## Explorations

You will need a box of toothpicks for the activity.

**1.** Start with 10 toothpicks. Can you arrange them into groups of equal numbers of toothpicks with no leftovers?

**2.** Repeat the above with 1, 2, 3, ..., 9 toothpicks and also with 11, 12, 13, ..., 20 toothpicks.

**3.** Classify the numbers 1, 2, 3, ..., 20 according to how each number can be arranged by copying and completing the following table:

| List I | List II | List III |
|---|---|---|
| The number 1:<br>1 group of<br>1 toothpick or<br>$1 = 1 \times 1$ | The number 2:<br>1 group of 2 toothpicks;<br>2 groups of 1 toothpick or<br>$2 = 1 \times 2 = 2 \times 1$<br><br>The number 3:<br>1 group of 3 toothpicks;<br>3 groups of 1 toothpick or<br>$3 = 1 \times 3 = 3 \times 1$ | The number 4:<br>1 group of 4 toothpicks; 4 groups of<br>1 toothpick; 2 groups of 2 toothpicks or<br>$4 = 1 \times 4 = 4 \times 1 = 2 \times 2$<br><br>The number 6:<br>1 group of 6 toothpicks; 6 groups of<br>1 toothpick; 2 groups of 3 toothpicks;<br>3 groups of 2 toothpicks or<br>$6 = 1 \times 6 = 6 \times 1 = 2 \times 3 = 3 \times 2$ |

## Factors and Multiples

From the above activity for the number 18, we have

$$18 = \mathbf{1} \times 18 = \mathbf{2} \times 9 = \mathbf{3} \times 6 = \mathbf{6} \times 3 = \mathbf{9} \times 2 = \mathbf{18} \times 1.$$

We call each of the numbers 1, 2, 3, 6, 9 and 18 a **factor** of 18. Conversely, we call 18 a **multiple** of each of the numbers 1, 2, 3, 6, 9 and 18. Clearly, when 18 is divided by any one of its factors, the remainder is zero. We say that 18 is **divisible** by 1, 2, 3, 6, 9 and 18.

## Example 1

*(a) List the factors of 60.* *(b) List the multiples of 5.*

### Solution

(a) We have $60 = 1 \times 60 = 2 \times 30 = 3 \times 20 = 4 \times 15 = 5 \times 12 = 6 \times 10$.
Thus, the factors of 60 are 1, 2, 3, 4, 5, 6, 10, 12, 15, 20, 30, 60.

(b) By multiplying 5 with 1, 2, 3, 4 and so on, we obtain the multiples of 5, i.e. the multiples of 5 are 5, 10, 15, 20, ....

# Prime Numbers and Composite Numbers

Refer to the investigation on page 3, did you notice that each of your List II numbers has exactly **two** different factors, 1 and itself? Such numbers are called **prime numbers**.

**A number which has only two different factors, 1 and the number itself, is a prime number.**

For example: 2, 3, 5, 7, 11, ...

Each of the List III numbers has more than two different factors. Such numbers are known as **composite numbers**.

**A number which has more than two different factors is a composite number.**

For example: 4, 6, 8, 9, 10, ...

### Thinking Time

The number 1 is neither a prime number nor a composite number. Why?

## Explorations

You may work on this activity with a partner.

**1.** Copy on a piece of paper the numbers from 1 to 100 inclusive and arrange them in 10 rows as shown below.

| | | | | | | | | | |
|---|---|---|---|---|---|---|---|---|---|
| 1 | 2 | 3 | 4 | 5 | 6 | 7 | 8 | 9 | 10 |
| 11 | 12 | 13 | 14 | 15 | 16 | 17 | 18 | 19 | 20 |
| 21 | 22 | 23 | 24 | 25 | 26 | 27 | 28 | 29 | 30 |
| 31 | 32 | 33 | 34 | 35 | 36 | 37 | 38 | 39 | 40 |
| 41 | 42 | 43 | 44 | 45 | 46 | 47 | 48 | 49 | 50 |
| 51 | 52 | 53 | 54 | 55 | 56 | 57 | 58 | 59 | 60 |
| 61 | 62 | 63 | 64 | 65 | 66 | 67 | 68 | 69 | 70 |
| 71 | 72 | 73 | 74 | 75 | 76 | 77 | 78 | 79 | 80 |
| 81 | 82 | 83 | 84 | 85 | 86 | 87 | 88 | 89 | 90 |
| 91 | 92 | 93 | 94 | 95 | 96 | 97 | 98 | 99 | 100 |

**(a)** Cross out the number 1.
**(b)** Circle the number 2 and cross out all the other multiples of 2.
**(c)** Circle the number 3 and cross out all the other multiples of 3.
**(d)** Circle the number 5 and cross out all the other multiples of 5.
**(e)** Circle the number 7 and cross out all the other multiples of 7.
**(f)** Continue the process until all numbers are either circled or crossed out.

**2.** Answer the following questions:
**(a)** What is the reason for crossing out the number 1?
**(b)** What are the circled numbers?
**(c)** What are the numbers crossed out?
**(d)** List the prime numbers that are less than 100.
**(e)** What is the smallest prime number?
**(f)** What is the largest prime number less than 100?
**(g)** What are the first 20 composite numbers?
**(h)** Is every odd number a prime number?
**(i)** Is every even number a composite number?

The above process of finding all the prime numbers less than a given number is called the **Sieve of Eratosthenes** in honour of a Greek mathematician, Eratosthenes.

**3.** There is a statement which says that "every even number greater than 2 can be expressed as the sum of two prime numbers". For example, $4 = 2 + 2$, $8 = 3 + 5$ and $12 = 7 + 5$. This statement is called **Goldbach's Conjecture**. It is a conjecture because it has not yet been proven.

For each of the following numbers, verify Goldbach's Conjecture by expressing the number as a sum of two prime numbers:
**(a)** 16 **(b)** 36
**(c)** 64 **(d)** 98

**Just for fun**

*37 and 73 are prime numbers with reversed digits. Can you name another pair of two-digit prime numbers with reversed digits?*

4. Prime numbers such as 5 and 7 that differ by 2 are called **twin primes**. Mathematicians believe that there is an infinite number of pairs of twin primes. 1 000 000 061 and 1 000 000 063 are twin primes. List five other pairs of twin primes.

### Thinking Time

Can the product of two prime numbers be
**(a)** an odd number; **(b)** an even number; **(c)** a prime number?

## Exercise 1a

1. Write down all the factors of each of the following.
   **(a)** 16 **(b)** 28 **(c)** 96
   **(d)** 100 **(e)** 120 **(f)** 210

2. Write down the first six multiples of the following numbers.
   **(a)** 4 **(b)** 7 **(c)** 9
   **(d)** 12 **(e)** 17 **(f)** 21

3. Identify the numbers below that have 18 as a factor.

   54, 126, 198, 240, 320

4. Identify the numbers below which are factors of 144.

   1, 2, 3, 4, 8, 9, 12, 14, 16, 32, 48, 144

5. Identify the multiples of 8 from the following numbers.

   14, 24, 32, 54, 56, 36, 72, 30, 64, 18, 40, 78, 96, 108, 120

6. Identify the numbers that have 224 as a multiple in the following.

   3, 4, 12, 14, 24, 28, 32, 36, 56

7. Use a calculator to find all the factors of the following numbers.
   **(a)** 480 **(b)** 600 **(c)** 960
   **(d)** 936

8. John has 48 orange-flavoured sweets and Susan has 45 lime-flavoured sweets.
   **(a)** John wishes to divide his sweets equally into bags. List all the possible ways he can do this. (For example, he can have 6 bags of 8 sweets.)
   **(b)** Susan also wishes to divide her sweets equally into bags. List all the possible ways she can do this.
   **(c)** Peter, their good friend, suggests that they combine the sweets and divide them equally into bags in such a way that each bag has equal number of orange-flavoured and lime-flavoured sweets. Explain how this can be done.

9. Determine whether each of the following is a prime number or a composite number:
   **(a)** 2 **(b)** 15 **(c)** 17
   **(d)** 21 **(e)** 27 **(f)** 29

10. Name the next five prime numbers after 30.

11. Find two prime numbers whose sum is an odd number.

## Explorations

We all know that if the digit of an integer in the ones place is 0, 2, 4, 6, 8, then the number is divisible by 2.

Or we say if the digit of a number in the ones place is divisible by 2, then the number is divisible by 2.

This makes it much easier to determine if a number is divisible by 2.
Can we also find easier ways to determine if a number is divisible by 3, 9, 11 without using a calculator?

1. Divisibility by 3.
   Is 1353 divisible by 3?
   (1) Use a calculator to check if 1353 is divisible by 3.
   (2) Calculate the sum of all digits of 1353, that is 1 + 3 + 5 + 3.
   Is the sum divisible by 3?
   (3) Repeat (1) and (2) for the numbers: 69 261, 324 126, 31 568.
   What conclusion can you get? Complete the following statement.
   A number is divisible by 3 if the sum of all the digits of the number is _____.

2. Divisibility by 9.
   Is 46 998 divisible by 9?
   (1) Use a calculator to check if 46 998 is divisible by 9.
   (2) Calculate the sum of all digits of 46 998.
   Is the sum divisible by 9?
   (3) Repeat (1) and (2) for the numbers: 493 101, 561 321, 390 152.
   What conclusion can you get? Complete the following statement.
   A number is divisible by 9 if the sum of all the digits of the number is _____ .

3. Divisibility by 11.
   Is 35 728 divisible by 11?
   (1) Calculate the sum of the digits in the odd places (ones, hundreds, ten thousands,......).
   (2) Calculate the sum of the digits in the even places (tens, thousands,......).
   (3) Calculate the difference between the two sums you have obtained in (1) and (2).
   Is the difference divisible by 11?
   (4) Repeat (1), (2), (3) for the numbers: 493 156, 325 413, 421 517.
   What conclusion can you get? Complete the following statement.
   A number is divisible by 11 if the difference between the sum of the digits in the odd places and the sum of the digits in the even places is _____ .

4. Palindromic numbers.
   A palindromic number is one which reads the same forwards or backwards. Some examples are: 121, 2332, 1234321, etc.

   Try to list more palindromic numbers with an even number of digits. Are they divisible by a common number? What is it?

**Thinking Time**

1. If a number is divisible by 8, must it also be divisible by 2? By 4? Explain your answer.

2. If a number is divisible by 3, must it also be divisible by 9? Explain your answer.

**Just for Fun**

*Han Sin, a Chinese general, devised a method to count the number of soldiers he had. First, he ordered his soldiers to form groups of 3 followed by groups of 5 and then groups of 7. In each case, he noted down the remainder. Using the three remainders, he was able to calculate the exact number of soldiers he had without doing the actual counting. Do you know how he did it?*

## Prime Factorisation

Every number (except 1) is either a prime number or a composite number. A composite number can be expressed as the product of two or more prime numbers, which are called **prime factors**. The **decomposition** of a composite number into prime factors is known as **prime factorisation**.

For example, $30 = 2 \times 3 \times 5$
and $60 = 2 \times 2 \times 3 \times 5$.

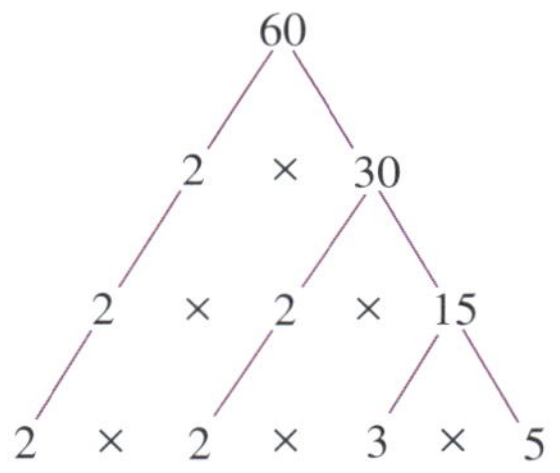


2, 3 and 5 are called the prime factors of 30. They are also the prime factors of 60. A **factor tree** can be used to express a composite number as a product of its prime factors.

The factor tree illustrates the prime factorisation of 60.

## Index Notation

To be brief, we write $5 \times 5$ as $5^2$ and read it as *5 squared* or *square of 5*.

$5 \times 5 \times 5$ is written as $5^3$ and is read as *5 cubed* or *the cube of 5*.

$5 \times 5 \times 5 \times 5$ is written as $5^4$ and is read as *5 to the power of 4*.

Similarly, $2 \times 2 \times 2 \times 2 \times 2 \times 2$ is written as $2^6$ and is read as *2 to the power of 6*.

In general, $\underbrace{a \times a \times \ldots \times a}_{n \text{ factors}}$ is written as $a^n$ and is read as *a to the power of n*.

This index notation $a^n$ gives us a more precise method of expressing the factors of a number.

For example, $12 = 2 \times 2 \times 3$ can be written as $12 = 2^2 \times 3$
$40 = 2 \times 2 \times 2 \times 5$ can be written as $40 = 2^3 \times 5$
$72 = 2 \times 2 \times 2 \times 3 \times 3$ can be written as $72 = 2^3 \times 3^2$

## Example 2

*Express 252 in prime factors.*

### Solution

**Method 1**

252
$2 \times 126$
$2 \times 2 \times 63$
$2 \times 2 \times 3 \times 21$
$2 \times 2 \times 3 \times 3 \times 7$

From the above factor tree,
we have $252 = 2 \times 2 \times 3 \times 3 \times 7$
$= 2^2 \times 3^2 \times 7$ (using index notation)

**Method 2**

| | |
|---|---|
| 2 | 252 |
| 2 | 126 |
| 3 | 63 |
| 3 | 21 |
| 7 | 7 |
| | 1 |

(Start by dividing the number by the smallest prime factor and continue dividing until you get 1 as shown.)

$\therefore\ 252 = 2 \times 2 \times 3 \times 3 \times 7$
$= 2^2 \times 3^2 \times 7$

## Explorations

1. The early Greeks called some numbers "perfect". A perfect number is one that is equal to the sum of its factors excluding the number itself. For example, the factors of 6 are 1, 2, 3 and 6. We exclude 6 and we have $1 + 2 + 3 = 6$. So 6 is a perfect number. Man has sought "perfect numbers" throughout the ages but found very few. Can you look up books in the library or from the internet for more perfect numbers?

2. If you wish to convey a friendship day greeting to someone to express sincere friendship, you can send a card on which is written something like this:
   Fandi – 2620
   Minghui – 2924
   You can explain that 2620 and 2924 are amicable numbers. Find out more about amicable numbers from your library books or from the internet and tell your friend what they are.

## Exercise 1b

1. Express the following using index notation:
   (a) $7 \times 7$
   (b) $2 \times 2 \times 5 \times 5$
   (c) $3 \times 7 \times 7 \times 7$
   (d) $5 \times 5 \times 11 \times 11 \times 11$
   (e) $2 \times 2 \times 13 \times 13 \times 31 \times 2$
   (f) $5 \times 5 \times 5 \times 19 \times 29 \times 19 \times 23 \times 29$

2. Express each of the following as a product of prime factors using index notation:
   (a) 28 (b) 48
   (c) 54 (d) 88
   (e) 108 (f) 144
   (g) 192 (h) 256

3. Complete the following factor trees:
   (a)

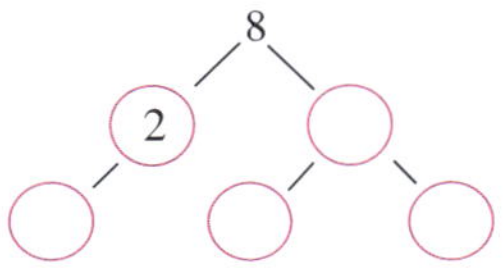


$8 = \_\_\_ \times \_\_\_ \times \_\_\_$

   (b)

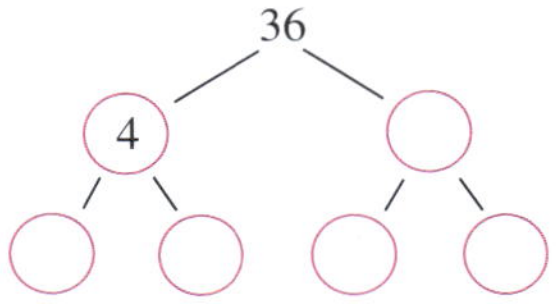


$36 = \_\_\_ \times \_\_\_ \times \_\_\_ \times \_\_\_$

   (c)

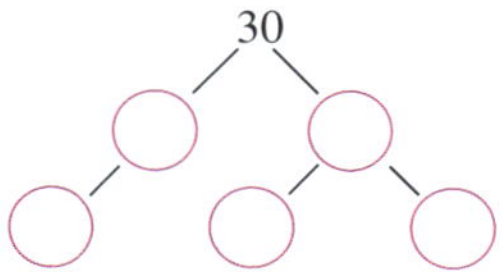


$30 = \_\_\_ \times \_\_\_ \times \_\_\_$

   (d)

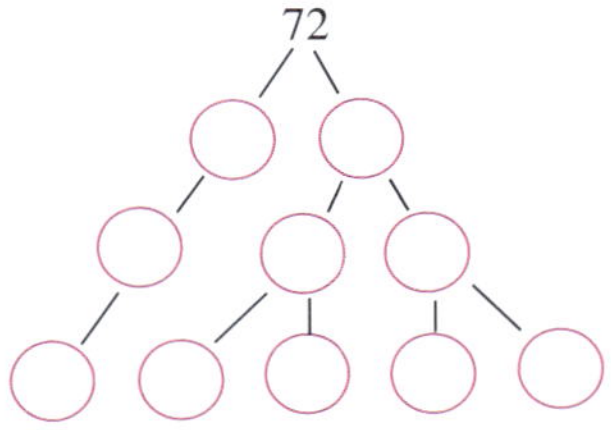


$72 = \_\_\_ \times \_\_\_ \times \_\_\_ \times \_\_\_ \times \_\_\_$

4. Factorise each of the following numbers into prime factors:
   (a) 100
   (b) 125
   (c) 147
   (d) 216
   (e) 225
   (f) 360
   (g) 567
   (h) 648

# Highest Common Factor (HCF)

Explorations

Suppose Mary, an art elective program student, is working on an assignment. She plans to cover a 30 cm by 36 cm sheet of paper completely with identical square patterns. Can you help her find the side of the largest possible square?

Step 1: Divide each side into groups of equal lengths. This is equivalent to finding the factors of 30 and 36. Making a complete list of the factors from the smallest to the largest, we have:

| | | | | | |
|---|---|---|---|---|---|
| The factors of 30 are | 1 | 2 | 3 | 5 | _____, _____, _____, _____ |
| The factors of 36 are | 1 | 2 | 3 | 4 | _____, _____, _____, _____, _____ |

Step 2: The common factors of 30 and 36 are _____, _____, _____ and _____. (The common factors are factors of both 30 and 36.)

Step 3: The largest common factor of 30 and 36 is _____.

Step 4: Hence the side of the largest possible square is _____.

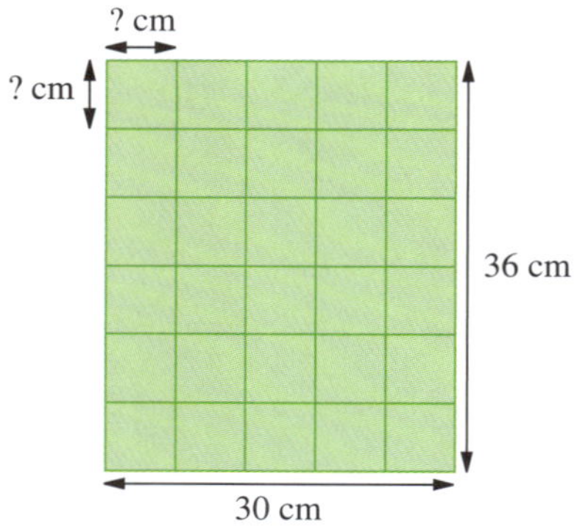


In Step 3 on page 11, the largest common factor of 30 and 36 is called the Highest Common Factor (HCF) of 30 and 36.

**Thinking Time**

The HCF of a set of numbers will be less than the numbers or equal to one of the numbers. Why?

Listing all the possible factors of numbers to find the HCF of the numbers as shown above can be tedious. The following provides two alternative methods:

**Method 1**

The prime factorisation of 30 and 36 is shown below:

$$30 = \boxed{2} \times \boxed{3} \times 5$$
$$36 = \boxed{2^2} \times \boxed{3^2}$$

$2 \times 3$ ← Identify the common factors.

$\therefore$ the HCF of 30 and 36 is $2 \times 3 = 6$.

**Method 2**

| | | | |
|---|---|---|---|
| Step 1 | 2 | 30 | 36 |
| Step 2 | 3 | 15 | 18 |
| Step 3 | | 5 | 6 ← 1 is their only common factor. Stop here. |

HCF = $\boxed{2 \times 3} = 6$

In step 1, we know that 30 and 36 are divisible by 2 because they are both even numbers.

In step 2, we divide 30 and 36 by 2 and write the quotients (15 and 18). We know that 15 and 18 are divisible by 3.

In step 3, we divide 15 and 18 by 3 and write the quotients (5 and 6).

Since 5 and 6 have no common factors except 1, we stop our computation.

We multiply the numbers listed on the left side of the above computation: $2 \times 3 = 6$.

$\therefore$ the HCF of 30 and 36 is 6.

## Example 3

*Find the HCF of 60, 180 and 210.*

### Solution

**Method 1**

$$60 = 2 \times 2 \times 3 \times 5 = \boxed{2^2} \times \boxed{3} \times \boxed{5}$$
$$180 = 2 \times 2 \times 3 \times 3 \times 5 = \boxed{2^2} \times \boxed{3^2} \times \boxed{5}$$
$$210 = 2 \times 3 \times 5 \times 7 = \boxed{2} \times \boxed{3} \times \boxed{5} \times 7$$

2 3 5 ← Identify the common factors.

$\therefore$ the HCF of 60, 180 and 210 is $2 \times 3 \times 5 = 30$.

**Method 2**

| | | | |
|---|---|---|---|
| 2 | 60 | 180 | 210 |
| 3 | 30 | 90 | 105 |
| 5 | 10 | 30 | 35 |
| | 2 | 6 | 7 |

← 2, 6 and 7 have no common factors except 1. Stop here.

HCF = $\boxed{2 \times 3 \times 5} = 30$

$\therefore$ the HCF of 60, 180 and 210 is 30.

## Exercise 1c

1. Find all the common factors of:
   **(a)** 6 and 9 **(b)** 15 and 18
   **(c)** 27 and 36 **(d)** 45 and 75

2. Find the HCF of the following:
   **(a)** 12 and 30 **(b)** 15 and 75
   **(c)** 16 and 40 **(d)** 24 and 64
   **(e)** 24 and 108 **(f)** 28 and 56
   **(g)** 36 and 243 **(h)** 324 and 128

3. Find the HCF of:
   **(a)** 27, 63 and 207
   **(b)** 84, 63 and 126
   **(c)** 48, 72 and 132
   **(d)** 30, 75, 90 and 135
   **(e)** 36, 168, 144 and 252
   **(f)** $2^3 \times 3^2 \times 5$ and $2^2 \times 3^4 \times 5^3$
   **(g)** $2 \times 5^2 \times 7$ and $2^3 \times 3^4 \times 5^3 \times 7^2$

4. James wants to cover a floor measuring 90 cm by 120 cm with square tiles of the same size. Given that he uses only whole tiles, find
   **(a)** the largest possible length of the side of each tile;
   **(b)** the number of tiles that are needed to cover the floor.

5. Paul has three pieces of rope with lengths of 140 cm, 168 cm and 210 cm. He wishes to cut the three pieces of rope into smaller pieces of equal length with no remainders.

**(a)** What is the greatest possible length of each of the smaller pieces of rope?

**(b)** How many of the smaller pieces of rope of equal length can he get altogether?

# Least Common Multiple (LCM)

## Explorations

Mary, the art elective program student, is working on a second assignment. She first designs a rectangular pattern measuring 9 cm by 12 cm. She then makes copies of the rectangular pattern. Next she uses the rectangular patterns to form a square. How many rectangular patterns does she need to form the smallest square? What is the length of a side of this square?

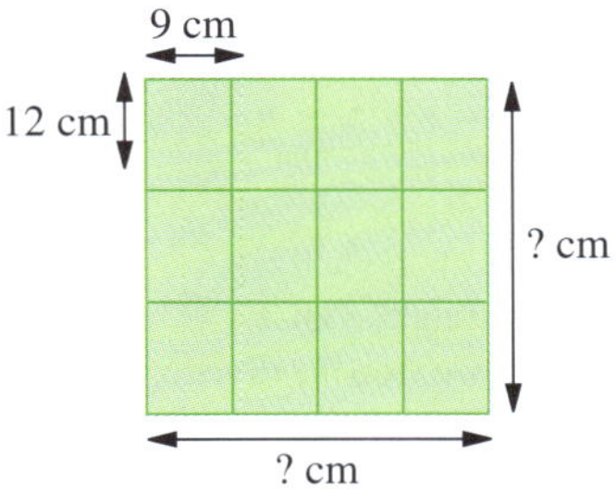


**Just For Fun**

*Each of the following letters represents a prime number. Find out what they are.*

```
      ab
  ×  ccb
  ------
     bdb
    ddb
   ddb
  ------
   dbdab
```

Step 1: Mary first considers the possible multiples of 9 and 12.

The multiples of 9 are 9 18 27 36 ___ ___ ___ ___ ___ ___ ___ ___ .. .. ..

The multiples of 12 are 12 24 36 ___ ___ ___ ___ ___ ___ ___ ___ ___ .. .. ..

Step 2: The first three multiples common to 9 and 12 are 36, _____ and _____.

Step 3: The smallest of all the common multiples of 9 and 12 is _____ and we call _____ the **Least Common Multiple (LCM)** of 9 and 12.

Step 4: In conclusion, we now know that the side of the smallest square is 36 cm. The number of rectangular patterns needed to create such a square is $4 \times 3 = 12$.

## Example 4

*Find the LCM of 30 and 36.*

**Solution**

Multiples of 30 = {30, 60, 90, 120, 150, 180, 210, ...}

Multiples of 36 = {36, 72, 108, 144, 180, 216, ...}

$\therefore$ the LCM of 30 and 36 is 180.

The above method of finding the LCM of two numbers is tedious. The following are two simpler methods:

**Method 1**

Using prime factorisation, we have

$$\begin{array}{rcccccl} 30 = & 2 & \times & 3 & \times & 5 & \\ 36 = & 2^2 & \times & 3^2 & & & \\ & \downarrow & & \downarrow & & \downarrow & \text{Choose the bigger number from each set.} \\ & 2^2 & \times & 3^2 & \times & 5 & \end{array}$$

$\therefore$ the LCM of 30 and 36 is $2^2 \times 3^2 \times 5 = 180$.

> **Notes**
>
> *2 is a factor of 30.*
> *$2^2$ is a factor of 36.*
> *$\therefore$ $2^2$ has to be a factor of the LCM of 30 and 36.*
> *Similary, $3^2$ and 5 are factors of the LCM of 30 and 36.*

**Method 2**

$$\begin{array}{r|rr} 2 & 30 & 36 \\ \hline 3 & 15 & 18 \\ \hline & 5 & 6 \end{array}$$

(Carry out the division as in the case of finding the HCF. Since 5 and 6 have no more common factor, we stop the division.)

$\longrightarrow$ We see that 2, 3, 5 and 6 are factors of 30 and 36. Therefore, multiply these numbers to obtain the LCM.

$\therefore$ the LCM of 30 and 36 is $2 \times 3 \times 5 \times 6 = 180$.

## Example 5

*Find the LCM of 18, 24 and 36.*

### Solution

Using prime factorisation, we have

$$\begin{aligned}
18 &= 2 \times 3 \times 3 &&= \boxed{2} \times \boxed{3^2} \\
24 &= 2 \times 2 \times 2 \times 3 &&= \boxed{2^3} \times \boxed{3} \\
36 &= 2 \times 2 \times 3 \times 3 &&= \boxed{2^2} \times \boxed{3^2}
\end{aligned}$$

$\downarrow$ $2^3$ $\quad \downarrow$ $3^2$ — Choose the largest number from each set.

$\therefore$ the LCM of 18, 24 and 36 is $2^3 \times 3^2 = 72$.

### Thinking Time

1. What do you notice about the LCM of a set of numbers?

2. Can the LCM be smaller than one of the numbers?

## Exercise 1d

1. Find the LCM of each of the following pairs of numbers:
   - **(a)** 3 and 7 **(b)** 6 and 9
   - **(c)** 7 and 21 **(d)** 12 and 9
   - **(e)** 24 and 18 **(f)** 65 and 135
   - **(g)** 100 and 75 **(h)** 144 and 36
   - **(i)** 250 and 125 **(j)** 63 and 490

2. Find the LCM of:
   - **(a)** 6, 9 and 15 **(b)** 14, 18 and 21
   - **(c)** 28, 44 and 68 **(d)** 65, 175 and 135
   - **(e)** 18, 12, 6 and 8 **(f)** 6, 12, 9 and 24

3. Find the LCM of each of the following pairs of numbers:
   - **(a)** $2^2 \times 3^3 \times 5^4$ and $2 \times 3^4 \times 5^3 \times 7$
   - **(b)** $2^3 \times 3^4 \times 5$ and $2^2 \times 3^3 \times 5^2$
   - **(c)** $2^2 \times 5 \times 7$ and $2^3 \times 3^2 \times 5^2 \times 11$

4. Find the HCF and LCM of each of the following:
   **(a)** 18 and 42 **(b)** 21 and 28
   **(c)** 26 and 39 **(d)** 140 and 210
   **(e)** 150 and 45 **(f)** 336 and 224

5. Two lighthouses flash their lights every 20 seconds and 30 seconds respectively. Given that they flash together at 8 p.m., when will they next flash together?

6. Three bells toll at intervals of 8 minutes, 15 minutes and 24 minutes respectively. If they toll together at 3 p.m., at what time will they next toll together again?

✶ 7. David was trying to sleep one night but there was too much noise around him. His clock ticked every 5 seconds; a tap was dripping every 7 seconds and his pet dog snored every 12 seconds.
He noticed on his clock that all three things happened together on the stroke of midnight.
   **(a)** After how many seconds would all three things happen together again?
   **(b)** How many times would all three things happen together again between midnight and one o'clock?

✶ 8. Two cars start to drive around a 2-km track at the same time. Car $X$ makes one lap every 80 s, while car $Y$ makes one lap every 60 s.
   **(a)** How long will it take for the two cars to be at their starting position again? Give your answer in minutes.
   **(b)** How long will it take the faster car to be ahead by 15 laps? Give your answer in hours.

# Squares and Square Roots

The area of a square of side 6 cm is given by

$6 \times 6 = 36 \text{ cm}^2$.

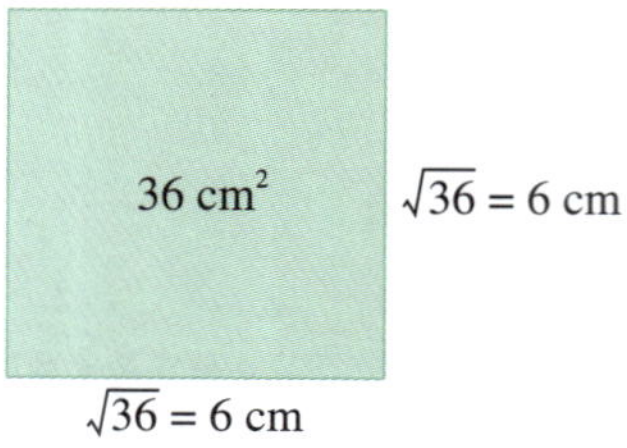


Therefore 36 is said to be the **square** of 6. We write $6^2 = 36$ and we read 'the square of 6 is 36' or simply **'6 squared is 36'**.

Clearly, to find the side of a square whose area is 36 cm², we find a positive number $x$ such that $36 = x \times x$ or $x^2$.

Now, $x = 6$ and we say that 6 is the positive **square root** of 36 and we write $\sqrt{36} = 6$.

Similarly,
$1 \times 1 = 1^2 = 1$ and $\sqrt{1} = 1$
$2 \times 2 = 2^2 = 4$ and $\sqrt{4} = 2$
$3 \times 3 = 3^2 = 9$ and $\sqrt{9} = 3$
$4 \times 4 = 4^2 = 16$ and $\sqrt{16} = 4$

**Notes**

*1, 4, 9 and 16 are perfect squares.*

*We use the symbol* $\sqrt{\phantom{x}}$ *to denote a square root.*

Notice that 1, 4, 9 and 16 are squares of whole numbers. These numbers are called **perfect squares**.

In general, if a number $y$ can be expressed as $y = x^2$, we say that $y$ is the square of $x$ and $x$ is a square root of $y$.

**Just for fun**

*Find the two-digit number which has the square of the sum of its digits equal to the number obtained by reversing its digits.*

## Example 6

*Find the positive square root of (a) 784; (b) 2025.*

**Solution**

Working: Use prime factorisation

(a)
$$\begin{aligned} 784 &= (2 \times 2) \times (2 \times 2) \times (7 \times 7) \\ &= (2 \times 2 \times 7)^2 \\ \sqrt{784} &= \sqrt{(2 \times 2 \times 7)^2} \\ &= \sqrt{28^2} \\ &= 28 \end{aligned}$$

| | |
|---|---|
| 2 | 784 |
| 2 | 392 |
| 2 | 196 |
| 2 | 98 |
| 7 | 49 |
| | 7 |

(b)
$$\begin{aligned} 2025 &= 5 \times 5 \times 3 \times 3 \times 3 \times 3 \\ &= (5 \times 3 \times 3)^2 \\ \sqrt{2025} &= \sqrt{(5 \times 3 \times 3)^2} \\ &= \sqrt{45^2} \\ &= 45 \end{aligned}$$

| | |
|---|---|
| 5 | 2025 |
| 5 | 405 |
| 3 | 81 |
| 3 | 27 |
| 3 | 9 |
| | 3 |

## Example 7

*Find the square of (a) 15; (b) 35.*

**Solution**

(a) $15^2 = 15 \times 15$
$= 225$

(b) $35^2 = 35 \times 35$
$= 1225$

**Thinking Time**

Can you calculate $55^2$ and $95^2$ mentally?

# Cubes and Cube Roots

## Explorations

The 1st, 2nd and 10th Cube Numbers are stated below:

$$1 \times 1 \times 1 = 1$$
$$2 \times 2 \times 2 = 8$$
$$10 \times 10 \times 10 = 1000$$

**1.** Write down the 3rd, 4th and 5th Cube Numbers.

**2.** **(a)** Find the square roots of the 3rd, 4th and 5th Cube Numbers.
**(b)** Which Cube Number has a whole number as its square root?
**(c)** Find another Cube Number which has a whole number as its square root.

**3.** In conclusion, some Cube Numbers have square roots which are whole numbers. The other Cube Numbers have square roots which are **not** whole numbers.
Explain whether or not the square root of the 16th Cube Number is a whole number.

Now,

$$1 \times 1 \times 1 = 1^3 = 1 \quad \text{and} \quad \sqrt[3]{1} = 1$$
$$2 \times 2 \times 2 = 2^3 = 8 \quad \text{and} \quad \sqrt[3]{8} = 2$$
$$3 \times 3 \times 3 = 3^3 = 27 \quad \text{and} \quad \sqrt[3]{27} = 3$$
$$4 \times 4 \times 4 = 4^3 = 64 \quad \text{and} \quad \sqrt[3]{64} = 4$$

**Notes**

*We use the symbol* $\sqrt[3]{\phantom{x}}$ *to denote a cube root.*

Notice that 1, 8, 27 and 64 are cubes of whole numbers. These numbers are called **perfect cubes**.

In general, if a number $y$ can be expressed as $y = x^3$, we say that $y$ is the cube of $x$ and $x$ is the cube root of $y$.

## Thinking Time

**1.** What is the value of $0^3$?

**2.** What is the value of $\sqrt[3]{0}$?

## Example 8

*Find the cube root of (a) 512; (b) 5832.*

> *Just for fun*
>
> *Find the two-digit number which has the sum of the cubes of its digits equal to three times itself.*

### Solution

(a) $512 = (2 \times 2 \times 2) \times (2 \times 2 \times 2) \times (2 \times 2 \times 2)$
$= 8 \times 8 \times 8$
$= 8^3$

$\therefore \quad \sqrt[3]{512} = \sqrt[3]{8^3} = 8$

Working

| | |
|---|---|
| 2 | 512 |
| 2 | 256 |
| 2 | 128 |
| 2 | 64 |
| 2 | 32 |
| 2 | 16 |
| 2 | 8 |
| 2 | 4 |
| | 2 |

(b) $5832 = (2 \times 3 \times 3) \times (2 \times 3 \times 3) \times (2 \times 3 \times 3)$
$= 18 \times 18 \times 18$
$= 18^3$

$\therefore \quad \sqrt[3]{5832} = \sqrt[3]{18^3} = 18$

| | |
|---|---|
| 2 | 5832 |
| 2 | 2916 |
| 2 | 1458 |
| 3 | 729 |
| 3 | 243 |
| 3 | 81 |
| 3 | 27 |
| 3 | 9 |
| | 3 |

## Exercise 1e

1. Find all the perfect squares between 50 and 150.

2. Copy the following numbers and circle those that are perfect squares:

   8, 16, 18, 25, 33, 49, 50, 72, 81, 100, 1, 125, 144, 200, 169, 111, 225, 400

3. Replace each box with the correct answer:
   **(a)** Given that $8 \times 8 = 64$, then $\sqrt{64} = \square$.
   **(b)** Given that $12 \times 12 = 144$, then $\sqrt{144} = \square$.
   **(c)** Given that $17 \times 17 = 289$, then $\sqrt{289} = \square$.
   **(d)** Given that $(3 \times 4 \times 5) \times (3 \times 4 \times 5) = 3600$, then $\sqrt{3600} = \square$.

4. Find the positive square root of each of the following numbers:
   **(a)** 36 **(b)** 81 **(c)** 196 **(d)** 256 **(e)** 441

5. Use a calculator to factorise the following (into prime numbers) and hence find the positive square root of each of the following numbers:
(a) 1156 (b) 1296 (c) 1764
(d) 9801 (e) 11 025 (f) 34 596

6. Replace each box with the correct answer:
(a) Given that $11 \times 11 \times 11 = 1331$, then $\sqrt[3]{1331} = \square$.
(b) Given that $19 \times 19 \times 19 = 6859$, then $\sqrt[3]{6859} = \square$.
(c) Given that $13^3 = 2197$, then $\sqrt[3]{2197} = \square$.
(d) Given that $(3 \times 4) \times (3 \times 4) \times (3 \times 4) = 1728$, then $\sqrt[3]{1728} = \square$.

7. Using a calculator, factorise each of the following numbers completely and hence find its cube root:
(a) 3375 (b) 4096 (c) 21 952
(d) 46 656 (e) 91 125 (f) 373 248

8. Find the area of a square of side 56 cm.

9. What is the length of a side of a square whose area is 2304 $cm^2$?

10. What is the volume of a cube of side 11 cm?

11. Given that the volume of a cube is 2744 $cm^3$, find the length of its edge.

## Mental Estimation

What is the square root of 48? What is the cube root of 65?

48 can be written as $48 = 4 \times 4 \times 3$.
65 can be written as $65 = 5 \times 13$.

Clearly, 48 and 65 cannot be expressed respectively as $a^2$ and $b^3$, where $a$ and $b$ are whole numbers. Thus, 48 is not a perfect square and 65 is not a perfect cube. $\sqrt{48}$ and $\sqrt[3]{65}$ are not whole numbers but decimals.

In this section, we are concerned with finding estimates of numbers such as $\sqrt{48}$ and $\sqrt[3]{65}$.

We observe that 48 is close to 49 which is a perfect square.

$\therefore \sqrt{48} \approx \sqrt{49} = 7$ (Is the exact value of $\sqrt{48}$ greater or less than 7?)

The above process can also be done mentally.

Similarly, 65 is close to 64 which is a perfect cube.

$\therefore \sqrt[3]{65} \approx \sqrt[3]{64} = 4$ (Is the exact value of $\sqrt[3]{65}$ greater or less than 4?)

## Example 9

*Estimate each of the following:*

*(a)* $29^2$ *(b)* $19^3$ *(c)* $401^2$

*(d)* $104^3$ *(e)* $\sqrt{99}$ *(f)* $\sqrt[3]{999}$

### Solution

(a) $29^2 \approx 30^2 = 900$

Is $29^2 < 900$ or $> 900$?

(b) $19^3 \approx 20^3 = 8000$

Is $19^3 < 8000$ or $> 8000$?

Similarly, (c) and (d) can be worked out in the same way.

(c) $401^2 \approx 400^2 = 160\ 000$

Is $401^2 < 160\ 000$ or $> 160\ 000$?

(d) $104^3 \approx 100^3 = 1\ 000\ 000$

Is $104^3 < 1\ 000\ 000$ or $> 1\ 000\ 000$?

(e) $\sqrt{99} \approx \sqrt{100} = \sqrt{10 \times 10} = 10$

Is $\sqrt{99} < 10$ or $> 10$?

(f) $\sqrt[3]{999} \approx \sqrt[3]{1000} = \sqrt[3]{10 \times 10 \times 10} = 10$

Is $\sqrt[3]{999} < 10$ or $> 10$?

**Notes**

*$30^2 = (3 \times 10) \times (3 \times 10)$*
*$= 3 \times 3 \times 10 \times 10$*
*$= 9 \times 100$*
*$= 900$*

*$20^3 = (2 \times 10) \times (2 \times 10) \times (2 \times 10)$*
*$= 2 \times 2 \times 2 \times 10 \times 10 \times 10$*
*$= 8 \times 1000$*
*$= 8000$*

# The Use of Calculators

We can use a scientific calculator to find the square, square root, cube and cube root of a number very easily. Below are some function keys for the purpose.

| | |
|---|---|
| $\sqrt{\phantom{x}}$ | square root key |
| $x^2$ | square key |
| $y^x$ | power key |
| $\sqrt[x]{\phantom{x}}$ | $x^{th}$ root key |

To find $\sqrt{25}$, press 25 $\boxed{\sqrt{\ }}$ $\boxed{=}$ and the display gives 5, the square root of 25.

For calculators with Direct Algebraic Logic (DAL), the sequence of pressing the keys is $\boxed{\sqrt{\ }}$ 25 $\boxed{=}$.

We shall show the sequence following the DAL in all our examples.

## Example 10

*Use your calculator to evaluate the following:*

(a) $14^2 + \sqrt[3]{2744} - \sqrt{529}$  (b) $\dfrac{\sqrt[3]{729} \times 39^2}{\sqrt{169}}$  (c) $\dfrac{\sqrt{65\,536} + 8^3}{11^2 - \sqrt[3]{15\,625}}$

### Solution

| | Sequence of pressing keys: | Final display |
|---|---|---|
| (a) | 14 $\boxed{x^2}$ $\boxed{+}$ 3 $\boxed{\sqrt[x]{\ }}$ 2744 – $\boxed{\sqrt{\ }}$ 529 $\boxed{=}$ | 187 |
| (b) | 3 $\boxed{\sqrt[x]{\ }}$ 729 $\boxed{\times}$ 39 $\boxed{x^2}$ ÷ $\boxed{\sqrt{\ }}$ 169 $\boxed{=}$ | 1053 |
| (c) | $\boxed{(}$ $\boxed{\sqrt{\ }}$ 65 536 $\boxed{+}$ 8 $\boxed{x^y}$ 3 $\boxed{)}$ $\boxed{\div}$ | |
| | $\boxed{(}$ 11 $\boxed{x^2}$ $\boxed{-}$ 3 $\boxed{\sqrt[x]{\ }}$ 15 625 $\boxed{)}$ $\boxed{=}$ | 8 |
| or | | |
| | 11 $\boxed{x^2}$ – 3 $\boxed{\sqrt[x]{\ }}$ 15 625 $\boxed{=}$ $\boxed{\text{STO}}$ | |
| | $\boxed{\sqrt{\ }}$ 65 536 $\boxed{+}$ 8 $\boxed{y^x}$ 3 $\boxed{=}$ $\boxed{\div}$ $\boxed{\text{RCL}}$ $\boxed{=}$ | 8 |

**Note:** Some calculators also have the $\boxed{\sqrt[3]{\ }}$ key for finding the cube root of a number. Try to use this key to work out the above calculations.

## Explorations

How to test whether a given number is a prime number?

**(a)** Is 221 a prime number?

Step 1: Use a calculator to find $\sqrt{221}$.

Step 2: List all the prime numbers less than $\sqrt{221}$.

Step 3: Use a calculator to divide 221 successively by the prime numbers listed in step 2.

You can stop the process when

**(a)** a prime number divides 221 exactly. If this happens, then 221 will not be a prime number (e.g. 25 is not a prime number because 5 divides 25 exactly).

**(b)** a prime number does not divide 221 exactly and the number in the display is less than the prime number. If this occurs, then 221 will be a prime number (e.g. 13 is a prime number since $13 \div 2 = 6.5 > 2$; $13 \div 3 = 4.3 > 3$; $13 \div 5 = 2.6 < 5$).

---

**Thinking Time**

Why do we test only the prime numbers less than the square root of the given number?

## Exercise 1f

**1.** Estimate mentally the following:

**(a)** $41^2$ **(b)** $58^2$
**(c)** $112^2$ **(d)** $32^3$
**(e)** $39^3$ **(f)** $98^3$
**(g)** $198^2$ **(h)** $301^3$

**2.** Give an estimate of each of the following mentally:

**(a)** $\sqrt{37}$ **(b)** $\sqrt[3]{26}$
**(c)** $\sqrt{63}$ **(d)** $\sqrt[3]{124}$
**(e)** $\sqrt{84}$ **(f)** $\sqrt{142}$
**(g)** $\sqrt[3]{1004}$ **(h)** $\sqrt{897}$

**3.** Evaluate the following using a calculator:

**(a)** $26^2$ **(b)** $37^2$
**(c)** $78^2$ **(d)** $99^2$
**(e)** $123^2$ **(f)** $13^3$
**(g)** $29^3$ **(h)** $34^3$
**(i)** $67^3$ **(j)** $109^3$
**(k)** $\sqrt{961}$ **(l)** $\sqrt{2209}$
**(m)** $\sqrt{3481}$ **(n)** $\sqrt{11\ 236}$
**(o)** $\sqrt{69\ 169}$ **(p)** $\sqrt[3]{4096}$
**(q)** $\sqrt[3]{68\ 921}$ **(r)** $\sqrt[3]{314\ 432}$
**(s)** $\sqrt[3]{753\ 571}$ **(t)** $\sqrt[3]{1\ 906\ 624}$
**(u)** $18^2 + 11^3 - \sqrt{484} + \sqrt[3]{4913}$
**(v)** $\sqrt{676} \times 9^3 - 17^2 + \sqrt{2704}$
**(w)** $24^3 \div \sqrt{4096} + \sqrt[3]{512} \times 44^2$
**(x)** $\sqrt[3]{1331} \times \sqrt{2916} - 42^3 \div 21^2$
**(y)** $\dfrac{7^3 \times \sqrt{576} + \sqrt[3]{512}}{\sqrt{7744} - 2^3}$

**4.** Determine whether each of the following is a prime number:

**(a)** 101 **(b)** 227
**(c)** 323 **(d)** 413

# SUMMARY

1. A **prime number** is a number which has only two different factors, 1 and the number itself. Prime numbers are 2, 3, 5, 7, 11, 13, 17, etc.

2. A **composite number** is a number which has more than two different factors. Composite numbers are 4, 6, 12, 15, 24, 32, etc.

3. A composite number can be expressed as the product of two or more prime numbers.

4. The process of expressing a composite number as the product of prime factors is called **prime factorisation**.

5. Index notation:
   In general, $\underbrace{a \times a \times \ldots \times a}_{n \text{ factors}}$ is written as $a^n$ and is read as $a$ to the power of $n$.

6. The largest of the factors common to two or more numbers is called the **Highest Common Factor** (HCF) of the numbers.

7. The smallest of the common multiples of two or more numbers is called the **Least Common Multiple** (LCM) of the numbers.

8. If a number $y$ can be expressed as $y = x^2$, we say that $y$ is the **square** of $x$ and $x$ is the **square root** of $y$.
   If $x$ is a whole number, then $y$ is a **perfect square**.

9. If a number $y$ can be expressed as $y = x^3$, we say that $y$ is the **cube** of $x$ and $x$ is the **cube root** of $y$.

# Review Examples 1

## Example 1

*Find the smallest value of k, when*
*(a) 264k is a perfect square,*
*(b) 1960k is a cube.*

### Solution

(a) $264 = 2^3 \times 3 \times 11$

For $2^2 \times 2 \times 3 \times 11 \times k$ to be a perfect square,

$k = 2 \times 3 \times 11$ i.e. $k = 66$.

| | |
|---|---|
| 2 | 264 |
| 2 | 132 |
| 2 | 66 |
| 3 | 33 |
| | 11 |

(b) $1960 = 2^3 \times 5 \times 7^2$

For $2^3 \times 5 \times 7^2 \times k$ to be a cube,

$k = 5^2 \times 7 = 175$.

| | |
|---|---|
| 2 | 1960 |
| 2 | 980 |
| 2 | 490 |
| 5 | 245 |
| 7 | 49 |
| | 7 |

## Example 2

*(a) If $1764 = (2 \times 3 \times 7)^P$, find the value of p.*
*(b) Given that 1764 and a number Q has an LCM of 8820 and a HCF of 36, find the number Q.*

### Solution

(a) $1764 = (2 \times 3 \times 7)^2$
$\therefore \quad p = 2$

| | |
|---|---|
| 2 | 1764 |
| 2 | 882 |
| 3 | 441 |
| 3 | 147 |
| 7 | 49 |
| | 7 |

(b) LCM $= 8820 = 2^2 \times 3^2 \times 7^2 \times 5$
HCF $= 36 = 2^2 \times 3^2$

$1764 = 2^2 \times 3^2 \times 7^2$

| | |
|---|---|
| 2 | 8820 |
| 2 | 4410 |
| 5 | 2205 |
| 3 | 441 |
| 3 | 147 |
| 7 | 49 |
| | 7 |

Obviously, $Q$ must contain all the factors of the HCF i.e. $2^2 \times 3^2$.

$Q$ must also contain the factor(s) of the LCM that is(are) not the factor(s) of the number 1764. In this case, that factor is 5.

$\therefore \quad Q = 2^2 \times 3^2 \times 5 = 180$

## Review Questions 1

**1.** State whether each of the following is true or false:

**(a)** The prime numbers between 1 and 20 are 2, 3, 5, 7, 9, 11, 13, 17 and 19.
**(b)** The HCF of 16, 20, 24 and 32 is 8.
**(c)** The smallest number that is divisible by 10, 15 and 20 is 60.
**(d)** 1725 when expressed as a product of prime factors is $23 \times 15 \times 5$.
**(e)** The HCF of $4 \times 3^2$, $3 \times 4^2$ and $2 \times 3 \times 5$ is $4^2 \times 3^2 \times 5$.
**(f)** $\sqrt{802\ 500}$ lies between 800 and 900.
**(g)** An approximation of $89^2$ is 8100 which is less than the exact value of $89^2$.

**2.** Find the HCF of each of the following:

**(a)** 144, 162
**(b)** 12, 18, 30
**(c)** 65, 78, 104

**3.** Find the LCM of each of the following:

**(a)** 12, 30
**(b)** 16, 18, 48
**(c)** 42, 63, 105

**4.** How many whole numbers are between each of the following pairs of numbers?

**(a)** $\sqrt{7}$ and $\sqrt{80}$
**(b)** $\sqrt[3]{7}$ and $\sqrt[3]{215}$
**(c)** $\sqrt[3]{18}$ and $\sqrt{120}$

**5.** Evaluate each of the following using a calculator:

**(a)** $\sqrt{27^2 + 36^2}$
**(b)** $\sqrt{136^2 - 64^2}$
**(c)** $29^2 + 19^3 + \sqrt{676} - \sqrt[3]{50\ 653}$
**(d)** $58^2 \div \sqrt[3]{24\ 389} \times \sqrt{3721} + 33^2$
**(e)** $\dfrac{26^3 \div \sqrt{2704} \times \sqrt[3]{42\ 875}}{25 \times \sqrt{1849} - 14^3 + 40^2 + 79}$.

★**6.** John, Peter and Paul were each given a piece of string of equal length. John cut his into equal lengths of 2 m; Peter cut his into equal lengths of 3 m; and Paul cut his into equal lengths of 5 m. If there was no remainder in each case, find the shortest length of string given to each of them.

∗ **7.** What is the shortest length which can be divided into 4 cm, 8 cm or 2 cm portions without remainders?

∗ **8.** Find two numbers if their LCM is 120 and their HCF is 4. (Give three possible answers.)

∗ **9.** Find the smallest value of $k$, when
**(a)** $280k$ is a perfect square,
**(b)** $882k$ is a cube.

∗ **10.** A three-digit number is the product of four prime numbers. Given that the three digits of the number are all prime and different and that the sum of its prime factors is 30, find the number.

*In this chapter, you will learn how to*

- *use negative numbers in practical situations;*
- *perform calculations with integers;*
- *represent numbers on the number line and order them;*
- *use the symbols >, <, ≥, ≤ and ≠.*

CHAPTER

# 2

# Integers

## Introduction

The main peak of the Jade Dragon Snow Mountain in China is at an altitude of 5596 metres above sea level. Riding in a cable car, the two young ladies reached the altitude of 4500 metres, from which they climbed another 150 metres to arrive at the Dragon Spruce Meadow. At a temperature of 10°C below freezing point, the ladies enjoyed having a photo taken with their fruit of labour.

In this chapter you will learn about negative numbers, which were first used by Indians for accounting purposes in the 6th and 7th centuries. If we take the altitude of the Dragon Spruce Meadow to be 0 metre altitude, then we use +946 m to indicate that the peak is 946 m above the Dragon Spruce Meadow and –150 m to indicate that the cable car terminal is 150 m below the Dragon Spruce Meadow. Similarly, we use –10°C below 0°C to represent a temperature that is 10°C below 0°C.

# Negative Numbers

The concept of **negative numbers** is derived from many of our real life situations. Here are some examples.

We normally record the temperature 10°C above zero as 10°C, and correspondingly, we record the temperature 10°C below zero as –10°C.

Sea level is usually taken as zero altitude. If a bird flies 50 m above zero altitude, we use 50 m to represent how high the bird is flying above sea level. Correspondingly, if a submarine dives 50 m underwater, we will use –50 m to represent its depth below sea level.

Tom and Johnson walk away from the same place but in opposite directions. If Tom walked 15 m heading towards the East, and Johnson walked 12 m heading towards the West, then we can use 15 m and –12 m to record how far and in which directions they walked respectively.

*Fig. 2.1*

In the above examples, 10, 50 and 15 are called positive numbers, and –10, –50 and –12 are called negative numbers.

In Mathematics, numbers with the sign "–" are called negative numbers. "–" is read as "negative".

**Thinking Time**

Can you give some other examples where both positive and negative numbers are used?

# Integers

The numbers 1, 2, 3, 4, ... are also called **positive integers**. The corresponding negative numbers –1, –2, –3, –4, ... are called **negative integers**. The positive integers and 0, together with the negative integers, are called integers. Zero is an integer but is neither positive nor negative.

# The Number Line

In Mathematics, it is often useful to represent integers by points on a line called the **number line**.

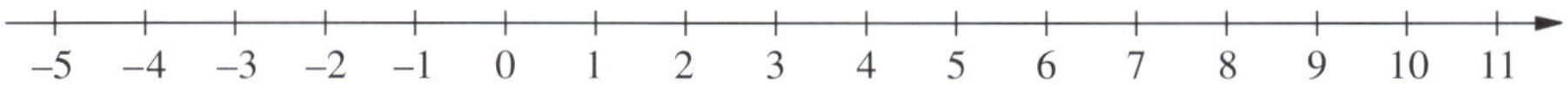


Draw a line. Choose any point on the line and label it 0. Starting with 0, mark off equal intervals of any suitable length towards two directions. Label the points marked 1, 2, 3, 4, ... on the right side and –1, –2, –3, –4, ... on the left side as shown in the figure above. The arrow on the extreme right indicates that the list of numbers continues in the same way indefinitely.

A number on the number line is always greater than any number to its left and smaller than any number to its right, i.e. 4 is greater than 3; –4 is smaller than –3, and –4 is smaller than +3.

The number line below shows that $a$ is greater than $b$ and $b$ is greater than $c$.

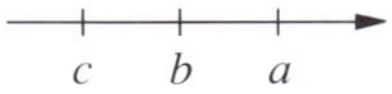


**Thinking Time**

Are all negative integers smaller than all positive integers? Why?

In Mathematics, we use the symbol '>' to denote 'is greater than' and the symbol '<' to denote 'is less than'. For example, we write '$7 > 4$' to express '7 is greater than 4' and we write '$-3 < 6$' to express '–3 is less than 6'.

We use the symbol '≠' to denote 'is not equal to', for example, we write '$a \neq 8$' for '$a$ is not equal to 8'.

We use the symbol '≥' to denote 'is greater than or equal to'. For example, we write '$a \geq 9$' for '$a$ is greater than or equal to 9'.

We use the symbol '≤' to denote 'is less than or equal to'. For example, we write '$b \leq -10$' for '$b$ is less than or equal to –10'.

## Example 1

*Draw a number line to represent the integers*
*(a) –2, –4, 6 and 8;* *(b) greater than –3.*

### Solution

(a) We use dots to indicate the numbers –2, –4, 6 and 8.

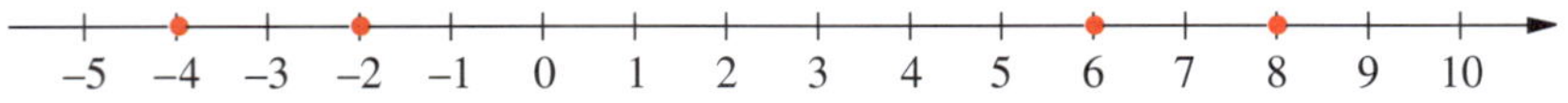


(b) We use dots to indicate the first few whole numbers that are greater than –3.

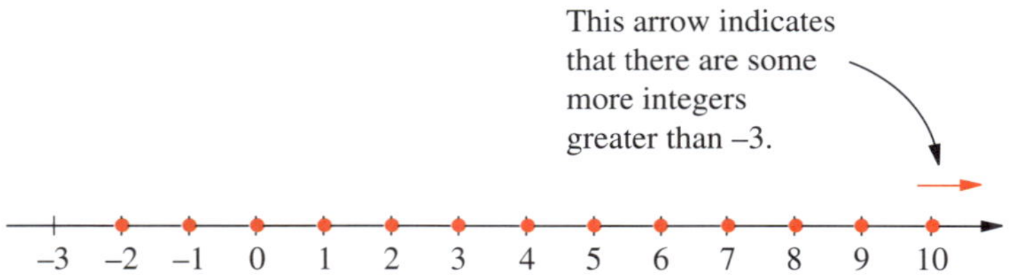


# Absolute Value of an Integer

Positions of –2 and 2 on the number line indicate that $-2 < 2$ but they are of the same **distance** from zero. –2 and 2 are said to have the same **absolute value**.

The absolute value of an integer is its distance from zero on the number line.

The absolute value of an integer is **always positive**.

We write the absolute value of an integer $x$ as $|x|$. Thus, $|-2| = |2| = 2$.

## Example 2

*Given the integers –8, 9 and –20,*
*(a) arrange the integers in ascending order;*
*(b) find the absolute values of the integers;*
*(c) arrange the absolute values of the integers in descending order.*

## Solution

(a) The required arrangement of the numbers is –20, –8, 9 since $-20 < -8 < 9$.

(b) Distances between –8, 9, –20 and zero on the number line are 8 units, 9 units and 20 units respectively.

$\therefore \quad |-8| = 8, \quad |9| = 9, \quad |-20| = 20.$

(c) The required arrangement of the absolute values is 20, 9, 8.

**Your Attention**

*20 is greater than 8, but –20 is smaller than –8.*

## Exercise 2a

**1.** Copy and complete the following:

**(a)** An ascent of –20 m means a descent of _____ m.

**(b)** A clockwise rotation of –90° means an anticlockwise rotation of _____°.

**(c)** Walking 9 km from East to West means walking _____ km from West to East.

**2.** Copy and complete the following:

**(a)** If –6 represents 6 m below sea-level, then +30 represents _____.

**(b)** If +40 represents depositing $40 in the bank, then a withdrawal of $35 is _____.

**(c)** If +60° represents rotating 60° clockwise, then –30° represents _____.

**(d)** If +45 represents a speed of 45km/h of a car moving to the East, then –45 represents _____.

**3.** Suppose time is measured in years and zero stands for the year 2000.

**(a)** What number stands for the year 1999?

**(b)** What number stands for the year 2002?

**(c)** What number stands for the year 1996?

**4.** Put the correct sign, < or >, between each pair of temperatures below.

**(a)** –4°, 4° **(b)** 8°, –2°

**(c)** –13°, –5° **(d)** –1°, 5°

**(e)** –3°, –4° **(f)** –20°, –19°

**5.** Translate each of the following into a mathematical statement using either a < or a > sign:

**(a)** A temperature of –5° is colder than a temperature of 12°.

**(b)** A gain of $200 is better than a loss of $120.

**(c)** A depth of 40 m below sea level is lower than a depth of 25 m below sea level.

**6.** Arrange the integers in each group in ascending order.

**(a)** –2, –6, 0, –60, –90

**(b)** 1, –1500, 2, –1

**7.** Use a number line to illustrate each of the following:

**(a)** –5, –2, 0, 5, –3

**(b)** ..., –10, –8, –6, –4, ...

**(c)** The set of integers greater than –4 and less than 2.

**(d)** The set of integers between –2 and +6.

**8.** Fill in the boxes with > or <:

**(a)** $8 \ \square \ -8$ **(b)** $-11 \ \square \ -6$

**(c)** $0 \ \square \ -2$ **(d)** $(-12)^2 \ \square \ -200$

**(e)** $-\sqrt{64} \ \square \ \sqrt[3]{125}$

**(f)** $\sqrt[3]{-27} \ \square \ -\sqrt{16}$

# Addition of Integers

You have learnt how to add positive integers. You can also add them using the number line. Below is an example.

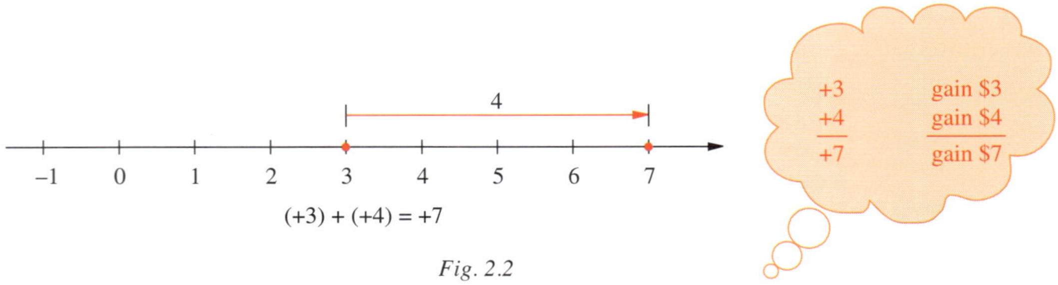


*Fig. 2.2*

Now, let us use the number line to add two negative integers e.g. –2 and –5. (See Fig. 2.3)

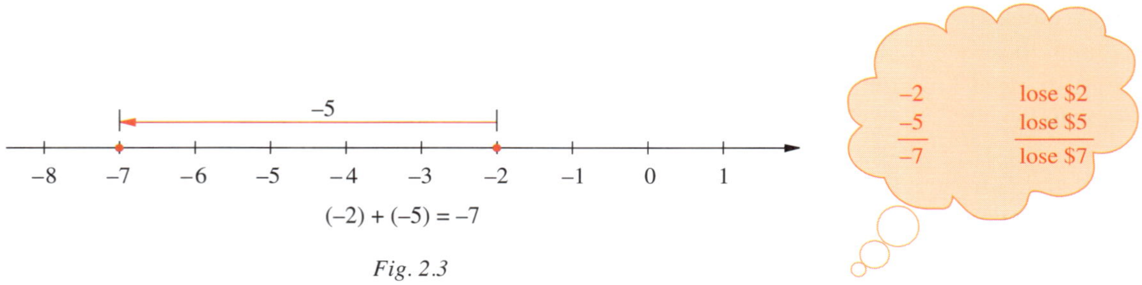


*Fig. 2.3*

The above discussion suggests the following rules for adding two numbers with the **same signs**.

**1.** To add two positive numbers, add their absolute values. The result is positive.

$$(+x) + (+y) = +(x + y)$$

**2.** To add two negative numbers, add their absolute values and place a negative sign before the result.

$$(-x) + (-y) = -(x + y)$$

## Example 3

*Find the sum of each of the following:*

*(a) (–3) + (–8)* *(b) (–71) + (–43)*

### Solution

(a) (–3) + (–8) = –(3 + 8) = –11

(b) (–71) + (–43) = –(71 + 43) = 114

## Explorations

Why do we say that the durian is the king of fruits? To find the answer, do the following additions and match the answer given outside each box with the correct letter.

C (–3) + (–5)
O (+2) + (+7)
I (–2) + (–11)
U (–8) + (–4)
T (+9) + (+6)
D (–11) + (–10)
L (–1) + (–19)
E (+25) + (+17)
S (–43) + (–16)

| | | | | | | | | | | | | | | |
|---|---|---|---|---|---|---|---|---|---|---|---|---|---|---|
| | | | | | | | | | | | | | | |
| –13 | +15 | | –13 | –59 | | –21 | +42 | –20 | –13 | –8 | –13 | +9 | –12 | –59 |

Now, let us add a positive integer and a negative integer.

Fig. 2.4 shows the addition of +6 and –2.

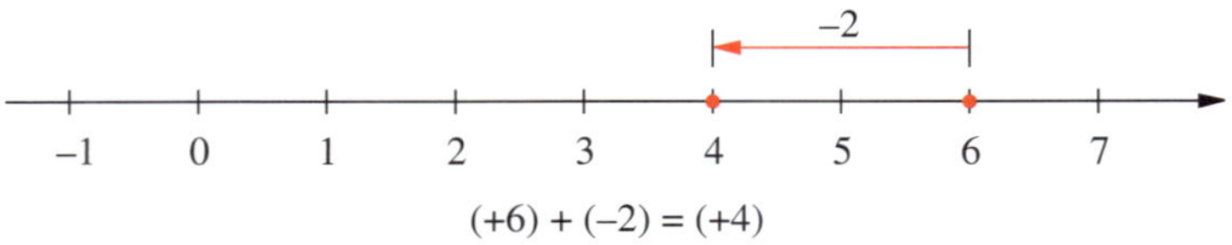


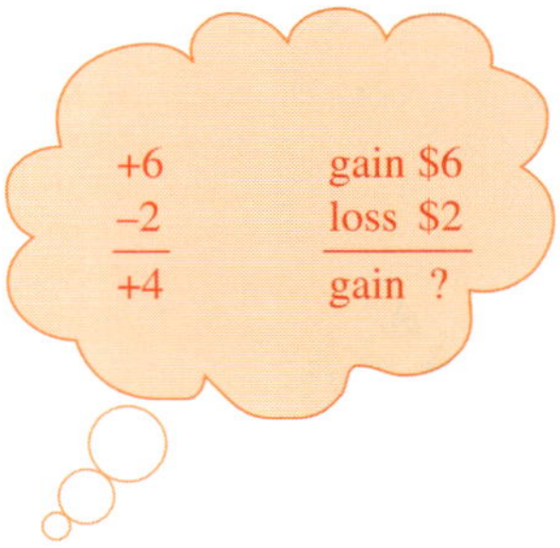


*Fig. 2.4*

Fig. 2.5 shows the addition of –7 and +2.

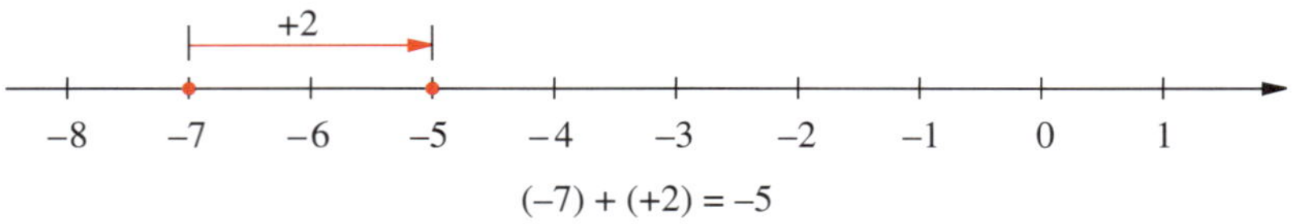


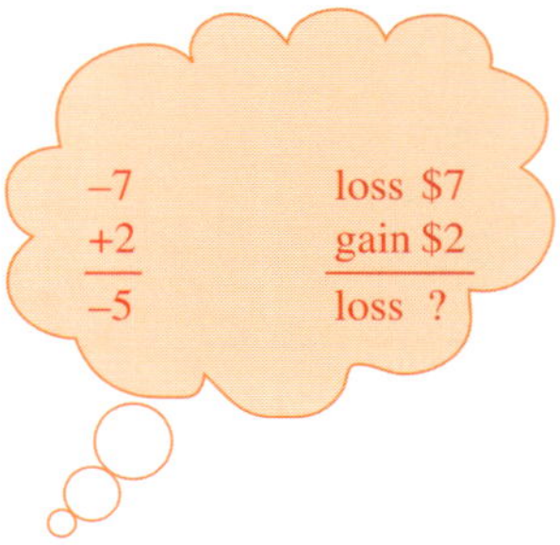


*Fig. 2.5*

The above discussion suggests the following rules for adding two numbers with the **different signs**.

To add a positive and a negative number, find the difference of their absolute values by subtracting the smaller absolute value from the larger absolute value. Place the sign of the number having the larger absolute value before the result.

For any $x > 0$, $y > 0$,

$(+x) + (-y) = +(x - y)$ if $x > y$
$(+x) + (-y) = -(y - x)$ if $y > x$
$(-x) + (+y) = -(x - y)$ if $x > y$
$(-x) + (+y) = +(y - x)$ if $y > x$

## Example 4

*Find the following sums:*
*(a) –6 + 18*
*(b) 23 + (–68)*

**Solution**

(a) $-6 + 18 = (-6) + (+18) = +(18 - 6) = +12 = 12.$ (+18 has a larger absolute value of 18)

(b) $23 + (-68) = (+23) + (-68) = -(68 - 23) = -45$ (–68 has a larger absolute value of 68)

## Example 5

*Find the following sums:*

*(a) $[-5 + (-9)] + [16 + (-21)]$*

*(b) $-13 + 14 + (-7)$*

**Solution**

(a) $[-5 + (-9)] + [16 + (-21)]$
$= [(-5) + (-9)] + [(+16) + (-21)]$
$= [-(5 + 9)] + [-(21 - 16)]$
$= (-14) + (-5) = -(14 + 5) = -19$

(b) $-13 + 14 + (-7)$
$= [(-13) + (+14)] + (-7)$
$= [+(14 - 13)] + (-7)$
$= (+1) + (-7) = -(7 - 1) = -6$

## Exercise 2b

1. Calculate the following.
   **(a)** $-5 + 13$ **(b)** $-11 + 19$
   **(c)** $14 + (-7)$ **(d)** $23 + (-12)$
   **(e)** $-37 + 22$ **(f)** $-45 + 19$
   **(g)** $25 + (-66)$ **(h)** $74 + (-89)$
   **(i)** $101 + (-200)$

2. Evaluate the following.
   **(a)** $-7 + (-11) + 9$
   **(b)** $-10 + 17 + (-21)$
   **(c)** $34 + (-18) + 9$
   **(d)** $81 + (-6) + (-62)$
   **(e)** $51 + 14 + (-100)$
   **(f)** $-27 + 71 + 12$
   **(g)** $[-7 + (-14)] + [-45 + 92]$
   **(h)** $[31 + (-48)] + [-16 + (-120)]$

3. The temperature of a piece of meat was $-8°C$ when it was taken out of a freezer. After several minutes of warming, its temperature rose by 17°C. What was the new temperature of the piece of meat?

4. A company's profits and losses for the first quarter of a certain year are as shown below.

| | |
|---|---|
| January: | $6000 (profit) |
| February: | $2000 (loss) |
| March: | $5500 (loss) |

   **(a)** Translate the company's performance for the first quarter into a mathematical statement.
   **(b)** How much profit or loss did the company make in the first quarter?

# Subtraction of Integers

We think of subtraction (–) as the **reverse** or **opposite** of addition (+).

On the number line, to **add** 5, we move 5 units to the **right**. Thus, to **subtract** 5, we move 5 units to the **left**. Recall also that to add –5, we move 5 units to the left. Thus, to **add** –5, we move 5 units to the **left** and to **subtract** –5, we move 5 units to the **right**.

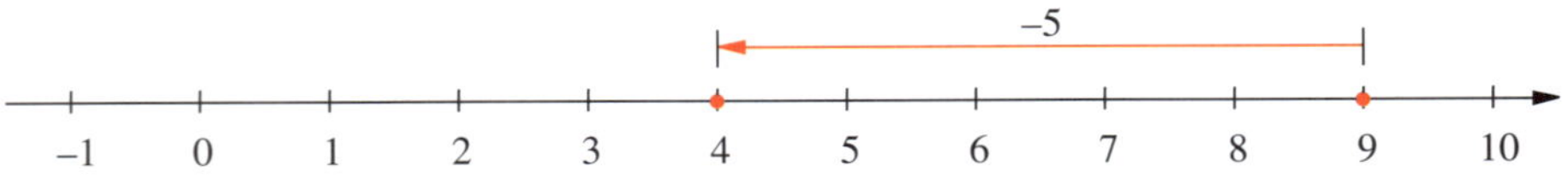


9 – 5 = 9 + (–5) = 4

*Fig. 2.6*

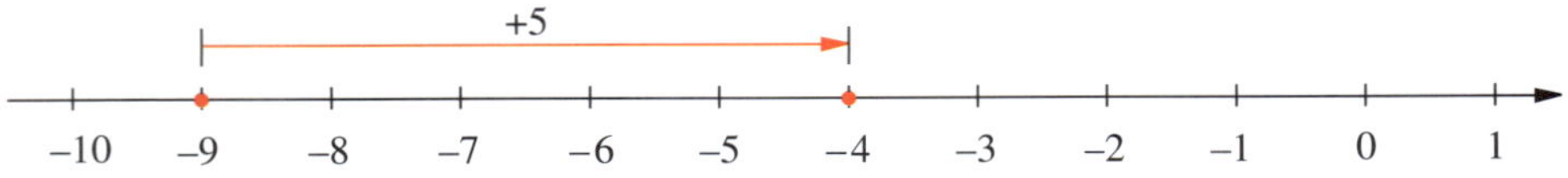


(–9) – (–5) = (–9) + (+5) = –4

*Fig. 2.7*

Fig. 2.6 represents both 9 – 5 and 9 + (–5).

Fig. 2.7 represents both (–9) – (–5) and (–9) + (+5).

Hence, 9 – 5 = 9 + (–5) = 4
(–9) – (–5) = (–9) + (+5) = –4

Similarly, (–9) – 5 = (–9) + (–5) = –14
9 – (–5) = 9 + 5 = 14.

The above discussion suggests the following rule for subtracting integers.

**To subtract integers, change the sign of the integer being subtracted and add according to the rules for addition of integers.**

$$x - y = x + (-y)$$

## Example 6

*Evaluate the following:*

*(a) 9 – 14* *(b) –12 – 6* *(c) –10 – (–7)* *(d) 4 – (–11)*

**Solution**

(a) 9 – 14 is the same as 9 – (+14). Change +14 to –14. Add 9 and –14.
∴ 9 – 14 = 9 + (–14) = –(14 – 9) = –5.

(b) –12 – 6 is the same as –12 – (+6). Change +6 to –6. Add –12 and –6.
∴ –12 – 6 = –12 + (–6) = –(12 + 6) = –18.

(c) Change –7 in –10 – (–7) to +7. Add –10 and +7.
∴ –10 – (–7) = –10 + (+7) = –(10 – 7) = –3.

(d) Change –11 in 4 – (–11) to +11 or simply 11. Add 4 and 11.
∴ 4 – (–11) = 4 + 11 = 15

## Example 7

*Do the following:*

*(a) 280 + (–120) – (–320) + 50*

*(b) –(–700) + (–500) – 130 + 70*

### Solution

(a) 280 + (–120) – (–320) + 50
= (280 – 120) + 320 + 50
= 160 + 320 + 50 = 530

(b) –(–700) + (–500) – 130 + 70
= 700 – 500 – 130 + 70
= 200 – 130 + 70
= 70 + 70 = 140

## Explorations

Why can't you have a conversation near the merlion? Do the following subtractions and match the answer given outside each box with the correct letter, and solve the riddle.

A (–5) – (+6)
T (+8) – (–5)
P (–4) – (–7)
O (+2) – (–10)
S 0 – (–4)
Y (–7) – (+12)
N (–2) – (+25)
E (–16) – (–9)
D (–1) – (0)
R (–6) – (–17)
U (+9) – (–13)
H (–4) – (+14)
W (–15) – (–21)

| | | | | | | | | | | | | | | | | | | | | | |
|---|---|---|---|---|---|---|---|---|---|---|---|---|---|---|---|---|---|---|---|---|---|
| –18 | –7 | +4 | +3 | +11 | –11 | –19 | +4 | +6 | –11 | +13 | –7 | +11 | +12 | –27 | –19 | +12 | +22 | +11 | –18 | –7 | –11 | –1 |

## Exercise 2c

1. Evaluate the following.
   **(a)** $-3 - (6 - 9)$
   **(b)** $-4 - (2 - 5)$
   **(c)** $-6 - (-3) - 1$
   **(d)** $4 - (-9) - 5$
   **(e)** $-2 - (-6) - 3$
   **(f)** $5 - 7 - (-10)$
   **(g)** $4 - (-8) - 6 - (-11)$
   **(h)** $-12 - 7 - (-9) - 3$
   **(i)** $15 - 20 - 13 - 32$

2. Find the values of the following expressions.
   **(a)** $-12 - (-24) + (-36)$
   **(b)** $-40 + 15 - 27 + 11$
   **(c)** $(-5 - 6) - (15 - 5)$
   **(d)** $-324 + 12 - 56$
   **(e)** $[-2 + (-10)] - [15 + (-20)]$
   **(f)** $146 - (-200) + (-100) - 150$
   **(g)** $-176 + (-123) - (-167) + 103$
   **(h)** $-14 + 26 - 37 - 45 + 56 - 67 - 71$
   **(i)** $19 - 27 + 34 + 43 - 58 - 66 - 76 + 81$

# Multiplication of Integers

## Explorations

Examine the following pattern and write down the next 2 lines.

$6 \times 11 + 6 = 8 \times 9$
$5 \times 10 + 6 = 7 \times 8$
$4 \times 9 + 6 = 6 \times 7$
$3 \times 8 + 6 = 5 \times 6$

________________

________________

Continuing the pattern, we get

$0 \times 5 + 6 = 2 \times 3$
$(-1) \times 4 + 6 = 1 \times 2$

For $(-1) \times 4 + 6 = 1 \times 2$ to be true, $(-1) \times 4$ must equal to $-4$.

Similarly, for the next line $(-2) \times 3 + 6 = 0 \times 1$ to be true,

$(-2) \times 3$ must equal to _____.

Observation: when a negative integer is multiplied by a positive integer, the result is _____.

Continue the pattern by filling in the blanks:

$(-2) \times 3 + 6 = 0 \times 1$
$(-3) \times 2 + 6 = (-1) \times 0$
$(-4) \times 1 + 6 = (-2) \times (\quad)$
$(\quad) \times (\quad) + 6 = (-3) \times (\quad)$

Write down your observations.

We already know the rule for multiplying positive integers using a number line. For example, $3 \times 2 = 6$.

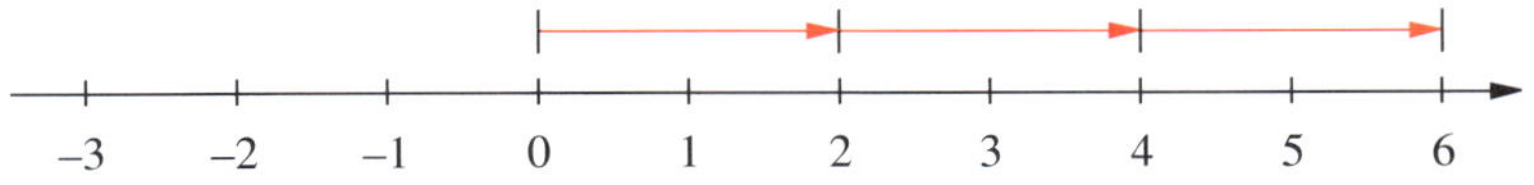


$3 \times 2 = 2 + 2 + 2 = 6$.

Similarly, for multiplying a positive integer by a negative integer, for example $3 \times (-2)$, we have

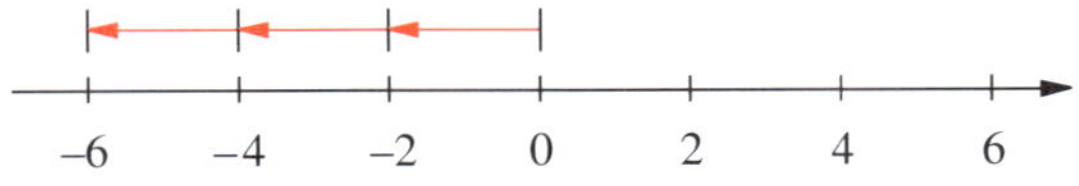


$3 \times (-2) = (-2) + (-2) + (-2) = -6 = -(3 \times 2)$.

We know $2 \times 3 = 3 \times 2 = 6$. Similarly, we have $(-2) \times 3 = 3 \times (-2) = -(3 \times 2) = -(2 \times 3)$.

**In general, if both $x$ and $y$ represent positive integers, then**

$$x \times (-y) = -(x \times y) \text{ or } (-x) \times y = -(x \times y).$$

**In other words, the product of a positive integer and a negative integer is a negative integer.**

**FOR YOUR INFORMATION**

*Scientists have found that the lowest possible temperature is –273 °C. The temperature is referred to as the absolute zero. The boiling points of some gases are given below:*

| | |
|---|---|
| *Hydrogen* | *–253 °C* |
| *Nitrogen* | *–196 °C* |
| *Oxygen* | *–183 °C* |

*If mixtures of the above three gases are cooled to –260 °C and then allowed to warm, which of the gases will escape first?*

## Example 8

*Evaluate the following:*

*(a)* $9 \times (-4)$

*(b)* $(-8) \times 6$

**Solution**

(a) $9 \times (-4) = -(9 \times 4) = -36$

(b) $(-8) \times 6 = -(8 \times 6) = -48$

We know that the product of two positive integers is a positive integer. Is the product of two negative integers a negative integer or a positive integer?

Let us consider $(-1) \times (-1)$ first. We know $1 \times (-1) = -1$. It is reasonable to have $(-1) \times (-1) = -(1 \times (-1)) = +1$.

Similarly for $(-2) \times (-3)$, we have $(-2) \times (-3) = (-1) \times 2 \times (-1) \times 3 = -(-1) \times 2 \times 3 = +2 \times 3$.

**In general, if $x$ and $y$ are any two positive integers, then**

$$x \times y = +(x \times y) \text{ or } (-x) \times (-y) = +(x \times y).$$

**In other words, the product of two negative integers is a positive integer.**

## Example 9

*Evaluate the following:*

*(a)* $(-8) \times (-3)$

*(b)* $(-7) \times (-13)$

**Solution**

(a) $(-8) \times (-3) = +(8 \times 3) = 24$

(b) $(-7) \times (-13) = +(7 \times 13) = 91$

## Example 10

*Evaluate the following:*

*(a)* $(-8) \times (4) - (-5) \times 3$

*(b)* $(-7 + 6) \times (-11)$

*(c)* $9 \times (-2) \times (-2) \times 10$

*(d)* $(-6)^2$

Solution

(a) $(-8) \times (4) - (-5) \times 3 = -32 - (-15)$
$= -32 + 15$
$= -17$

*Notes*

*$-(-15)$ means $+(-1)(-15)$*

(b) $(-7 + 6) \times (-11) = (-1) \times (-11)$
$= 11$

(c) $9 \times (-2) \times (-2) \times 10 = (-18) \times (-2) \times 10 = 36 \times 10 = 360$

(d) $(-6)^2 = (-6) \times (-6) = 36$

*Notes*

*Since $(-6)^2 = 36$*

*So $-6$ is also a square root of 36*

## Division of Integers

We know division is the inverse operation of multiplication. For example, $3 \times 2 = 6$, $6 \div 3 = 2$ and $6 \div 2 = 3$ and similarly as $3 \times (-2) = -6$, we naturally have $(-6) \div 3 = -2 = -(6 \div 3)$ and $(-6) \div (-2) = 3$.

**In general, if $x$ and $y$ are any two positive integers, then**

**(1) $(-x) \div y = -(x \div y) = x \div (-y)$.**
**(2) $(-x) \div (-y) = +(x \div y)$.**

We know that if $x$ is a positive integer, $0 \times x = 0$.

Therefore, $0 \div x = 0$. Similarly, as $0 \times (-x) = 0$, we say $0 \div (-x) = 0$.

Thinking Time

Can we divide an integer by 0?

### Example 11

*Evaluate the following:*
*(a) $-100 \div (-20)$* *(b) $625 \div (-25)$* *(c) $0 \div (-14)$*

Solution

(a) $-100 \div (-20) = 100 \div 20 = 5$
(b) $625 \div (-25) = -(625 \div 25) = -25$
(c) $0 \div (-14) = 0$

## Exercise 2d

Evaluate the following:

**(a)** $3 \times (-7)$
**(b)** $(-3) \times (-2)$
**(c)** $(-20) \times (-14)$
**(d)** $(-4)^2$
**(e)** $-4 \times (-9)$
**(f)** $(-4) \div 2$
**(g)** $(-122) \div (-2)$
**(h)** $275 \div (-5)$
**(i)** $(-3) \times (-4) \times (-8) \times (-2)$
**(j)** $(-3) \times 0 \times (-8) \times (-5) \times 4$

# Rules for Operating on Integers

## (I) Commutative Law

We know that:

| Addition | Multiplication |
|---|---|
| e.g. $3 + 5 = 8 = 5 + 3$<br>$2 + (-5) = -3 = (-5) + 2$<br>$(-3) + (-4) = -7 = (-4) + (-3)$ | e.g. $3 \times 5 = 15 = 5 \times 3$<br>$2 \times (-3) = -6 = (-3) \times 2$<br>$(-3) \times (-4) = 12 = (-4) \times (-3)$ |

In general, if $x$ and $y$ represent two integers, then
$x + y = y + x$, this is called the commutative law of addition.
$x \times y = y \times x$, this is called the commutative law of multiplication.

**Thinking Time**

Do you think subtraction and division of integers are commutative?

e.g. Is $8 - 5 = 5 - 8$?
Is $9 \div 3 = 3 \div 9$?

## (II) Associative Law

We also know that:

| Addition | Multiplication |
|---|---|
| e.g. $(2 + (-5)) + 7 = -3 + 7 = 4$<br>$2 + ((-5) + 7) = 2 + 2 = 4$<br>$\therefore (2 + (-5)) + 7 = 2 + ((-5) + 7)$ | e.g. $(2 \times (-5)) \times 7 = (-10) \times 7 = -70$<br>$2 \times ((-5) \times 7) = 2 \times (-35) = -70$<br>$\therefore (2 \times (-5)) \times 7 = 2 \times ((-5) \times 7)$ |

In general, if $x$, $y$ and $z$ represent three integers, then
$(x + y) + z = x + (y + z)$, this is called the associative law of addition.
$(x \times y) \times z = x \times (y \times z)$, this is called the associative law of multiplication.

Thinking Time

Do you think subtraction and division of integers are associative?

e.g. Is $(9 - 3) - 4 = 9 - (3 - 4)$?
Is $(6 \div 2) \div 3 = 6 \div (2 \div 3)$?

## (III) Distributive Laws

John did 5 hours and 9 hours of community work in the first and second half of the year respectively. He was awarded 4 points for each hour of community work. Let us find the total points John received for the whole year.

The points he received can be calculated in two ways as shown below:

**(1)** $4 \times (5 + 9) = 4 \times 14 = 56$

**(2)** $(4 \times 5) + (4 \times 9) = 20 + 36 = 56$

Clearly, $4 \times (5 + 9) = (4 \times 5) + (4 \times 9)$

In general, for any three integers $x$, $y$ and $z$, we have $x \times (y + z) = x \times y + x \times z$. This is called the Distributive Law of Multiplication over Addition.

We also have the Distributive Law of Multiplication over Subtraction, that is for any three integers $x$, $y$ and $z$, we have

$$x \times (y - z) = x \times y - x \times z.$$

Let us look at more examples:

$$-2 \times (-3 + 4) = -2$$
$$-2 \times (-3) + (-2) \times 4 = -2$$
$$\therefore \quad -2 \times (-3 + 4) = -2 \times (-3) + (-2) \times 4$$

$$(-3 - 4) \times (-2) = 14$$
$$-3 \times (-2) - 4 \times (-2) = 14$$
$$\therefore \quad (-3 - 4) \times (-2) = -3 \times (-2) - 4 \times (-2)$$

Just For Fun

*If a cashier first scans in an item costing \$3 and then an item costing \$5, the total amount obtained must be the same as that obtained if the item costing \$5 is scanned in first.*
*In our everyday life, there are things that we must do in a certain order. For example, we have to put on our socks before we put on our shoes.*

*Can you give examples of things that we can do in any order and things that we must do in a definite order?*

The rules for the order of operations on **integers** are the same as those for whole numbers.

For example,

(a) $-3 \times (5 - 3) = -3 \times 2$
$= -6$
(Simplify the expression within the brackets first.)

(b) $-3 \times [-15 + (7 - 2)] = -3 \times (-15 + 5)$
$= -3 \times (-10)$
$= 30$
(Simplify the expression within the innermost pair of brackets first.)

(c) $-28 + 12 - 9 = (-28 + 12) - 9$
$= -16 - 9 = -25$
(Work from left to right.)

(d) $-125 \div 5 \times (-10) = (-125 \div 5) \times (-10)$
$= (-25) \times (-10) = 250$
(Work from left to right.)

(e) $-12 + (-3) \times 4 - 35 \div (-7) = -12 + [(-3) \times 4] - [35 \div (-7)]$
$= -12 + (-12) - (-5)$ (Do multiplication and division first.)
$= -24 + 5 = -19$

(f) $2 - (-3)^2 = 2 - (9) = -(9 - 2)$
$= -7$
$[(-3)^2 = (-3) \times (-3) = 9]$

## Exercise 2e

**1.** Copy and complete the following by filling in an appropriate operation symbol in each box:

**(a)** $(-5) \square 3 = 3 \square (-5) = -15$
**(b)** $23 \square (-11) = -11 \square 23 = 12$
**(c)** $-64 \square (-36) = -36 \square (-64) = -100$
**(d)** $-6 \square (-15) = -15 \square (-6) = 90$

**2.** Replace each $\square$ with an integer.

**(a)** $(-3) \times (\square + 8) = (-3) \times (-28) + (-3) \times 8 = \square$
**(b)** $(-16) \times 12 + 6 \times 12 = (-16 + 6) \times \square = \square$
**(c)** $(-5) \times (-4 - \square) = [(-5) \times (-4)] - [(-5) \times (-14)] = \square$
**(d)** $(-11) \times (8 - 9) = [(-11) \times \square] - [(-11) \times \square] = \square$
**(e)** $(-11 \times 10) - (-9 \times 10) = [-11 - (-9)] \times \square = \square$
**(f)** $[-15 - (-5)] \times (-9) = [\square \times (-9)] - [(-5) \times \square] = \square$

**★3.** Evaluate the following:

**(a)** $5 \times 2 - (-3)$
**(b)** $5 \times [3 \times (-2) - 10]$
**(c)** $\sqrt{10 - 3 \times (-2)}$
**(d)** $[3 - (-2)]^3$
**(e)** $24 \times (-2) \times 5 \div (-6)$
**(f)** $(16 - 24) - (57 - 77) \div (-2)$
**(g)** $[(12 - 18) \div 3 - 5] \times (-4)$
**(h)** $160 \div (-40) - 20 \div (-5)$
**(i)** $2 \times (-2) + (-2) \times (-3) + 2 \times (-3)$
**(j)** $\{[(-15 + 5) \times 2 + 8] - 32 \div 8\} - (-7)$
**(k)** $[3 \times (-3) - 4 \times (-2) + 5 \times 2] + [-2 \times (-3) + 8 \times (-2) - 8 \times 2]$

## SUMMARY

1. **Integers** are: ..., –4, –3, –2, –1, 0, 1, 2, 3, 4, ...

2. **Addition of Integers**

   **(a)** For any two negative integers $-x$ and $-y$,

$$-x + (-y) = -(x + y).$$

   **(b)** For a positive integer $x$ and a negative integer $-y$,

$$x + (-y) = x - y \quad \text{if } x > y \quad \text{and} \quad x + (-y) = -(y - x) \quad \text{if } y > x.$$

3. **Subtraction of Integers**

   For any two integers $a$ and $b$, $a - b = a + (-b)$.

4. **Multiplication of Integers**

   For any two positive integers $x$ and $y$,

   **(a)** $x \times (-y) = -(x \times y)$ and $(-x) \times y = -(x \times y)$,

   **(b)** $x \times y = +(x \times y)$ and $(-x) \times (-y) = +(x \times y)$.

5. **Division of Integers**

   For any two positive integers $x$ and $y$,

   **(a)** $0 \div x = 0$ and $0 \div (-x) = 0$,

   **(b)** $(-x) \div y = -(x \div y)$ and $x \div (-y) = -(x \div y)$,

   **(c)** $x \div y = +(x \div y)$ and $-x \div (-y) = +(x \div y)$.

## Review Examples 1

### Example 1

*The average of –30, –12 and a third number is –5. What is the third number?*

**Solution**

The sum of the three numbers = (–5) × 3 = –15

∴ –15 = The third number + (–30) + (–12)

∴ The third number = (–15) – (–30) – (–12)

= –15 + 30 + 12

= 27

## Example 2

*Evaluate each of the following without the use of a calculator*

*(a)* $2 \times (-5) - 7 \times (-3) - (-2)^2$ *(b)* $(-5)^3 - (-2)^3 + 3 \times (-4)^2$

### Solution

(a) $2 \times (-5) - 7 \times (-3) - (-2)^2 = -10 + 21 - (4)$
$= -10 + 21 - 4 = 7$

(b) $(-5)^3 - (-2)^3 + 3 \times (-4)^2 = -125 -(-8) + 3(16)$
$= -125 + 8 + 48$
$= -69$

## Review Questions 2

1. Evaluate each of the following:
   **(a)** $13 - (-54)$
   **(b)** $(-74) - (-46)$
   **(c)** $11 + (-33)$
   **(d)** $-12 - 88$
   **(e)** $[-13 + (-15)] + (-8)$
   **(f)** $500 - (-200) - 210 - 100$
   **(g)** $777 - (-111) - (-222) + 20$
   **(h)** $-(-129) + (-119) - 301 + 12$

2. Evaluate each of the following:
   **(a)** $(-4) \times (-5) \times (-6)$
   **(b)** $(-3 + 6) \times (-4)$
   **(c)** $(-3 - 5) \times (-3 - 4)$
   **(d)** $(-3 - 15) \div (-6)$
   **(e)** $4 \times (-5) \div (-2)$
   **(f)** $-5 \times 6 - 18 \div (-3)$
   **(g)** $2 \times (-3)^2 - 3 \times 4$
   **(h)** $[-3 \times (-2)] \times (2 - 5)^2$
   **(i)** $3 \times 5^2 - 2 \times (-3) \times 2$

3. Evaluate the following:
   **(a)** $(-2)^2 - (-2 \times 3) + (2 \times 3^2)$
   **(b)** $5 \times (-2)^3 + (-4)^2 \times (-3)$
   **(c)** $(-4)^2 \div (-8) + 3 \times (-2)^3$
   **(d)** $(-2)^3 - \left(-2 \times \frac{1}{2}\right) + 3(-1)^2$
   **(e)** $4 \times 3^2 \div (-6) - (-1)^3 \times (-3)^2$
   **(f)** $-2 \times (-2)^3 \times (-2) \times 3 + (-2) \times 3 \times (-1)^2$

4. Complete the following:

   **(a)**
$$\begin{array}{r} \square\;2\;\square \\ \times \quad \square\;7 \\ \hline 2\;2\;\square\;8 \\ \square\;6\;\square\;0 \\ \hline 1\;\square\;4\;6\;\square \end{array}$$

   **(b)**
$$\begin{array}{r} \square\;0\;\square\;\square\;3 \\ - \quad \square\;7\;2\;\square \\ \hline 7\;7\;7 \end{array}$$

   **(c)**
$$\begin{array}{r} \square\;\square \\ 28\,\overline{)\,1\;\square\;\square\;4} \\ 1\;\square\;\square\phantom{\;4} \\ \hline \square\;\square\;4 \\ \square\;\square\;4 \\ \hline 0 \end{array}$$

5. Fill in each of the □ with a digit so as to make the long division correct.

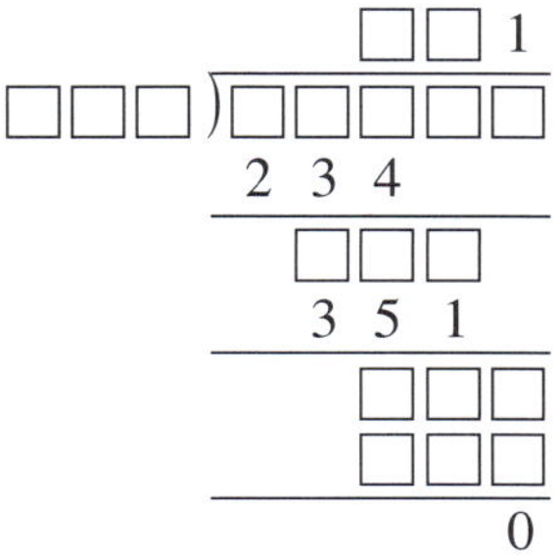


6. Fill in the numbers 1, 2, 3, 4, 5, 6, 7, 8 and 9 so that the sum of the numbers on each side of the triangle is 17.

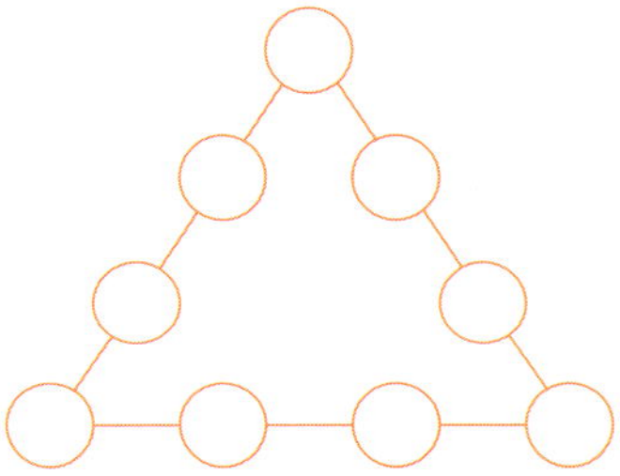

7. Insert +, –, ×, ÷ and brackets to make the following sentences true. The first one has been done for you.

**(a)** $(3 + 3) \div 3 - 3 \div 3 = 1$

**(b)** 3 3 3 3 3 = 2

**(c)** 3 3 3 3 3 = 3

**(d)** 3 3 3 3 3 = 4

**(e)** 3 3 3 3 3 = 5

**(f)** 3 3 3 3 3 = 6

**(g)** 3 3 3 3 3 = 7

**(h)** 3 3 3 3 3 = 8

**(i)** 3 3 3 3 3 = 9

**(j)** 3 3 3 3 3 = 10

***In this chapter, you will learn how to***

- ***recognise rational numbers;***
- ***perform calculations with rational numbers;***
- ***convert rational numbers into decimals;***
- ***convert decimals into rational numbers;***
- ***calculate rational numbers and real numbers using calculators.***

CHAPTER

# 3

# Rational Numbers

## Introduction

A good picture requires sufficient light to enter the camera and fall on the film. The amount of light is controlled by the size of the aperture (an opening in front of the camera) and the duration it remains open. The aperture varies in size, indicated by a sequence of numbers: 1, 4, 2, 2.8, 4, 5.6, 8, 11, 16 and 22, while the duration follow a sequence of fractions of a second:

$$\frac{1}{2}, \frac{1}{4}, \frac{1}{8}, \frac{1}{15}, \frac{1}{30}, \frac{1}{60}, \frac{1}{125}, \frac{1}{250}, \frac{1}{500}, \frac{1}{1000}, \frac{1}{2000} \text{ and } \frac{1}{4000}.$$

The above numbers are rational numbers.

The picture was taken with an aperture of size 5.6, left opened for $\frac{1}{250}$ of a second. The result is a sharp, frozen image of the moving basketball player. A longer duration of, say $\frac{1}{60}$ may lead to a blurred picture which may be a deliberate attempt to show action.

In this chapter, we will learn more about rational numbers.

# Rational Numbers

When we add two whole numbers, the result is always a whole number. However, if we subtract one whole number from another, do we always obtain a whole number? We know that $5 - 3 = 2$ and $3 - 5 = -2$.

not a whole number, it is an integer

**Notes**

*Whole numbers are 0, 1, 2, 3, ......*

Thus subtraction of whole numbers does not always produce a whole number. This explains the need to include negative numbers to form the set of integers.

**Thinking Time**

What will you get when adding or subtracting integers?

When we multiply two integers, we always obtain an integer. However, if we divide one integer by another, do we always obtain an integer?

We know that $6 \div 2 = 3$ and $2 \div 6 = \frac{1}{3}$.

not an integer, it is a fraction.

Thus, the division of two integers does not necessarily result in an integer. Hence, there is a need to include fractions to form a bigger set of numbers known as the set of **rational numbers**.

**For Your Information**

***Rational Numbers*** *are the set of numbers that includes integers and fractions.*

*The word 'rational' comes from the word 'ratio'. Thus 'irrational' means 'not ratio'.*

**A rational number is a number that can be expressed in the form $\frac{a}{b}$, where $a$ and $b$ are integers and $b \neq 0$.**

From the definition of a rational number, we see that fractions such as $-\frac{3}{4}$, $\frac{1}{2}$ and $\frac{5}{3}$ are rational numbers. If we replace $b$ in $\frac{a}{b}$ by the integer 1, we have $\frac{a}{b} = \frac{a}{1} = a$. Hence, the integers $6 = \frac{6}{1}$, $-5 = \frac{-5}{1}$, $-10 = \frac{-10}{1}$, $0 = \frac{0}{1}$ and so on are rational numbers.

**Thinking Time**

Is $\frac{-3}{4} = \frac{3}{-4} = -\frac{3}{4}$?

**For Your Information**

*$\pi$ is not a rational number. $\frac{22}{7}$ is only an approximate value for $\pi$.*

# Ordering of Rational Numbers

Fig. 3.1 shows the fractions $\frac{1}{2}$, $\frac{3}{4}$ and $\frac{13}{12}$ on the number line.

*Fig. 3.1*

**Notes**

*Divide the distance between 0 and 1 into 4 equal parts.*

From the number line, we know that $\frac{3}{4}$ is greater than $\frac{1}{2}$ and $\frac{13}{12}$ is greater than $\frac{12}{12}$ i.e. 1.

Where would you mark $\frac{1}{4}$, $\frac{7}{10}$ and $\frac{5}{4}$ on the above number line?

## Example 1

*Arrange the rational numbers $\frac{3}{4}$, $\frac{5}{9}$ and $\frac{7}{12}$ in ascending order.*

**Notes**

*Descending order means from the largest to the smallest.*

**Solution**

**Method 1:**

The rational numbers $\frac{3}{4}$, $\frac{5}{9}$ and $\frac{7}{12}$ have different denominators. To compare the rational numbers, we convert them to equivalent rational numbers with the same denominator.

The LCM of 4, 9 and 12 is 36.

$$\frac{3}{4} = \frac{3 \times 9}{4 \times 9} = \frac{27}{36}; \quad \frac{5}{9} = \frac{5 \times 4}{9 \times 4} = \frac{20}{36}; \quad \frac{7}{12} = \frac{7 \times 3}{12 \times 3} = \frac{21}{36}.$$

$$\frac{20}{36} < \frac{21}{36} < \frac{27}{36} \quad \text{or} \quad \frac{5}{9} < \frac{7}{12} < \frac{3}{4}.$$

Hence, the arrangement of the rational numbers in ascending order is $\frac{5}{9}, \frac{7}{12}, \frac{3}{4}$.

**Method 2:**

Express $\frac{3}{4}$, $\frac{5}{9}$ and $\frac{7}{12}$ in decimals,

$\frac{3}{4} = 0.75$, $\frac{5}{9} = 0.55$, $\frac{7}{12} = 0.58$

$\therefore \quad 0.55 < 0.58 < 0.75$

Hence $\frac{5}{9} < \frac{7}{12} < \frac{3}{4}$.

## Exercise 3a

1. Which of the three rational numbers is the largest?

   (a) $\frac{1}{17}, \frac{6}{17}, \frac{4}{17}$ (b) $\frac{3}{4}, \frac{7}{8}, \frac{3}{16}$

2. Peter, the centre forward on the basketball team, is $1\frac{7}{12}$ m tall. His rival for the position is $1\frac{5}{18}$ m tall. Who is more likely to be selected for the position if they are equal in all aspects except in height?

3. Mary, Susan and Joan went fishing. Mary caught a sea bass that weighed $2\frac{11}{13}$ kg; Susan caught one that weighed $2\frac{17}{18}$ kg and Joan caught one that weighed $2\frac{13}{14}$ kg. Who caught the heaviest sea bass?

4. Arrange the following rational numbers in ascending order:

   (a) $\frac{11}{12}, \frac{5}{8}, \frac{3}{4}$ (b) $\frac{7}{11}, \frac{15}{22}, \frac{14}{33}$

5. Arrange the following rational numbers in descending order:

   (a) $\frac{3}{4}, \frac{4}{5}, \frac{7}{10}, \frac{11}{12}$ (b) $\frac{7}{9}, \frac{5}{6}, \frac{13}{18}, \frac{2}{3}$

# Addition and Subtraction of Rational Numbers

In adding or subtracting rational numbers with different denominators, we must first express the rational numbers in the same denominator. We always use the LCM of the denominators as the common denominator.

### Example 2

*Evaluate (a)* $3\frac{8}{15} + 1\frac{5}{8}$; *(b)* $6\frac{1}{16} - 1\frac{17}{24}$.

**Solution**

(a) $3\frac{8}{15} + 1\frac{5}{8} = (3 + 1) + \left(\frac{8}{15} + \frac{5}{8}\right)$ (Add the whole numbers and rational numbers separately.)

$= 4 + \frac{64 + 75}{120}$

$= 4 + \frac{139}{120} = 4 + 1\frac{19}{120} = 5\frac{19}{120}$

**Notes**

*The LCM of 15 and 8 is 120.*

(b) $6\frac{1}{16} - 1\frac{17}{24} = (5 - 1) + \left(1\frac{1}{16} - \frac{17}{24}\right)$

$= 4 + \left(\frac{17}{16} - \frac{17}{24}\right)$

$= 4 + \frac{51 - 34}{48}$

$= 4\frac{17}{48}$

Notes

*The LCM of 16 and 24 is 48.*

Thinking Time

Do you know why we have written $6\frac{1}{16}$ as $5 + 1\frac{1}{16}$ instead of $6 + \frac{1}{16}$?

## Exercise 3b

1. Evaluate the following:
   (a) $3\frac{3}{5} + \frac{1}{5}$
   (b) $2\frac{1}{7} + 3\frac{5}{8}$
   (c) $2\frac{7}{100} + 1\frac{3}{40}$
   (d) $7\frac{1}{5} - 3\frac{3}{10}$
   (e) $9\frac{2}{5} - 5\frac{1}{3}$

2. Evaluate the following:
   (a) $\frac{4}{7} + \frac{5}{21} + \frac{11}{42}$
   (b) $\frac{5}{8} + \frac{21}{10} - \frac{17}{20}$
   (c) $4\frac{2}{3} + 1\frac{3}{5} - 1\frac{1}{4}$
   (d) $4\frac{3}{5} - 1\frac{2}{15} - 2\frac{7}{9} - \frac{2}{45}$

3. On Thursday, Jean ran $1\frac{17}{20}$ km. On Friday, she ran $2\frac{8}{15}$ km. Find the total distance she ran in the two days.

4. John has $9\frac{5}{16}$ litres of paint. After using $4\frac{11}{12}$ litres for painting a room, how much paint has he left?

5. During the school's spring cleaning day, Robin and his classmates spent $1\frac{1}{4}$ hours and $1\frac{1}{12}$ hours cleaning two classrooms assigned to them respectively.
   (a) Find the total time they spent.
   (b) Which classroom took them longer to clean?
   (c) How much longer did they spend in cleaning one classroom than the other?

6. Mary had $6\frac{3}{4}$ cups of flour. She used $2\frac{7}{12}$ cups of flour in one recipe and $2\frac{13}{24}$ cups of flour in another.
**(a)** How much flour did she use altogether?
**(b)** How much flour had she left?

7. The police conduct regular anti-secret society operations among youths. On a Friday night, the officers spent $\frac{3}{4}$ hour, $1\frac{1}{2}$ hours and $2\frac{1}{4}$ hours checking on youths at three youth hangouts respectively. Find the total time the officers spent on their operation.

# Multiplication and Division of Rational Numbers

A rule for the multiplication of rational numbers.

$$\frac{a}{b} \times \frac{c}{d} = \frac{a \times c}{b \times d}, \text{ where } a, b, c, d \text{ are integers and } b \neq 0, d \neq 0.$$

A rule for the division of rational numbers.

$$\frac{a}{b} \div \frac{c}{d} = \frac{a}{b} \times \frac{d}{c} = \frac{a \times d}{b \times c}, \text{ where } a, b, c, d \text{ are integers and } b \neq 0, c \neq 0, d \neq 0.$$

## Example 3

*Evaluate* (a) $2\frac{2}{3} \times \frac{5}{14}$; (b) $\frac{24}{55} \div \frac{-8}{11}$; (c) $1\frac{4}{13} \times \left(-7\frac{4}{5}\right) \div 11\frac{1}{3}$.

*Just for Fun*

*$\frac{1}{3}$ of a number of animals are cows, $\frac{1}{4}$ are sheep. $\frac{1}{5}$ are horses, $\frac{1}{6}$ are deer and 4 are dogs. How many animals are there altogether?*

**Solution**

(a) $2\frac{2}{3} \times \frac{5}{14} = \frac{\cancel{8}^{4}}{3} \times \frac{5}{\cancel{14}_{7}}$

$= \frac{4 \times 5}{3 \times 7} = \frac{20}{21}$

(b) $\frac{24}{55} \div \frac{-8}{11} = \frac{24}{55} \times \frac{-11}{8}$

$= -\frac{3}{5}$

(c) $1\frac{4}{13} \times \left(-7\frac{4}{5}\right) \div 11\frac{1}{3} = \frac{17}{13} \times \left(\frac{-39}{5}\right) \div \frac{34}{3}$

$= \frac{17}{13} \times \frac{-39}{5} \times \frac{3}{34}$

$= -\frac{9}{10}$

## Exercise 3c

1. Evaluate the following, expressing your answers in the simplest form:

   (a) $3\frac{1}{2} \times 4\frac{4}{5} \times \frac{5}{14}$

   (b) $5\frac{1}{4} \div 2\frac{4}{5} \div 1\frac{7}{9}$

   (c) $3\frac{1}{9} \times \left(-3\frac{3}{5}\right) \div 2\frac{1}{10}$

2. A school has $6\frac{3}{4}$ kg of detergent in stock. During the 'Use Your Hands' campaign, each class will be given $\frac{3}{8}$ kg of detergent. There are 28 classes in the school.
   **(a)** What fraction of the school will be supplied with the detergent in stock?
   **(b)** How much detergent will be required altogether for the whole school?
   **(c)** How much more detergent does the school need to order?
   **(d)** If the school gives out the detergent in stock to the 15 lower secondary classes first,
   **(i)** how much detergent will be given out;
   **(ii)** how much detergent in stock will be left?

3. **(a)** Last year, Peter spent a total of $8\frac{1}{6}$ hours on community service. Visits to old folks' homes made up $\frac{4}{7}$ of the total time. How much time did he spend in visiting old folks' homes?
   **(b)** This year, Peter plans to spend $1\frac{1}{5}$ of his time spent last year on community service.
   **(i)** Find the time he will spend on community service.
   **(ii)** How much more time will he spend this year than last year?

*4. Five boys Alan, Ben, Chris, Den and Eng share a sum of money.
   Alan takes $\frac{1}{5}$ of the sum of money.
   After Alan has taken his share, Ben takes $\frac{1}{3}$ of the remaining money.
   After Ben has taken his share, Chris takes $\frac{1}{4}$ of the remaining money.
   After Chris has taken his share, Den takes $\frac{1}{7}$ of the remaining money.
   After Den has taken his share, Eng takes all the remaining money.
   What fraction of the sum of money is Eng's share?

# Arithmetical Operations on Rational Numbers

The rules for the operations on rational numbers are the same as the rules for whole numbers.

## Example 4

*Evaluate the following:*

*(a)* $\frac{1}{3} \times \left(\frac{1}{2} + \frac{1}{4}\right) \div \frac{1}{6}$

*(b)* $\left(\frac{3}{4} \times \frac{1}{3} - \frac{1}{12}\right) \div \frac{1}{2}$

### Solution

(a) $\frac{1}{3} \times \left(\frac{1}{2} + \frac{1}{4}\right) \div \frac{1}{6}$

$= \frac{1}{3} \times \left(\frac{2+1}{4}\right) \div \frac{1}{6}$ (Calculate the expression within the brackets first. Work from left to right.)

$= \frac{1}{3} \times \frac{3}{4} \times \frac{6}{1}$

$= \frac{3}{2} = 1\frac{1}{2}$

(b) $\left(\frac{3}{4} \times \frac{1}{3} - \frac{1}{12}\right) \div \frac{1}{2}$

$= \left(\frac{1}{4} - \frac{1}{12}\right) \div \frac{1}{2}$ (Within brackets, do multiplication before subtraction.)

$= \left(\frac{3-1}{12}\right) \times \frac{2}{1}$

$= \frac{2}{12} \times 2 = \frac{1}{3}$

## Explorations

Can you spot the errors in each of the following?

(a) $2 = \frac{8}{4} = \frac{8}{2+2} = \frac{8}{2} + \frac{8}{2} = 4 + 4 = 8$

(b) $\frac{3+4}{3} = \frac{3}{3} + 4 = 1 + 4 = 5$

(c) $\frac{2}{3} + \frac{4}{5} = \frac{10}{15} + \frac{12}{15} = \frac{10+12}{15+15} = \frac{22}{30} = \frac{11}{15}$

## Exercise 3d

**1.** Evaluate the following, expressing your answers in the simplest form.

**(a)** $\left(\frac{4}{5} + \frac{1}{3}\right) \div \frac{2}{3}$

**(b)** $\frac{1}{3} \times \left(\frac{1}{3} + \frac{3}{4}\right) \div \frac{5}{2}$

**(c)** $\frac{1}{3} \times \left(\frac{1}{4} - \frac{1}{12} + \frac{1}{2}\right)$

**(d)** $\left(\frac{1}{2} + \frac{1}{3}\right) \div \left(\frac{2}{3} \times \frac{1}{8}\right)$

**(e)** $\frac{2}{3} \times \frac{1}{4} - \frac{1}{12} \div \frac{1}{2}$

**2.** Evaluate the following. Express each answer in its lowest terms.

**(a)** $\frac{5}{8} - \left(-\frac{1}{8}\right)$

**(b)** $-\frac{13}{15} + \left(-\frac{2}{15}\right)$

**(c)** $2\frac{1}{5} - 3\frac{3}{10}$

**(d)** $-\frac{1}{2} + \frac{1}{4} - \frac{2}{3}$

**(e)** $-\frac{1}{10} + \left(-\frac{4}{5}\right) - \left(-\frac{1}{4}\right)$

**(f)** $-\frac{1}{2} - \frac{1}{4} - \left(-\frac{9}{10}\right)$

**(g)** $\frac{1}{3} + \left(\frac{-1}{2}\right)^2$

**3.** Evaluate the following. Express each answer in its lowest terms.

**(a)** $-5 \times \left(-\frac{9}{10}\right)$

**(b)** $\frac{4}{9} \times \left(-\frac{3}{10}\right)$

**(c)** $\frac{3}{4} \div \left(\frac{-2}{3}\right)^2$

**(d)** $\frac{11}{12} \times \left(-\frac{23}{33} + \frac{7}{11}\right)$

**(e)** $\frac{3}{5} \times \left(-\frac{1}{4} - \frac{1}{6}\right) \div \left(-\frac{7}{3} + \frac{5}{4}\right)$

# Problem Solving Involving Rational Numbers

## Example 5

*Four boys shared a certain sum of money. The first received* $\frac{1}{6}$ *of it, the second* $\frac{1}{4}$ *and the third* $\frac{1}{2}$. *If the fourth boy received \$5, how much was the sum of money shared?*

### Solution

**Use a model**

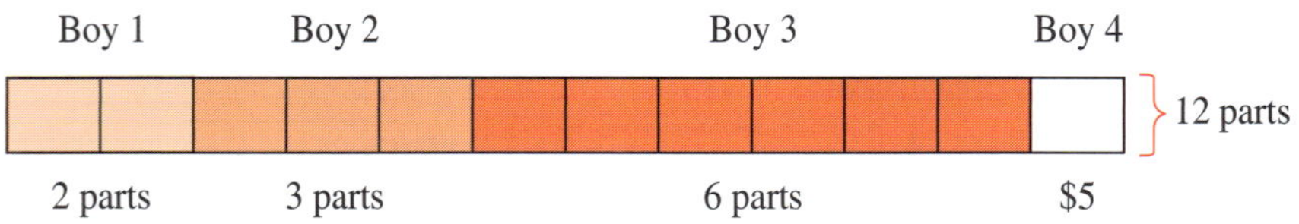


From the model, the fourth boy shared 1 part of a total of 12 parts.

$\therefore$ 1 part = \$5 and thus, 12 parts = 12 × \$5 = \$60.

The sum of money shared was \$60.

Alternatively, the first three boys together received $\frac{1}{6} + \frac{1}{4} + \frac{1}{2} = \frac{2 + 3 + 6}{12} = \frac{11}{12}$ of the money. The fourth boy thus received $\left(1 - \frac{11}{12}\right) = \frac{1}{12}$ of the money.

$\frac{1}{12}$ of the money = \$5 or $\frac{1}{12}$ × the money = \$5.

$\therefore$ the money = $\$5 \div \frac{1}{12} = \$5 \times 12 = \$60$.

## Example 6

*Peter has 45 English and Chinese books. $\frac{4}{5}$ of the English books and $\frac{3}{4}$ of the Chinese books are fiction. The total number of fiction books he has is 35.*

*How many of his fiction books are in English?*

### Solution

**Method 1: Make a list**

| | |
|---|---|
| Fraction of English fiction books | $\frac{4}{5}, \frac{8}{10}, \frac{12}{15}, \frac{16}{20}, \mathbf{\frac{20}{25}}, \frac{24}{30}, \ldots$ |
| Fraction of Chinese fiction books | $\frac{3}{4}, \frac{6}{8}, \frac{9}{12}, \frac{12}{16}, \mathbf{\frac{15}{20}}, \frac{16}{24}, \ldots$ |

Study the fractions in bold from the lists.

The sum of the numerators = 20 (number of English fiction) + 15 (number of Chinese fiction)
= 35 (number of fiction Peter has)

The sum of the denominators = 25 + 20 = 45 (Total number of books Peter has)

$\therefore$ 20 of his fiction books are in English.

**Method 2: Use a model**

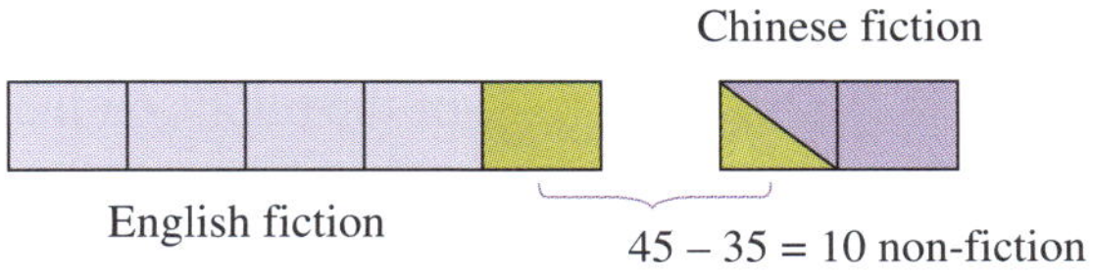


Rearrange the rectangles and triangles as shown below.

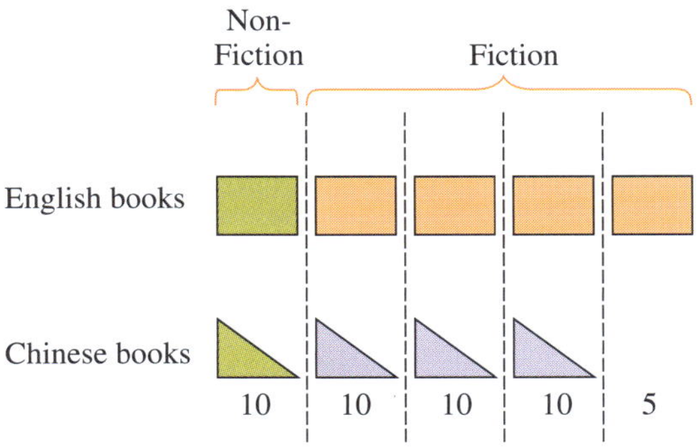


Each pair of rectangle and triangle represents 10 books.

A single rectangle represents 45 – 4 × 10 = 5 books.

$\therefore$ The number of English fiction books = 4 × 5 = 20.

## Exercise 3e

1. James uses $\frac{1}{3}$ of his land for growing durians, $\frac{1}{4}$ for bananas, $\frac{3}{8}$ for guavas and the remaining 9 hectares for mangoes. What is the total area of his land?

2. A sum of money is shared among three brothers. The eldest receives $\frac{7}{13}$ of it and the next receives $\frac{2}{3}$ of the remainder. If the youngest brother receives \$6, find the sum of money shared.

3. There are 42 pupils in a class. $\frac{3}{4}$ of the boys and $\frac{2}{3}$ of the girls travel to school by bus. The total number of boys and girls who travel to school by bus is 30.
   **(a)** How many boys are there in the class?
   **(b)** How many girls travel to school by bus?

4. The Mathematics Club in a school has 53 members. $\frac{5}{7}$ of the girl members and $\frac{3}{5}$ of the boy members are from lower secondary. The total number of lower secondary members is 35. How many members of the club are boys?

5. Mary has two tanks of fish. If she transfers 15 fish from the larger tank to the smaller tank, then the number of fish in the smaller tank will be $\frac{5}{7}$ of the number of fish in the larger tank. Given that there are 35 fish in the smaller tank originally, find the number of fish in the larger tank before the transfer of fish.

6. As part of the 'Learning Journey' program, 73 pupils travelled on two buses, one air-conditioned and one non-air-conditioned, to the Singapore Discovery Centre. Three-fifths of the pupils on the air-conditioned bus were girls. There were 17 boys in the non-air-conditioned bus. The number of girls on the two buses were equal. How many pupils were there on the air-conditioned bus?

# Terminating and Recurring Decimals

## Explorations

1. Divide 1 by 8. Does the division process come to an end?
2. Divide 1 by 3. Does the division process come to an end? If it does not, do you observe anything interesting? Can you suggest a way of writing the answer?
3. Divide 7 by 12. Does any digit keep repeating itself? Suggest a way to write the answer.
4. Divide 13 by 99. How many digits keep repeating? Suggest a way to write the answer.
5. Divide 41 by 333. How many digits keep repeating? Suggest a way to write the answer.
6. Divide 1 by 7. Suggest a way to write the answer.

There is an end to the division process when we divide 1 by 8, i.e., $1 \div 8 = 0.125$. Therefore, 0.125 is called a **terminating** decimal.

When dividing 1 by 3, the digit 3 repeats itself and we get

$$1 \div 3 = \frac{1}{3} = 0.333\ ... = 0.\dot{3}.$$

$0.\dot{3}$ is called a **recurring decimal**. The dot above '3' indicates that '3' is the repeating digit. Similarly, we write

$$\frac{7}{12} = 0.583\,333\ ... = 0.58\dot{3}$$

$$\frac{41}{333} = 0.123\,123\,123\ ...$$
$$= 0.\dot{1}2\dot{3}$$

$$\frac{13}{99} = 0.13\,13\,13\ ... = 0.\dot{1}\dot{3}$$

$$\frac{1}{7} = 0.142\,857\,142\,857\ ...$$
$$= 0.\dot{1}42\,85\dot{7}$$

Some numbers such as $\frac{5}{27}$ ($= 0.185\,185\,185\ ...$) cannot be written as decimal with a finite number of decimal places, but result in a number of digits that repeat infinitely. These are called **recurring decimals**.

**Just for fun**

*Would you like to order a pizza? Just call $62\frac{35}{99}$?*

**Notes**

*It is more convenient to write $0.\dot{1}\dot{2}\dot{3}$ as $0.\dot{1}2\dot{3}$ and $0.\dot{1}\dot{4}\dot{2}\,\dot{8}\dot{5}\dot{7}$ as $0.\dot{1}42\,85\dot{7}$. Hence, the first dot and the last dot mark the beginning and end of a repeating block of digits.*

## Exercise 3f

1. Express the following fractions as recurring decimals:

   (a) $\frac{2}{3}$  (b) $\frac{5}{11}$

   (c) $\frac{29}{33}$  (d) $\frac{5}{22}$

   (e) $\frac{4}{7}$  (f) $\frac{35}{72}$

2. Express each of the following as a decimal and indicate whether it is recurring or non-recurring:

   (a) $\frac{4}{9}$  (b) $\frac{7}{3}$

   (c) $\frac{61}{90}$  (d) $\frac{59}{99}$

   (e) $\frac{97}{125}$  (f) $\frac{13}{44}$

# Use of Calculators on Real Numbers

All the numbers on the number line form the set of **real numbers**.

The relationship between the various sets of numbers we have discussed is shown in Fig. 3.2.

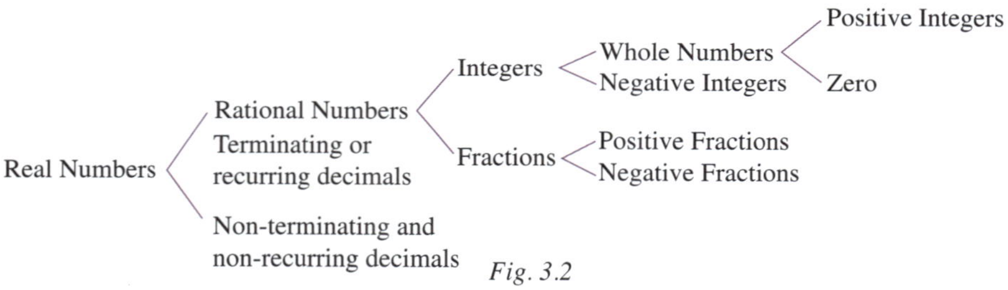


*Fig. 3.2*

**Notes**

*There are some real numbers which are not rational numbers for example, $\sqrt{2}$, $\sqrt{5}$ and $\pi$. They are called irrational numbers.*

In primary schools, we have learnt how to do operations involving whole numbers, fractions and decimals using a calculator. We have also learnt how to use the $\sqrt{\ }$ key, $\sqrt[x]{\ }$ and $y^x$ key. Now we shall have more practice with these and use three more operating keys to evaluate real numbers.

| | |
|---|---|
| +/– | Sign Change Key |
| $\pi$ | Pi Key, which is 3.141 592 654 correct to 9 decimal places. |
| $a^{b}/_{c}$ | Fraction Key |

For example,

(a) to find $-65 + 47 \times (-79)$, press [+/–] 65 [+] 47 [×] [+/–] 79 [=] to get –3778,

(b) to find $-3\frac{2}{3} \div \left(-1\frac{5}{6}\right)$, press [+/–] 3 [$a^b/_c$] 2 [$a^b/_c$] 3 [÷] [+/–] 1 [$a^b/_c$] 5 [$a^b/_c$] 6 [=] to get 2,

(c) to find $-23.4 \div \pi$, press [+/–] 23.4 [÷] [$\pi$] [=] to get –7.448 451 33.

Like $\pi$, real numbers which are not rational numbers, such as $\sqrt{2}$, $\sqrt{3}$, $\sqrt{5}$, $\sqrt[3]{2}$, $\sqrt[3]{3}$, $\sqrt[3]{4}$, cannot be expressed as a terminating or recurring decimal. A calculator provides approximate values of these numbers.

For example, to find $\sqrt{2}$, press [$\sqrt{\ }$] 2 [=] to get 1.414 213 562,

to find $\sqrt{3}$, press [$\sqrt{\ }$] 3 [=] to get 1.732 050 808,

to find $\sqrt[3]{4}$, press 3 [$\sqrt[x]{\ }$] 4 [=] to get 1.587 401 052,

to find $\sqrt[3]{5}$, press 3 [$\sqrt[x]{\ }$] 5 [=] to get 1.709 975 947.

As your calculator might be different, please check your calculator manual before using it.

## Example 7

*Evaluate each of the following using a calculator:*

(a) $\dfrac{3.2^3 + 4.3^2}{\sqrt{47.5} - 2.74}$ (b) $\dfrac{\frac{2}{5} + \frac{3}{4}}{1\frac{2}{3} + \frac{3}{4} - \frac{5}{6}}$

**Solution**

| The steps: | Final display: |
| --- | --- |
| (a) 3.2 [$y^x$] 3 [+] 4.3 [$x^2$] [=] [÷] [(] [$\sqrt{\ }$] 47.5 – 2.74 [)] [=] | 12.345 303 24 |
| or [(] 3.2 [$y^x$] 3 [+] 4.3 [$x^2$] [)] ÷ [(] [$\sqrt{\ }$] 47.5 – 2.74 [)] [=] | |
| (b) 1 [$a^b/_c$] 2 [$a^b/_c$] 3 [+] 3 [$a^b/_c$] 4 [–] 5 [$a^b/_c$] 6 [=] [STO] 2 [$a^b/_c$] 5 [+] 3 [$a^b/_c$] 4 [=] [÷] [RCL] [=] | $\frac{69}{95}$ |

## Example 8

*Use a calculator to evaluate (a)* $\dfrac{\sqrt{3}}{2}$, *(b)* $\dfrac{1}{\sqrt{7}-1}$ *and (c)* $\dfrac{\sqrt{5}+\sqrt{2}}{\sqrt{2}-\sqrt{5}}$ *correct to 2 decimal places.*

### Solution

| Sequence of pressing keys: | Final display: |
|---|---|
| (a) [√] 3 [÷] 2 [=] | 0.866 025 403 |
| (b) 1 [÷] [(] [√] 7 [−] 1 [)] [=] | 0.607 625 218 |
| (c) [√] 5 + [√] 2 [=] [÷] [(] [√] 2 − [√] 5 [)] [=] | −4.44 151 844 |

The answers are thus (a) 0.87, (b) 0.61 and (c) –4.44.

## Exercise 3g

1. State whether each of the following is a rational number:
   (a) $\dfrac{1}{5}$ (b) $-4$
   (c) 0.6 (d) $\sqrt{6}$
   (e) $\sqrt{9}$ (f) 0
   (g) $2\pi$ (h) $\sqrt[3]{8}$
   (i) $\dfrac{22}{7}$ (j) $\sqrt{1000}$

2. State whether each of the following statements is true or false:
   **(a)** An integer is a rational number.
   **(b)** A rational number is an integer.
   **(c)** The set of real numbers consists of positive numbers, negative numbers and zero.

3. Use a calculator to compute each of the following:
   **(a)** $-5165 + 2844 + 8416$
   **(b)** $-4715 \times (-78)$
   **(c)** $-29\ 187 \div 69$
   **(d)** $597 \times (-57) - 4648 \div (-83)$
   **(e)** $-\dfrac{2}{21} + \dfrac{1}{43}$
   **(f)** $-\dfrac{47}{98} \times \dfrac{518}{77}$
   **(g)** $-\dfrac{51}{23} \div \left(-\dfrac{48}{62}\right)$
   **(h)** $\dfrac{-\dfrac{7}{19} + \dfrac{5}{18}}{-\dfrac{6}{13} - \left(-\dfrac{11}{7}\right)}$
   ***(i)** $\dfrac{\left(-\dfrac{4}{7}\right)^2 - \left(-\dfrac{2}{5}\right)^3}{-\sqrt{\dfrac{64}{625}} \div \sqrt[3]{-\dfrac{8}{125}}}$

**4.** Use a calculator to evaluate the following. Give your answers correct to 2 decimal places.

**(a)** $\sqrt{129}$

**(b)** $\sqrt[3]{81}$

**(c)** $-\dfrac{\pi^2}{4}$

**(d)** $\sqrt{\dfrac{45}{8}}$

**(e)** $\sqrt{5} - \sqrt{3}$

**(f)** $\sqrt{14^2 + 19^2}$

**(g)** $\pi \times (79.67)^2$

**(h)** $\dfrac{1}{3} \times \pi \times (43.6)^2 \times 56.9$

**✶(i)** $\dfrac{\sqrt[3]{12} - \sqrt{6}}{\sqrt{50} - \sqrt[3]{111}}$

**✶(j)** $\sqrt{\dfrac{46^2 + 83^2 - 65^2}{2 \times 46 \times 83}}$

## SUMMARY

**1. Rational Numbers**

A rational number is a number which can be expressed in the form $\frac{a}{b}$, where $a$ and $b$ are integers and $b \neq 0$. A rational number can be expressed as a non-recurring or recurring decimal.

**2. Real Numbers**

- Real Numbers
  - Rational Numbers / Terminating or recurring decimals
    - Integers
      - Whole Numbers
        - Positive Integers
        - Zero
      - Negative Integers
    - Fractions
      - Positive Fractions
      - Negative Fractions
  - Non-terminating and non-recurring decimals

# Review Examples 3

## Example 1

*Each month, the Lee's family spends $\frac{5}{18}$ of its income on food, $\frac{4}{15}$ on housing loan, $\frac{1}{24}$ on clothes and $\frac{7}{36}$ on transport. If there is \$790 left, calculate the monthly income.*

**Solution**

Fraction of income left $= 1 - \frac{5}{18} - \frac{4}{15} - \frac{1}{24} - \frac{7}{36}$

$= \frac{79}{360}$

i.e. $\frac{79}{360} \times$ Monthly Income = \$790

$\therefore$ Monthly Income = \$790 $\div \frac{79}{360}$

= \$3600

## Example 2

*Evaluate* $\left(1 - \frac{1}{3}\right) \times \left(1 - \frac{1}{4}\right) \times \left(1 - \frac{1}{5}\right) \times \ldots \times \left(1 - \frac{1}{100}\right)$.

**Solution**

$\left(1 - \frac{1}{3}\right) \times \left(1 - \frac{1}{4}\right) \times \left(1 - \frac{1}{5}\right) \times \ldots \times \left(1 - \frac{1}{99}\right)\left(1 - \frac{1}{100}\right)$

$= \frac{2}{\cancel{3}} \times \frac{\cancel{3}}{\cancel{4}} \times \frac{\cancel{4}}{5} \times \ldots \times \frac{98}{\cancel{99}} \times \frac{\cancel{99}}{100}$

$= \frac{2}{1} \times \frac{1}{100}$

$= \frac{1}{50}$

# Review Questions 3

1. Evaluate the following, expressing your answers in the simplest form:

(a) $2\frac{1}{3} + 1\frac{5}{12}$

(b) $4\frac{5}{7} - 2\frac{5}{21}$

(c) $8\frac{1}{3} - 1\frac{1}{9} - \frac{5}{18} - 2\frac{5}{6}$

(d) $2\frac{17}{27} - \frac{1}{18} + \frac{61}{72} - \frac{1}{54}$

2. Calculate the following without using a calculator:

(a) $\frac{6}{15} \times \frac{8}{9} \div \frac{8}{45}$

(b) $\dfrac{\frac{3}{5} - \frac{1}{2}}{\frac{2}{5}}$

(c) $\frac{3}{4} - \left(\frac{2}{3} \times \frac{1}{3} + \frac{1}{9}\right) \div \frac{4}{9}$

(d) $\frac{3}{49} \div \frac{1}{7} + \left(\frac{1}{5} + \frac{7}{10}\right) \times \frac{4}{21}$

(e) $\frac{1}{2} - \frac{1}{3} - \frac{2}{7} \times \frac{21}{24}$

(f) $\frac{1}{2} \times \frac{3}{5} \times \left(\frac{7}{9} + \frac{1}{3}\right)$

3. Use a calculator to evaluate each of the following, giving your answer correct to 2 decimal places if it is not exact:

(a) $\left(1\frac{1}{4}\right)^3 \times \frac{6}{25} \div \frac{5}{6}$

(b) $\dfrac{9.35^2 + 96.1^3}{65.9^3 - 38.4^2}$

★(c) $\dfrac{\frac{4}{7} + \frac{2}{9}}{2\frac{3}{5} - 1\frac{5}{8} + 5\frac{3}{4}}$

(d) $522.76 \times 647.15 \div 322.11 + 631.7 \div 524.38$

(e) $\frac{4}{9} \times \left(\frac{5}{8} + \frac{5}{6} - \frac{2}{3}\right) \div \frac{2}{5} \times \left(\frac{3}{4} + \frac{1}{5}\right)$

4. Which of the two rational numbers $\frac{1993}{1994}$ and $\frac{1995}{1996}$ is larger? Explain the reason without making the denominators of the two fractions the same or converting the fractions into decimals.

5. We can write $\frac{1}{6}$ as the sum of two fractions with numerators equal to 1. For example, $\frac{1}{6} = \frac{1}{12} + \frac{1}{12}$ and $\frac{1}{6} = \frac{1}{7} + \frac{1}{42}$.

There are three other ways of writing $\frac{1}{6}$ in the same form. Can you find them?

6. Calculate $\left(5\frac{1}{3} + \dfrac{2\frac{1}{4} - 1\frac{1}{5}}{1\frac{1}{3} + 1\frac{1}{2}}\right) \times \dfrac{1\frac{1}{3} + 1\frac{1}{2}}{5\frac{1}{3} \times \left(1\frac{1}{3} + 1\frac{1}{2}\right) + 2\frac{1}{4} - 1\frac{1}{5}}$. [Hint: observe carefully before using your calculator.]

7. In a boys' school, $\frac{5}{8}$ of the boys play football and $\frac{4}{7}$ play rugby. If every boy plays at least one of the two games, find the fraction of the boys who play both.

*In this chapter, you will learn*

- *how to round off numbers and measures to a specified degree of accuracy;*
- *how to make estimates of numbers and measures;*
- *the concept of rounding and truncation errors.*

# Estimation

CHAPTER

# 4

# and Approximation

## Introduction

Can you estimate the number of oranges in the picture shown? Many a time we have to make estimation and approximation. A methodical way of making estimations is a far better method than making a wild guesses.

# Estimation and Rounding

In our daily life, we often need to use estimation when getting a precise answer is impossible, unnecessary, or inconvenient.

Estimation often involves rounding. In Mathematics we **round off** a number to the nearest. For example, we round off 3.824 to 3.82 and round off 2.815 to 2.82.

**Notes**

*In rounding, we may round up, round down, or round off to the nearest. For example, we round up 850 to 1000 when we budget for a trip that will cost at least $850. At the car park, the car park charges are often rounded up. For example, you park for 1 hour and 50 minutes, you pay $2 at a parking rate of 50 cents for every half hour. At some supermarkets, the bill is round down to the nearest 5 cents. For example, if your bill is $12.03, you pay $12.00 and if your bill is $12.09, you pay $12.05.*

## Example 1

*Estimate the cost of 8 copies of a textbook at $10.49 each.*

### Solution

To estimate the cost of 8 copies of the textbook, we may choose to round $10.49 to the nearest 10 cents and obtain the result as shown below. The actual cost is also provided for comparison.

| Estimation cost: | | Actual cost: | |
|---|---|---|---|
| | $10.50 | | $10.49 |
| | × 8 | | × 8 |
| | $84.00 | | $83.92 |

For a quick estimate, which can be done mentally, round $10.49 to the nearest dollar and the estimate is $10 × 8 = $80.

## Explorations

You may carry out this activity individually.

It is useful to estimate the total sum of a bill to avoid overpayment.

For example you can estimate the total amount of a supermarket receipt. This is done by rounding the cost of each item to the nearest 50 cents and keeping a running total mentally from the first item to the last item.

The table below is drawn up from a supermarket receipt. Observe how the estimation is done.

| Sale | Actual Cost ($) | Estimated Cost ($) | Running total ($) |
|---|---|---|---|
| P Butter | 4.50 | 4.50 | 4.50 |
| B/Sardine | 3.50 | 3.50 | 8.00 |
| Cod Fish | 1.30 | 1.50 | 9.50 |
| M+M Plain | 0.60 | 0.50 | 10.00 |
| Almond Cho | 2.85 | 3.00 | 13.00 |
| Hi Lo Milk | 2.85 | 3.00 | 16.00 |
| F/P Kaya | 1.60 | 1.50 | 17.50 |
| F/Spread | 2.85 | 3.00 | 20.50 |
| Tapioca Crisp | 2.40 | 2.50 | 23.00 |
| Subtotal | 22.45 | | |
| Total | 22.45 | | |

$\therefore$ The estimated total amount = $23.00

You notice that the estimated total is very close to the actual total.

1. Collect supermarket receipts with at least 5 items. Without seeing the actual total, estimate the total amount by rounding the cost of each item and adding them mentally. Compare your estimate with the actual total.
2. Exchange the receipts with other students and repeat the process.
3. Choose a receipt with more than 10 items. Use it to start a friendly competition among your classmates. Give a round of applause to the winner, the student who takes the shortest time to estimate the total mentally.

## Example 2

*Make an estimate and pick the nearest answer in each of the following cases:*

| | | | | | |
|---|---|---|---|---|---|
| (a) $\dfrac{3902}{23\,839}$ | (i) 0.02 | (ii) 0.2 | (iii) 2 | (iv) 20 | (v) 200 |
| (b) $\dfrac{4.19 \times 0.0309}{0.0222}$ | (i) 0.006 | (ii) 0.06 | (iii) 0.6 | (iv) 6 | (v) 60 |
| (c) $\dfrac{52.41 \times 0.044}{0.001\,18}$ | (i) 20 | (ii) 200 | (iii) 2000 | (iv) 20 000 | (v) 200 000 |
| (d) $\sqrt{990}$ | (i) 10 | (ii) 30 | (iii) 100 | (iv) 300 | (v) 1000 |
| (e) $\sqrt{\dfrac{8.05 \times 24.78}{1.984}}$ | (i) 0.1 | (ii) 1 | (iii) 10 | (iv) 100 | (v) 1000 |

### Solution

(a) $\dfrac{3902}{23\,839}$ is roughly $\dfrac{4000}{20\,000}$, i.e., 0.2. Therefore the answer is (ii).

(b) $\dfrac{4.19 \times 0.0309}{0.0222}$ is roughly $\dfrac{4 \times 0.03}{0.02}$, i.e., 6. Therefore the answer is (iv).

(c) $\dfrac{52.41 \times 0.044}{0.001\ 18}$ is roughly $\dfrac{50 \times 0.04}{0.001}$, i.e., 2000. Therefore the answer is (iii).

(d) $\sqrt{990}$ is roughly $\sqrt{900}$, i.e., 30. Therefore, the answer is (ii).

(e) $\sqrt{\dfrac{8.05 \times 24.78}{1.984}}$ is roughly $\sqrt{\dfrac{8 \times 25}{2}}$, i.e., $\sqrt{100}$ which is equal to 10. Therefore the answer is (iii).

## Exercise 4a

**1.** During a sale, one kilogram of fish was sold for \$4.95. Estimate how many kilograms of fish you could buy with \$20.

**2.** Without doing an exact calculation, determine whether you can afford all the items below if you have only \$30.

— 1 two-kilogram bottle of corn oil for \$6.95.
— 5 cans of peach at \$1.95 per can.
— 300 g of beef at \$1.02 per 100 g.
— 24 packets of recombined milk at \$2.85 for 6.

**3.** Given below are the prices of three brands of soap. Determine which brand is the cheapest.

| Brand | No. of bars | Net weight of each bar | Price |
|---|---|---|---|
| *A* | 3 | 100 g | \$1.30 |
| *B* | 6 | 100 g | \$2.35 |
| *C* | 4 | 125 g | \$2.85 |

**4.** Estimate each of the following mentally and choose the correct answer in each case:

**(a)** $3.14 \times 80.5 =$
**(i)** 2.5277 **(ii)** 25.277
**(iii)** 252.77 **(iv)** 2527.7

**(b)** $91.44 \div 0.36 =$
**(i)** 2.54 **(ii)** 25.4
**(iii)** 254 **(iv)** 2540

**(c)** $\sqrt{917} =$
**(i)** 3.0282 **(ii)** 30.282
**(iii)** 302.82 **(iv)** 3028.2

**(d)** $\sqrt[3]{998} =$
**(i)** 9.993 **(ii)** 31.591
**(iii)** 3.159 **(iv)** 99.93

**(e)** $\dfrac{4927}{9835} =$
**(i)** 0.5 **(ii)** 5
**(iii)** 50 **(iv)** 500

**5.** Estimate each of the following and pick the closest answer in each case:

**(a)** $4831.9 \times 229.78 =$
**(i)** 10 000 **(ii)** 100 000
**(iii)** 1 000 000 **(iv)** 10 000 000

**(b)** $17\ 913 \times 0.963 =$
**(i)** 180 **(ii)** 1800
**(iii)** 18 000 **(iv)** 180 000

**(c)** $\dfrac{52.47 \times 0.083}{0.001\ 98} =$
**(i)** 200 **(ii)** 2000
**(iii)** 20 000 **(iv)** 200 000

**(d)** $\dfrac{2857 \times (0.5)^2}{0.0049} =$
**(i)** 1500 **(ii)** 15 000
**(iii)** 150 000 **(iv)** 1 500 000

**(e)** $\sqrt{0.0815} =$
**(i)** 0.09 **(ii)** 0.03
**(iii)** 0.9 **(iv)** 0.3

# Approximations in Measurements and Accuracy

In most of our daily situations we do not need to use highly sensitive measuring devices. How accurate our measurements are depends on what we need the information for. For example if we use a compass to guide us from one end of the school to the other, it would not be a serious error if we are 1° off course. However, 1° off course on a journey from the earth to the moon will mean an error of 644 000 km!

Besides the errors arising from the use of different instruments, human error is another source of error. In an athletics meet, the time for the first placing of a 100-m race given by two timekeepers may be slightly different. That is why in a school athletics meet there are usually two or more timekeepers for the first few placings.

In fact, all physical measurements such as mass, length, time, area and volume can never be absolutely accurate. They are only approximations. The accuracy of a measurement depends on the measuring instruments and the person taking the measurement. Both of them can never be absolutely accurate.

**For Your Information**

*Light travels approximately 870 000 times faster than sound and sound travels approximately 33 times faster than the best runner in the world.*

## Explorations

You can do this activity with a partner.

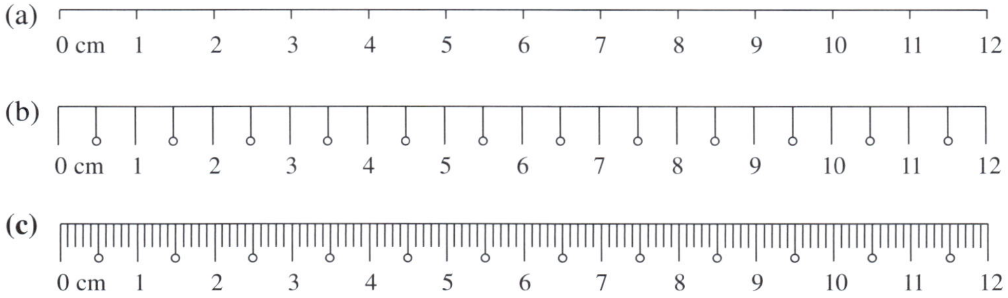


*Fig. 4.1*

**1.** Make a photocopy of Fig. 4.1 and paste it on a piece of vanguard sheet. Cut out the three strips and use them as rulers for this activity.

**2.**

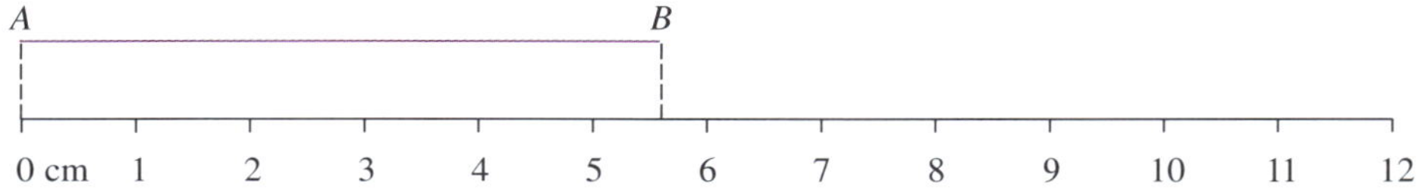


*Fig. 4.2*

In Fig. 4.2, we say that the length of $AB$ is **6 cm to the nearest cm** as the end point is nearer to 6 cm.

**(a)** Draw three lines of lengths between 5 cm and 10 cm each.

(**b**) Use the first ruler (Fig. 4.1(a)) to measure the lengths of the line drawn by your partner.
(**c**) What are the lengths of the lines to the nearest cm?

**3.** Fig. 4.3 shows the same line *AB* measured using the second ruler (Fig. 4.1(b)).

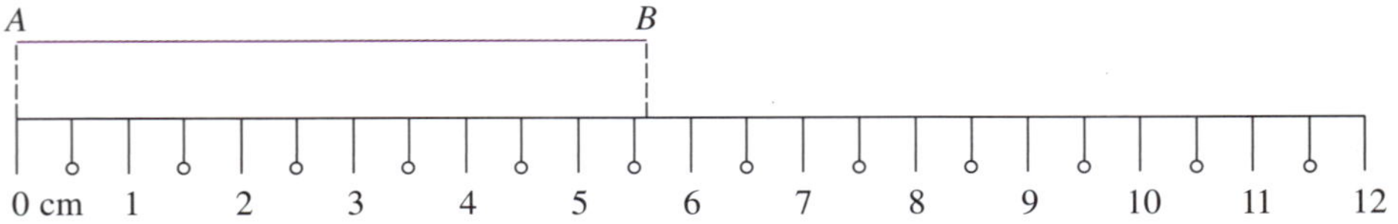


*Fig. 4.3*

Notice that the end point *B* is nearer to 5.5 cm. We say that the length of *AB* is **5.5 cm to the nearest 0.5 cm**. Do you agree that we achieve a greater accuracy by using the second ruler to measure the length of *AB*.
(**a**) Use your second ruler to measure the lengths of your lines.
(**b**) What are their lengths to the nearest 0.5 cm?

**4.** If we divide each centimetre (cm) into 10 equal parts like the third ruler (Fig. 4.1 (c)), we can measure *AB* even more accurately.

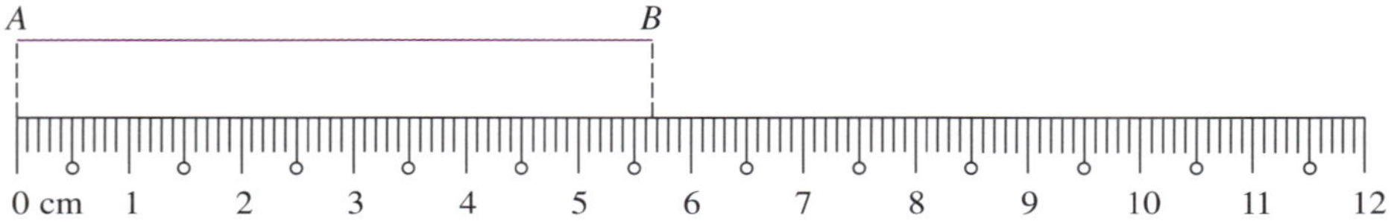


*Fig. 4.4*

Fig. 4.4 shows that the end point *B* lies between the sixth-tenth and seventh-tenth of a centimetre mark. However, it is nearer to 5.6 cm. We say that the length of *AB* is **5.6 cm to the nearest** $\mathbf{\frac{1}{10}}$ **or 0.1 cm**.

(**a**) Use your third ruler to measure the lengths of the lines you have drawn.
(**b**) What are their lengths to the nearest 0.1 cm?

**5.** Make an estimation of the following to the nearest 0.5 cm and use your ruler to check your estimation.

| | Estimation | By Measurement |
|---|---|---|
| (**a**) The length of the thumb of your partner. | ______ | ______ |
| (**b**) The length of the ear of your partner. | ______ | ______ |
| (**c**) The circumference of the wrist of your partner. | ______ | ______ |
| (**d**) The circumference of the neck of your partner. | ______ | ______ |

# Rounding Off a Number to a Given Number of Decimal Places

Sometimes it is impractical or meaningless to have too many decimal places in a decimal. For example, it is absurd to give the speed of a car as 48.234 56 km/h. Here, we will learn how to round off a number to a desired number of decimal places.

## Example 3

*Round off (a) 8.2, (b) 8.9 and (c) 8.5 to the nearest whole number.*

### Solution

(a) The number line above shows that 8.2 lies between 8 and 9 but is closer to 8.
$\therefore$ $8.2 \approx 8$ (correct to the nearest whole number)

(b) The number line shows that 8.9 lies closer to 9 than to 8.
$\therefore$ $8.9 \approx 9$ (correct to the nearest whole number)

(c) Similarly, it shows that 8.5 is exactly halfway between 8 and 9. For such cases, it is common to round up.
$\therefore$ $8.5 \approx 9$ (correct to the nearest whole number)

> **Notes**
> *The symbol $\approx$ represents approximation. For example, when we say that 8.1 is approximately equal to 8, we can write $8.1 \approx 8$.*

The following are steps to follow when rounding off a decimal:

1. Include one extra digit for consideration.
2. Simply drop the extra digit if it is less than 5. If it is 5 or more, add 1 to the previous digit before dropping the extra digit.

## Example 4

*Write the following numbers correct to (i) the nearest whole number, (ii) 2 decimal places and (iii) 3 decimal places:*

*(a) 9.7168* *(b) 19.2147* *(c) 0.825 14*

### Solution

(a) (i) $9.7168 \approx 10$ (↑ 7: This digit is more than 5.)

(ii) $9.7168 \approx 9.72$ (↑ 6: This digit is more than 5.)

(iii) $9.7168 \approx 9.717$ (↑ 8: This digit is more than 5.)

(b) (i) $19.2147 \approx 19$ — ↑ This digit is less than 5.

(ii) $19.2147 \approx 19.21$ — ↑ This digit is less than 5.

(iii) $19.2147 \approx 19.215$ — ↑ This digit is more than 5.

(c) (i) $0.825\ 14 \approx 1$ — ↑ This digit is more than 5.

(ii) $0.825\ 14 \approx 0.83$ — ↑ This digit is 5.

(iii) $0.825\ 14 \approx 0.825$ — ↑ This digit is less than 5.

## Example 5

*Using a calculator, express*

*(a)* $\frac{5}{14}$ *as a decimal correct to 3 decimal places, and*

*(b)* $\frac{14}{37}$ *as a decimal correct to 4 decimal places.*

### Solution

(a) $\frac{5}{14} = 0.357\ 14...$

$\therefore \quad \frac{5}{14} \approx 0.357$

(correct to 3 decimal places)

(b) $\frac{14}{37} = 0.378\ 37...$

$\therefore \quad \frac{14}{37} \approx 0.378\ 4$

(correct to 4 decimal places)

## Exercise 4b

**1.** Write the following correct to **(i)** the nearest whole number and **(ii)** 2 decimal places:

**(a)** 5.424 67 **(b)** 15.824
**(c)** 7.862 **(d)** 130.829

**2.** Write the following correct to 3 decimal places:

**(a)** 712.8926 **(b)** 0.002 72
**(c)** 0.8274 **(d)** 7.024 489

**3.** Express $\frac{2}{17}$ as a decimal and give your answer correct to 3 decimal places.

**4.** Express the following fractions as decimals correct to 2 decimal places:

**(a)** $\frac{4}{19}$ **(b)** $\frac{7}{11}$
**(c)** $\frac{9}{14}$ **(d)** $\frac{11}{15}$

# Accuracy and Significant Figures

The accuracy of a measurement is indicated by the number of figures or digits, called **significant figures** or **digits**, that it contains. Suppose that a line $PQ$, with an actual length of 5.01 cm, is measured using ruler 1 (Fig. 4.1(a)) and ruler 3 (Fig. 4.1(c)) respectively. The measurements of $PQ$ would be 5 cm (to the nearest cm) and 5.0 (to the nearest 0.1 cm) respectively.

We say that

5 cm is the length of $PQ$ (measured) correct to **one** significant figure

and 5.0 cm is the length of $PQ$ correct to **two** significant figures.

Estimation can be done by rounding a number to a specified decimal place. It can also be done by rounding a number to a specified number of significant figures. Consider a HDB flat priced at \$185 500. To get a quick idea of the affordability of the flat, we may estimate the price by rounding it to \$200 000, an approximation of \$185 500 to one significant figure.

# Rounding a Number to a Given Number of Significant Figures

We have the following **rules** for rounding a number to a given number of significant figures:

**1.** Consider the place values of the number from left to right, starting with the first non-zero figure. Include one extra figure for consideration.

**2.** If the extra figure is less than 5, drop the extra figure and all other following figures to the right. Use zeros to keep the place values if necessary (e.g. correct to 4 significant figures, 2.040 45 = 2.040 and not 2.04; correct to 3 significant figures, 0.400 127 = 0.400 and not 0.4.)

**3.** If the extra figure is 5 or more, add 1 to the previous figure before dropping the extra figure and all other following figures. Use zeros to keep the place values if necessary.

To find the number of significant figures in a number we have the following rules:

1. The following figures in a number are significant:
   **(a)** All non-zero figures (e.g. 7.12 has three significant figures).
   **(b)** All zeros between significant figures (e.g. 2003 has four significant figures).
   **(c)** All zeros at the end of a decimal (e.g. 22.300 has five significant figures).

2. The following figures in a number are not significant:
   **(a)** All zeros at the beginning of a decimal less than 1 (e.g. 0.000 325 has three significant figures).
   **(b)** All zeros at the end of a number may or may not be significant. It depends on how the estimation is made (e.g. 20 000 has one significant figure if the estimation is made correct to the nearest 10 000).

3. In general, we start counting significant figures by the first non-zero number.

For example,

**1.** 0.060 52 ≈ 0.06 (correct to 1 significant figure).

**2.** 0.06052 ≈ 0.061 (correct to 2 significant figures).

**3.** 0.003 824 ≈ 0.004 (correct to 1 significant figure).

**4.** 0.003 824 ≈ 0.0038 (correct to 2 significant figures).

**5.** 0.003 824 ≈ 0.003 82 (correct to 3 significant figures).

**6.** 2.0057 ≈ 2.0 (correct to 2 significant figures).

**7.** 2.0057 ≈ 2.01 (correct to 3 significant figures).

**8.** 2.0057 ≈ 2.006 (correct to 4 significant figures).

**9.** 4276 ≈ 4000 (correct to 1 significant figure).

**10.** 4276 ≈ 4300 (correct to 2 significant figures).

**11.** 4276 ≈ 4280 (correct to 3 significant figures).

**12.** 40 004 ≈ 40 000 (correct to 1 significant figure).
(if the estimation is made, correct to the nearest 10 000)

**13.** 40 004 ≈ 40 000 (correct to 2 significant figures).
(if the estimation is made, correct to the nearest 1000)

**14.** 40 004 ≈ 40 000 (correct to 3 significant figures).
(if the estimation is made, correct to the nearest 100)

**15.** 40 004 ≈ 40 000 (correct to 4 significant figures).
(if the estimation is made, correct to the nearest 10)

**16.** 0.834 ≈ 0.83 (correct to 2 significant figures).

**17.** 0.600 27 ≈ 0.600 (correct to 3 significant figures).

**18.** 0.059 002 ≈ 0.059 00 (correct to 4 significant figures).

Notes

*The zeros in examples 1–5 are not significant.*

*The zeros in examples 6–8 are significant.*

## Example 6

*Estimate the following, giving your answers correct to 1 significant figure:*

(a) $\dfrac{74.97}{2.52}$ (b) $\sqrt{\dfrac{12.02 \times 24.99}{3.001}}$

> *Problem Solving Tips*
>
> *When we are asked to estimate to 1 significant figure, we normally estimate to 2 significant figures in the working and then round off to 1 significant figure for the final answer. Similarly, for 2 significant figures, work with 3 significant figures before rounding off and so on.*

### Solution

(a) $\dfrac{74.97}{2.52} \approx \dfrac{75}{2.5} = \dfrac{750}{25} = 30$

$\therefore \quad \dfrac{74.97}{2.52} \approx 30$ (correct to 1 significant figure)

(b) $\sqrt{\dfrac{12.02 \times 24.99}{3.001}} \approx \sqrt{\dfrac{12 \times 25}{3}} = \sqrt{100} = 10$

$\therefore \quad \sqrt{\dfrac{12.02 \times 24.99}{3.001}} \approx 10$ (correct to 1 significant figure)

## Example 7

*(a) Express 0.061 54 correct to*
*(i) three decimal places and state the number of significant figures in the result;*
*(ii) three significant figures and state the number of decimal places in the result.*

*(b) Express 12.2057 correct to*
*(i) two decimal places and state the number of significant figures in the result;*
*(ii) five significant figures and state the number of decimal places in the result.*

### Solution

(a) (i) 0.061 54 = 0.062 (correct to three decimal places)
0.062 has two significant figures.

(ii) 0.061 54 = 0.0615 (correct to three significant figures)
0.0615 has four decimal places.

(b) (i) 12.2057 = 12.21 (correct to two decimal places)
12.21 has four significant figures.

(ii) 12.2057 = 12.206 (correct to five significant figures)
12.206 has three decimal places.

## Exercise 4c

1. Round off the following:
   (a) 456 g to the nearest 10 g
   (b) 722 g to the nearest 100 g
   (c) 3.27 cm to the nearest cm
   (d) 123.452 cm to 1 decimal place
   (e) 18.2 to the nearest whole number
   (f) 31.256 m to the nearest 10 m
   (g) 12.35 cm to the nearest 0.1 cm
   (h) 4325 pupils to the nearest 100 pupils
   (i) 845 km to the nearest 10 km
   (j) 22.58 mm to the nearest $\frac{1}{10}$ mm

2. State the number of significant figures in each of the following:

| | |
|---|---|
| (a) 15.0 | (b) 27.3 |
| (c) 30 756 | (d) 4.02 |
| (e) 63 | (f) 9.5 |
| (g) 30.08 | (h) 48.20 |
| (i) 6 000 000 | (j) 0.0283 |

3. Express the following numbers correct to the number of significant figures indicated within the brackets:

| | |
|---|---|
| (a) 3.084 | (2) |
| (b) 1.483 56 | (4) |
| (c) 0.003 46 | (1) |
| (d) 16.047 | (1) |
| (e) 3.141 59 | (2) |
| (f) 0.574 38 | (2) |
| (g) 0.056 78 | (3) |
| (h) 217.006 | (5) |
| (i) 3.598 | (3) |
| (j) 15.7037 | (4) |

4. (a) Estimate the value of $\frac{7.94}{2.01}$ correct to 1 significant figure.
   (b) Use your result to estimate the value of $\frac{79\ 400}{0.000\ 201}$.

5. Estimate the value of $\frac{21.83 \times 0.498}{220.1}$, giving your answer correct to 1 significant figure.

6. Express 28.136 275 correct to two significant figures. How many decimal places are there in the result?

7. Express 0.211 087 94 correct to three significant figures. Write down the number of decimal places in the result.

8. Express 0.008 345 7 correct to four decimal places. State the number of significant figures in the answer.

9. Express 117.964 8 correct to two decimal places. How many significant figures are there in the answer?

## Rounding and Truncation Errors

With the extensive use of calculators in our daily life, we notice that the simple four-function calculators and the scientific calculators do not give exact answers as they use numeric computing. The calculators usually display up to 10 digits. The number of digits stored and used for calculation depends on the capacity of the calculator. These approximations will result in the rounding or truncation errors especially in working with very large or very small numbers.

## Explorations

You need a simple four-function calculator with a $\sqrt{\ }$ function and a scientific calculator.

Do this activity with a partner.

| No. | Event | Answer obtained by using four-function calculator | Answer obtained by using scientific calculator | Questions |
|---|---|---|---|---|
| 1 (i)<br>(ii) | Divide 2 by 3.<br>Multiply answer in (i) by 4. | _____<br>_____ | _____<br>_____ | Do your calculators display different answers? Why do you think this is so? |
| 2 (a)(i)<br>(a)(ii) | Key in $\sqrt{\sqrt{7}}$ and square the answer twice<br>Subtract 7 from the answer in (a)(i). | _____<br>_____ | _____<br>_____ | 2(a)(i) Does the calculator display the expected answer 7?<br>(ii) Does the calculator display the expected answer 0? Why? |
| 2 (b)(i)<br>(b)(ii) | Key in $\sqrt{\sqrt{\sqrt{\sqrt{7}}}}$ and square the answer 4 times.<br>Subtract 7 from the answer in (b)(i). | _____<br>_____ | _____<br>_____ | 2(b)(i) Does the calculator display the expected answer 7?<br>(ii) Does the calculator display the expected answer 0? Why? |
| 2 (c)(i)<br>(c)(ii) | Key in $\sqrt{\sqrt{\sqrt{\sqrt{\sqrt{\sqrt{7}}}}}}$ and square the answer 6 times.<br>Subtract 7 from the answer in (c)(i). | _____<br>_____ | _____<br>_____ | (i) Do you get the same expected answer as in 2(a) and 2(b)?<br>(ii) Why are there differences in the answers in 2(a), (b) or (c)? |
| 3 (i)<br>(ii) | Add 999 888 888 to 0.007 and then subtract 999 888 888.<br>Repeat (i) replacing 0.007 with 0.993. | _____<br>_____ | _____<br>_____ | (i) Does the calculator display the expected answer 0.007?<br>(ii) Why do the calculators give us these kind of answers? |

In view of the above results, you will notice that

1. The answer for the division is either 0.66...67 (rounding) or 0.66...666 (truncation). The answer for the product is either 2.66...67 or 2.66...64. This gives you an idea whether the calculator stores and displays the same number of digits or whether it stores and uses more digits for computation than it displays.
2. An error will occur, either it stabilises or diverges, at a certain value depending on the types of calculators used.
3. This is a case where manual calculation is faster and more accurate than using a calculator. For certain scientific calculators, the very large number is displayed as the answer when adding a very large number to a very small number. You may try this example again, key in 999 999 999 and then add 0.008.

In conclusion, if a problem requires an answer correct to $n$ significant figures, you should either

(i) store your intermediate working in your calculators, or

(ii) round up to at least $(n + 1)$ significant figures if the question requires to write down intermediate working.

(iii) When we use a calculator, errors may arise due to rounding and truncation as well as possible human errors in keying in the numbers or the sequence we key in. It is wise to make a rough estimation of the result. If the result looks ridiculous, check the answer again.

## Example 8

*A rectangular tank, 21.35 cm by 9.86 cm by 17.4 cm, is completely filled with water. Find*
*(a) the volume of the water in the rectangular tank,*
*(b) the length of the side of a cubical tank that has the same volume as the rectangular tank.*
*Correct all answers in 3 significant figures.*

### Solution

(a) Volume of the rectangular tank
$= 21.35 \times 9.86 \times 17.4 \approx 3662.8914$ (store this value in the calculator)
$\approx 3660 \text{ cm}^3$ (correct to 3 significant figures)

(b) To find the length of the side of the cubical tank, take cube root of the stored value or 3663 (correct to 4 significant figures)

$\sqrt[3]{3662.8914} = 15.414\,922\,57$ or $\sqrt[3]{3663} = 15.415\,074\,91$

$\therefore$ The length is 15.4 cm (correct to 3 significant figures)

## Explorations

Do this investigation with a partner using different types of calculators and make your conclusion.

1. (i) Divide 2 by 7,
   (ii) Multiply your answers in (i) by 7.

2. (a) (i) Key in $\sqrt[3]{\sqrt[3]{2}}$ and then cube the answer twice.
   (ii) Subtract 2 from the answer displayed by the calculator in (a)(i). What is the result?

   (b) (i) Key in $\sqrt[3]{\sqrt[3]{\sqrt[3]{\sqrt[3]{2}}}}$ and then cube the answer 4 times.
   (ii) Subtract 2 from the answer displayed by the calculator in (b)(i). What is the result?

   (c) Do you get the expected result of 0 in both the above cases?

## Exercise 4d

Use a calculator to evaluate questions 1 to 7, giving your answers correct to 3 significant figures.

1. $(0.2187)^3$
2. $\sqrt[3]{0.086\ 42}$
3. $\sqrt{25.6^2 + 17.89^2}$
4. $\sqrt{\dfrac{1976 \times (14.98)^2}{(59.87)^2}}$
5. $\dfrac{99.85 \times \sqrt{63.8}}{24.81}$
6. $\sqrt[3]{\dfrac{79.97 \times 24.999}{2.007}}$
7. $999 \times 998\ 998 - 0.009$
8. The internal dimensions of an open concrete tank are 1.89 m long, 0.77 m wide and 1.23 m high. Find the capacity of the tank in litres. If the concrete side walls of the tank are 0.1 m thick, find also, in cubic metres, the volume of concrete used. Give all answers correct to 3 significant figures.
9. Find the area of a field measuring 28.99 m by 32.76 m. What is the cost of spraying insecticide on this field if the cost is $25.50 per square metre? Give all answers correct to 4 significant figures.
10. A rectangular tank, 7.41 cm by 8.52 cm by 36.9 cm is completely filled with oil. Find
    (a) the volume of oil in the rectangular tank,
    (b) the length of the side of a cubical tank if the oil is to be stored in a cubical tank.

    Leave all answers correct to 4 significant figures.

## SUMMARY

**Rules for rounding a number to a given number of significant figures:**

**(a)** Count the given number of significant figures from left to right, starting with the first non-zero figure. Include one extra figure for consideration.

**(b)** If the extra figure is less than 5, drop the extra figure and all other following figures. Use zeros to keep the place value if necessary.

**(c)** If the extra figure is 5 or more, add 1 to the previous figure before dropping the extra figure and all other following figures. Use zeros to keep the place value if necessary.

**Rules for determining the number of significant figures:**

**(a)** The following figures in a number are significant:
- **(i)** All non-zero figures.
- **(ii)** All zeros between significant figures.
- **(iii)** All zeros at the end of a decimal.

**(b)** The following figures in a number are not significant:
- **(i)** All zeros at the beginning of a decimal less than 1.
- **(ii)** All zeros at the end of a whole number may or may not be significant. It depends on how the estimation is made.

## Review Examples 4

### Example 1

*The length of the circumference of a circle is given by $l = 2\pi r$ and its area is given by $A = \pi r^2$, where r is its radius. Find*

*(a) the approximate length of the circumference of a circle whose radius is 997 cm. (Giving your answer to the nearest whole mumber.)*
*(b) the approximate area of a circle whose radius is 11.09 m. (Giving your answer to the nearest whole number.)*
*[Take π as 3.1416.]*

**Solution**

(a) The approximate length of the circumference
$= 2 \times 3.1416 \times 997$
$= 6264.35$ cm
$= 63$ m (to the nearest whole number)

(b) The approximate area
$= 3.1416 \times 11.09^2$
$= 386.38\ \text{m}^2$
$= 386\ \text{m}^2$ (to the nearest whole number)

### Example 2

*Estimate, correct to 1 significant figure, the value of (a)* $\dfrac{97.85 \times \sqrt{63.8}}{24.79}$, *(b)* $\dfrac{4870 \times 1227 + 968 \times 4870}{1936 \times 0.49}$

**Solution**

(a) $$\frac{97.85 \times \sqrt{63.8}}{24.79} \approx \frac{100 \times \sqrt{64}}{25} = 4 \times 8 \approx 30$$

(b) $$\frac{4870 \times 1227 + 968 \times 4870}{1936 \times 0.49} \approx \frac{5000 \times 1000 + 1000 \times 5000}{2000 \times 0.5} = \frac{2000 \times 5000}{2000 \times 0.5} = 10\,000$$

# Review Questions 4

1. Estimate each of the following mentally and pick the correct answer in each case:
   **(a)** $4.07 \times 6.998 =$
   **(i)** 0.2848 **(ii)** 2.848
   **(iii)** 28.48 **(iv)** 284.8

   **(b)** $29.7 \div 5.03 =$
   **(i)** 0.5905 **(ii)** 5.905
   **(iii)** 59.05 **(iv)** 590.5

   **(c)** $\sqrt{899} + \sqrt[3]{1029} =$
   **(i)** 4.008 **(ii)** 40.08
   **(iii)** 400.8 **(iv)** 4008

   **(d)** $\dfrac{6.003 - 5.12 \times 0.992}{5.97 \div 3.103} =$
   **(i)** 0.048 **(ii)** 0.48
   **(iii)** 4.8 **(iv)** 48

2. Estimate each of the following and pick the closest answer in each case:
   **(a)** $4962.8 \times 312.93 =$
   **(i)** 1 500 000 **(ii)** 150 000
   **(iii)** 15 000 **(iv)** 1500

   **(b)** $\dfrac{9.1034 - 7.9902}{(10.0123)^2} =$
   **(i)** 0.001 **(ii)** 0.01
   **(iii)** 0.1 **(iv)** 1.0

   **(c)** $\dfrac{45.117 \div 8.97}{4.983 \times 0.002\ 104} =$
   **(i)** 50 **(ii)** 500
   **(iii)** 5000 **(iv)** 50 000

   **(d)** $\dfrac{3.9897 \times 2.1005}{(0.0199)^3} =$
   **(i)** 1 000 000 **(ii)** 100 000
   **(iii)** 10 000 **(iv)** 1000

3. Express the following correct to the number of decimal places or significant figures indicated within the brackets:
   **(a)** 0.085 67 (3 decimal places)
   **(b)** 0.085 67 (3 significant figures)
   **(c)** 5.096 (3 significant figures)
   **(d)** 726 990 (2 significant figures)
   **(e)** 0.058 76 (2 decimal places)

4. Estimate, correct to 1 significant figure, the value of 52.976 03 – 31.321 86.

5. Estimate, correct to 1 significant figure, the value of
   **(a)** $\dfrac{79.81}{1.62}$; **(b)** $\dfrac{66.4}{0.0319}$.

6. Estimate the number of loaves of bread required to serve 40 pupils so that each pupil will have 3 meals with 1.8 servings for each meal, assuming that 5 servings can be made from 1 loaf of bread.

7. Estimate the number of classroom desks that you would need to stack up to reach the ceiling of your classroom.

✶8. Find the approximate value of $\dfrac{31.98 \div 8.03}{48.109 - 29.989 \times 0.995}$, giving your answer correct to 1 significant figure.

✶9. Estimate the value of $20.02 \times 9.99 - 6.112 \times \dfrac{16.027}{(1.977)^3}$ correct to 2 significant figures.

✶10. Estimate, correct to 1 significant figure, the value of $\sqrt{136.05 - (2.985 + 7.001)^2}$.

✶11. A Singaporean has assets of \$5 billion (\$5 000 000 000). If he spends \$10 every second, how long will it take for him to spend all his money?

✶12. One million seconds is about
   **(A)** 3 days **(B)** 12 days
   **(C)** 3 months **(D)** 1 year
   **(E)** 2 years

*In this chapter, you will learn how to:*

- *use letters to represent numbers;*
- *express basic arithmetical processes algebraically;*
- *substitute numbers for letters in formulae and expressions;*
- *manipulate simple algebraic expressions;*
- *manipulate linear algebraic expressions with fractional coefficients;*
- *factorise algebraic expressions.*

CHAPTER

# 5

# Fundamental Algebra

## Introduction

Have you ever received SMS (Short Message System) from your classmates that read "how r u hav u done ur math" or "OMG late g2g ttul"? Nowadays many of us use short forms in our SMS. We also use other short forms to write our e-mails and many stenographers or secretaries use shorthand to draft their letters or take minutes during meetings.

Just as we need to use shorthand or short forms to make our work easier in our everyday situations, mathematicians had over the years developed a system called ALGEBRA which uses symbols and letters of the alphabet to replace numbers.

In primary school, you were introduced to some simple concepts of algebra. You have learned how to represent an unknown number by a letter, how to simplify algebraic expressions and to find the value of an algebraic expression by substitution. We shall have a quick revision of these before proceeding to more challenging problems.

## Notations in Algebra

The signs +, –, ×, ÷, =, etc, are used in algebra as in arithmetic.

1. In arithmetic, $5 + 4 = 9$ means that the sum of 5 and 4 is equal to 9. In algebra, $x + y = z$ means that the sum of two numbers represented by $x$ and $y$ is equal to the number represented by $z$.

   If $x = 4$, $y = 3$, then $z$ represents 7.
   If $x = 4$, $z = 8$, then $y$ represents 4.

   Similarly, $x - y = z$ means that the difference between two numbers represented by the letters $x$ and $y$ is equal to the number represented by the letter $z$.

2. In arithmetic, $5 \times 4 = 20$ means that the product of 5 and 4 is 20. In algebra $a \times b = c$ means that the product of two numbers $a$ and $b$ is equal to the number represented by the letter $c$. Normally we write $a \times b = c$ simply as $ab = c$, omitting the multiplication sign. Similarly $x = a \times b \times c$ is simply written as $x = abc$.

3. In arithmetic, $36 \div 4 = 9$ means that 36 divided by 4 gives 9. In algebra $a \div b = c$ means that the number represented by $a$ is divided by the number represented by the letter $b$ to give the result represented by the letter $c$. Also $a \div b$ is normally written as $\frac{a}{b}$.

4. In arithmetic, $3 \times 3 \times 3 \times 3$ may be written as $3^4$, and $7^5$ means $7 \times 7 \times 7 \times 7 \times 7$. In algebra $a \times a \times a \times a$ is written as $a^4$ and $y^5$ means $y \times y \times y \times y \times y$.

**Notes**

*If $x + y = z$ and $z$ represents 9, then $x$ and $y$ may represent any pair of numbers whose sum is 9, for example, 4 and 5, 7 and 2, 1.7 and 7.3 etc.*

**Notes**

*(1) When we multiply y by 3, we normally write it as 3y and **not** y3.*

*(2) When we multiply x by 1, we write it simply as x and **not** 1x or x1.*

**Thinking Time**

(1) Can we write $6 \times 7$ simply as 67? Explain your answer clearly.

(2) Is $3a^2 = (3a)^2$? Explain your answer.

## Example 1

*Write an algebraic expression for each of the following:*

*(a) Add 2x to 3y.*

*(b) Subtract 5x from 12.*

*(c) Multiply u by 3v.*

*(d) Divide 3k by 7x.*

*(e) Subtract the product of x and 3y from the sum of p and q.*

### Solution

(a) $2x + 3y$

(b) $12 - 5x$

(c) $u \times 3v = 3uv$

(d) $3k \div 7x = \dfrac{3k}{7x}$

(e) $p + q - x \times 3y = p + q - 3xy$

## Example 2

*Ali is three times as old as Beng and Beng is five years older than Chandra. If Beng is x years old, write an algebraic expression for each of the following:*

*(a) Ali's age today.*

*(b) Chandra's age today.*

*(c) Ali's age in 5 years' time.*

*(d) The sum of Ali's and Beng's ages in 2 years' time.*

*(e) The sum of Beng's and Chandra's ages 4 years ago.*

### Solution

(a) Ali's age is $3x$ years old.

(b) Chandra's age is $x - 5$ years old.

(c) Ali will be $3x + 5$ years old.

(d) In 2 years' time, Ali will be $3x + 2$ years old and Beng will be $x + 2$ years old. The sum of their ages is $(3x + 2) + (x + 2) = 4x + 4$.

(e) Four years ago, Beng was $x - 4$ years old and Chandra was $x - 5 - 4$, i.e., $x - 9$ years old. The sum of their ages then was $(x - 4) + (x - 9) = 2x - 13$ years old.

**For Your Information**

*In 820 A.D., the Muslim mathematician, Al-Khwarizmi, wrote the algebraic text entitled Al-Jabr Wa-Al Muqabala (the science of cancellation and reduction). A Latin translation of this text became known in Europe under the title Al-Jabr. Thus the Arabic word for reduction, Al-Jabr, became the word algebra. Al-Khwarizmi was known as the "Father of Algebra".*

## Example 3

*Write an algebraic expression for each of the following:*

*(a) The sum of three consecutive integers, of which x is the middle integer.*
*(b) The product of two consecutive odd integers, of which x is the smaller integer.*
*(c) The total cost of x 10¢ stamps and 25 y¢ stamps.*
*(d) The cost of each cup if the total cost of 5 cups is $y.*

**Solution**

(a) The integer after $x$ is $x + 1$ and the one before $x$ is $x - 1$. The sum of the three consecutive integers is $(x - 1) + x + (x + 1) = 3x$.

(b) The greater of the two consecutive odd integers is $x + 2$. The product of the two integers is $x \times (x + 2)$ or $x(x + 2)$.

(c) The cost of $x$ 10¢ stamps = $(x \times 10)$¢ = $10x$¢.
The cost of 25 $y$¢ stamps = $(25 \times y)$¢ = $25y$¢.
The total cost = $(10x + 25y)$¢.

(d) The cost of each cup = $\$y \div 5 = \$\frac{y}{5}$.

**Explorations**

**Nine, A Mysterious Number**

Nine is a mysterious number. Do you know that 9 is hidden in the birthday of every notable person in the world, past or present? Take George Washington, America's first President, who was born on the 22nd February 1732. Let us write his birthday as the number 2221732. Now rearrange the digits of this number to obtain another number, say 3227212. Compute the difference of these two numbers by subtracting the smaller from the larger, as follows:

$$\begin{array}{r} 3\,2\,2\,7\,2\,1\,2 \\ -\ 2\,2\,2\,1\,7\,3\,2 \\ \hline 1\,0\,0\,5\,4\,8\,0 \\ \hline \end{array}$$

Add the digits of the resulting number until we get a single digit number. In our example, we have

$$1005480 \rightarrow 1+0+0+5+4+8+0 \rightarrow 18 \rightarrow 1+8 \rightarrow 9. \text{ We obtain 9!}$$

If you carry out the above calculations using the birthday of any other notable person in the world, you are going to get 9. Does this suggest a certain mysterious relationship between the birthdays of all world famous persons and the number 9?

Try working it out, using the birthdays of every student in your class. Are you surprised by the results? Can you explain why?

# Exercise 5a

1. Write an algebraic expression for each of the following:
   (a) Add $2x$ to 14.
   (b) Subtract 14 from $5a$.
   (c) Multiply 4 by $2k$.
   (d) Divide $8x$ by $24y$.
   (e) Add $2x$ to twice $3y$.
   (f) Subtract $5x$ from half of $y$.

2. Translate each of the following word expressions into an algebraic expression:
   (a) The sum of a number $2x$ and a number $y$.
   (b) The product of 7 and a number $k$.
   (c) Fifteen subtracted from twice the number $t$.
   (d) Three times the number $u$ decreased by four.
   (e) Eight more than half of a number $v$.
   (f) The total value of $h$ 50-cent coins and $k$ 2-dollar notes in dollars.
   (g) The total cost of buying $x$ sets of stamps to commemorate the 50th Anniversary of the Inter-Religious Organisation (IRO) at $1.82 per set, and $y$ sets of the Rabbit Zodiac series at $2.22 per set.

3. Write an algebraic expression for each of the following:
   (a) The cost of $x$ litres of petrol at $1.70 per litre.
   (b) Three times the variable $x$ divided by the sum of 3 and $k$.
   (c) Five times a number which is 3 more than $h$.
   (d) One quarter of a number which is 4 less than $m$.
   (e) The total number of eggs in $k$ cartons where each carton contains $n$ eggs.
   (f) The total distance travelled by a car for $x$ hours at an average speed of $z$ km/h.

4. Mary is $x$ years old. Write an algebraic expression for each of the following:
   (a) Three times Mary's age next year.
   (b) Five times Mary's age six years ago.
   ✶(c) The present age of Mary's aunt if her aunt is four times as old as Mary will be 2 years from now.
   ✶(d) The present age of Mary's niece if her niece is 3 years less than one-third Mary's age 5 years ago.

## Evaluation of Algebraic Expressions

Consider the algebraic expressions: (a) $4x$ (b) $5x + 7$

The algebraic expression in (a) will take on different values for different values of $x$ given.

For example, when $x = 2$, $4x = 4 \times 2 = 8$
$x = 3$, $4x = 4 \times 3 = 12$
$x = 5$, $4x = 4 \times 5 = 20$ and so on.

**Notes**

*A **variable** is a changing quantity, usually denoted by a letter in algebraic equations and expressions, that might have any one of a range of possible values.*

Since the value of $4x$ varies according to the value given to $x$, $x$ is called a **variable**.

In the term $4x$, the constant factor 4 is called the **coefficient** of the term. Thus, the coefficient of $x$ in $5\boldsymbol{x}$ is 5, the coefficient of $xy$ in $7\boldsymbol{xy}$ is 7, the coefficient of $abc$ in $23\boldsymbol{abc}$ is 23, and so on.

What is the coefficient of $x^2$ in $25x^2$?

The algebraic expression in (b) will take on different values for different values of $x$ given.

For example, when $x = 2$, $5x + 7 = 5 \times 2 + 7 = 17$
$x = 3$, $5x + 7 = 5 \times 3 + 7 = 22$
$x = 5$, $5x + 7 = 5 \times 5 + 7 = 32$ and so on.

Notice that the value of the algebraic expression depends on $x$; the numeral 7 always remains unchanged. We call this numeral a **constant term** or simply a **constant**.

## Example 4

*Evaluate $2x + 5y$ when*

*(a) $x = 3$ and $y = 5$;* *(b) $x = 2$ and $y = -1$;* *(c) $x = -2$ and $y = 3$.*

**Solution**

(a) When $x = 3$ and $y = 5$, $2x + 5y = 2(3) + 5(5) = 6 + 25 = 31$.

(b) When $x = 2$ and $y = -1$, $2x + 5y = 2(2) + 5(-1) = 4 - 5 = -1$.

(c) When $x = -2$ and $y = 3$, $2x + 5y = 2(-2) + 5(3) = -4 + 15 = 11$.

## Example 5

*Evaluate the following when $a = 3$, $b = 4$ and $c = 6$:*

*(a) $a(2c - b)$* *(b) $ab[(3b - c) - a]$*

*(c) $(a + b)(c - b) + ab$* *(d) $\dfrac{a + b}{c} + \dfrac{a + c}{c - b}$*

**Solution**

(a) $a(2c - b) = 3(2 \times 6 - 4) = 3(12 - 4) = 3(8) = 24$

(b) $ab[(3b - c) - a] = 3 \times 4[(3 \times 4 - 6) - 3]$
$= 12[(12 - 6) - 3]$
$= 12[6 - 3] = 12(3) = 36$

(c) $(a + b)(c - b) + ab = (3 + 4)(6 - 4) + 3 \times 4$
$= 7 \times 2 + 12 = 14 + 12 = 26$

(d) $\dfrac{a + b}{c} + \dfrac{a + c}{c - b} = \dfrac{3 + 4}{6} + \dfrac{3 + 6}{6 - 4} = \dfrac{7}{6} + \dfrac{9}{2}$
$= \dfrac{7 + 9 \times 3}{6} = \dfrac{7 + 27}{6} = \dfrac{34}{6} = 5\dfrac{2}{3}$

**For Your Information**

*When a ship is in distress, it sends out an S.O.S. signal for help. This signal was first invented by Samuel Morse in 1837 for use with his telegraph. It was adopted in 1912 for international telegraphic use and is widely known as the "Morse code". The Morse code is a system of dots, dashes and spaces. A combination of dots. (one-unit length) and dashes — (three-unit length) make up the system of codes. The distance between two single characters is a space of one-unit length and the distance between two alphabets is a space of three-unit length. The space between two words is given a distance of seven-unit length.*

*The following are some of the words used in the Morse code:*

| — | .... | .— | —. | —.— |
|---|---|---|---|---|
| *T* | *H* | *A* | *N* | *K* |

| —.— — | — — — | ..— |
|---|---|---|
| *Y* | *O* | *U* |

*Can you find out the symbols for all the alphabets?*

*With the advancement of new technologies the Morse code had served its usefulness and it is now no longer in use.*

## Exercise 5b

**1.** What bird produces the largest egg in the world?

Copy the code box and evaluate the following given that $a = 2$, $b = -3$, $c = 4$, $d = -1$ and $e = 0$.

| | | | |
|---|---|---|---|
| $C$ | $2a + 5$ | $I$ | $a(2b + 3c)$ |
| $E$ | $3a - b + d$ | $T$ | $2a^2 - 3b^2 + 2$ |
| $R$ | $c - 3d + 2e$ | $O$ | $2e - 3d^2 + 5c$ |
| $H$ | $3ac \times \frac{1}{4}d$ | $S$ | $8ae \div 2bdc$ |

Put the letter corresponding to the number in the box.

| | | | | | | | | | |
|---|---|---|---|---|---|---|---|---|---|
| | | | | | | | | | |
| −17 | −6 | 8 | 17 | 0 | −17 | 7 | 12 | 9 | −6 |

**2.** Why do the students like the first day of a particular month?

Copy the code box and evaluate the following given that $a = 2$, $b = -3$, $c = 4$, $d = 5$ and $e = -6$.

| | |
|---|---|
| $R$ | $3a - 3(2c - e)$ |
| $O$ | $4(a - 3b) - 5c$ |
| $I$ | $4c - (a - 2b - e)$ |
| $P$ | $9c - 3(2d + c)$ |
| $L$ | $7e - 5b^2 + 4ac$ |
| $A$ | $5abe - 4(e + c)^2$ |
| $F$ | $\frac{a}{c} + \frac{b}{e}$ |

Put the letter corresponding to the number in the box.

| | | | | | | | | |
|---|---|---|---|---|---|---|---|---|
| | | | | | | | | |
| 164 | −6 | −36 | 2 | −55 | 1 | 24 | 24 | −55 |

# Some Rules in Algebra

1. In algebra, terms of the same kind, called **like terms**, can be combined into a single term; added to or subtracted from one another. For example,

   (a) $3a + 5a = 8a$ (b) $7b - 3b = 4b$

   (c) $2a + 5b + 4a + 8b = (2a + 4a) + (5b + 8b) = 6a + 13b$

   (d) $9c + 7d - 4c - 5d = (9c - 4c) + (7d - 5d) = 5c + 2d$

   (e) $7a + 9b - 5a - 4b + 2a - b = (7a - 5a + 2a) + (9b - 4b - b)$
   $= 4a + 4b$

2. In multiplication and division, the coefficients and the variables are multiplied or divided separately. For example,

   (a) $3 \times 6a = 3 \times 6 \times a = 18a$

   (b) $3(a - b) = 3 \times a - 3 \times b = 3a - 3b$

   (c) $14m \div 7 = \frac{14m}{7} = 2m$

## Example 6

*Simplify the following expressions:*

(a) $3x + 8x - 2x$ (b) $4x + 2y + 7x - y$

(c) $3a - 2b + c - a + 5b$

**Solution**

(a) $3x + 8x - 2x = (3x + 8x) - 2x$
$= 11x - 2x$
$= 9x$

(b) $4x + 2y + 7x - y = 4x + 7x + 2y - y$ (group the like terms)
$= 11x + y$

(c) $3a - 2b + c - a + 5b = 3a - a - 2b + 5b + c$ (group the like terms)
$= 2a + 3b + c$

## Exercise 5c

1. Simplify each of the following.
   - **(a)** $7a + 2a$
   - **(b)** $5b + 3b + 5$
   - **(c)** $3a + 6a + c$
   - **(d)** $5h + 3k + 2h + k$
   - **(e)** $7p + 4q - 2p + q$
   - **(f)** $7u + 3v - 2u + 5v$
   - **(g)** $5p + 5 - 3p - 2$
   - **(h)** $4a - 6b - 3a$
   - **(i)** $2c + 4d - 5d$

2. Simplify the following expressions:
   - **(a)** $5n \times 12$
   - **(b)** $-2 \times 6a$
   - **(c)** $-4 \times (-2k)$
   - **(d)** $2 \times \frac{3m}{4}$
   - **(e)** $-\frac{2}{5} \times 20n$
   - **(f)** $-\frac{3}{7}a \times \left(-4\frac{2}{3}\right)$
   - **(g)** $51u \div 17$
   - **(h)** $(-11c) \div 121$
   - **(i)** $(-27v) \div (-3)$
   - **(j)** $45 \div (-15d)$ $(d \neq 0)$

# Use of Brackets in Simplification

When brackets occur in an algebraic expression, the rules by which operations are performed apply exactly as in arithmetic:

(a) Simplify the expression within the brackets first.

(b) When an expression contains more than one pair of brackets, simplify the expression within the innermost pair of brackets first, for example,

$$[2c - 4(c - 1)] + c = [2c - 4c + 4] + c = 4 - 2c + c = 4 - c.$$

(c) If an expression within a pair of brackets is multiplied by a number, **each term** within the brackets must be multiplied by that number when the brackets are removed. We call this **expansion** by using the distributive law, for example,

$$4(a - 2b + 3c) = 4a - 8b + 12c$$

## Example 7

*Expand each of the following:*

*(a)* $5(2x + y)$ *(b)* $3(2a + b) - 2(4a - 3b)$

*(c)* $3[3x - 2(x - 4)]$ *(d)* $7h - 2[3(h - 2) - 2(h - 5)]$

**For Your Information**

1) $a(b + c) = ab + ac$
2) $a(b - c) = ab - ac$
3) $-a(b + c) = -ab - ac$
4) $-a(b - c) = -ab + ac$
5) $a(-b + c) = -ab + ac$
6) $a(-b - c) = -ab - ac$
7) $-a(-b + c) = ab - ac$
8) $-a(-b - c) = ab + ac$

**Solution**

(a) $5(2x + y)$
$= 5 \times 2x + 5 \times y$
$= 10x + 5y$

(b) $3(2a + b) - 2(4a - 3b) = 6a + 3b - 8a + 6b$
$= 6a - 8a + 3b + 6b$
$= -2a + 9b$

(c) $3[3x - 2(x - 4)] = 3[3x - 2x + 8] = 3[x + 8]$
$= 3x + 24$

(d) $7h - 2[3(h - 2) - 2(h - 5)] = 7h - 2[3h - 6 - 2h + 10]$
$= 7h - 2(3h - 2h - 6 + 10) = 7h - 2(h + 4)$
$= 7h - 2h - 8 = 5h - 8$

## Example 8

*Simplify* (a) $3a + 5b - 3c - 2b + 7c - a + 4c - 6b + 8a$;
(b) $3x + 2(x + 4) - (2x - 3) + 5x - 7$;
(c) $2[4p - 3(m + p)]$.

### Solution

(a) $3a + 5b - 3c - 2b + 7c - a + 4c - 6b + 8a$
$= (3 - 1 + 8)a + (5 - 2 - 6)b + (-3 + 7 + 4)c = 10a - 3b + 8c$

(b) $3x + 2(x + 4) - (2x - 3) + 5x - 7 = 3x + 2x + 8 - 2x + 3 + 5x - 7$
$= (3 + 2 - 2 + 5)x + (8 + 3 - 7)$
$= 8x + 4$

(c) $2[4p - 3(m + p)] = 2(4p - 3m - 3p)$
$= 2(p - 3m)$
$= 2p - 6m$

## Exercise 5d

**1.** Expand each of the following:
**(a)** $3(x + 2)$ **(b)** $4(k - 5)$
**(c)** $6(4 - m)$ **(d)** $2(3x + 7)$
**(e)** $8(2y - 5)$ **(f)** $3(2x + 5y)$
**(g)** $5(x - 4z)$ **(h)** $8(3h - 4k)$
**(i)** $-3(x + 6)$ **(j)** $-4(y - 6)$

**2.** Simplify the following expressions:
**(a)** $5(a + 2b) - 3b$
**(b)** $4u - 3(2u - 5v)$
**(c)** $-2a - 3(a - b)$
**(d)** $7m - 2n - 2(3n - 2m)$
**(e)** $-3(2h - k) + 4(k - 3h)$
**(f)** $-4(a - 3b) - 5(a - 3b)$
**(g)** $5(3p - 2q) - 2(3p + 2q)$
**(h)** $a + 3(2a - 3b + c) + 7c$
**(i)** $(x + y) - 2(3x - 4y + 3)$
**(j)** $3(p - 2q) - 4(2p - 3q - 5)$

**3.** Write down the simplest forms for the following:
**(a)** $\frac{1}{2}\left[2x + \frac{1}{2}(4x - 12)\right]$
**(b)** $\frac{2}{5}[12p - (5 + 2p)]$
**(c)** $a - [b - (c + d)]$
**(d)** $4[2x + 5 - (3x - 2)]$
**(e)** $2[(3p - 2q) - (p - q)]$
**(f)** $-2\{3a - 4[a - (2 + a)]\}$
**(g)** $5\{3c - [d - 2(c + d)]\}$
**(h)** $(w + r) - 4(3t - 2w) + 6(t - 2r)$
**(i)** $\frac{1}{2}(4x - 2y) - \frac{2}{3}(9x - 3y) - 2(x - y)$
**(j)** $-4[5(2x + 3y) - 4(x + 2y)]$

# Addition and Subtraction of Algebraic Expressions

## Example 9

*Find the sum of* $2a + 3b - 4c$, $3a - 2b$, $-4a + 5c$ *and* $a - 2b + c$.

**Solution**

The expressions are arranged so that the like terms are grouped in the same columns. Then, each column is added.

$$\begin{array}{rrrrrrr} & 2a & + & 3b & - & 4c \\ & 3a & - & 2b & & \\ - & 4a & & & + & 5c \\ + & a & - & 2b & + & c \\ \hline & 2a & - & b & + & 2c \end{array}$$

**Notes**

*Since there is no term involving c in 3a – 2b, a space is left. Similarly, a space is left in –4a + 5c for the column for b.*

$\therefore$ the sum is $2a - b + 2c$.

Alternatively, $(2a + 3b - 4c) + (3a - 2b) + (-4a + 5c) + (a - 2b + c)$
$= 2a + 3b - 4c + 3a - 2b - 4a + 5c + a - 2b + c$
$= 2a - b + 2c$

## Example 10

*Subtract* $3a + 4b - 2c - 5d$ *from* $2a + 5b - d$.

**Solution**

Again, the expressions are arranged in order. The expression to be subtracted is placed below the other expression and like terms are grouped in the same columns.

$$\begin{array}{r} 2a + 5b \qquad - \ d \\ -\,(3a + 4b - 2c - 5d) \\ \hline -\ \ a + \ \ b + 2c + 4d \end{array}$$

$\therefore$ the result is $-a + b + 2c + 4d$.

This shows that if the expressions are written down as in arithmetic, the result is obtained by changing the sign of each term in the lower line and then adding. Alternatively, if we use brackets, we write

$(2a + 5b - d) - (3a + 4b - 2c - 5d)$
$= 2a + 5b - d - 3a - 4b + 2c + 5d$
$= -a + b + 2c + 4d$

or

$$\begin{array}{r} 2a + 5b \qquad - \ d \\ +\ -3a - 4b + 2c + 5d \\ \hline -\ a + \ \ b + 2c + 4d \end{array}$$

## Exercise 5e

1. Find the sum of each of the following expressions:
   (a) $4a + 6b + 5c$, $-3a - 9b$, $a + 3b - 4c$
   (b) $p + 2q - 4r$, $2p - 3q + 5r$, $3p - q + 2r$
   (c) $5x - 4y$, $6y - 7z$, $3z - 4x$
   (d) $9p + 12q - 3r - 4s$, $-8q + 4s$, $-7p + q + 2r$, $p + 4r - 5s$

2. Subtract
   (a) $a - 2b + 6c$ from $3a + 3b - 4c$;
   (b) $2q - 3r - s$ from $p - 4q - 6r$;
   (c) $8a - 3b + 5c - 2d$ from $10a - b - 4c - 6d$;
   (d) $7p + 6q - 5r$ from $10p - 7q + 8r$

3. Simplify the following expressions:
   (a) $(-2a + 7b) - (a + 4b)$
   (b) $(3a + 4b - 5c) + (2a - 7b - 6c) + (8a - 5b + 9c)$
   (c) $2(a + b - 3c) - 4(a - b + c) + 5a$
   (d) $5(b + a - 6c) - 7(c - b + 6a)$
   (e) $3(a - 5c) - 4(b - a) + 3(c - b)$
   (f) $6(a - 3b + 5c) - 4(5b - c + 2a) - 5(2a - 4c + 3b)$
   (g) $8(3a - 4b + c) + 5(2a - 3b + c) - 3(2c - 9a + 7b)$
   (h) $9(2a - 7c + 4b) - 4(b - c) - 7(-c - 4b)$

## Linear Algebraic Expressions with Fractional Coefficients

In $\frac{2x - 3}{4}$ and $\frac{5x - 7}{6}$, the numerators can be placed within brackets.

Thus, $\frac{2x - 3}{4}$ can be written as $\frac{(2x - 3)}{4}$ or $\frac{1}{4}(2x - 3)$ and $\frac{5x - 7}{6}$ can be written as $\frac{(5x - 7)}{6}$ or $\frac{1}{6}(5x - 7)$.

The procedure for simplifying such algebraic expressions is similar to that of simplifying ordinary fractions.

## Example 11

*Simplify the following expressions:*

(a) $\dfrac{2x-3}{5} + \dfrac{x-3}{3}$ (b) $\dfrac{2x-7}{3} - \dfrac{2x-6}{9}$

(c) $\dfrac{x+y}{2} + \dfrac{3x-y}{5} - \dfrac{7(2x-4)}{6}$

**Notes**

$-\dfrac{2x-6}{9} = -\dfrac{(2x-6)}{9}$

*The numerator is placed within brackets.*

### Solution

(a) The LCM of 5 and 3 is 15.

$$\begin{aligned}&\frac{2x-3}{5} + \frac{x-3}{3}\\ &= \frac{3(2x-3)}{15} + \frac{5(x-3)}{15}\\ &= \frac{3(2x-3)+5(x-3)}{15}\\ &= \frac{6x-9+5x-15}{15}\\ &= \frac{11x-24}{15}\end{aligned}$$

(b) The LCM of 3 and 9 is 9.

$$\begin{aligned}&\frac{2x-7}{3} - \frac{2x-6}{9}\\ &= \frac{3(2x-7)}{9} - \frac{(2x-6)}{9}\\ &= \frac{3(2x-7)-(2x-6)}{9}\\ &= \frac{6x-21-2x+6}{9}\\ &= \frac{4x-15}{9}\end{aligned}$$

(c)
$$\begin{aligned}\frac{x+y}{2} + \frac{3x-y}{5} - \frac{7(2x-4)}{6} &= \frac{15(x+y)+6(3x-y)-35(2x-4)}{30}\\ &= \frac{15x+15y+18x-6y-70x+140}{30}\\ &= \frac{-37x+9y+140}{30}\end{aligned}$$

## Exercise 5f

**1.** Simplify the following algebraic fractions:

**(a)** $\dfrac{x}{2} + \dfrac{2x}{5}$

**(b)** $\dfrac{x}{3} - \dfrac{x}{4}$

**(c)** $\dfrac{2x}{7} + \dfrac{x+1}{5}$

**(d)** $\dfrac{3x}{8} - \dfrac{x+2}{4}$

**(e)** $\dfrac{2a+3}{6} - \dfrac{a+2}{4}$

**(f)** $\dfrac{3y-1}{4} - \dfrac{2y-3}{6}$

**(g)** $\dfrac{x}{2} + \dfrac{x+2}{3} + \dfrac{x-3}{4}$

**(h)** $\dfrac{x-4}{5} + \dfrac{2x-5}{3} - \dfrac{2(x+1)}{6}$

**(i)** $\dfrac{a+2}{3} - \dfrac{3a+1}{4} + \dfrac{5+4a}{6}$

**(j)** $2x + \dfrac{x-4}{9} - \dfrac{2x-5}{3}$

**2.** Simplify the following algebraic fractions:

**(a)** $\dfrac{x}{5} + \dfrac{2x-4}{7}$

**(b)** $\dfrac{2x+7}{3} - \dfrac{6x-3}{5}$

**(c)** $\dfrac{4x+1}{5} + \dfrac{3x-1}{2}$

**(d)** $\dfrac{2x-7}{4} - \dfrac{x-6}{7}$

**(e)** $\dfrac{3(x-2)}{4} - \dfrac{4(2x-3)}{5}$

**(f)** $\dfrac{2(x+3)}{5} - \dfrac{1}{2} + \dfrac{3x-4}{4}$

# Factorisation

In primary school, we learnt that in $21 = 3 \times 7$, 3 and 7 are called the factors of 21. Similarly, in $3(x + y) = 3x + 3y$, 3 and $(x + y)$ are called the factors of $3x + 3y$. In $2 + 4x + 12z = 2(1 + 2x + 6z)$, the factors are 2 and $(1 + 2x + 6z)$. Can you name the factors in $4x + 2 = 2(2x + 1)$ and $3x + 9b + 15y = 3(x + 3b + 5y)$?

The process of writing an algebraic expression as a **product** of two or more other algebraic expressions, which are called the factors of the first algebraic expression, is called **factorisation**.

**Notes**

*For the rest of this chapter, all letters, like a, b, c, p, q, r, s except x, y, z, represent constants.*

## Example 12

*Factorise the following:*

*(a)* $5x - 15$ *(b)* $3ax + 12ay$ *(c)* $dx + dy + dz$

**Solution**

(a) $5x - 15 = 5 \times x - 5 \times 3$
$= 5(x - 3)$

(b) $3ax + 12ay$
$= 3 \times a \times x + 3 \times 4 \times a \times y$
$= 3 \times a(x + 4 \times y)$
$= 3a(x + y)$

(c) $dx + dy + dz = d(x + y + z)$

**Your Attention**

*Always check that no other factor can be taken out from within the brackets in your final answer. If you write $3ax + 12bx = 3(ax + 4bx)$, your factorisation is not complete as x is also a factor of the expression.*

## Exercise 5g

1. Factorise the following expressions where possible.
   **(a)** $4x + 12$  **(b)** $6x - 18$  **(c)** $-10x + 25$  **(d)** $16 - 24x$

2. Factorise the following expressions where possible.
   **(a)** $18a - 54ay$  **(b)** $28x + 56y$  **(c)** $2px - 3py$  **(d)** $3a + 6ax - 9az$
   **(e)** $4x + 8y + 16z$  **(f)** $4m - 6my - 18mz$

## Example 13

*Factorise each of the following:*
*(a)* $h(2x + 3) + 2(2x + 3)$  *(b)* $2h(5x + 2) - (5x + 2)$
*(c)* $3a(2y - z) + 4b(2y - z)$  *(d)* $5x(2a + b - 3c) - 6y(2a + b - 3c)$

### Solution

(a) In $h(2x + 3) + 2(2x + 3)$, $(2x + 3)$ is a common factor.
$\therefore\ h(2x + 3) + 2(2x + 3) = (2x + 3)(h + 2)$.

(b) $2h(5x + 2) - (5x + 2)$
$= 2h(5x + 2) - 1(5x + 2)$
$= (2h - 1)(5x + 2)$

(c) In $3a(2y - z) + 4b(2y - z)$, $(2y - z)$ is a common factor.
$\therefore\ 3a(2y - z) + 4b(2y - z)$
$= (3a + 4b)(2y - z)$
You will be able to identify the common factors after some practice.

(d) $5x(2a + b - 3c) - 6y(2a + b - 3c)$
$= (2a + b - 3c)(5x - 6y)$

**Notes**

*Note that*
$-(5k + 2) = (-1)(5k + 2)$

## Exercise 5h

1. Factorise each of the following:
   **(a)** $a(3x + 2) + 2(3x + 2)$
   **(b)** $3a(2y + 3) + h(2y + 3)$
   **(c)** $a(5x + y) - c(5x + y)$
   **(d)** $6h(3x + 5y) + (3x + 5y)$
   **(e)** $3k(z - 3) - h(z - 3)$

2. Factorise each of the following:
   **(a)** $a(x + y) + 3(x + y)$
   **(b)** $2h(x + 3y) - k(x + 3y)$
   **(c)** $6a(x - 2y) + 5(x - 2y)$
   **(d)** $3x(5k + h) - y(5k + h)$
   **(e)** $x(2a + b) + 3y(b + 2a)$

3. Factorise each of the following:
   **(a)** $5y(m + 3n) + 10z(m + 3n)$
   **(b)** $3a(x - 3) - 6k(x - 3)$
   **(c)** $(x + y)(a + b) - (y + z)(a + b)$
   **(d)** $(2y - 3z)(p + q) + (y - z)(p + q)$
   **(e)** $3x(2a + 3b - 4c) - 2y(2a + 3b - 4c)$

# Factorisation by Grouping

Sometimes, it is possible to find a common factor by grouping terms of an expression.

*For example,* $kx - ky + bx - by = k(x - y) + b(x - y)$
$= (x - y)(k + b)$

At other times, it may be necessary to re-group the terms of an expression in order to find the common factor.

*For example,* $dx + cy + cx + dy = (dx + dy) + (cx + cy)$
$= d(x + y) + c(x + y)$
$= (d + c)(x + y)$

Factorisation may be simplified by changing the signs of the factors.

*For example,* $4(a - b) + x(b - a) = 4(a - b) - x(a - b)$
$= (a - b)(4 - x)$

**Notes**

$(b - a) = -(a - b)$
$(q - p) = -(p - q)$

## Example 14

*Factorise the following:*
*(a)* $3px - 6qx - 2py + 4qy$ *(b)* $ax - ay - by + bx$

**Solution**

(a) $3px - 6qx - 2py + 4qy$
$= (3px - 2py) - (6qx - 4qy)$
$= p(3x - 2y) - 2q(3x - 2y)$
$= (p - 2q)(3x - 2y)$

(b) $ax - ay - by + bx$
$= (ax - ay) - (by - bx)$
$= a(x - y) - b(y - x)$
$= a(x - y) + b(x - y)$
$= (x - y)(a + b)$

**Notes**

$-(y - x) = (x - y)$

Alternatively we can rearrange the terms as
$ax - ay - by + bx$
$= (ax - ay) + (bx - by)$
$= a(x - y) + b(x - y)$
$= (x - y)(a + b)$

**Problem Solving Tips**

*To factorise an expression, the most important thing is to rearrange the terms to get a common factor.*

## Exercise 5i

Factorise the following where possible. Where there are no factors, state so.

1. $ax - 5a + 4x - 20$
2. $12hy + 20h - 15 - 9y$
3. $6az - 15z + 5 - 2a$
4. $ax + bx + ay + by$
5. $ay + by - bz - az$
6. $kx + hy - hx - ky$
7. $3ky - 4hx + 12hy - kx$
8. $6hx + 12ky + 9kx + 8hy$
9. $24vx - 15uy - 20vy + 18ux$
10. $14cx + 10dy - 4cy - 35xd$

## SUMMARY

1. In algebra, we use symbols, e.g. $a$, $x^2$ and $xy$, to represent numbers and variables. We add or subtract the like terms by adding or subtracting the coefficients, e.g. $2a + 5a = 7a$ and $7b - 3b = 4b$. We do not add the coefficients of unlike terms, so adding $3x$ and $4y$ gives $3x + 4y$.

2. When an expression of arithmetic operations contains brackets, work with the expressions within the brackets first. (If there are brackets within brackets, work with the innermost pair of brackets first.)

3. If an expression within brackets is multiplied by a number, each term within the brackets must be multiplied by that number when the brackets are removed.

4. Factorisation of algebraic expressions can be done by
   **(a)** extracting common factors from all terms of the given expressions,
   **(b)** grouping terms in such a manner that the new terms obtained have a common factor.

# Review Examples 5

## Example 1

*(a) If* $x = 3$, $y = -4$ *and* $z = 2$, *evaluate* $\dfrac{3x - 5y^2 - 2xyz}{\dfrac{x}{y} - \dfrac{y^2}{z}}$

*(b) Simplify* *(i)* $15x - [3x - \{2y - 4x - 5(y - x)\} + 7y]$

*(ii)* $\dfrac{3(x + y)}{5} - \dfrac{2(3x - y)}{3}$

## Solution

(a) $\dfrac{3x - 5y^2 - 2xyz}{\dfrac{x}{y} - \dfrac{y^2}{z}} = \dfrac{3(3) - 5(-4)^2 - 2(3)(-4)(2)}{\dfrac{3}{-4} - \dfrac{(-4)^2}{2}}$

$= \dfrac{9 - 80 + 48}{-\dfrac{3}{4} - 8} = \dfrac{-23}{-8\dfrac{3}{4}} = 2\dfrac{22}{35}$

(b) (i) $15x - [3x - \{2y - 4x - 5(y - x)\} + 7y]$

$= 15x - [3x - \{2y - 4x - 5y + 5x\} + 7y]$

$= 15x - [3x - 2y + 4x + 5y - 5x + 7y]$

$= 15x - 2x - 10y$

$= 13x - 10y$

(ii) $\dfrac{3(x + y)}{5} - \dfrac{2(3x - y)}{3} = \dfrac{9(x + y) - 10(3x - y)}{15}$

$= \dfrac{9x + 9y - 30x + 10y}{15} = \dfrac{-21x + 19y}{15}$

## Example 2

*Factorise each of the following completely.*

*(a)* $2x(h - 3k) - 6y(h - 3k)$

*(b)* $3ax - 6ay - 8by + 4bx$

*(c)* $2hx - 6kx + 24ky - 8hy$

## Solution

(a) $2x(h - 3k) - 6y(h - 3k)$
$= (h - 3k)(2x - 6y)$
$= 2(h - 3k)(x - 3y)$

(b) $3ax - 6ay - 8by + 4bx$
$= 3ax - 6ay + 4bx - 8by$
$= 3a(x - 2y) + 4b(x - 2y)$
$= (x - 2y)(3a + 4b)$

(c) $2hx - 6kx + 24ky - 8hy$
$= 2hx - 8hy - 6kx + 24ky$
$= 2h(x - 4y) - 6k(x - 4y)$
$= (x - 4y)(2h - 6k)$
$= 2(x - 4y)(h - 3k)$

## Review Questions 5

**1.** Given that $a = -2$ and $b = 7$, evaluate the following expressions:
**(a)** $4a + 5b$ **(b)** $2a^2$
**(c)** $3a - 4b$ **(d)** $a(b - a)$
**(e)** $b - a^2$ **(f)** $(b - a)^2$

**2.** Given that $a = \dfrac{1+b}{1-b}$, calculate the value of $a$ when $b = -3$, giving your answer as a fraction in its lowest terms.

**3.** Given that $\dfrac{1}{v} + \dfrac{1}{u} = \dfrac{1}{f}$, find the value of $f$ when $v = 10$ and $u = 15$.

**4.** Simplify each of the following expressions:
**(a)** $4[e - 3\{f - 6(f - e)\}]$
**(b)** $-3 + m - \{2 - (m - 4)\}$
**(c)** $10 + \dfrac{1}{3}[4k - (18 + 7k)]$
**(d)** $1 - 3(1 + x) + \{2 - (4x - 7)\}$
**(e)** $18 - [10 - x - (9 - x)]$
**(f)** $-2[3x - (4 - 5) - (6 - 8)x]$
**(g)** $\dfrac{1}{2}\left[16x - \dfrac{2}{3}(6x - 12)\right]$
**(h)** $a - [b - \{c - (d + e)\}]$
**(i)** $9[5a - 2\{3a - 7 - 2a)\}]$
**(j)** $5x - [2x - \{3x - 3(x - 2y) + y\}]$

**5.** Simplify each of the following algebraic fractions:
**(a)** $\dfrac{x}{2} + \dfrac{x}{4} - \dfrac{x}{6}$
**(b)** $\dfrac{2x}{3} + \dfrac{5-x}{4}$
**(c)** $\dfrac{x+1}{5} + \dfrac{2x-1}{3}$
**(d)** $\dfrac{2y-3}{5} + \dfrac{y-5}{6}$
**(e)** $\dfrac{3a-5}{4} - \dfrac{2a-7}{3}$
**(f)** $3a - \dfrac{4-3a}{5} - \dfrac{a-4}{6}$
**(g)** $\dfrac{a+b}{3} - \dfrac{b+c}{2} + \dfrac{4a-c}{5}$
**(h)** $\dfrac{2(3a+b)}{5} + \dfrac{4(2a-b)}{3}$
**(i)** $\dfrac{6x-y}{5} + \dfrac{3x-4y}{10} - \dfrac{5(x-2)}{6}$
**(j)** $\dfrac{4(x-5)}{7} - \dfrac{5(x-y)}{6} + \dfrac{7x-y}{21}$

**6.** Factorise the following:

**(a)** $4x - 8(y - 2z)$
**(b)** $5x + 10(b + c)x$
**(c)** $(a + b)(x + y) - (a + b)(y + z)$
**(d)** $5a(2x + y) - 3(2x + y)$
**(e)** $(2a - 3b)(x + y) + (a - b)(x + y)$
**(f)** $ax + by - ay - bx$
**(g)** $8ax - 4a - 2x + 1$
**(h)** $6cx + 4dy - 3cy - 8dx$
**(i)** $2ax - 21dy + 7dx - 6ay$
**(j)** $5mx - 10m - 3x + 6$

**7.** If $a = 3$, $b = -4$ and $c = -2$, evaluate each of the following:

**(a)** $\dfrac{3a - b}{2c} + \dfrac{3a - c}{c - b}$

**(b)** $\dfrac{2c - a}{3c + b} - \dfrac{5a + 4c}{c - a}$

**(c)** $\dfrac{a + b + 2c}{3c - a - b} - \dfrac{5c}{4b}$

**(d)** $\dfrac{b - c}{3c + 4b} \div \left(\dfrac{bc}{a} + \dfrac{ac}{b}\right)$

**8.** At a famous "roti prata" shop, for every two people who ordered egg prata, there were five people who ordered plain prata.

**(a)** If $a$ people ordered egg prata, how many people ordered plain prata?
**(b)** If $b$ people ordered plain prata, how many people ordered egg prata?
**(c)** If there were a total of $c$ people in the shop, how many of them ordered egg prata?

**9.** The average salary of $m$ male employees and $f$ female employees of a company is $\$A$. If the average salary of the male employees is $\$B$, find an expression for the average salary of the female employees.

**10.** A collection of coins contains only 10-cent and 5-cent coins. There are $x$ 5-cent coins in the collection. Write an algebraic expression for each of the following:

**(a)** The total value of the 5-cent coins.
**(b)** The total value of the 10-cent coins if there are three times as many 10-cent as 5-cent coins.
**(c)** The total value of the coins if for every three 10-cent coins there are five 5-cent coins.

## Revision Exercise I No. 1

1. (a) What is the sum of $4x$ dollars and $4x$ cents? Express your answer in cents.

   (b) Simplify $2x - \{[2x - 3(2y - 3x) - (5y - 2x) - 9y] - 3x\}$.

   (c) Find the HCF and LCM of 12, 15 and 18.

2. Estimate the following mentally, giving your answer correct to 1 significant figure:

   (a) $101 \times \sqrt{81}$  (b) $\sqrt[3]{26} \times 502 \div 49$

   (c) $\sqrt{65} \times \sqrt[3]{63} \div 17$

3. Simplify

   (a) $15a \div 5$;  (b) $4c \times 7c$;
   (c) $a \times 5 \times b$;  (d) $a \times 5 \times 2 + c$;
   (e) $\frac{2}{3} \times 21a^2$;  (f) $15x - 3 \times 5x$;
   (g) $8a + 2 \times 6a$;  (h) $4x + 8x \div 2$.

4. A family of 2 adults and 4 children visited Sentosa during the June school holidays. An adult ticket cost \$18.90 and a child-ticket cost \$12.80 during the promotional period. If the family enjoyed the promotional price,

   (a) what was the total cost of the tickets for the family?
   (b) what would be the change if a \$100-note was used to purchase the tickets?

5. A teacher distributes 255 pencils, 425 sheets of graph paper and 595 sheets of writing paper equally to a group of students.

   (a) Calculate the smallest and largest possible numbers of students in the group.
   (b) Calculate the smallest and largest numbers of pencils, graph paper and writing paper each student can receive.

6. State whether each of the following statements is true or false:

   (a) The first four prime numbers are 1, 2, 3 and 5.
   (b) Every prime number is a rational number.
   (c) 3.141 59 is an irrational number.
   (d) $\left(2\frac{3}{4} + 1.25\right)$ is an integer.

7. Simplify the following expression:
   $5h - 3[4k - 6m + 2\{3h - (2h + 3k)\}]$

8. (a) Estimate the value of $\frac{\sqrt{2505} \times 8.705}{4.98 \times 2.907}$ giving your answer as a whole number.

   (b) Given that $6.5 \times 232 = 1508$, find the value of
   (i) $0.065 \times 2.32$
   (ii) $1.508 \div 650$
   showing your working clearly.

9. (a) One-fifth of a plank is sawn off and three-eighths of the remaining piece is then thrown away. What fraction of the original plank remains?

   (b) Lamp posts along one side of a street are 6 m apart. If they extend for one and a half kilometres, find the number of lamp posts along the street.

   (c) An apple costs $x$ cents. An orange costs $y$ cents more than an apple. Express the cost of 8 apples and 12 oranges in terms of $x$ and $y$.

10. Copy and complete the following using one of the symbols >, = or <:

    (a) 5 ☐ 2
    (b) –5 ☐ –2
    (c) $\frac{1}{3}$ ☐ 0.33
    (d) –7 ☐ 14 ÷ (–2)

## Revision Exercise I No. 2

1. (a) Find the LCM of the following:
   (i) 56, 63
   (ii) 8, 56, 140
   (iii) 35, 45, 55

   (b) Find the HCF of the following:
   (i) 32, 48, 72
   (ii) 54, 90, 240

**2. (a)** If $x = \dfrac{\frac{1}{2}k}{h^2 + \frac{g}{5y^2}}$, find the value of $x$ when $y = 3$, $h = -2$, $k = -1$ and $g = 4$.

**(b)** A ribbon is 14 m long. Fourteen pieces, each of length 25 cm, are cut from it. The remaining piece is cut into equal lengths of 40 cm. How many pieces of length 40 cm are there? What is the length of the piece that is left?

**3.** The numerals –2, –1, 1, 2, 3, 4, 5 and 6 are written on eight separate cards, with one number on each card.

**(a)** List the pairs of cards that have a sum of 4.

**(b)** List the pairs of cards that have a product of 2.

**(c)** List the groups of three cards that have a sum of 10.

**4.** On a certain morning, the temperature in London was –14°C, the temperature in Hong Kong was 8°C and the temperature in Singapore was 24°C.

**(a)** Find the difference in temperature between London and Singapore.

**(b)** The temperature in Shanghai was mid-way between the temperatures of London and Hong Kong. Find the temperatures of Shanghai on that morning.

**5. (a)** Estimate the value of

**(i)** $\dfrac{64.11}{1.62}$;

**(ii)** 56.967 13 – 34.230 67;

correct to 1 significant figure.

**(b)** Use your calculator to evaluate

**(i)** $4(25 - 4\pi)$;

**(ii)** $\dfrac{\frac{2}{7} + \sqrt{15} \times \sqrt[3]{3}}{\sqrt{5} \div \frac{11}{2}}$;

**(iii)** $\dfrac{10.92^3 \times \sqrt{7.42}}{5.68^2} + \sqrt[3]{\dfrac{5.16 \times 6.42}{1.56 \times 4.72}}$;

giving each answer correct to 3 decimal places.

**6.** $\frac{4}{9}$ of the passengers in an MRT train are men, $\frac{2}{5}$ of them are women and the rest are children. If there are 72 women, find

**(a)** the total number of passengers in the train;

**(b)** how many more men than children there are in the train.

**7. (a)** Write down the difference between $3y$ minutes and $25y$ seconds, giving your answer in seconds.

**(b)** Cane sugar is sold at \$2.88 for a 900 g pack or at \$1.65 for a 500 g pack. Which is more expensive and by how much per kilogram?

**8.** Given that $a = 1$, $b = 2$, $c = 0$ and $d = -3$, evaluate

**(a)** $\dfrac{a^2bd}{3a - d}$; **(b)** $\dfrac{d^2 + bc}{b + a}$;

**(c)** $a^2 + b^2 + d^2$; **(d)** $a^3 + b^3 + d^3$.

**9. (a)** Express $\frac{5}{16}$ as a decimal.

**(b)** Change 0.225 into a fraction in its lowest term.

**(c)** Express 9.906 correct to

**(i)** 2 decimal places;

**(ii)** 2 significant figures.

## Revision Exercise I No. 3

**1.** Replace each ☐ with '>' or '<',

**(a)** $\frac{4}{7}$ ☐ $\frac{3}{7}$ **(b)** $1\frac{2}{5}$ ☐ $2\frac{1}{5}$

**(c)** –5 ☐ –4 **(d)** $\frac{7}{12}$ ☐ $\frac{7}{13}$

**(e)** 0.54 ☐ 0.5399

**(f)** –0.001 ☐ 0.0001

2. Evaluate each of the following, giving your answer as a fraction in its lowest terms.

(a) $\dfrac{\frac{3}{4}-3\frac{1}{2}}{2\frac{1}{3}\times\left(-\frac{3}{2}\right)+\frac{1}{4}}$

(b) $\left(\frac{-3}{5}\right)\times\left(2\frac{1}{2}\right)-2\times\left(-1\frac{1}{3}\right)^2$

(c) $\sqrt{\frac{7}{8}\times\frac{3}{4}\div 1\frac{1}{6}}$

(d) $\dfrac{2\frac{3}{4}-\frac{3}{8}}{\frac{8}{9}\times\frac{3}{4}\div\frac{4}{17}}$

3. (a) Express (i) 48 and (ii) 324 as a product of prime factors.

(b) Find the HCF of
(i) 49, 63; (ii) 36, 54, 75.

(c) Find the LCM of
(i) 56, 72; (ii) 21, 35, 42.

4. (a) Mrs Lee was given $1200 every month for the household expenses. She used $\frac{17}{24}$ of it to buy food for the family, $\frac{5}{7}$ of the remainder for clothes and other expenses. She put the remaining money into the bank as savings for the family. Find the amount she put into the bank.

(b) Arrange the fractions $-\frac{3}{4}, -\frac{5}{6}, -\frac{11}{12}, -\frac{8}{9}$ in ascending order.

(c) Arrange the fractions $-\frac{2}{5}, -\frac{1}{4}, -\frac{3}{8}, -\frac{1}{3}$ in descending order.

5. (a) Simplify each of the following expressions:
(i) $x + 2(3x - y) - 5(y - x)$
(ii) $-5(2x - z) + 2z - x$

(b) Given that $a = 2$, $b = -1$ and $c = 5$, evaluate
(i) $3(c - b) + 2bc$;
(ii) $3a^2 - 2bc + b^3$.

6. (a) An apple costs $a$ cents while a pear costs $b$ cents more than an apple. Find, in terms of $a$ and $b$, the cost of 5 apples and 7 pears.

(b) Write the following using mathematical symbols:
(i) 7 is greater than 5 but less than 10.
(ii) $a$ lies between $b$ and $c$ where $a$, $b$ and $c$ are numbers.

7. Four girls, $A$, $B$, $C$ and $D$, share a box of chocolates among themselves. $A$ takes $\frac{1}{4}$ of the chocolates, $B$ takes $\frac{3}{8}$ of the remainder and $C$ takes $\frac{5}{9}$ of the left-overs. What fraction of the chocolates does $D$ have?

8. (a) Evaluate
$-20 + 2[-38 - 20 \times (-8) + 5 - (-10 - 20)]$.

(b) Find the HCF and LCM of $3^2 \times 5$, $3 \times 5^2$ and $2^3 \times 3 \times 5$.

(c) Express 3136 as a product of prime factors and hence, find the value of $\sqrt{3136}$.

9. (a) If $37.56 \div 2.31 = 16.26$, find the value of
(i) $0.3756 \div 23.1$
(ii) $1.626 \times 23\,100$
showing your working clearly.

(b) There are seven consecutive even numbers. If the largest number is $x$, write an algebraic expression for the middle number. If the middle number is 48, find the value of the smallest number.

10. Given that $a = 5$, $b = 6$ and $c = -1$, evaluate

(a) $3a + b + c$; (b) $\frac{ab}{2c}$;

(c) $2a + 3b - 4c$; (d) $ab - bc$.

## Revision Exercise I No. 4

1. Round off the number
(a) 5.3352 to 2 decimal places;
(b) 0.090 38 to 2 significant figures;
(c) 4972 to the nearest hundred;
(d) 12 097 to the nearest ten.

**2.** Simplify
**(a)** $(x - 3y + 4z) - (x - y - 2z)$;
**(b)** $\frac{1}{2}(x - y) + \frac{1}{3}(2x - 5y) - \frac{1}{5}(6x - 5y)$.

**3.** Given that $x = 2$, $y = -3$, $z = 4$ and $k = -1$, evaluate each of the following:
**(a)** $xy - 3zk + 2x$ **(b)** $y(2z - 3kx - xz)$
**(c)** $\frac{3x - 4z}{4y - 5k}$ **(d)** $\frac{4k}{3x - y + z}$

**4. (a)** David cycles $x$ km in 3 hours. If he maintains the same average speed, how far can he cycle in $12y$ minutes?

**(b)** A water pump can fill a tank at the rate of 750 litres per hour. A tank has a capacity of 9000 litres. Find the time needed to fill $\frac{7}{15}$ of the tank.

**5.** The entrance fee to an amusement park during a promotional period was $4.75 for an adult and $2.85 per child. Mr Ong took all his little nephews to the park and paid a total of $30.40 as entrance fee. How many children did Mr Ong bring?

**6. (a)** Find the LCM and HCF of
**(i)** 12, 16, 32;
**(ii)** 20, 24, 140.

**(b)** Which of the fractions, $-\frac{1}{2}$, $-\frac{3}{7}$, $-\frac{2}{5}$ and $-\frac{4}{9}$, is the largest?

**(c)** Which of the fractions, $-\frac{5}{6}$, $-\frac{3}{5}$, $-\frac{2}{9}$ and $-\frac{3}{10}$, is the smallest?

**7.** Write an algebraic expression for each of the following:
**(a)** The distance a car travels in $x$ hours if its average speed is 70 km/h.
**(b)** The number of 22-cent stamps you can buy with $\$x$.
**(c)** The number of grams in $k$ kilograms.

**8. (a)** Express 5832 as a product of prime factors and hence, find the cube root of 5832.

**(b)** Simplify
$25 - (-3) + [12 \times (-3) + 15 - (-35)] \div 7$.

**(c)** Simplify $\frac{2x - 1}{3} - \frac{2(x - 2)}{5}$.

**9.** If $x = 3$, $y = 2$ and $z = -1$, find the value of $\frac{2z^2 - 5x}{xz - y}$.

**10.** Mrs Goh bought 4 bags of rice at $5.28 per bag, 5 packets of biscuits at $1.25 per packet and 2.8 kg of meat at $7.40 per kg. Assuming that GST was absorbed by the retailer, calculate the total amount she spent. What would be the change if she paid for all these with a $50-note?

## Revision Exercise I No. 5

**1. (a)** Simplify $\frac{7x - 1}{2} + \frac{2x + 3}{6} - \frac{5x - 4}{2}$.

**(b)** 3 cars leave Town $A$ for Town $B$ on a straight road. Car $A$ will sound a horn after every 60 m, Car $B$ will sound a horn after every 80 m and Car $C$ will sound a horn after every 180 m. After how many metres will the 3 cars sound their horns at the same time?

**2. (a)** Tickets to an Arts festival concert are priced at $35 each for seats at the front rows and at $15 each for seats at the back rows. There are 12 front rows with 26 seats in each row and 18 back rows with 28 seats in each row. If $\frac{5}{6}$ of all the front row seats and $\frac{6}{7}$ of the back row seats are sold, how much is the total in ticket sales?

**(b)** Use your calculator to evaluate each of the following, giving your answer correct to 4 significant figures.

**(i)** $\frac{28.75^2 - \sqrt[3]{45.67}}{84.4 + \sqrt{75.6}}$

**(ii)** $\frac{56.34^4}{384.7^2 - 46.5^3}$

**(iii)** $46.9^2 - 15.3 \times \sqrt{47.4}$

**(iv)** $\sqrt{5487} \div (48.3 - \sqrt[3]{784})$

3. (a) Arrange the following numbers in descending order

$$\frac{22}{7},\ 3.142,\ \frac{47}{15},\ 3\frac{3}{20}$$

(b) Find the HCF and LCM of the following.
(i) 48, 64 (ii) 54, 63, 84

4. Given that $x = 5$, $y = -4$, $z = -2$, $h = 0$ and $k = -1$, evaluate each of the following:
(a) $5x - 3yk + 4hz$ (b) $xy(2x - 4k + xhz)$
(c) $\dfrac{5x - 3z}{xhk - 5yk}$ (d) $\dfrac{2k}{xh - yz}$

5. (a) Find the HCF and LCM of $(2 \times 5 \times 7)$, $(3^3 \times 5)$ and $(2^2 \times 3^3 \times 5)$.

(b) List the whole numbers between $\sqrt[3]{65}$ and $\sqrt{65}$.

(c) What is the value of the number that is mid-way between –24 and 6?

6. (a) Find the value of $x^2 - 4x + 2$ when $x$ is
(i) 0; (ii) 1;
(iii) $\frac{1}{2}$; (iv) –1.

(b) Simplify
(i) $5a - b - (4a + 3b)$;
(ii) $2(m - 2n) - 7m$.

7. (a) Express 2304 as a product of prime factors and hence, find the value of $\sqrt{2304}$.

(b) Arrange the fractions $-\frac{2}{3}$, $-\frac{4}{9}$, $-\frac{5}{7}$ and $-\frac{8}{11}$ in descending order.

8. The table below shows a man's deposits and withdrawals in dollars. If he had $520 in his savings account initially, calculate his balance after his withdrawals and last deposit of $75.

| Deposits (in $) | Withdrawals (in $) |
|---|---|
| 65 | 44 |
| 30 | 134 |
| 20 | 25 |
| 40 | 55 |
| 60 | 50 |
| 75 | |

9. Subtract $2\frac{1}{2}$ from the sum of $3\frac{1}{6}$ and $4\frac{5}{8}$.

10. (a) Similar bars of soap are sold in packs of 3 for $2 or packs of 8 for $5.20. Find the difference in the price for 48 bars of soap.

(b) Mr Tan gave $240 to his wife, $\frac{2}{5}$ of the remainder of his money to his son and kept the rest. If he had $195 left, how much money did he have originally?

(c) The temperature in Arrowtown for six days are –4°C, –12°C, –18°C, 2°C, 4°C and –8°C. Find the average temperature for these six days.

*In this chapter, you will learn*

- ***how to recognise simple patterns from various number sequences;***
- ***how to continue a given number sequence and find its general term;***
- ***about problem-solving heuristics.***

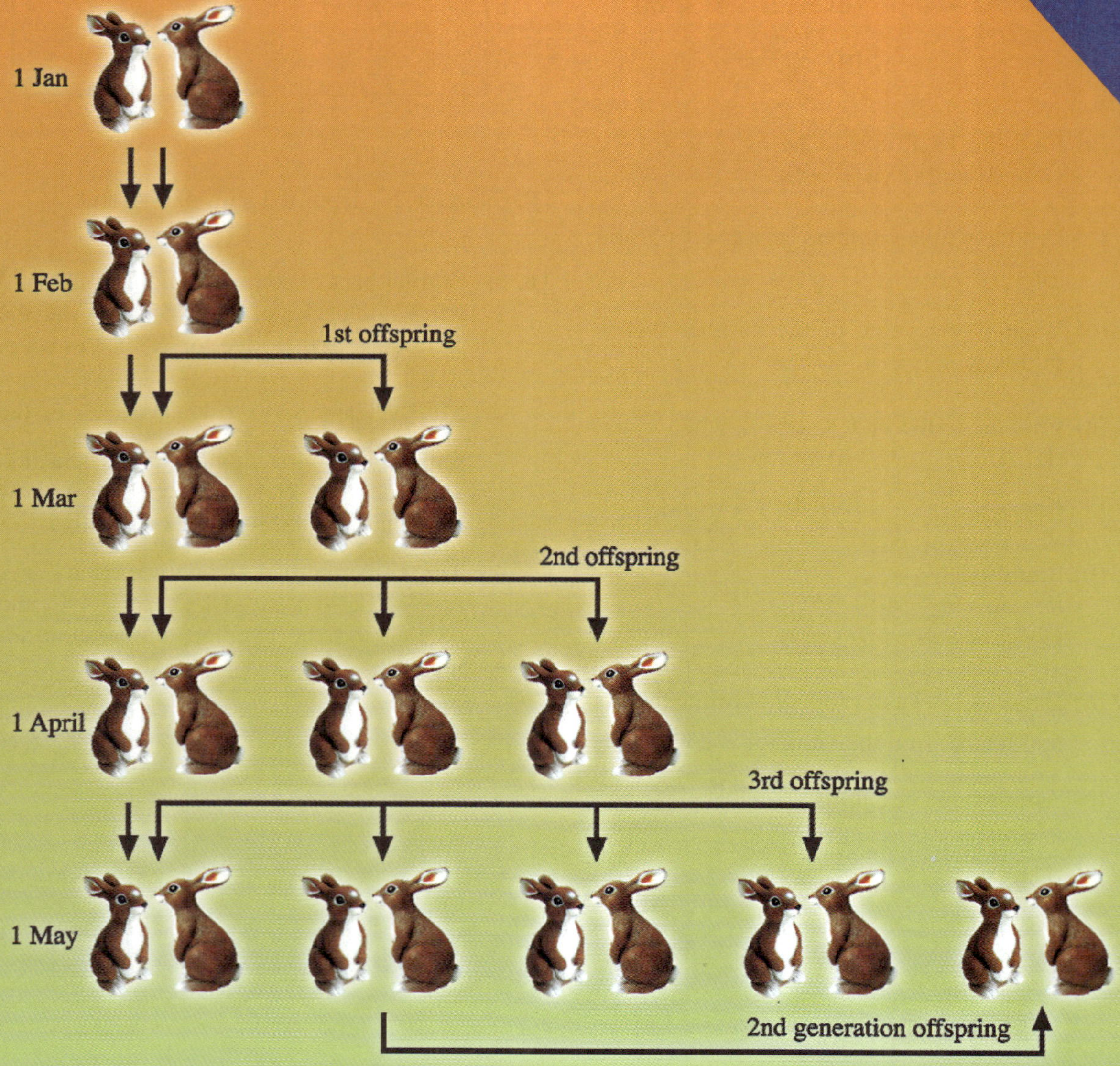

CHAPTER

# 6

# Number Sequences

## Introduction

Suppose a pair of newborn rabbits takes one month to mature, and starts to reproduce a pair of offsprings two months after they are born. They reproduce a pair each month after that.

If we started with a pair of newborn rabbits on 1 January 2000, how many pairs of rabbits will there be on 1 January 2010? This problem will lead to an interesting and useful number sequence.

# Number Sequences

Consider the following natural numbers:

$$1, 2, 3, 4, 5, 6, 7, 8, 9, 10, \ldots$$

We know that the numbers which come after 10 are 11, 12, 13, 14 and so on. We are able to continue writing down the numbers because each successive number follows the preceding one according to a specific rule. We say that the natural numbers form a **number sequence**. The numbers in a sequence are the **terms** of the sequence.

For the sequence of natural numbers, the rule is: start with 1, then add 1 to each term to get the next term.

Here are other examples of number sequences:

(a) Sequence of even numbers

$$2, 4, 6, 8, 10, 12, \ldots$$

**Rule**: Start with 2, then add 2 to each term to get the next term or multiply each term of the sequence 1, 2, 3, 4, 5, 6, ... by 2.

(b) Sequence of odd numbers

$$1, 3, 5, 7, 9, 11, \ldots$$

**Rule**: Start with 1, then add 2 to each term to get the next term or subtract 1 from each term of the sequence 2, 4, 6, 8, 10, 12, ...

(c) Sequence of powers of 2

$$1, 2, 4, 8, 16, 32$$

**Rule**: Start with 1, then multiply each term by 2 to get the next term.

(d) Sequence of squares

$$1, 4, 9, 16, 25, 36, \ldots$$

or

$$1^2, 2^2, 3^2, 4^2, 5^2, 6^2, \ldots$$

**Rule**: Square each term of the sequence 1, 2, 3, 4, 5, 6, ...

Try to find the pattern of the sequence below and state the rule for the pattern yourself.

$$1, 8, 27, 64, 125, 216, \ldots$$

## Example 1

*For each sequence, state a rule and write the next three terms:*

*(a) 3, 8, 13, 18 ...*

*(b) 38, 32, 26, 20 ...*

*(c) 2, 6, 18, 54 ...*

*(d) 128, 64, 32, 16 ...*

> **Just for fun**
>
> *Fill in the two missing numbers in the sequence below:*
>
> *1, 4, 9, 61, 52, ___, 94, ___, 18, 1, 121, ...*
>
> *State a rule for this sequence.*

### Solution

(a) **Rule**: Add 5 to each term to get the next term.
The next three terms are 23, 28 and 33.

(b) **Rule**: Subtract 6 from each term to get the next term.
The next three terms are 14, 8 and 2.

(c) **Rule**: Multiply each term by 3 to get the next term.
The next three terms are 162, 486 and 1458.

(d) **Rule**: Divide each term by 2 to get the next term.
The next three terms are 8, 4 and 2.

## Explorations

Carry out the activity with a partner.

1. **Fibonacci Sequence**

**(a)** The number sequence resulting from the problem described at the beginning of the chapter: 1, 1, 2, 3, 5, ... is known as a **Fibonacci sequence**. Write down the next two terms and figure out the rule for generating the next number in the sequence.

**(b)** Using the rule for generating a Fibonacci sequence, write down the next four terms of the Fibonacci sequence starting with

**(i)** 0, 1 **(ii)** 4, 2 **(iii)** 3, 3.

**(c)** The second and the third terms of a Fibonacci sequence are 3 and 10 respectively. Write down the first 8 terms of the sequence.

**(d)** Consider any three consecutive terms of the Fibonacci sequence 1, 1, 2, 3, 5, .... Find the product of the first and third terms and the square of the middle term. Does the product differ from the square by 1? Check whether this is true for any other three consecutive terms.

**(e)** Consider the pattern:

$$1^2 + 1^2 = 1 \times 2$$
$$1^2 + 1^2 + 2^2 = 2 \times 3$$
$$1^2 + 1^2 + 2^2 + 3^2 = 3 \times 5$$
$$1^2 + 1^2 + 2^2 + 3^2 + 5^2 = 5 \times 8$$

Write down the next three lines.

**2. Pascal's Triangle**

**(a)** The following triangle of numbers is called **Pascal's triangle**.

1
1 1
1 2 1
1 3 3 1
1 4 6 4 1
⋮

Each term is obtained by adding the two terms immediately above. For example, the number 2 in the third row is obtained by adding 1 and 1 which are above. The first 4 in the fifth row is obtained by adding 1 and 3 which are immediately above. (See the diagram below)

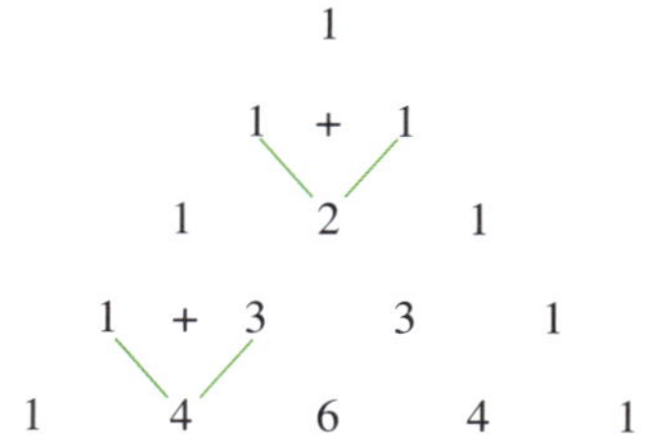


Write down the next two rows of the triangle.

**(b)** The table of numbers below is formed in the same way as the Pascal's triangle.

**(i)** Write down the next six rows.

**(ii)** Find the sum of the terms of each diagonal shown.

**(iii)** State a rule for the sequence of numbers formed by these diagonal sums.

**(iv)** Write down the next five diagonal sums.

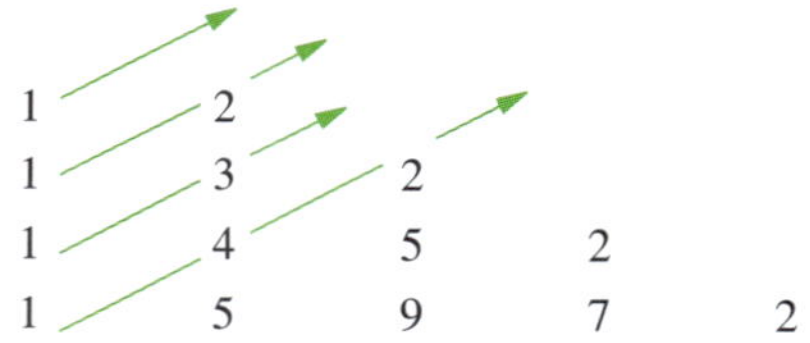


**(c)** In the diagram below, diagonals are drawn starting at the number from the left of each row.

**(i)** Find the sum of the terms along each diagonal.

**(ii)** Do you find the Fibonacci sequence in the Pascal's triangle?

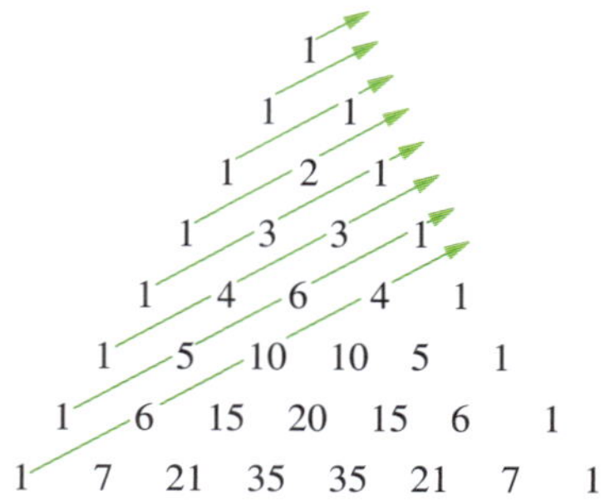

## Exercise 6a

1. Identify a rule and complete the following number sequences:
   (a) 2, 5, 8, 11, ______, ______, ______
   (b) 0, 10, 20, 30, ______, ______, ______
   (c) 52, 59, 66, 73, ______, ______, ______
   (d) 80, 72, 64, 56, ______, ______, ______
   (e) 37, ______, 55, 64, ______, ______
   (f) 59, ______, 51, 47, ______, ______

2. State a rule and write the next three terms of each sequence.
   (a) 1, 3, 9, 27, ...
   (b) 6, 12, 24, 48, ...
   (c) 1600, 800, 400, 200, ...
   (d) 4, 12, 36, 108, ...

3. For each of the following sequences, state a rule and write down the next two terms:
   (a) 14, 19, 24, 29, ...
   (b) 28, 39, 50, 61, ...
   (c) 73, 67, 61, 55, ...
   (d) 99, 90, 81, 72, ...
   (e) 15, 30, 60, 120, ...
   (f) 2187, 729, 243, 81, ...

4. Identify a rule and complete the following:
   (a) 1, 3, 6, 10, ______, ______, ______
   (b) 17, 22, 27, 32, ______, ______, ______
   (c) 50, 45, 44, 39, 38, ______, ______, ______
   (d) 1, 1, 2, 3, 5, 8, 13, ______, ______, ______

## General Term in a Number Sequence

The sequence of even numbers 2, 4, 6, 8, 10, 12, ... can be rewritten as

$$2 \times \mathbf{1}, 2 \times \mathbf{2}, 2 \times \mathbf{3}, 2 \times \mathbf{4}, 2 \times \mathbf{5}, 2 \times \mathbf{6}, ..., 2 \times \boldsymbol{n}, ...$$

The expression $2 \times n$ or $2n$ is the **formula** for the ***n*th term** or the **general term** of the number sequence. By varying the values of the letter $n$ in the formula, we obtain corresponding values of the formula $2n$ and thus generate terms of the number sequence. The letter $n$ is a **variable**.

Similarly, the sequence of odd numbers 1, 3, 5, 7, 9, 11, ... can be rewritten as

$$2 \times \mathbf{1} - 1, 2 \times \mathbf{2} - 1, 2 \times \mathbf{3} - 1, 2 \times \mathbf{4} - 1, 2 \times \mathbf{5} - 1, ..., 2\boldsymbol{n} - 1, ....$$

The formula for the $n$th term of the sequence of odd numbers is $2n - 1$.

Clearly, the formula for the $n$th term of the number sequence, $1 = 1^2$, $4 = 2^2$, $9 = 3^2$, $25 = 5^2$, $36 = 6^2$, ... is $n^2$.

# Example 2

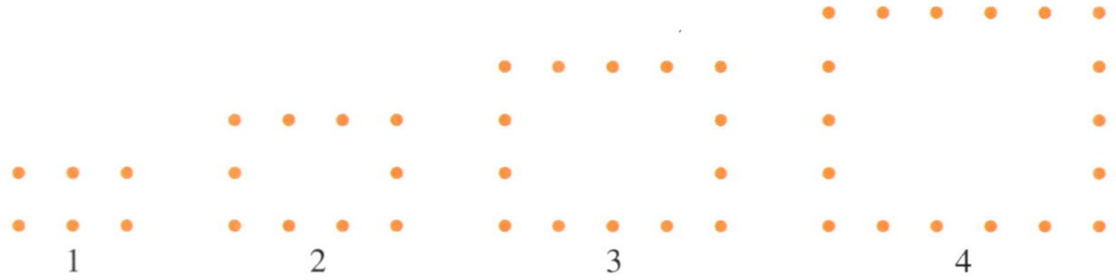


*The diagram shows the first four of a sequence of figures. Figures 1 and 2 contain 6 dots and 10 dots respectively. The sequence continues as shown in Figures 3, 4 and so on. Let n denote the figure number and d the corresponding number of dots.*

*(a) Count the number of dots in each of the Figures 3 and 4 and write down the next 2 terms of the number sequence 6, 10, ....*
*(b) Find a formula that connects n and d, i.e., a formula for the nth term of the sequence in (a).*
*(c) Using the formula in (b), find*
*(i) the number of dots there will be in Figure 30;*
*(ii) the figure that has 42 dots.*

**Solution**

(a) The next two terms are 14 and 18 respectively.

(b) Notice that

$6 = 4 \times \mathbf{1} + 2,$ (Figure **1**)
$10 = 4 \times \mathbf{2} + 2,$ (Figure **2**)
$14 = 4 \times \mathbf{3} + 2,$ (Figure **3**)
$18 = 4 \times \mathbf{4} + 2,$ (Figure **4**)

$\therefore$ for Figure $n$, the number of dots, $d = 4 \times \boldsymbol{n} + 2 = 4n + 2$.

(c) (i) To find the number of dots there will be in Figure 30, we find the value of $d$ when $n$ has a value 30, i.e. when $n = 30$.

Replacing $n$ by 30 in the formula $d = 4n + 2$, we have
$d = 4 \times 30 + 2 = 122.$

$\therefore$ there are 122 dots in Figure 30.

(ii) By writing $42 = 4 \times 10 + 2$ and comparing with the formula $d = 4 \times n + 2$, we have $n = 10$ when $d = 42$.

$\therefore$ the figure that has 42 dots is Figure 10.

## Example 3

*Consider the pattern:*

$$\begin{aligned} 2 &= 1 \times 2 \\ 6 &= 2 \times 3 \\ 12 &= 3 \times 4 \\ 20 &= 4 \times 5 \\ &\vdots \\ 110 &= k \times (k + 1) \\ &\vdots \end{aligned}$$

*Write down (a) the 8th line in the pattern; (b) the value of k.*

**Solution**

(a) From the pattern, the right-hand side of the 8th line is $8 \times 9$.

$\therefore$ the 8th line is $72 = 8 \times 9$.

(b) Since $110 = 10 \times 11 = 10 \times (10 + 1)$,

$\therefore$ $k = 10$.

## Example 4

*(a) Draw the next two patterns and complete the table below.*

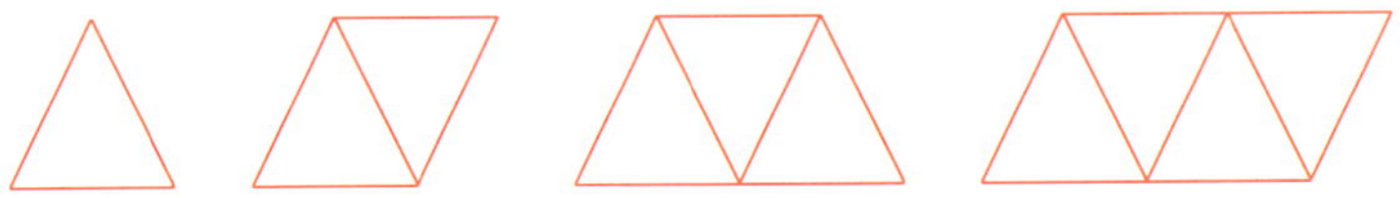

| *Pattern Number* | *1* | *2* | *3* | *4* | *5* | *6* | ... | *n* |
|---|---|---|---|---|---|---|---|---|
| *No. of lines* | $2 \times \mathbf{1} + 1 = 3$ | $2 \times \mathbf{2} + 1 = 5$ | $2 \times \mathbf{3} + 1 = 7$ | $2 \times \mathbf{4} + 1 = 9$ | | | ... | |
| *No. of triangles* | **1** | **2** | **3** | **4** | | | ... | |

*(b) How many lines and triangles are there in the*

*(i) 10th pattern,* *(ii) 121st pattern?*

*(c) Write down the formula connecting the number of lines (L) and the number of triangles (T) in a pattern.*

## Solution

(a) The next two patterns are shown below.

| Pattern Number | ... | 5 | 6 | ... | $n$ |
|---|---|---|---|---|---|
| No. of lines | ... | $2 \times \mathbf{5} + 1 = 11$ | $2 \times \mathbf{6} + 1 = 13$ | ... | $2 \times n + 1$ |
| No. of triangles | ... | **5** | **6** | ... | $n$ |

(b) (i) The number of lines in the 10th pattern = $2 \times 10 + 1 = 21$
The number of triangles in the pattern = 10

(ii) The number of lines in the 121st pattern = $2 \times 121 + 1 = 243$
The number of triangles in the pattern = 121

(c) From (a), the number of lines L = $2 \times n + 1 = 2n + 1$
and the number of triangles T = $n$.
The formula connecting L and T is L = 2T + 1.

## Exercise 6b

**1.** Consider the pattern:

$$1 + 3 = 4 = 2^2 = (1 + 1)^2$$
$$1 + 3 + 5 = 9 = 3^2 = (2 + 1)^2$$
$$1 + 3 + 5 + 7 = 16 = 4^2 = (3 + 1)^2$$
$$1 + 3 + 5 + 7 + 9 = 25 = 5^2 = (4 + 1)^2$$
$$\vdots$$
$$1 + 3 + 5 + \ldots + a = b = c^2 = (d + 1)^2$$
$$\vdots$$

**(a)** Write down the fifth and sixth lines in the pattern.
**(b)** Write down the 11th line in the pattern.
**(c)** Given that $b = 169$, find the values of $a$, $c$ and $d$.

2. Consider the pattern:

$$2 + 1^2 = 3$$
$$2 + 2^2 = 6$$
$$2 + 3^2 = 11$$
$$2 + 4^2 = 18$$
$$\vdots$$
$$2 + x^2 = 66$$
$$\vdots$$

**(a)** Write down the 6th line in the pattern.
**(b)** Find the value of $x$.

3. Consider the pattern:

$$1^2 - 0^2 = 1 = 1 + 0$$
$$2^2 - 1^2 = 3 = 2 + 1$$
$$3^2 - 2^2 = 5 = 3 + 2$$
$$4^2 - 3^2 = 7 = 4 + 3$$
$$\vdots$$
$$m^2 - n^2 = 81 = m + n$$
$$\vdots$$

**(a)** Write down the 10th line in the pattern.
**(b)** Find the value of $598^2 - 597^2$.
**(c)** Find the values of $m$ and $n$.

4. Consider the pattern:

$$\frac{1 \times 2}{2} + (1 - 1)^2 = 1$$
$$\frac{2 \times 3}{2} + (2 - 1)^2 = 4$$
$$\frac{3 \times 4}{2} + (3 - 1)^2 = 10$$
$$\frac{4 \times 5}{2} + (4 - 1)^2 = 19$$
$$\vdots$$
$$\frac{110}{2} + (p - 1)^2 = q$$
$$\vdots$$

**(a)** Write down the 11th line in the pattern.
**(b)** Find the values of $p$ and $q$.

5.

Diagram 1 Diagram 2 Diagram 3 Diagram 4

A sequence of diagrams consisting of shaded and unshaded small triangles is shown above. Diagrams 1 and 2 contain 3 and 6 shaded triangles respectively. The sequence continues as shown in Diagrams 3 and 4 and so on. Let $n$ denote the diagram number and $t$ the corresponding number of shaded triangles.

**(a)** By counting the number of shaded triangles in each of the Diagrams 3 and 4, write down the next 2 terms of the number sequence 3, 6, ....

**(b)** Find a formula that connects $n$ and $t$.

**(c)** Using the formula in (b), find

**(i)** the number of shaded triangles there will be in Diagram 50;

**(ii)** the diagram that has 87 shaded triangles.

**6.**

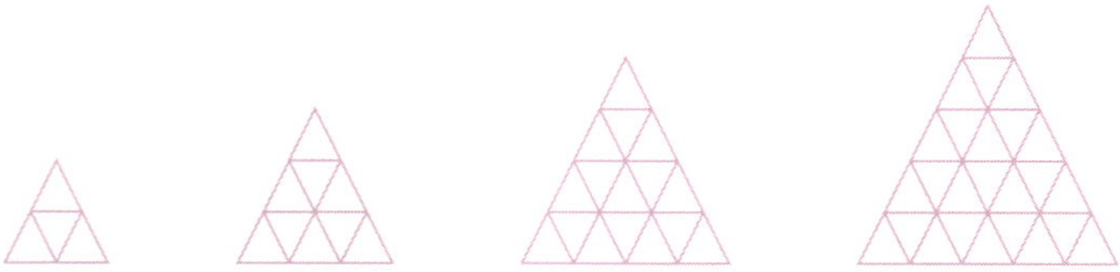

Figure 1 Figure 2 Figure 3 Figure 4

The above shows the first four of a sequence of figures. Figures 1 and 2 contain 4 and 9 small triangles respectively. The sequence continues as shown in Figures 3 and 4 and so on. Let $N$ denote the figure number and $T$ the corresponding number of small triangles.

**(a)** By counting the number of small triangles in Figures 3 and 4, write down the next 2 terms of the number sequence 4, 9, ....

**(b)** Find a formula that connects $N$ and $T$.

**(c)** Using the formula in (b), find

**(i)** the number of small triangles there will be in Figure 9;

**(ii)** the figure that has 121 small triangles.

# Problem Solving

So far, in this chapter, you have learnt how to

(i) recognise simple patterns from various number sequences;

(ii) generalise the terms in the number sequences following the pattern;

(iii) associate pattern of figures with sequences of numbers.

In the following activities, you shall explore various **heuristics** or strategies of solving problems.

## Explorations

You may do the first problem with a partner and the second problem with four others in a group.

**Problem 1**

How many logs are there in the stack shown below?

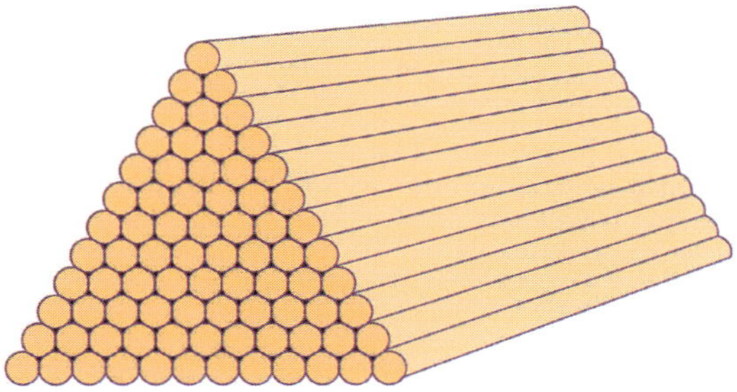

**Step 1: Simplify the problem and use a model**

Use coins to model the different numbers of layers of logs from the top starting with 1, 2, 3 and 4 layers as shown below.

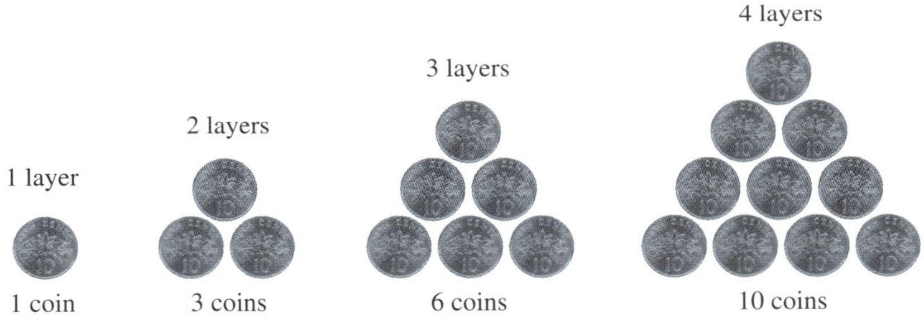


**Step 2: Look for a pattern**

**(a)** Use the piles of coins in the figure above to complete the following:

1 layer $= 1 \qquad\qquad\qquad = 1$

2 layers $= 1 + \square \qquad\qquad = 3$

3 layers $= 1 + 2 + \square \qquad = 6$

4 layers $= 1 + \square + 3 + \square = 10$

**(b)** Does the following rule completely describe the sequence?

**Rule**: The $r$th term of the sequence is obtained by adding the first $r$ natural numbers so that

the 5th term $= 1 + 2 + 3 + 4 + 5 = 15$,

the 6th term $= 1 + 2 + 3 + 4 + 5 + 6 = 21$,

the 7th term $= 1 + 2 + 3 + 4 + 5 + 6 + 7 = 28$, and so on.

**(c)** Write down the 8th, 9th and 10th terms of the sequence.
**(d)** How many layers of logs are there in the stack?
**(e)** Which term of the sequence corresponds to the number of logs in the stack?
**(f)** Find the number of logs in the stack.

**Alternatively:**

The sequence of numbers 1, 3, 6, 10 can be written as follows:

1 layer : $1 = \frac{1 \times 2}{2} = \frac{1 \times (1 + 1)}{2}$

2 layers : $3 = \frac{2 \times 3}{2} = \frac{2 \times (2 + 1)}{2}$

3 layers : $6 = \frac{3 \times 4}{2} = \frac{3 \times (3 + 1)}{2}$

4 layers : $10 = \frac{4 \times 5}{2} = \frac{4 \times (4 + 1)}{2}$

**(a)** Do you notice an emerging pattern showing how the number of logs is connected to the number of layers in the stack?

**(b)** Can you state the rule for the pattern?

**(c)** Using the rule, find the number of logs when the number of layers in the stack is
**(i)** 8; **(ii)** 9; **(iii)** 10.

**(d)** Use the rule to find the number of logs in the stack shown. Does your answer agree with that obtained earlier?

**(e)** As a challenge, find the number of layers in a stack of 820 logs.

**(f)** Multiply each number in the sequence by 8 and add 1. What do you notice?

**(g)** Choose any two consecutive numbers in the sequence and add them. What do you notice?

**(h)** Select any two consecutive numbers in the sequence. Square each of them and add the squares. Do you get another number in the same sequence?

**Problem 2**

John invites his Chinese, Malay and Indian friends for a Chinese New Year party. They shake hands to greet one another. Each person shakes hands with every other person exactly once. How many total handshakes are there if there are 20 people including John attending the party?

**Step 1: Simplify the problem and act it out**

Have two of you shake hands. Next, three of you shake hands exactly once with one another. Each time, count the number of handshakes. Repeat this with four of you and five of you respectively.

**Step 2: Use tabulation and look for a pattern**

**(a)** Tabulate your results by copying and completing the table below.

| Number of people ($n$) | **2** | **3** | **4** | **5** |
|---|---|---|---|---|
| Number of handshakes ($N$) | 1 | 3 | | |

**(b)** Write down the formula connecting $n$, the number of people and $N$, the number of handshakes.

**(c)** Use the formula to find how many handshakes there are when 20 people shake hands with one another exactly once.

**Alternatively**, in **Step 1** above, you can use **diagrams**.
The first two diagrams, involving two people, $A$ and $B$, and three people, $A$, $B$ and $C$, respectively, are shown below.

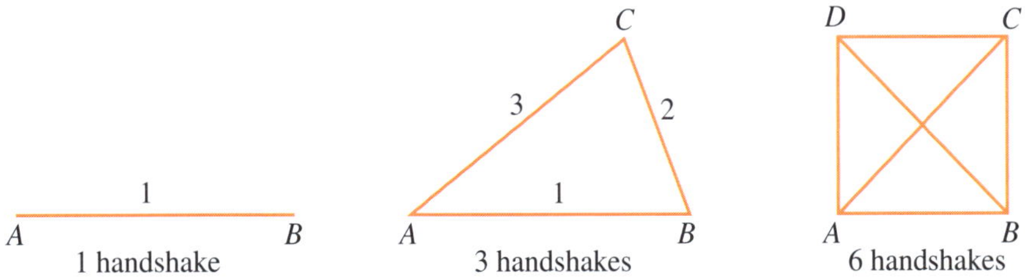


Add the next two diagrams, involving 5 people, $A$, $B$, $C$, $D$ and $E$, and 6 people, $A$, $B$, $C$, $D$, $E$ and $F$, respectively.

---

From the above activities, we see that a problem may be solved in more than one way and in solving the above problems, we have used different strategies or heuristics like:

- simplifying the problem
- using a model
- using tabulation
- acting it out
- looking for a pattern
- drawing a diagram

Some other strategies you will find useful are:

- changing your point of view
- working backwards
- making a supposition
- solving part of the problem
- solving a simpler problem
- eliminating the unlikely possibilities
- making an organised list
- writing an equation
- using trial-and-error
- thinking of a related problem

**Just For Fun**

*Use only 4 lines to join the 9 points without lifting your pencil.*

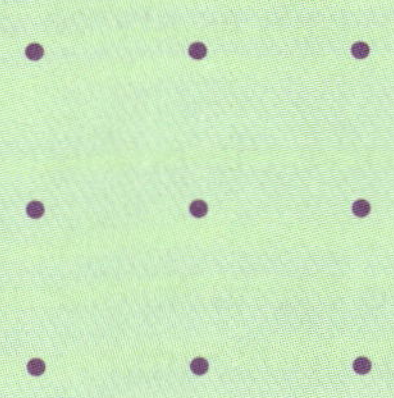

In the next example, we shall consider two of the above problem-solving strategies, i.e., drawing a diagram and changing your point of view. We shall introduce some of the other strategies in later chapters.

## Example 5

*Suppose in the S League soccer tournament, 8 teams qualify for the knock-out round, if the organisers adopt a single-elimination method for the round, i.e., winners play against winners until 1 team is left, what is the total number of matches played? How many matches would be played if there are n teams generally?*

### Solution

Draw a diagram showing the progress of the tournament as shown below.

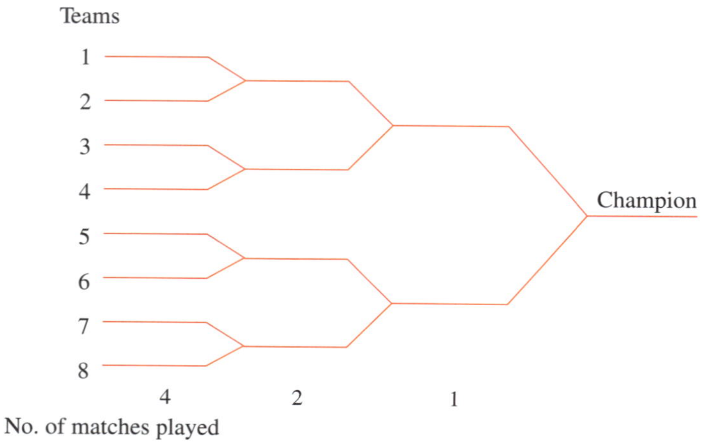


**Just for fun**

*The diagram shows 12 matchsticks arranged to form 6 equilateral triangles. Can you rearrange these 12 matchsticks to form 8 equilateral triangles of the same size?*

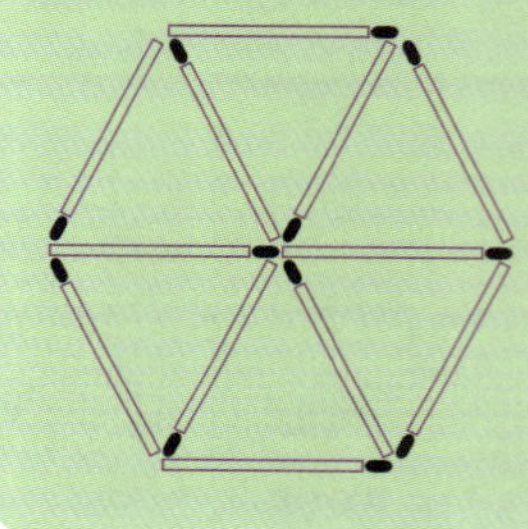

From the diagram, the number of matches played = 4 + 2 + 1 = 7.

Can we solve the second part of the question as above? It seems difficult because it is impossible to draw a diagram as above when the number of teams is variable and not a fixed number.

To solve the problem, we have to change our point of view.

Since 1 team is eliminated in each game played, we need $n - 1$ games to eliminate $n - 1$ teams to leave 1 champion team.

## Exercise 6c

**1.** How many segments will there be when 49 points divide a given line segment?

**(a)** Study the figures below and complete the table that follows.

*Figure 1* *Figure 2* *Figure 3* *Figure 4*

| No. of points | 1 | 2 | 3 | 4 | 5 | 6 |
|---|---|---|---|---|---|---|
| No. of segments | 1 + 1 = 2 | 2 + 1 = 3 | | | | |

**(b)** Hence, write down the number of segments there will be when 49 points divide the given line segment.

**(c)** How many points are needed to divide a given line segment into 101 segments?

**2.** 20 people go to a restaurant for a buffet dinner. They requested to be seated at the same table. The restaurant has only small square tables that can be joined end to end to form a large long table. If each small square table can seat only one people on each side, how many of such small tables are needed to seat this group of people?

**(a)** Study the diagram below and complete the tables (i) and (ii) that follow.

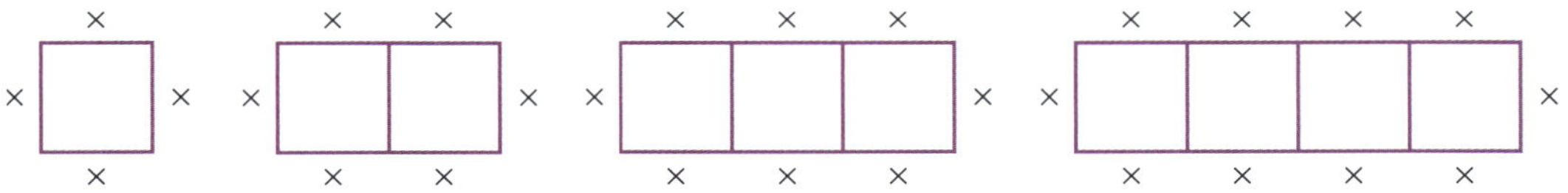

*Figure 1* *Figure 2* *Figure 3* *Figure 4*

**(i)**

| No. of people | 4 | 6 | 8 | 10 | 12 | 14 |
|---|---|---|---|---|---|---|
| No. of tables | $\frac{4-2}{2} = 1$ | $\frac{6-2}{2} = 2$ | | | | |

**(ii)**

| No. of tables | 1 | 2 | 3 | 4 | 5 | 6 |
|---|---|---|---|---|---|---|
| No. of people | 2(1) + 2 = 4 | 2(2) + 2 = 6 | | | | |

**(b)** How many tables will be needed to seat
**(i)** 20 people; **(ii)** 30 people?

**(c)** How many people can be seated if there are
**(i)** 22 tables; **(ii)** 36 tables?

**3.** The diagram shows a line segment, $AB$, on which 18 points ($P_1$, $P_2$, ..., $P_{18}$) are marked.

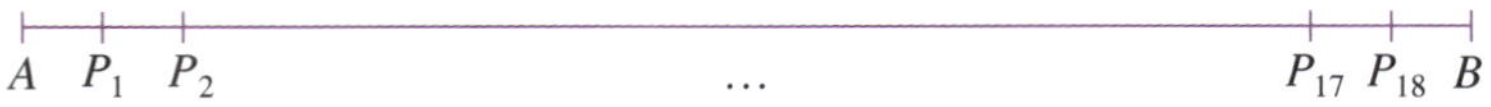


*Fig. 6.1*

**(a)** Study the diagram below and complete the table that follows.

*Figure 1* *Figure 2* *Figure 3* *Figure 4*

| No. of points on the line segment $AB$ (including the points $A$ and $B$) | 2 | 3 | 4 | 5 | 6 | 7 |
|---|---|---|---|---|---|---|
| No. of possible line segments | $\frac{2 \times (2-1)}{2} = 1$ | $\frac{3 \times (3-1)}{2} = 3$* | | | | |

* The three line segments are $AP_1$, $P_1B$ and $AB$.

**(b)** What is the total number of possible line segments in Fig. 6.1?

**4.** The following diagrams show a sequence of tiling patterns. The shape of each tile is a 1 cm by 1 cm square.

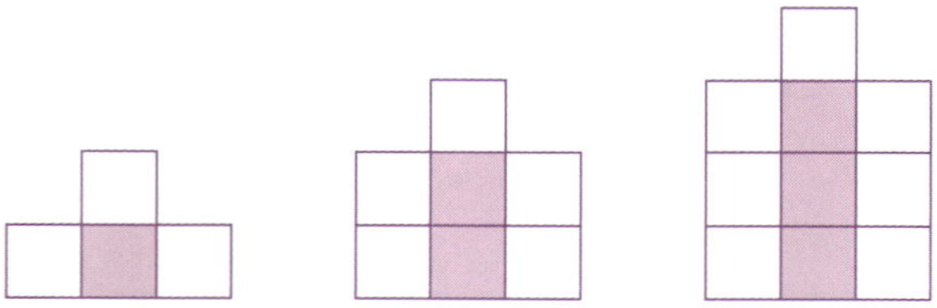

Figure 1 Figure 2 Figure 3

**(a)** Copy and complete the table below:

| Figure | 1 | 2 | 3 | 4 | 5 | 6 |
|---|---|---|---|---|---|---|
| Number of black squares ($b$) | 1 | 2 | | | | |
| Number of white squares ($w$) | 3 | 5 | | | | |
| Area of the whole figure ($b + w$) | 4 | 7 | | | | |
| Perimeter of the whole figure (cm) | 10 | 12 | | | | |

(b) Find the number of white squares in figure 9.

(c) Find the perimeter of figure 9.

(d) Find the number of white squares in Figure $n$.
(Give your answer in terms of $n$.)

(e) Find the perimeter of the whole figure in Figure $n$.
(Give your answer in terms of $n$.)

**5.** A man was trying to swim to a buoy placed at a distance of 200 m out in the sea. It took him 1 min to swim 20 m. Then a wave pushed him back 10 m and he rested for about 1 min before swimming again. He continued in this way for the rest of the journey. How long would it take the man to swim to the buoy?

**6.** Consider the pattern:

$$\frac{2}{1 \times 2 \times 3} = \frac{1}{1} - \frac{2}{2} + \frac{1}{3}$$
$$\frac{2}{2 \times 3 \times 4} = \frac{1}{2} - \frac{2}{3} + \frac{1}{4}$$
$$\frac{2}{3 \times 4 \times 5} = \frac{1}{3} - \frac{2}{4} + \frac{1}{5}$$
$$\vdots$$
$$\frac{2}{p(p+1)(p+2)} = \frac{1}{p} - \frac{2}{p+1} + \frac{1}{p+2}$$
$$\vdots$$

(a) Write down the eighth line in the pattern.

(b) Use the above pattern to find the value of $\frac{1}{10} - \frac{2}{11} + \frac{1}{12}$.

(c) Find the value of $p$ such that $\frac{2}{7980} = \frac{1}{p} - \frac{2}{p+1} + \frac{1}{p+2}$.

**Just for Fun**

*Fill in the next line in the following sequence and explain your answer.*

*3*
*1 3*
*1 1 1 3*
*3 1 1 3*
*1 3 2 1 1 3*
*1 1 1 3 1 2 2 1 1 3*
⋮

7. **(a)** Write down the next two terms of the sequence,

**(i)** 3, 5, 7, 9,... **(ii)** 8, 12, 16, 20,...

**(iii)** 4, 12, 24, 40, 60,... **(iv)** 5, 13, 25, 41, 61,...

**(b)** Consider the following pattern carefully:

$$3^2 + 4^2 = 5^2$$
$$5^2 + 12^2 = 13^2$$
$$7^2 + 24^2 = 25^2$$
$$9^2 + 40^2 = 41^2$$
$$11^2 + 60^2 = 61^2$$
$$\vdots$$

Write down the next two lines in the pattern with the help of the number sequences in part (i).

## SUMMARY

1. A **number sequence** is a set of numbers arranged in such a way that each successive number follows the preceding one according to a specific rule. The numbers in a sequence are known as the **terms** of the sequence.

2. Some heuristics for problem solving are:
   - change your point of view
   - make an organised list
   - work backwards
   - make a supposition
   - use trial-and-error
   - think of a related problem
   - simplify the problem
   - eliminate the unlikely possibilities
   - draw a diagram
   - look for a pattern
   - write an equation
   - solve a simpler problem
   - solve part of the problem
   - use tabulation
   - use a model
   - act it out

# Review Examples 6

## Example 1

*Consider the following sequence of subtraction of fractions.*

$$\frac{1}{1} - \frac{1}{2} = \frac{1}{2}$$

$$\frac{1}{2} - \frac{1}{3} = \frac{1}{6}$$

$$\frac{1}{3} - \frac{1}{4} = \frac{1}{12}$$

$$\vdots \quad \vdots \quad \vdots$$

$$\frac{1}{x} - \frac{1}{x+1} = \frac{1}{132}$$

*(a) Write down the $5^{th}$ and $6^{th}$ lines of the sequence.*
*(b) Find the value of $x$.*

### Solution

(a) The $5^{th}$ line is $\frac{1}{5} - \frac{1}{6} = \frac{1}{5 \times 6} = \frac{1}{30}$

The 6th line is $\frac{1}{6} - \frac{1}{7} = \frac{1}{6 \times 7} = \frac{1}{42}$

(b) We need to find $x$ such that $x \times (x + 1) = 132$.
We know that $10 \times 11 = 110$ and
$11 \times 12 = 132$
$\therefore \quad x = 11.$

# Review Questions 6

**1.** State the rule and write down the next three terms in the following sequences:

**(a)** 1, 7, 13, 19, ...
**(b)** 12, 15, 21, 30, ...
**(c)** 0, 2, 2, 4, 6, 10, ...
**(d)** 101, 97, 93, 89, ...
**(e)** 5, 9, 13, 17, ...
**(f)** 3, 6, 11, 18, ...

**2.** Fill in each box with an appropriate number. State the rule you use to obtain your answer.

**(a)** 1, 3, □, 27, 81
**(b)** □, 12, 24, 48, 96
**(c)** 2, 4, 7, 11, □
**(d)** 1, 5, 6, 11, 17, □
**(e)** 41, 40, 38, 35, 31, □

**3.** Consider the pattern:

$$\begin{aligned} 11 - 2 &= 3^2 \\ 1\,111 - 22 &= 33^2 \\ 111\,111 - 222 &= 333^2 \\ &\vdots \\ x - y &= 33\,333\,333^2 \end{aligned}$$

**(a)** Write down the 5th line in the pattern.
**(b)** Find the values of $x$ and $y$.

**4.** **(a)** Consider the pattern:

$$\begin{aligned} 1 &= \frac{1 \times (1+1)}{2} \\ 1 + 2 = 3 &= \frac{2 \times (2+1)}{2} \\ 1 + 2 + 3 = 6 &= \frac{3 \times (3+1)}{2} \\ 1 + 2 + 3 + 4 = 10 &= \frac{4 \times (4+1)}{2} \\ &\vdots \\ 1 + 2 + 3 + 4 + \ldots + k = 45 &= \frac{k \times (k+1)}{2} \\ &\vdots \end{aligned}$$

**(i)** Write down the 7th line in the pattern.
**(ii)** Find the value of $k$.

**(b)** Consider the pattern:

$$\begin{aligned} 1^3 = \quad 1 &= 1^1 \\ 1^3 + 2^3 = 1 + 8 = \quad 9 &= 3^2 \\ 1^3 + 2^3 + 3^3 = 1 + 8 + 27 = \quad 36 &= 6^2 \\ 1^3 + 2^3 + 3^3 + 4^3 = 1 + 8 + 27 + 64 = 100 &= 10^2 \\ &\vdots \\ 1^3 + 2^3 + 3^3 + \ldots + x^3 = 1 + 8 + 27 + \ldots + y = z &= 36^2 \\ &\vdots \end{aligned}$$

**(i)** Write down the 6th line in the pattern.
**(ii)** Making use of the pattern in (a), find the values of $x$, $y$ and $z$.

**5.** Using toothpicks, we can form patterns consisting of different layers of triangles as shown in the diagrams below.

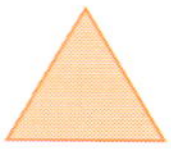
1 layer

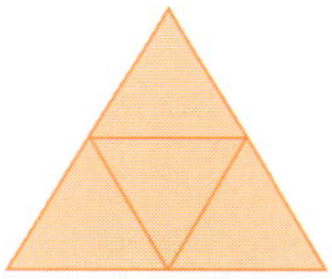
2 layers

3 layers

The numbers of toothpicks used form the following pattern:

| Number of layers | Number of toothpicks |
|---|---|
| 1 | $1 \times 3 + 3 \times 0$ |
| 2 | $2 \times 3 + 3 \times 1$ |
| 3 | $3 \times 3 + 3 \times 3$ |
| 4 | $4 \times 3 + 3 \times 6$ |
| 5 | $5 \times 3 + 3 \times 10$ |
| ... | ... |
| $k$ | $k \times 3 + 3 \times \dfrac{k \times (k-1)}{2}$ |
| ... | ... |

How many toothpicks are needed to form

**(a)** an 8-layer triangular pattern;

**(b)** a 20-layer triangular pattern and a 40-layer triangular pattern?

**6.** Study the following diagrams and then complete the table below.

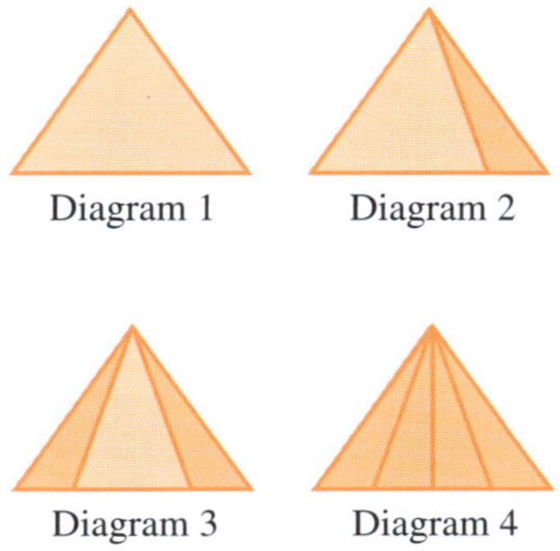


| Diagram | 1 | 2 | 3 | 4 | 5 | 6 |
|---|---|---|---|---|---|---|
| Number of triangles | 1 | 3 | 6 | | | |

**(a)** How many triangles will there be in Diagram 7?

**(b)** How many triangles will there be in Diagram $n$? Give your answer in terms of $n$.

*In this chapter, you will learn how to*

- *solve simple algebraic equations;*
- *construct simple linear equations from given situations and solve these equations;*
- *solve simple inequalities.*

CHAPTER

# 7

# Algebraic Equations and Simple Inequalities

## Introduction

Albert Einstein derived the simple yet elegant formula, $E = mc^2$, to calculate the amount of energy released when a quantity of matter is destroyed. This idea has been used by scientists to develop the atomic bomb. An atomic explosion produces a 'mushroom shaped' cloud which may look majestic but is actually very destructive and harmful.

## Open Sentences

Consider the following sentences and state whether each of them is true or false:

(a) 5 is greater than 4.
(b) 4 is a factor of 32.
(c) $4 + 5 = 7$
(d) London is an island.

Clearly, we can conclude that sentences (a) and (b) are true while sentences (c) and (d) are false.

Now, consider the following sentences:

(e) $7 + \square = 13$.
(f) 3 is a factor of $\triangle$.
(g) $\bigcirc$ is the capital of Malaysia.

We cannot say whether sentences (e), (f) and (g) are true or false because it is not given in the sentence what the symbols $\square$, $\triangle$ and $\bigcirc$ stand for. We call such sentences **open sentences** and the symbols $\square$, $\triangle$ and $\bigcirc$ **unknowns** or **variables**. An open sentence is a sentence which contains one or more unknowns. An open sentence can be true or false depending on what we replace the unknown(s) in the sentence with.

Normally, we use letters such as $a$, $b$, $c$, $x$, $y$ and $z$ to represent unknowns. Open sentences that include numbers, variables and operation symbols in mathematics are called **mathematical sentences**. In particular, (e) and (f) are called open mathematical sentences.

In Example 2 of Chapter 6, we have shown that the number of dots in the $n$th figure of the pattern is $d = 4n + 2$. This is also an open mathematical sentence.

## Simple Equations

An open mathematical sentence which contains an equal sign "=" is called an equation. The following are some simple equations:

(a) $x - 5 = 7$ (b) $2x + 7 = 26$ (c) $\frac{x + 5}{2} = 3x - 2$ (d) $x^2 + x = 6$

Equations like $3x - 5 = x + 7$ which contain only one unknown or variable are called equations in one unknown.

To solve an equation means to find the value of the unknown so that the equation becomes a true or correct sentence. The value found is called the **solution** of the equation.

Let us recall Example 2 of Chapter 6 again. To find the figure in which the number of dots is 34, we have to find the value of $n$ such that $4n + 2 = 34$. This is to solve the equation $4n + 2 = 34$.

## Example 1

*What does x stand for if (a) $x + 5 = 11$; (b) $x - 6 = 14$?*

### Solution

By observation, we deduce that

(a) $x$ stands for 6 because $6 + 5 = 11$.

(b) $x$ stands for 20 because $20 - 6 = 14$.

## Example 2

*What does x stand for if (a) $3x = 18$; (b) $\frac{x}{5} = 7$?*

### Solution

By observation, we deduce that

(a) $x$ stands for 6 because $3 \times 6 = 18$.

(b) $x$ stands for 35 because $\frac{35}{5} = 7$.

## Exercise 7a

1. Find the solution of each of the following equations by observation:

   **(a)** $3a - 4 = a$ **(b)** $50 - 5a = 10$

   **(c)** $\frac{24}{a} = 3$ **(d)** $\frac{1}{2}a - \frac{1}{3}a = 2$

   **(e)** $\frac{3}{4}a - 3 = \frac{1}{2}a$ **(f)** $\frac{1}{3}a - 3 = 0$

   **(g)** $1.5a + 2 = 5$ **(h)** $0.5a - 1 = 4$

2. If $a$ is an integer, find the possible solutions, if any, for each of the following by observation:

   **(a)** $a^2 = 4$ **(b)** $9 - a^2 = 0$

   **(c)** $a^2 + 16 = 0$ **(d)** $\sqrt{a} = 5$

   **(e)** $\sqrt[3]{a} = 3$ **(f)** $\sqrt[3]{a} = -2$

   **(g)** $\sqrt[3]{a + 4} = 4$ **(h)** $\sqrt{2a + 1} = 7$

# Solving Simple Equations

We can use the idea of a balance to help us solve equations. Consider the case where the contents of the two scale pans balance each other. They will remain balanced if equal weights are added to both sides or if equal weights are taken away from both sides.

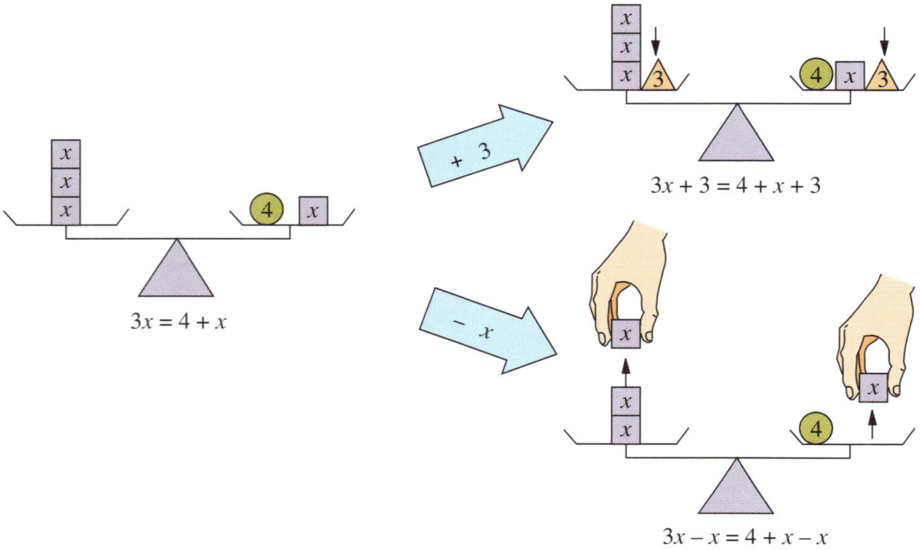


What will happen to the scales below if the weights of the contents of both sides are trebled or halved?

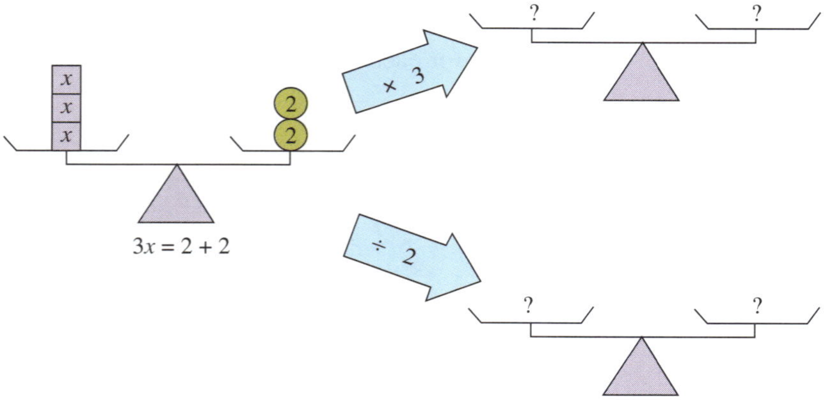


*You are given a bag of marbles, all of which are of the same shape and appearance. All the marbles are also of the same weight except one of which is heavier than the rest. You are to use a balance to make exactly two weighings to pick out the heavier marble from the rest of the marbles. Describe clearly how you can do it and explain it to your neighbour if the bag contains exactly*

*(a) 6*
*(b) 7*
*(c) 8*
*(d) 9 marbles.*

The above discussion leads to the following rules:

An equation remains balanced when

1. equal numbers are added to each side;

2. equal numbers are subtracted from each side;
3. each side is multiplied by equal numbers;
4. each side is divided by equal numbers except zero.

We shall use the above rules to solve simple equations.

## Example 3

*Solve the following equations:*

*(a)* $x + 3 = 14$ *(b)* $y - 5 = 8$ *(c)* $5x = 25$ *(d)* $\frac{1}{3}y = 4$

**Solution**

(a)
$$x + 3 = 14$$
$$x + 3 - 3 = 14 - 3 \quad \text{(Subtract 3 from both sides)}$$
$$\therefore \quad x = 11$$

(b)
$$y - 5 = 8$$
$$y - 5 + 5 = 8 + 5 \quad \text{(Add 5 to both sides)}$$
$$\therefore \quad y = 13$$

(c)
$$5x = 25$$
$$\frac{5x}{5} = \frac{25}{5} \quad \text{(Divide both sides by 5)}$$
$$\therefore \quad x = 5$$

(d)
$$\frac{1}{3}y = 4$$
$$\frac{1}{3}y \times 3 = 4 \times 3 \quad \text{(Multiply both sides by 3)}$$
$$\therefore \quad y = 12$$

> **Notes**
> *It is not necessary to write down the statements in the brackets when you are solving the equations.*

## Example 4

*Solve the following equations:*

*(a)* $3x + 2 = 8$ *(b)* $4a - 3 = 13$

**Solution**

(a)
$$3x + 2 = 8$$
$$3x + 2 - 2 = 8 - 2 \quad \text{(Subtract 2 from both sides)}$$
$$3x = 6$$
$$\frac{3x}{3} = \frac{6}{3} \quad \text{(Divide both sides by 3)}$$
$$\therefore \quad x = 2$$

> **Notes**
> *Substitute the answer back into the equation to check if the equation remains balanced: Left-hand side (LHS) = 3(2) + 2 = 6 + 2 = 8 = Right-hand side (RHS)*

(b) $4a - 3 = 13$

$4a - 3 + 3 = 13 + 3$ (Add 3 to both sides)

$4a = 16$

$\dfrac{4a}{4} = \dfrac{16}{4}$ (Divide both sides by 4)

$\therefore \quad a = 4$

**Notes**

*LHS = 4(4) – 3 = 16 – 3 = 13 = RHS*

## Example 5

*Solve the following equations:*

*(a)* $8.3 - 3x = 2.6$

*(b)* $3(2y - 3) = 9$

**Solution**

(a) $8.3 - 3x = 2.6$

$8.3 - 3x - 8.3 = 2.6 - 8.3$ (Subtract 8.3 from both sides)

$-3x = -5.7$

$\dfrac{-3x}{-3} = \dfrac{-5.7}{-3}$ (Divide both sides by –3)

$\therefore \quad x = 1.9$

**Your Attention**

*It is always a good practice to check your answer by substituting the answer back into the original question.*

(b) **Method 1**

$3(2y - 3) = 9$

$6y - 9 = 9$

$6y - 9 + 9 = 9 + 9$ (Add 9 to both sides)

$6y = 18$

$y = \dfrac{18}{6}$ (Divide both sides by 6)

$\therefore \quad y = 3$

**Method 2**

$3(2y - 3) = 9$

$2y - 3 = \dfrac{9}{3}$ (Divide both sides by 3)

$2y - 3 + 3 = 3 + 3$ (Add 3 to both sides)

$2y = 6$

$\therefore \quad y = \dfrac{6}{2} = 3$ (Divide both sides by 2)

## Exercise 7b

1. Solve the following equations:

   (a) $a + 5 = 17$
   (b) $c + 5 = -2$
   (c) $8 + x = 15$
   (d) $k + 0.7 = 2.7$
   (e) $y - 7 = -3$
   (f) $k - 1.6 = -0.4$
   (g) $4a = -20$
   (h) $-12e = 36$
   (i) $\frac{1}{4}x = 3$
   (j) $\frac{3}{4}x = -9$

2. Solve the following equations:

   (a) $3x - 2 = 4$
   (b) $4 + 9x = 22$
   (c) $10 - 7a = 3$
   (d) $2 - 5h = -18$
   (e) $11 - 2k = 17$
   (f) $5x - \frac{3}{4} = 2\frac{1}{4}$
   (g) $3\frac{1}{2} - 2x = 1\frac{1}{4}$
   (h) $3\frac{1}{4} + 4x = 1\frac{3}{4}$
   (i) $2h + 1.3 = 2.8$
   (j) $3x - 1.7 = 4.3$

3. Solve the following equations:

   (a) $3(x + 3) = 15$
   (b) $4(h - 2) = 28$
   (c) $6(3 - x) = 3$
   (d) $3(2x + 5) = 18$
   (e) $3(5 - 4x) = 21$
   (f) $2(2 - 0.3x) = 1.4$
   (g) $3(2x - 3.4) = 2.4$
   (h) $7(2.8 - 4x) = 2.8$
   (i) $3(3h - 1.2) = 5.7$

## Further Examples on Equations

### Example 6

*Solve the following equations:*

*(a)* $5x + 3 = 3x - 7$

*(b)* $5(x + 3) + 2(x + 1) = 5 - 4x$

**Solution**

(a)
$$5x + 3 = 3x - 7$$
$$5x + 3 - 3x = 3x - 7 - 3x \quad \text{(Subtract } 3x \text{ from both sides)}$$
$$2x + 3 = -7$$
$$2x + 3 - 3 = -7 - 3 \quad \text{(Subtract 3 from both sides)}$$
$$2x = -10$$
$$\frac{2x}{2} = \frac{-10}{2} \quad \text{(Divide both sides by 2)}$$
$$\therefore \quad x = -5$$

(b)
$$5(x + 3) + 2(x + 1) = 5 - 4x$$
$$5x + 15 + 2x + 2 = 5 - 4x$$
$$7x + 17 = 5 - 4x$$
$$7x + 17 + 4x = 5 - 4x + 4x \quad \text{(Add } 4x \text{ to both sides.)}$$
$$11x + 17 - 17 = 5 - 17 \quad \text{(Subtract 17 from both sides.)}$$
$$11x = -12$$
$$\frac{11x}{11} = -\frac{12}{11} \quad \text{(Divide both sides by 11.)}$$
$$\therefore \quad x = -1\frac{1}{11}$$

Can you check to see if the solution obtained is correct?

## Example 7

*Solve the equation* $\frac{x-1}{3} = \frac{2x+5}{7}$.

**Solution**

$$\frac{x-1}{3} = \frac{2x+5}{7}$$
$$7(x - 1) = 3(2x + 5) \quad \text{(Multiply both sides by the LCM of 3 and 7 i.e. 21)}$$
$$7x - 7 = 6x + 15$$
$$7x - 7 + 7 = 6x + 15 + 7 \quad \text{(Add 7 to both sides)}$$
$$7x - 6x = 6x + 22 - 6x \quad \text{(Subtract } 6x \text{ from both sides)}$$
$$\therefore \quad x = 22$$

**Notes**

*Check:*
*LHS*
$$= \frac{x-1}{3} = \frac{22-1}{3} = \frac{21}{3} = 7$$
*RHS*
$$= \frac{2x+5}{7} = \frac{2(22)+5}{7} = \frac{49}{7}$$
$$= 7$$

## Example 8

*Solve the equation* $1\frac{4}{5}x - \frac{3}{5} = 1\frac{1}{5}x + 10\frac{1}{5}$.

**Solution**

$$1\frac{4}{5}x - \frac{3}{5} = 1\frac{1}{5}x + 10\frac{1}{5}$$

$$1\frac{4}{5}x - \frac{3}{5} + \frac{3}{5} = 1\frac{1}{5}x + 10\frac{1}{5} + \frac{3}{5} \qquad \left(\text{Add } \frac{3}{5} \text{ to both sides}\right)$$

$$1\frac{4}{5}x - 1\frac{1}{5}x = 1\frac{1}{5}x + 10\frac{4}{5} - 1\frac{1}{5}x \qquad \left(\text{Subtract } 1\frac{1}{5}x \text{ from both sides}\right)$$

$$\frac{3}{5}x = 10\frac{4}{5}$$

$$x = 10\frac{4}{5} \times \frac{5}{3} \qquad \left(\text{Multiply both sides by } \frac{5}{3}\right)$$

$$\therefore \quad x = 18$$

**Notes**

*Check:*

$$LHS = 1\frac{4}{5}x - \frac{3}{5} = 1\frac{4}{5}(18) - \frac{3}{5} = 31\frac{4}{5}$$

$$RHS = 1\frac{1}{5}x + 10\frac{1}{5} = 1\frac{1}{5}(18) + 10\frac{1}{5} = 31\frac{4}{5}$$

## Example 9

*Solve the following equations:*

*(a)* $\frac{8}{2x - 3} = 4$ *(b)* $\frac{x - 3}{x + 4} = \frac{2}{5}$

**Solution**

(a)

$$\frac{8}{2x - 3} = 4$$

$$\frac{8}{2x - 3} \times (2x - 3) = 4(2x - 3) \qquad (\text{Multiple both sides by } (2x - 3))$$

$$\therefore \quad 8 = 4(2x - 3)$$

$$2 = 2x - 3 \qquad (\text{Divide both sides by } 4)$$

$$2 + 3 = 2x$$

$$\therefore \quad x = 2\frac{1}{2}$$

(b)

$$\frac{x - 3}{x + 4} = \frac{2}{5} \qquad (\text{Multiply both sides by } 5(x + 4))$$

$$5(x + 4) \times \frac{x - 3}{x + 4} = 5(x + 4) \times \frac{2}{5}$$

$$5(x - 3) = 2(x + 4)$$

$$5x - 15 = 2x + 8$$

$$5x - 2x = 8 + 15$$

$$3x = 23$$

$$x = \frac{23}{3} = 7\frac{2}{3}$$

## Exercise 7c

1. Solve the following equations:

   (a) $3a = 2a + 4$

   (b) $4x = 3x - 5$

   (c) $4x = 15 - x$

   (d) $8x + 3 = 5x + 12$

   (e) $3y + 4 = 8 - 5y$

   (f) $2x - 3 = 5x - 6$

   (g) $7 + 2x = 6x - 5$

   (h) $9h - 7 = 13h + 9$

   (i) $3k + 1 = 7k - 15$

   (j) $6x + 5 = 9x - 19$

2. Solve the following equations:

   (a) $2(c - 4) = 3(c - 2)$

   (b) $3(2a + 3) = 4a + 3$

   (c) $8(x - 5) = 4(3x - 3)$

   (d) $2(3a - 4) = 5(a + 6)$

   (e) $3(2x + 5) = 4(x - 3)$

   (f) $6(2x + 5) = 2(5x - 3)$

   (g) $9(2x + 7) = 11(x + 14)$

   (h) $5(7x - 3) = 14(2x - 2)$

3. Solve the following equations:

   (a) $\frac{a}{3} + 3 = 4$

   (b) $\frac{h}{4} + 6 = -4$

   (c) $\frac{x}{5} - 2 = 3$

   (d) $\frac{y}{6} - \frac{1}{2} = 3\frac{1}{2}$

   (e) $3 - \frac{x}{4} = 2$

   (f) $4 + \frac{y}{3} = -2$

   (g) $4 - \frac{2x}{3} = -2$

   (h) $6 - \frac{3x}{5} = 12$

   (i) $\frac{5 - 2x}{7} = 2$

   (j) $\frac{5 + 4x}{9} = -1$

4. Solve the following equations:

   (a) $x + \frac{x}{5} = 12$

   (b) $\frac{2}{5} = 2(3x - 7)$

   (c) $\frac{3}{5}x - \frac{3}{7} = \frac{5}{7}$

   (d) $2\frac{1}{2}y = 10 - 1\frac{2}{3}y$

   (e) $\frac{1}{2}a - 2 = \frac{3}{5}a + 4$

   (f) $m + 2 = \frac{2 - m}{3} - 2$

   (g) $\frac{k}{2} - \frac{k}{5} - \frac{k}{6} = 2$

   (h) $\frac{5x}{2} = \frac{4}{3} + 2x$

   (i) $\frac{2x - 1}{5} + \frac{x + 3}{8} = 0$

   (j) $\frac{2x + 3}{4} - \frac{x - 5}{6} = 0$

5. Solve the following equations:

**(a)** $\frac{3}{x} = \frac{2}{5}$

**(b)** $\frac{12}{x + 3} = 2$

**(c)** $\frac{15}{2x - 1} = 3$

**(d)** $\frac{1}{2} = \frac{1}{y + 2} - 1$

**(e)** $\frac{42}{2x - 5} - 3 = 11$

**(f)** $\frac{2x + 1}{3x - 5} = \frac{4}{7}$

**(g)** $\frac{5}{a - 2} = \frac{7}{a + 6}$

**(h)** $\frac{2}{7x - 3} = \frac{3}{9x - 5}$

## Formulae

A **formula** expresses a rule in algebraic terms. It uses variables to write instructions for performing a calculation. For example, to find the area of a rectangle, we use the formula $A = l \times b$ or $lb$. We have used the letter $A$ to represent the area, $l$ the length and $b$ the breadth. This is an algebraic expression of the rule. In words, this rule can be written as: "The area of a rectangle is equal to the product of its length and breadth".

If $P$ represents the perimeter of the rectangle, then we have $P = 2(l + b)$. If the values of $l$ and $b$ are known, we can find the values of $A$ and $P$.

For example, if $l = 4$ and $b = 3$, then

$$A = 4 \times 3 = 12$$
$$P = 2(4 + 3) = 14$$

We can also find the value of $l$ if $A$ and $b$ are given. This is an example of the process of replacing letters by numbers and this process is called **substitution**.

## Example 10

*The formula for the volume V of a cuboid is V = lbh where l is the length, b is the breadth and h is the height. Find the volume of the cuboid where*

*l = 5 cm, b = 4 cm and h = 3 cm;*

### Solution

Substitute $l = 5$, $b = 4$ and $h = 3$ into $V = lbh$.

$$\begin{aligned} V &= lbh \\ &= 5 \times 4 \times 3 = 60 \end{aligned}$$

$\therefore$ volume of the cuboid = 60 $cm^3$

## Example 11

*If* $a = \dfrac{b}{c - b}$, *find (a) a when b = 5 and c = 7; (b) b when a = 3 and c = 10.*

### Solution

(a) Substituting $b = 5$ and $c = 7$ into $a = \dfrac{b}{c - b}$, we get

$$\begin{aligned} a &= \frac{5}{7 - 5} \\ &= \frac{5}{2} \\ &= 2\frac{1}{2} \end{aligned}$$

(b) Substituting $a = 3$ and $c = 10$ into $a = \dfrac{b}{c - b}$, we get

$$\begin{aligned} 3 &= \frac{b}{10 - b} \\ 3(10 - b) &= b \\ 30 - 3b &= b \\ 30 &= 4b \\ \therefore \quad b &= 7\frac{1}{2} \end{aligned}$$

## Exercise 7d

Take the value of $\pi$ as $3\frac{1}{7}$, where necessary.

**1.** If $V = \frac{1}{3}Ah$, find $V$ when $A = 43$ and $h = 6$.

**2.** If $F = \frac{9c}{5} + 32$, find $F$ when $c = 30$.

**3.** If $S = 4\pi r^2$, find $S$ when $r = 10\frac{1}{2}$.

**4.** If $T = \pi(R^2 - r^2)$, find $T$ when $R = 4$ and $r = 3$.

**5.** If $a = \frac{y^2 - xz}{5}$, find $a$ when $x = 4$, $y = 7$ and $z = 6$.

**6.** If $k = \frac{x + y}{3}$, find $x$ when $k = 12$ and $y = 4$.

**7.** If $t = \frac{v - u}{a}$, find $a$ when $t = 1$, $u = 1\frac{2}{3}$ and $v = 3\frac{1}{2}$.

**8.** If $U = \pi(r + h)$, find $r$ when $U = 16\frac{1}{2}$ and $h = 2$.

**9.** If $v^2 = u^2 + 2gs$, find $s$ when $v = 20$, $u = 10$ and $g = 10$.

**10.** If $\frac{m}{x + q} = N$, find $q$ when $m = 9$, $x = 2$ and $N = 1\frac{4}{5}$.

**11.** If $n - y = \frac{4y - n}{m}$, find $n$ when $y = 3$ and $m = 7$.

**12.** If $\frac{a}{b} + e = \frac{c}{b}$, find $c$ when $a = 4$, $b = 12$ and $e = -\frac{1}{6}$.

✶**13.** If $y + b = \frac{ay + c}{b}$, find $c$ when $y = 12$, $a = 14$ and $b = 3$.

✶**14.** If $\frac{1}{a} = \frac{1}{b} + \frac{1}{c} + \frac{1}{d}$, find $c$ when $a = 2$, $b = 3$ and $d = 5$.

✶**15.** If $\frac{m(ny - x^2)}{z} + n = 5n$, find $y$ when $n = 4$, $x = 2$, $m = 6$ and $z = 8$.

✶**16.** If $c = \frac{a}{b} - \frac{d - e}{f - d}$, find $f$ when $a = 2$, $b = 3$, $c = 4$, $d = 5$ and $e = 6$.

# Construction of Formulae

To construct a formula, choose letters to represent the quantities. Usually, the first letter of the word is used. Then express the rule in algebraic terms. For example, the sum, $S$ kg, of the weights of two boys, one weighing $m$ kg and the other $n$ kg, is expressed as $S = m + n$.

## Example 12

*Find a formula for the sum (S) of any three consecutive even numbers.*

**Solution**

Let the smallest number be $n$.

The next number is $n + 2$ and the biggest number is $n + 4$.

$S = n + (n + 2) + (n + 4)$

$S = n + n + 2 + n + 4$

$\therefore \quad S = 3n + 6$ or $3(n + 2)$

## Exercise 7e

**1.** Using the letters suggested, construct a simple formula in each case:

**(a)** The sum ($S$) of three numbers $a$, $b$ and $c$.
**(b)** The product ($P$) of two numbers $x$ and $y$.
**(c)** The difference ($D$) between the ages of two boys; one being $a$ years old and the other $e$ years old.
**(d)** The area ($A$) of a semicircle whose radius is $r$.
**(e)** The cost ($\$C$) of $m$ eggs at 12 cents each.
**(f)** The average age ($A$) of 4 boys whose ages are $m$, $n$, $p$ and $q$ years.
**(g)** The vertical angle ($x°$) of an isosceles triangle whose base angle is $y°$.
**(h)** The time ($T$) in minutes for a train journey of $a$ hours $b$ minutes.
**(i)** The total cost ($\$T$) of $d$ chairs at $\$p$ each and $c$ tables at $\$q$ each.
*****(j)** A car travels $x$ km/h for $p$ km and makes another journey at $y$ km/h for $q$ km. Find
**(i)** the total distance ($r$ km) the car travels;
**(ii)** the formula for the average speed.

# Writing Algebraic Expressions

## Example 13

*How many days are there in (a) 5 weeks; (b) n weeks?*

**Solution**

There are 7 days in a week.

(a) There are $(7 \times 5) = 35$ days in 5 weeks.

(b) There are $(7 \times n) = 7n$ days in $n$ weeks.

## Example 14

*A girl is now 10 years old. How old will she be (a) 5 years later; (b) t years later?*

*How old was she (c) 3 years ago; (d) x years ago?*

*What are the possible values of x? Can x be 15?*

**Solution**

The girl becomes 1 year older every year.

One year later, she will be $(10 + 1) = 11$ years old.

(a) 5 years later, she will be $(10 + 5) = 15$ years old.
(b) $t$ years later, she will be $(10 + t)$ years old.
(c) 3 years ago, she was $(10 - 3) = 7$ years old.
(d) $x$ years ago, she was $(10 - x)$ years old.

The possible values of $x$ are 10, 9, 8, 7, 6, 5, 4, 3, 2, 1 and 0.
$x$ cannot be 15 since the girl is now only 10 years old.

## Example 15

*A car travels at a speed of 55 km/h. How far will it travel in (a) 2 hours; (b) t hours?*

**Solution**

In 1 hour, the car travels 55 km.

(a) In 2 hours, the car will travel $(55 \times 2)$ km, i.e., 110 km.
(b) In $t$ hours, the car will travel $(55 \times t)$ km, i.e., $55t$ km.

## Exercise 7f

1. How many grams are there in 5 kg? How many grams are there in $x$ kg?

2. What is the cost of 6 magazines at \$4 each? What is the cost of $p$ magazines at \$4 each? Find also the cost of $p$ magazines at \$$q$ each.

3. Four tennis balls have a total mass of $m$ kg. Find the mass of each ball.

4. If a horse runs at $b$ km/h, how far can it go in 2 hours if it keeps the same speed?

5. How many minutes are there in $m$ hours?

6. How many weeks are there in $y$ days?

7. If Dan has $x$ dollars, how many marbles can he buy if each marble costs five cents?

8. Write the number which is half as big as $b$.

9. Find the time taken by a cyclist to travel 21 km if he is travelling at $v$ km/h.

10. A shopkeeper buys an armchair for \$$a$ and then sells it at a profit of \$$b$. What is the selling price of the armchair?

11. A boy is $b$ years old and his father is 6 times as old as him. Find the father's age. Find also the sum of their ages in $y$ years' time.

12. Mrs Jones's age is equal to the sum of the ages of her two daughters. If the younger daughter is $x$ years old and the elder is 4 years older, how old is Mrs Jones?

13. Find the three consecutive numbers in which $n$ is the middle number.

14. A motorist drives for 5 hours at $u$ km/h and for 3 hours at $v$ km/h. Find the total distance travelled.

# Problem Solving with Algebra

One of the most powerful mathematical tools to solve problems is by **using an equation and solving it**. We shall look at this problem-solving method more closely. (Alternative methods will be shown too if they are shorter and easier than the method of using an equation).

> **Just for fun**
>
> *The sum of the ages of two brothers is 11. One is 10 years older than the other. How old are the brothers?*

## Setting up Equations

In order to solve word problems in mathematics, we often need to translate them into mathematical equations first. First, consider these problems:

1. "When I subtract 3 from a number, the result is the same as if I had halved the number. Find the number."

2. The sum of two consecutive odd numbers is 64. Find the two numbers.

For the first question if we let $x$ represent the number, we can write the given question mathematically as:

$$x - 3 = \frac{1}{2}x$$

For the second question, let $x$ be the smaller of the two consecutive odd numbers, then the larger one will be $x + 2$. We can then write the mathematical equation as:

$$x + (x + 2) = 64$$

## Exercise 7g

**1.** In each of the following, let $x$ denote the unknown. Derive an equation involving $x$:

**(a)** When a certain number is increased by 7, the result is 18.

**(b)** When a number is decreased by 2 and the result multiplied by 3, the final result is 24.

**(c)** When 5 is subtracted from a certain number and the result multiplied by 7, the final result is 63.

**(d)** When a certain number is subtracted from 24 and the result divided by 5, the final result is 4.

**(e)** The sum of three consecutive numbers is 63.

**(f)** One number is bigger than the other number by 3 and the sum of these two numbers is 43.

**(g)** Six times of a certain number is 16 more than twice the number.

**(h)** The length of a rectangle is 5 m more than its width and the perimeter of the rectangle is 32 m.

**(i)** The length of a rectangle is twice its width and the perimeter is 54 m.

**2.** Peter has five times as much money as David. If Peter gives \$28 to David, both of them will have equal amounts of money. How much money did Peter have at the beginning?

**3.** There are a total of 225 pupils in Secondary one. If the number of pupils who pay their school fees through GIRO scheme is 14 times the number of pupils who do not, find the number of pupils who did not join the scheme.

**4.** \$4800 is divided among three brothers $A$, $B$ and $C$. $A$ receives three times as much as $B$ and $C$ receives twice as much as $B$. If $B$ receives \$$x$, form an equation in $x$.

**5.** Three wallets and two handbags cost \$450 and a handbag costs twice as much as a wallet. If a wallet costs \$$x$, form an equation in $x$.

## Example 16

A man is now 3 times as old as his son. In 10 years' time, the sum of their ages will be 76. How old was the man when his son was born?

> **Just for fun**
>
> *If 10 cats can catch 10 mice in 10 minutes, how many cats are required to catch 100 mice in 100 minutes?*

### Solution

To see the problem clearly, first we draw a model.

*Age now*

*Age when the son was born*

*Age in 10 years' time from now*

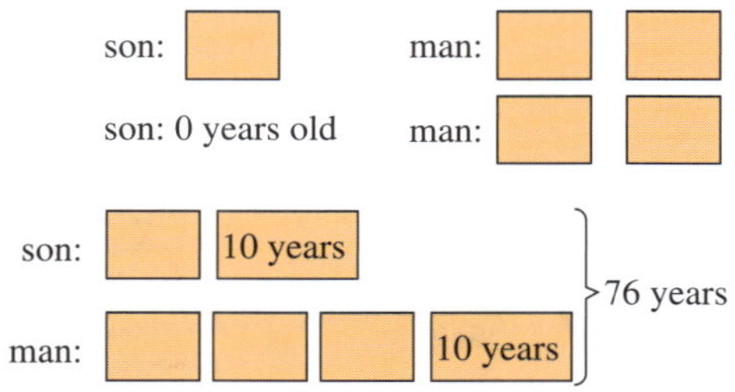


Let the present age of the son be $x$ years old.

The man is now $3x$ years old.

In 10 years' time, the son will be $(x + 10)$ years old.

In 10 years' time, the man will be $(3x + 10)$ years old.

$$\begin{aligned}(x + 10) + (3x + 10) &= 76\\ 4x + 20 &= 76\\ 4x &= 56\\ x &= 14\end{aligned}$$

Hence, the man is now $(3 \times 14) = 42$ years old.

$\therefore$ the man was $(42 - 14) = 28$ years old when his son was born.

## Example 17

*Find three consecutive odd numbers whose sum is 57.*

### Solution

**Strategy: Use an equation**

Let the smallest of the three consecutive odd numbers be $x$.

The other odd numbers are $x + 2$ and $x + 4$.

$$\begin{aligned} x + (x + 2) + (x + 4) &= 57 \\ 3x + 6 &= 57 \\ 3x &= 51 \\ x &= 17 \end{aligned}$$

$\therefore$ the three consecutive odd numbers are 17, 19 and 21.

Alternatively, we can find the solution by **making a systematic list**. We shall start with 11, 13, 15, ... etc. Do you think it is necessary to start the list with 1, 3, 5, ...?

11, 13, 15, (17), (19), (21), 23, 25, 27, ...

The three circled odd numbers add up to 57.

$\therefore$ the three consecutive odd numbers are 17, 19 and 21.

Notes

*Check: 17 + 19 + 21 = 57*

## Example 18

*A woman buys 50 eggs for \$6.60. Some cost 12 cents each and the rest 14 cents each. How many of each kind of eggs has she bought?*

Solution

**Strategy: Use an equation**

Let $x$ be the number of eggs she buys at 12 cents each.
No. of eggs bought at 14 cents each = $(50 - x)$.
Cost of $x$ eggs at 12 cents each = $12x$ cents.
Cost of $(50 - x)$ eggs at 14 cents each = $(50 - x) \times 14$ cents

$$\begin{aligned} 12x + 14(50 - x) &= 660 \\ 12x + 700 - 14x &= 660 \\ 40 &= 2x \\ x &= 20 \end{aligned}$$

$\therefore$ no. of eggs bought at 12 cents each = 20.
$\therefore$ no. of eggs bought at 14 cents each = 50 – 20 = 30.

Notes

*Check:*
*20(12¢) + 30(14¢)*
*= \$2.40 + \$4.20*
*= \$6.60*

## Example 19

*John walked for 45 minutes at the rate of 3 km/h and then ran for half an hour at a certain speed. At the end of that time he was 6 km away from the starting point. How fast did he run?*

**Solution**

**Strategy: Use an equation**

Let the speed at which he ran be $x$ km/h.

Distance = Speed × Time

$\therefore$ the distance he walked = $\left(\frac{45}{60} \times 3\right)$ km = $\frac{9}{4}$ km

The distance he ran = $\left(\frac{1}{2} \times x\right)$ km = $\frac{x}{2}$ km

$$\frac{9}{4} + \frac{x}{2} = 6$$

$$\frac{9 + 2x}{4} = 6$$

(Remember that when an equation is divided or multiplied by a number, every term must be divided or multiplied by that number.)

$$9 + 2x = 24$$

$$2x = 15$$

$$x = \frac{15}{2} = 7\frac{1}{2}$$

$\therefore$ John ran at a speed of $7\frac{1}{2}$ km/h.

## Exercise 7h

1. What are the four consecutive numbers whose sum is 50?
2. If a number is trebled, it gives the same result as when 28 is added to it. What is the number?
3. When a number is added to another number 5 times as large, the result is 24. What is the first number?
4. Joe and Ahmad had 80 marbles altogether. Ahmad has 4 times as many marbles as Joe. How many marbles does each boy have?
5. Arasoo is 4 years older than Peter. Ali is 2 years younger than Peter. If the sum of their ages is 47, what are their respective ages?
6. Lilian and Susan share \$30 between themselves. If Susan gets twice as much as Lilian, find each girl's share.
7. When loaded with bricks, a lorry weighs 11 600 kg. If the bricks weigh three times as heavy as the empty lorry, find the weight of the bricks.
8. When a number is added to two-thirds of itself, the result is 45. Find the number.
9. The numerator of a fraction is 5 less than the denominator. If 1 is added to both the numerator and the denominator, the fraction would become $\frac{2}{3}$. Find the fraction.

**10.** Mooncakes with double egg yolks cost 50 cents more than those with only a single egg yolk. Mrs Tan bought 6 mooncakes with double egg yolks and 5 with only a single egg yolk. If she paid $36 for the 11 mooncakes, how much does a mooncake with double egg yolks cost?

**11.** Tom, Dick and Harry share $256. Dick's share is four times as much as Tom's and Tom's share is one-third of Harry's. How much is each of their share?

***12.** Meng Kuang had saved a small sum of money from his weekly pocket money. After receiving a total of $108 as hong bao money from his relatives during the Chinese New Year, he decided to donate one-fifth of his money to the Community Chest of Singapore. His other siblings also donated a total of $148 to the Community Chest. If their total contribution to the Community Chest was $200, how much did Meng Kuang save originally?

***13.** A group of boys had to choose between playing soccer and badminton. The number of boys who chose soccer was three times that who chose badminton. Getting 12 boys who chose soccer to play badminton would make the number of players for each game equal. Find the number whose chose badminton originally.

# Inequalities

So far, we have learned about equations and their solutions, we shall now learn another important concept, inequalities and solving simple inequalities.

We have learned the ordering of numbers on the number line earlier. A number on the number line is ***greater*** than any number on its **left** and ***smaller*** than any number on its **right**. We use the symbol '>' to represent 'is greater than' and the symbol '<' to represent 'is smaller than'.

For example,

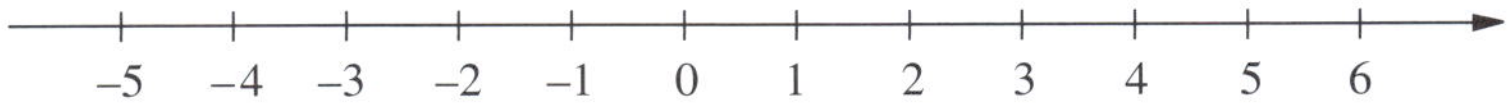


Since 5 lies on the right of 2, 5 is greater than 2 and we write $5 > 2$ or $2 < 5$. Similarly –1 is greater than –3 as –1 lies on the right of –3 and we write $-1 > -3$ or $-3 < -1$.

Consider any two numbers, $x$ and $y$, on the number line. One and only one of the following statements is true:

**(a)** $x > y$ **(b)** $x = y$ **(c)** $x < y$

This is known as the **Law of Trichotomy**.

## Exercise 7i

**1.** Copy and fill in the blanks with '<', '=' or '>' to make each of the following statements true.

**(a)** $9$ ______ $3$

**(b)** $7$ ______ $13$

**(c)** $-3$ ______ $2$

**(d)** $3^2$ ______ $9$

**(e)** $-4$ ______ $-2$

**(f)** $-6 \times 0$ ______ $7 \times 0$

**(g)** $0.4^2$ ______ $0.2$

**(h)** $(-5)^2$ ______ $(-4)^2$

**(i)** $(-0.5)^2$ ______ $(-0.4)^2$

**(j)** $\sqrt{6}$ ______ $\sqrt{8}$

**(k)** $\sqrt{0.6}$ ______ $\sqrt{0.8}$

**(l)** $3 \times 0$ ______ $-24 \times 0$

**(m)** $(0.3)^2$ ______ $(-0.3)^2$

**(n)** $(0.2)^2$ ______ $(0.2)^3$

**(o)** $(1.2)^2$ ______ $(1.2)^3$

**(p)** $\sqrt{0.8}$ ______ $\sqrt[3]{0.8}$

# Properties of Inequalities

## Explorations

Consider the inequality $8 > 6$, which we know is true.

1. What will happen when we multiply both sides of the inequality by 2?

   Left side $= 8 \times 2 = 16$
   Right side $= 6 \times 2 = 12$
   $16 > 12$
   That is, $8 \times 2 > 6 \times 2$.

   The inequality is still true when we multiply both sides of it by 2.

Let consider more cases.

2. What will happen when we multiply both sides of $8 > 6$ by 5?

   Left side = $8 \times 5 = 40$
   Right side = $6 \times 5 = 30$
   $40 \square 30$

   That is, $8 \times 5 \square 6 \times 5$

   The inequality is still true when we multiply both sides of it by 5.

3. What will happen when we multiply both sides of $8 > 6$ by $\frac{1}{2}$?

   Left side = $8 \times \frac{1}{2} = 4$

   Right side = $6 \times \frac{1}{2} = 3$

   $4 \square 3$

   That is, $8 \times \frac{1}{2} \square 6 \times \frac{1}{2}$

   The inequality $\square$ when we multiply both sides of it by $\frac{1}{2}$.

4. What will happen when we multiply both sides of $8 > 6$ by $\frac{1}{3}$?

   Left side = $8 \times \frac{1}{3} = \frac{8}{3}$

   Right side = $6 \times \frac{1}{3} = \frac{6}{3}$

   $\frac{8}{3} \square \frac{6}{3}$

   That is, $8 \times \frac{1}{3} \square 6 \times \frac{1}{3}$

   The inequality $\square$ when we multiply both sides of it by $\frac{1}{3}$.

   From the above, what conclusion can we get?

   The inequality $8 > 6$ $\square$ when we multiply both sides of it by a positive number.

Now, let's consider more inequalities: $-2 > -5$, $-3 < 2$, $4 < 9$.

As in the above. Observe what happens when we multiply both sides of inequality by 2, 5, $\frac{1}{2}$ and $\frac{1}{3}$ respectively. What conclusion can you get?

Check more cases to confirm your conclusion.

---

Generally, we can

**multiply both sides of an inequality by a positive number without changing the inequality sign.**

$\frac{a}{b} = a \times \frac{1}{b}$ is true for any real number $a$, $b$, $b \neq 0$.

Based on this fact, can you deduce the rule for dividing both sides of an inequality by a positive number corresponding to those in part 3?

Summarising all the above, we have, for all real numbers $x$, $y$ and $z$:

(1) If $x > y$ and $z > 0$, then $xz > yz$.
(2) If $x < y$ and $z > 0$, then $xz < yz$.
(3) If $x > y$ and $z > 0$, then $\frac{x}{z} > \frac{y}{z}$.
(4) If $x < y$ and $z > 0$, then $\frac{x}{z} < \frac{y}{z}$.

# Equations and Inequalities

Let us consider $x$ students attending a lecture.

(a) If there are 200 students, we write $x = 200$. This is an **equation**. It has only one answer, i.e., $x$ has only one value.

(b) If there are less than 200 students, we write $x < 200$.
This is an **inequality**. It has many answers, i.e., $x$ can have any numerical value ranging from 0 to 199 inclusive. In this example, negative values of $x$ have no real meaning.

For an inequality with one unknown, $x$, all values of $x$ that satisfy the inequality are called the solutions of the inequality. For example,

(a) some of the solutions of the inequality $x > -3$ are $-2, -1, 0, 1, 2, 3$, etc.;
(b) some of the solutions of the inequality $x < 3$ are $2, 1, 0, -1, -2, -3$, etc.

Notice that there are infinitely many possible solutions to each of the above inequalities.

## Example 20

*Solve the following inequalities.*

*(a)* $\frac{1}{2}x < 3$ *(b)* $2x > 18$

**Solution**

(a)
$$\frac{1}{2}x < 3$$
$$\left(\frac{1}{2}x\right) \times 2 < 3 \times 2 \quad \text{(Multiply both sides by 2)}$$
$$\therefore \quad x < 6$$

(b)
$$2x > 18$$
$$\frac{2x}{2} > \frac{18}{2} \quad \text{(Divide both sides by 2)}$$
$$\therefore \quad x > 9$$

## Example 21

*A chicken pie costs 70 cents. What is the maximum number of chicken pies that one can buy with \$5?*

**Solution**

Let the number of chicken pies that one can buy be $x$.
We require that $70x < 500$
$$x < \frac{500}{7}$$
$$x < 7\frac{1}{7}$$
$\therefore$ the maximum number of chicken pies that one can buy is 7.

## Exercise 7j

1. Solve the following inequalities.

   (a) $4x > 28$

   (b) $\frac{1}{5}y < 7$

   (c) $\frac{2}{3}h > \frac{7}{9}$

   (d) $0.5a > 1.4$

   (e) $3z < -18.3$

2. A bus can ferry a maximum of 32 pupils. How many buses are needed to ferry 440 pupils?

## SUMMARY

1. To solve an equation, we can

   (a) add the same number to each side;

   e.g. if $x - 5 = 7$,
   then $(x - 5) + 5 = 7 + 5$

   (b) subtract the same number from each side;

   e.g. if $x + 4 = 15$
   then $(x + 4) - 4 = 15 - 4$

   (c) multiply each side by the same number;

   e.g. if $\frac{1}{3}x = 8$
   then $3\left(\frac{1}{3}x\right) = 3(8)$

   (d) divide each side by the same number, except 0.

   e.g. if $3x = 21$
   then $\frac{3x}{3} = \frac{21}{3}$

2. To solve an inequality, we can multiply or divide both sides of an inequality by a positive number without having to change the inequality sign.

   If $x > y$, then $xa > ya$ and $\frac{x}{a} > \frac{y}{a}$, $(a > 0)$

# Review Examples 7

## Example 1

*Elvin and Carol together have \$84. If Elvin gives Carol \$32, then Carol will have twice as much money as Elvin. How much money do Elvin and Carol have originally?*

### Solution

Let the amount of money Elvin has originally be $\$x$.

$\therefore$ Carol originally has $\$(84 - x)$.

After giving \$32 to Carol,

Elvin now has $\$(x - 32)$ and Carol has $\$(84 - x + 32)$

We have
$$\begin{aligned} 2(x - 32) &= 84 - x + 32 \\ 2x - 64 &= 116 - x \\ 3x &= 180 \\ \therefore \quad x &= 60 \end{aligned}$$

$\therefore$ Elvin originally has \$60 and Carol has \$24.

## Example 2

*The numerator of a fraction is 2 less than the denominator. If 3 is subtracted from both the numerator and the denominator, the fraction becomes $\frac{3}{4}$. Find the fraction.*

### Solution

Let the numerator be $x$, $\therefore$ the denominator is $(x + 2)$.

$\therefore$ the fraction is $\dfrac{x}{x+2}$

Now
$$\begin{aligned} \frac{x - 3}{(x + 2) - 3} &= \frac{3}{4} \\ 4(x - 3) &= 3(x - 1) \\ 4x - 12 &= 3x - 3 \\ 4x - 3x &= 12 - 3 \\ \therefore \quad x &= 9 \text{ and} \\ x + 2 &= 11 \end{aligned}$$

$\therefore$ the fraction is $\dfrac{9}{11}$.

# Review Questions 7

1. Solve the following equations:
   (a) $\frac{8-x}{3} = \frac{2x+3}{5}$
   (b) $\frac{x+2}{6} = \frac{2x+7}{4}$
   (c) $\frac{x}{2} = 5 + \frac{x}{3}$
   (d) $\frac{9}{4x} - 5 = 4$
   (e) $\frac{2x+1}{6} - \frac{x-3}{4} = \frac{5}{12}$
   (f) $\frac{x+3}{7} - \frac{2(x-4)}{3} = 1$
   (g) $\frac{3x-4}{5} - \frac{x+1}{4} = \frac{2x-4}{3}$
   (h) $\frac{x}{3} + \frac{x-3}{4} - \frac{2x-7}{2} = 0$

2. Solve the following inequalities:
   (a) $3x < 27$
   (b) $\frac{2}{3}k < 14$

3. A man has a number of ducks costing \$10 each and three times as many chickens costing \$6 each. If the total cost of ducks and chickens is \$420, find the number of chickens the man has.

4. Three families, *A*, *B*, and *C*, share 480 kg of rice. *B* gets twice as much as *A*, and *C* gets half as much as *B*. How much does each family get?

5. A shopkeeper bought some eggs at 15 cents each. Six of them were broken while the rest were sold at 20 cents each. If he made a profit of \$4.80, find how many eggs he bought.

6. Tiong Beng worked part time at a fast food restaurant during the school holidays. He was paid the normal hourly rate of \$2.80 per hour. Overtime rate is one and a half times the normal rate. He received \$173.60 for 54 hours of work. How many hours did he work overtime?

7. Peter has 25 sweets and Lilian has 55. How many sweets must Peter give Lilian so that Lilian will have 4 times as many sweets as Peter?

8. With a fixed amount of money \$M, Mr Phua can use it to pay the wages of a local worker for 40 days or to pay the wages of an unskilled foreign worker for 60 days. If he employs one local worker and one unskilled foreign worker for a renovation project, how many days will the \$M be enough to pay for these two workers?

9. Drawing pencils cost 8 cents each and coloured pencils cost 11 cents each. Two dozen assorted pencils cost \$2.16. How many coloured pencils are there?

10. A man travels from *A* to *B* at 4 km/h and from *B* to *A* at 6 km/h. The total journey takes 45 minutes. Find the distance travelled.

11. A number is 5 times another number. By adding 8 to each number, the first number becomes only 3 times the second. What are the two numbers?

12. The sum of two numbers is 45. 4 times the first number is 9 less than 5 times the second. What are the numbers?

*13. A man normally takes 5 hours to travel at a certain speed from city *A* to city *B*. One day, he increases his speed by 4 km/h and finds that the journey from *A* to *B* takes half an hour less than the normal time. Find his normal speed.

∗**14.** The sum of the ages of Fandi and Ahmad is 38. Seven years ago, Fandi was three times as old as Ahmad. Find their present ages.

∗**15.** Aramugan has enough money to buy 24 apples. If the price of each apple is reduced by 5 cents, he will be able to buy an extra 6 apples with the same sum of money. Find the original cost of each apple.

∗**16.** A tank can be fully filled with water using a pipe that fills 20 litres in a minute. A bigger pipe that can fill 25 litres in a minute will take one minute less to fill the same tank. How many minutes does the smaller pipe take to fill the tank?

∗**17.** Mary is twice as old as John and half as old as Bob. In 22 years' time, Bob will be twice as old as John. How old is Mary?

∗**18.** Alice and Belinda start off simultaneously from two towns to meet one another. If Alice travels 2 km/h faster than Belinda, they would meet in 3 hours. If Belinda travels 1 km/h slower and Alice's speed is two-thirds of her previous speed, they would meet in 4 hours. How far apart are the two towns?

∗**19.** Given that $A$, $B$, $C$ and $D$ are whole numbers such that $A + B = 8$, $B + C = 11$, $B + D = 13$ and $C + D = 14$, find the values of $A$, $B$, $C$ and $D$.

∗**20.** Given that $A$, $B$, $C$ and $D$ are whole numbers such that $A \times B = 8$, $B \times C = 28$, $C \times D = 63$ and $B \times D = 36$, find the values of $A$, $B$, $C$ and $D$.

*In this chapter, you will learn how to*

- *find the perimeter and area of simple geometrical figures;*
- *solve problems involving these figures and figures related to them.*

CHAPTER

# 8

# Perimeter and Area of Simple Geometrical Figures

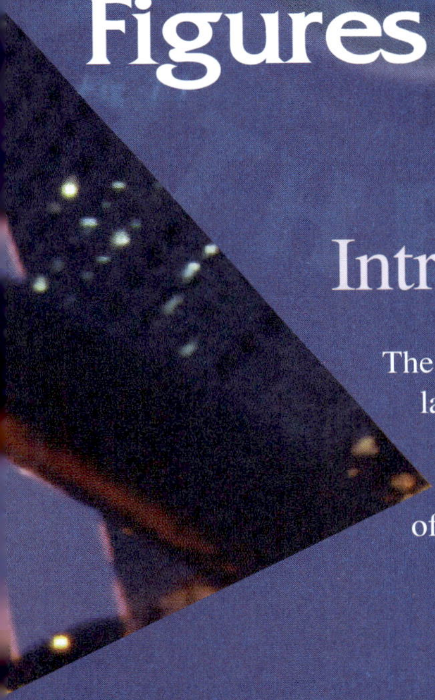

## Introduction

The picture shows the famous "Fountain of Wealth", the world's largest fountain, located at Suntec City in Singapore. It has attracted many visitors from all over the world. The bronze fountain has a circular ring of perimeter 66 m amd a base area of 1683 $m^2$.

In primary school, you learned how to find the areas and perimeters of squares, rectangles, triangles and circles. We shall have a quick revision of these before proceeding to find the areas of parallelograms and trapeziums.

The following are some important results that were covered in primary school.

| **Name** | **Figure** | **Area ($A$)** | **Perimeter ($P$)** |
|---|---|---|---|
| Square | 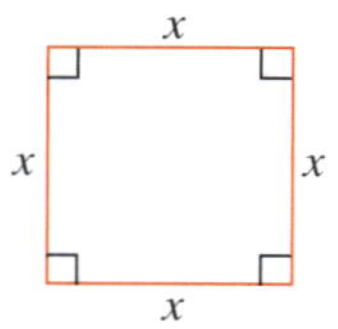  | $A = x^2$ | $P = 4x$ |
| Rectangle | 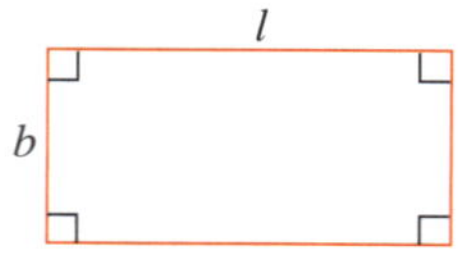  | $A = lb$ | $P = 2l + 2b$ |
| Triangle | 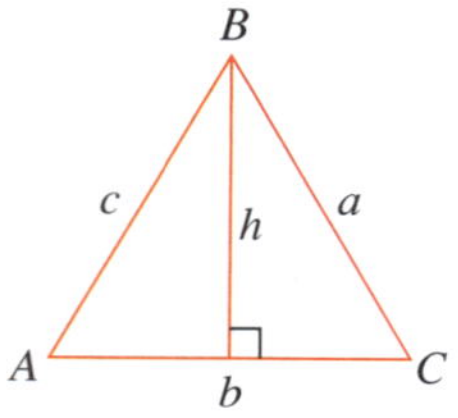  | $A = \frac{1}{2}bh$ | $P = a + b + c$ |
| Circle | 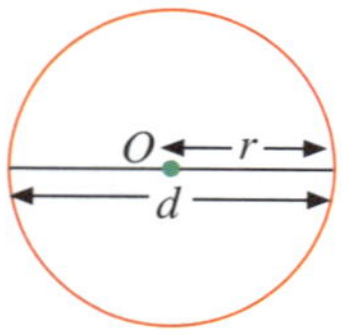  | $A = \pi r^2$ | $P = 2\pi r$ or Circumference $= 2\pi r$ $= \pi d$ |

## Units of Area

We often use the following metric units to measure area.

**Square centimetre (cm$^2$):** We usually use unit squares to compare areas. A square of side 1 cm is used as a standard unit. We call this unit area 1 square centimetre (1 cm$^2$). See Fig. 8.1.

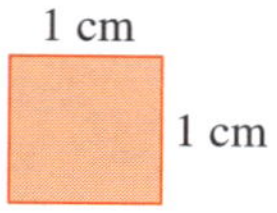


*Fig. 8.1*

FOR YOUR Information

*The metric system of units was first introduced in the nineteenth century. It is essentially a simple system based on the decimal system. As such, it allows easy conversion from one unit to another. The metric system of units involves measurements of length, area, mass, capacity and volume; all these units are related through the decimal notation.*

**Square millimetre ($mm^2$):** In Fig. 8.2, each small square has an area of 1 square millimetre (1 $mm^2$).

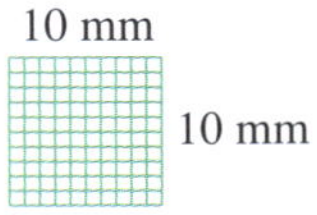


*Fig. 8.2*

$$1 \text{ cm} = 10 \text{ mm}$$
$$\therefore \quad 1 \text{ cm}^2 = 10 \text{ mm} \times 10 \text{ mm} = 100 \text{ mm}^2$$

Square millimetres are used to measure the areas of very small shapes.

**Square metre ($m^2$):** The square metre is used to measure the area of large surfaces such as the floor area of a flat.

$$1 \text{ m} = 100 \text{ cm}$$
$$\therefore \quad 1 \text{ m}^2 = 100 \text{ cm} \times 100 \text{ cm} = 10\,000 \text{ cm}^2$$

**Hectare (ha):** The hectare is used to measure large land areas such as farms, housing estates, etc.

$$1 \text{ ha} = 10\,000 \text{ m}^2$$

**Square km ($km^2$):** The square kilometre is used to measure the area of a very large surface such as the area of a country.

$$1 \text{ km} = 1000 \text{ m}$$
$$\therefore \quad 1 \text{ km}^2 = 1000 \text{ m} \times 1000 \text{ m}$$
$$= 1\,000\,000 \text{ m}^2 = 100 \text{ ha}$$

FOR YOUR INFORMATION

*The largest freshwater lake in the world is Lake Superior, one of the Great Lakes of North America. It covers an area of 82 350 $km^2$, roughly 130 times the size of Singapore.*

## Explorations

The British used feet, inches, yards, furlongs, miles, etc. to measure length, and acre to measure area. Find out what these units are and compare them with the SI units (International System of Units). In the SI units, the units of mass, length and time are the kilogram, meter and second respectively. Can you also find out other SI units?

## Example 1

*Express (a) 975 $cm^2$ in $m^2$; (b) 2.65 $km^2$ in $m^2$; (c) 5 $mm^2$ in $cm^2$.*

### Solution

(a)

$$975 \text{ cm}^2 = \_\_\_\_\_ \text{ m}^2$$

$$1 \text{ cm} = \frac{1}{100} \text{ m}$$

$$1 \text{ cm}^2 = \frac{1}{100} \text{ m} \times \frac{1}{100} \text{ m}$$

$$= \frac{1}{10\,000} \text{ m}^2$$

$$\therefore \quad 975 \text{ cm}^2 = 975 \times \frac{1}{10\,000} \text{ m}^2$$

$$= 0.0975 \text{ m}^2$$

(b)

$$2.65 \text{ km}^2 = \_\_\_\_\_ \text{ m}^2$$

$$1 \text{ km} = 1000 \text{ m}$$

$$1 \text{ km}^2 = 1000 \times 1000 \text{ m}$$

$$= 1\,000\,000 \text{ m}^2$$

$$\therefore \quad 2.65 \text{ km}^2 = 2.65 \times 1\,000\,000 \text{ m}^2$$

$$= 2\,650\,000 \text{ m}^2$$

(c)

$$5 \text{ mm}^2 = \_\_\_ \text{ cm}^2$$

$$1 \text{ mm} = \frac{1}{10} \text{ cm}$$

$$1 \text{ mm}^2 = \frac{1}{10} \text{ cm} \times \frac{1}{10} \text{ cm}$$

$$= \frac{1}{100} \text{ cm}^2$$

$$\therefore \quad 5 \text{ mm}^2 = 5 \times \frac{1}{100} \text{ cm}^2 = 0.05 \text{ cm}^2$$

## Example 2

*The perimeter of a rectangle is 22 cm and its breadth is 4 cm. Find its area.*

### Solution

We can **draw a diagram** and then **form an equation** to solve the problem.

Let the length of the rectangle be $x$ cm.

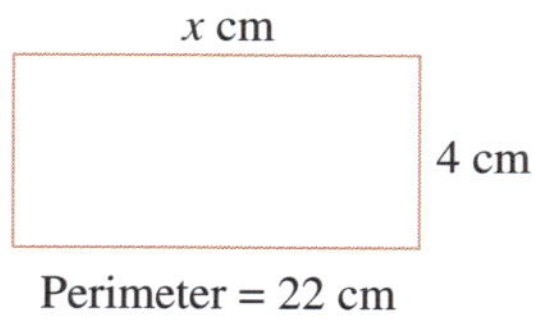


Then

$$2(x + 4) = 22$$

$$2x + 8 = 22$$

$$2x = 14$$

$$x = 7$$

$\therefore$ the area of the rectangle = (7 × 4) $cm^2$ = 28 $cm^2$

## Example 3

*The width of a rectangle is 5 cm less than its length, and its perimeter is 46 cm. Find its width and its area.*

**Solution**

Let the width of the rectangle be $x$ cm.

Then the length is $x + 5$ cm.

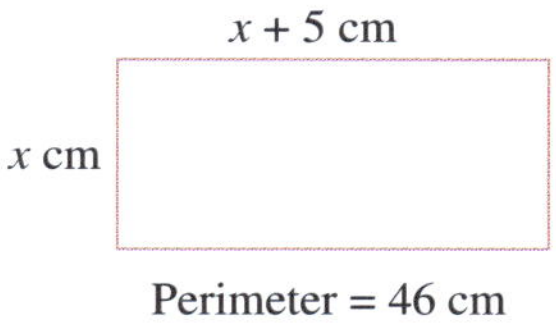


$$\begin{aligned} \text{Perimeter} = 2[x + (x + 5)] \text{ cm} &= 46 \text{ cm} \\ 4x + 10 &= 46 \\ \therefore \quad 4x &= 36 \\ x &= 9 \end{aligned}$$

$\therefore$ Its width is 9 cm and its length is 14 cm.

$\therefore$ Its area $= (9 \times 14)\ \text{cm}^2$
$= 126\ \text{cm}^2$

## Example 4

*A rectangular field is 13 m long and 10 m wide. It has a cement path $3\frac{1}{2}$ m wide around it. What is the area of the cement path?*

**Solution**

Area of the field and cement path $= (20 \times 17)\ \text{m}^2 = 340\ \text{m}^2$

Area of the field $= (13 \times 10)\ \text{m}^2 = 130\ \text{m}^2$

$\therefore$ area of cement path $= (340 - 130)\ \text{m}^2 = 210\ \text{m}^2$

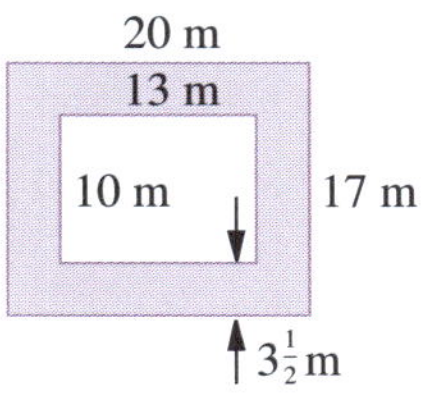

## Example 5

*In the figure below, the sides AB and DC of the quadrilateral ABCD are both perpendicular to the diagonal AC. Given AB = 2 cm, DC = 8 cm and the area of $\triangle ABC = 6$ cm², calculate*

*(a) the length of AC;*

*(b) the area of the quadrilateral ABCD.*

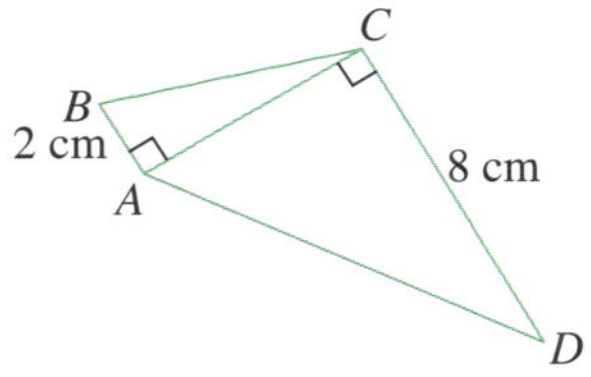


**Solution**

(a) Area of $\triangle ABC = \frac{1}{2} \times AB \times AC$

$$\therefore \quad \frac{1}{2} \times AB \times AC = 6$$

$$\frac{1}{2} \times 2 \times AC = 6$$

$$\therefore \quad AC = 6 \text{ cm}$$

(b) Area of $\triangle ACD = \frac{1}{2} \times AC \times CD$

$$= \left(\frac{1}{2} \times 6 \times 8\right) \text{ cm}^2 = 24 \text{ cm}^2$$

$\therefore$ the area of the quadrilateral $ABCD = (6 + 24)$ cm² $= 30$ cm²

> **Just for fun**
>
> *You are given 4 rectangular pieces of wood. Two of these measure 4 cm by 3 cm while the other two measure 13 cm by 1 cm. Use these 4 pieces of wood to enclose an area as large as you possibly can.*

## Example 6

*In the figure on the right, AB = 17 cm, BC = 25 cm, AP = 8 cm, BQ = 5 cm, CR = 14 cm and CS = 3 cm. Find the area of the shaded region.*

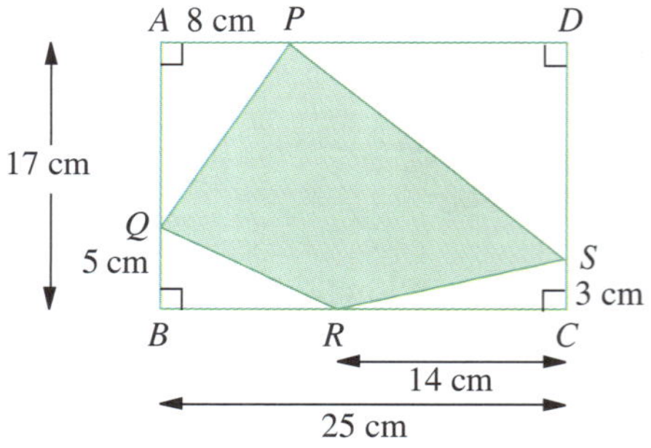


**Solution**

Area of shaded region $PQRS$ = area of $ABCD$ – area of triangles $(APQ + PDS + CSR + BQR)$

Area of $ABCD = (17 \times 25)$ cm² $= 425$ cm²

Area of $\triangle APQ = \frac{1}{2} \times AP \times AQ = \left[\frac{1}{2} \times 8 \times (17 - 5)\right]$ cm² $= 48$ cm²

Area of $\triangle PDS = \frac{1}{2} \times PD \times DS = \left[\frac{1}{2} \times (25 - 8) \times (17 - 3)\right]$ cm² $= 119$ cm²

Area of $\triangle SRC = \frac{1}{2} \times RC \times SC = \left(\frac{1}{2} \times 14 \times 3\right)$ cm² $= 21$ cm²

Area of $\triangle BRQ = \frac{1}{2} \times BR \times BQ = \left[\frac{1}{2} \times (25 - 14) \times 5\right]$ cm² $= 27.5$ cm²

$\therefore$ Area of $PQRS = (425 - 48 - 119 - 21 - 27.5)$ cm² $= 209.5$ cm²

## Example 7

*The area of a circle is 78.5 cm². Calculate the circumference of the circle. (Take $\pi = 3.142$)*

**Solution**

$$\begin{aligned} \text{Area of circle} &= \pi r^2 \\ 78.5 &= 3.14r^2 \\ r^2 &= \frac{78.5}{3.14} = 25 \end{aligned}$$

Since $5^2 = 25$

so $r = 5$

Since $\pi = 3.142$ and radius = 5 cm,

we have circumference of circle $= 2\pi r$

$= 2(3.142)(5)$ cm

$= 31.4$ cm

**Just for fun**

*Which of the following shaded figures has the greatest area? The squares are of the same length and the curved lines are all arcs of circles.*

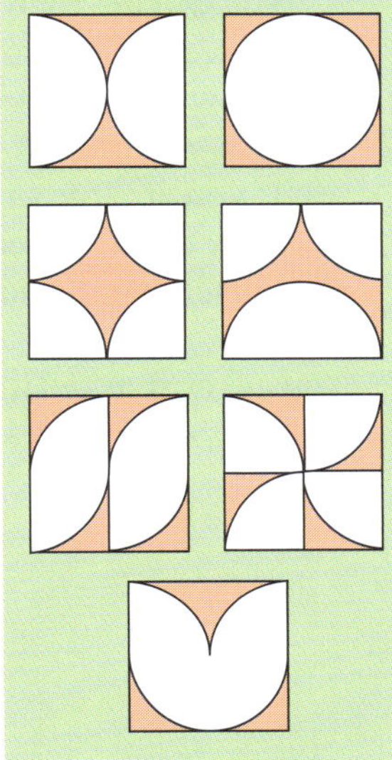

## Example 8

*The diameter of the wheel of a car is 0.35 m. Find the number of revolutions made by the wheel per minute when the car is travelling at 33 km/h. (Take $\pi = 3.142$)*

**Solution**

In 60 minutes, the car travels $(33 \times 1000)$ m.

In 1 minute, the car travels $\frac{33 \times 1000}{60}$ m.

$$\begin{aligned} \text{Number of revolutions made per minute} &= \frac{\text{distance travelled}}{\text{circumference of wheel}} \\ &= \frac{33 \times 1000}{60} \times \frac{1}{\pi d} \\ &= \frac{33 \times 1000}{60} \times 3.142 \times \frac{1}{0.35} \\ &= 4937 \end{aligned}$$

**Just for fun**

*The circumference of the Earth's equator is approximately 40 000 km. An imaginary belt of 40 000 km will fit the equator nicely. If we increase the length of the belt by 1 m, will it be possible for a cat to squeeze through? How far above the surface of the equator will the belt be?*

## Explorations

(1) You are asked to design a rectangular sheep pen of area 144 m$^2$ on an open field using wire fencing. You can work in a group of three.

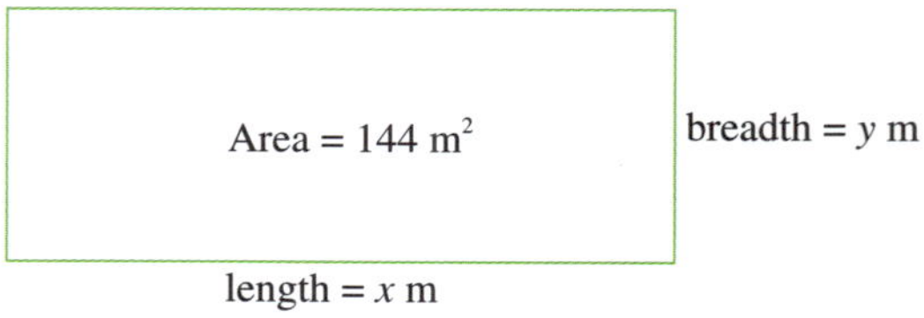


(a) What are the dimensions of the rectangular pen you will use for it to have the smallest perimeter?
(b) How do you justify that your answer is correct?
(c) Show your working clearly with appropriate use of mathematical terms and concepts.
(d) Can you make a general conclusion from this investigation?
(e) If you are free to choose shapes other than rectangle, what shape will give an area of 144 m$^2$ and have the least perimeter?
(f) Present your findings to your class.

(2) You are now asked to design another rectangular sheep pen with an area of 144 m$^2$ by using a straight hedge as one of the sides of the rectangular sheep pen.

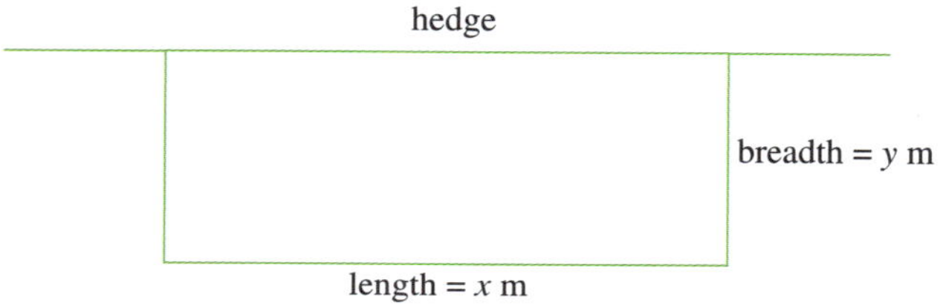


(a) What are the dimensions of the rectangular pen you will use for it to have the smallest perimeter?
(b) Is your answer very different from (1)?
(c) Show your working clearly with appropriate use of mathematical terms and concepts.
(d) Can you make a general conclusion from this investigation?
(e) If you are free to choose shapes other than rectangle, what shape will give an area of 144 m$^2$ and have the least perimeter?
(f) Present your findings to your class.

(3) There are two straight hedges in the farm that could be used as two adjacent sides of a rectangular sheep pen. You are to use wire fencing to enclose an area of 144 m$^2$ and using the least amount of wire fencing.

(a) What are the dimensions of the rectangular pen you will use for it to have the smallest perimeter?
(b) Discuss the difference in the answers for (1), (2) and (3).
(c) Can you make a general conclusion from this investigation?
(d) If you are free to choose shapes other than rectangle, what shape will give an area of 144 m$^2$ and have the least perimeter?
(f) Present your findings to your class and answer any questions that your friends may ask.

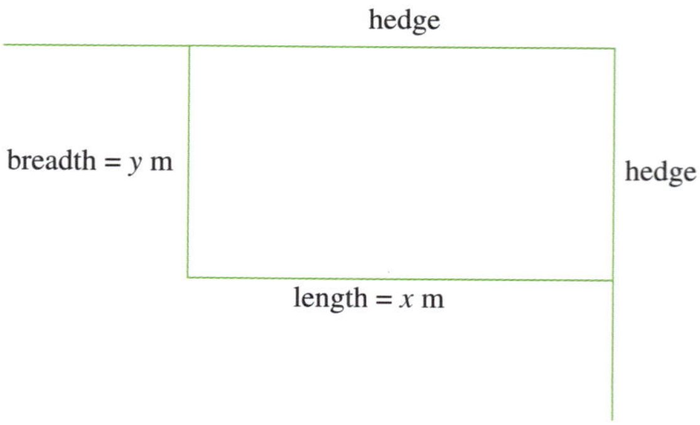

To help you make a good write up of the report and do the presentation, you may find the following scoring tool or rubric useful.

| Category | 4 or Excellent | 3 or Very Good | 2 or Good | 1 or Unsatisfactory |
|---|---|---|---|---|
| Use of problem solving heuristics | Use effective heuristics to solve all the problems completely. | Use effective heuristics to solve most of the problems. | Use effective heuristics to solve only some of the problems. | Unable to use effective heuristics to solve most of the problems. |
| Use of math terminology | Use correct terminology all the time to make it easy to follow and understand. | Use correct terminology most of the time to make it fairly easy to follow and understand. | Use correct terminology most of the time but it is at times not easy to follow and understand. | There is little use of correct terminology and it is difficult to follow and understand. |
| Presentation | The presentation is clear. Able to capture the attention of the audience. | The presentation is clear but contains minor errors. Able to capture the attention of most of the audience. | The presentation is not complete and contains several serious errors. Unable to capture the attention of most of the audience. | The presentation contains serious errors. Students don't seem to understand what is required. Unable to capture the attention of most of the audience. |

## Exercise 8a

Use the value of $\pi$ given by your calculator.

1. Copy and fill in the missing numbers:
   - **(a)** $8.5\ \text{cm}^2 =$ ______ $\text{m}^2$
   - **(b)** $2.5\ \text{mm}^2 =$ ______ $\text{cm}^2$
   - **(c)** $6.3\ \text{m}^2 =$ ______ $\text{cm}^2$
   - **(d)** $40.6\ \text{cm}^2 =$ ______ $\text{mm}^2$
   - **(e)** $44.4\ \text{km}^2 =$ ______ ha
   - **(f)** $3.1$ ha $=$ ______ $\text{m}^2$
   - **(g)** $53.7\ \text{m}^2 =$ ______ $\text{km}^2$
   - **(h)** $0.28\ \text{km}^2 =$ ______ $\text{m}^2$
   - **(i)** $53\ 200\ \text{mm}^2 =$ ______ $\text{m}^2$
   - **(j)** $69\ 450\ \text{cm}^2 =$ ______ ha

2. Copy and complete the following table below for each circle:

| | (a) | (b) | (c) | (d) |
|---|---|---|---|---|
| *Radius* | 10 m | | | |
| *Diameter* | | | | 3.6 m |
| *Circumference* | | 176 mm | | |
| *Area* | | | 616 $\text{cm}^2$ | |

3. Find the perimeter and area of each of the following figures. All dimensions are given in cm and the circular portions are semicircles or quadrants.

**(a)**

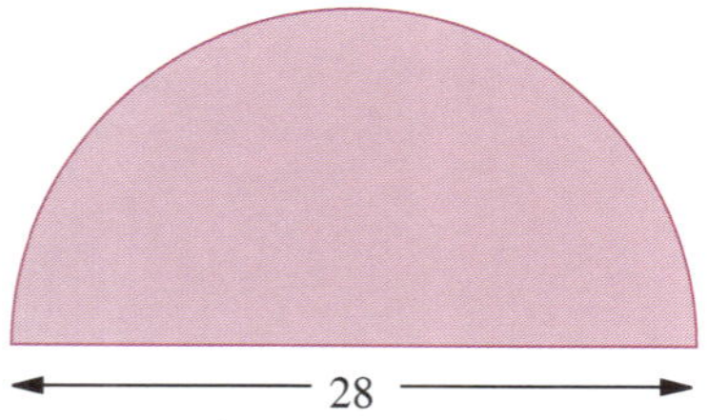


**(b)**

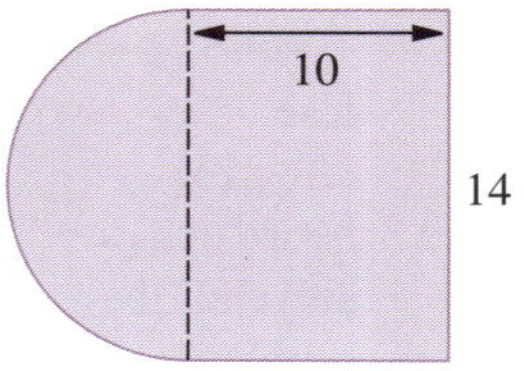


**(c)**

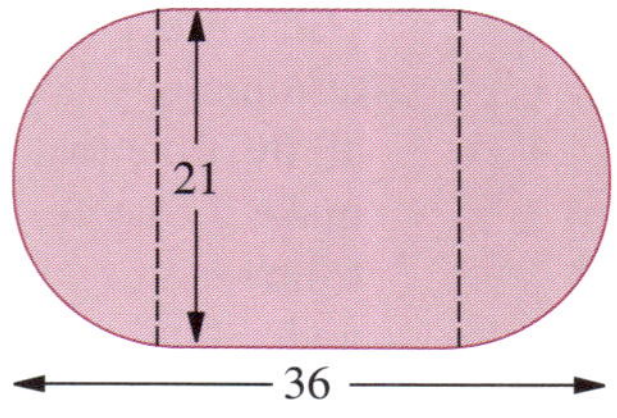


*(d)

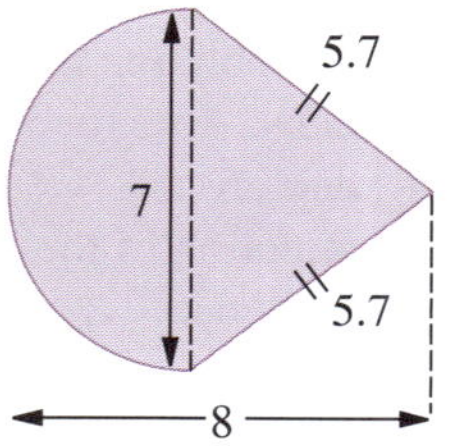


**(e)**

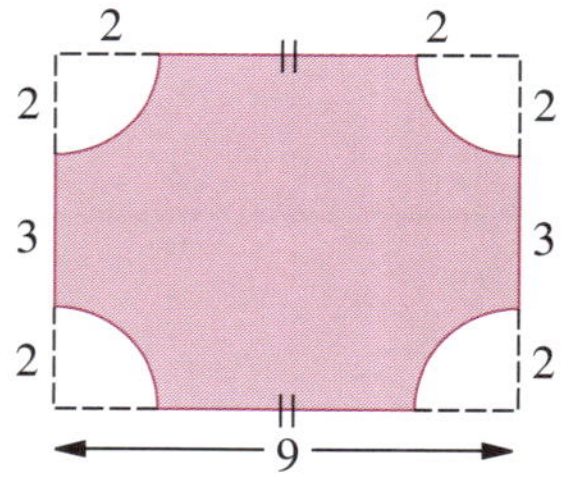


**(f)**

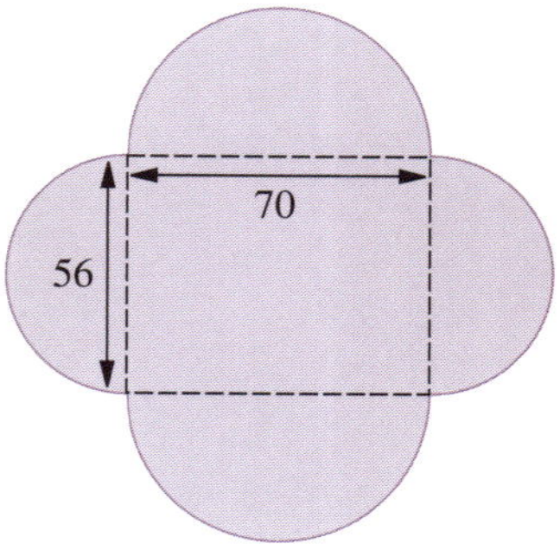

4. Find the area, in square centimetres, of a rectangular strip of board 3.28 m long and 75 mm wide.

5. A square cardboard of side 1.2 m has a 10 cm wide border round the exterior of three of its sides. Find the area of the border.

6. A paper box without a lid is 25 cm long, 16 cm wide and 5 cm deep. How many square centimetres of paper have been used to make the box?

7. Find the cost required to carpet a hall 8 m by 5.5 m if a rectangular section 2 m by $1\frac{1}{2}$ m is taken out to provide for the fireplace and the carpet costs \$52.50 per $m^2$.

8. A swimming pool 25 m by 10 m has a concrete border all round. Find the area of the concrete border if it is 2.5 m wide at the sides and 5 m at the ends.

9. If the minute hand of a big clock is 1.12 m long, find the rate at which its tip is moving in centimetres per minute.

★10. Find the total area of cardboard used in making a match box, complete with the sliding portion, 4 cm long, 2.5 cm wide and 1.2 cm deep (ignore the thickness of the cardboard.)

11. Find, in hectares, the area of the figure shown below. Give your answer correct to 2 decimal places.

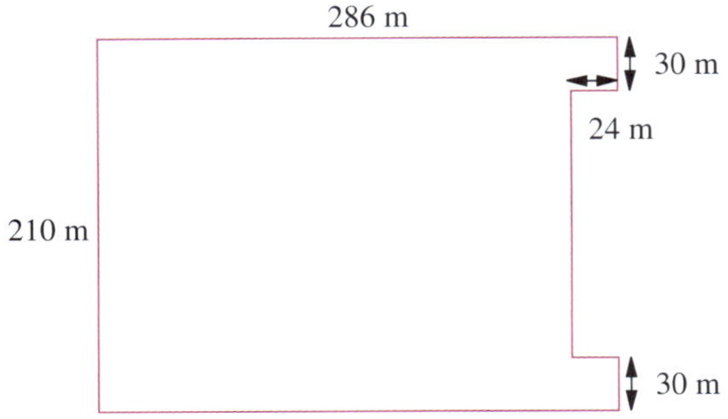


12. The length of a rectangle is 8 cm more than its width. If its perimeter is 56 cm, find its length and its area.

13.

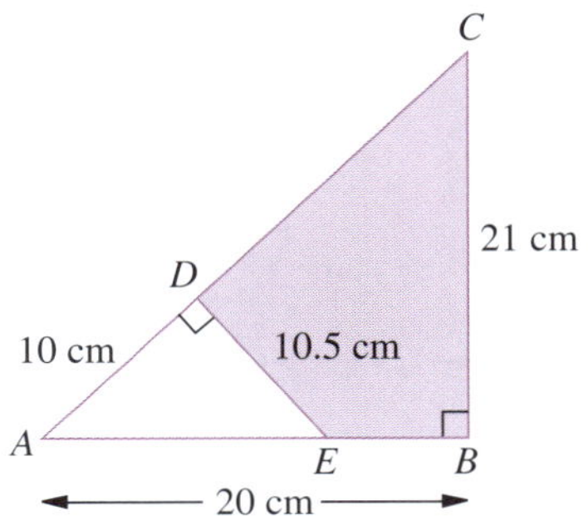


In the diagram, $AB = 20$ cm, $BC = 21$ cm, $AD = 10$ cm and $DE = 10.5$ cm. Angles $ABC$ and $ADE$ are right angles. If $\triangle ADE$ is removed from $\triangle ABC$, what is the area of the shaded region that remains?

14. For questions (a) to (c), refer to the figure below in which $QS$ is perpendicular to $PR$ and $PK$ is perpendicular to $QR$.
   **(a)** Find the area of $\triangle PQR$ if $PR = 17$ cm and $QS = 12$ cm.
   **(b)** Find the area of $\triangle PQS$ if $QS = 7$ cm, $PR = 14$ cm and $SR = 9$ cm.
   **(c)** Find $QS$ if $PR = 14$ cm and the area of $\triangle PQR = 147\ cm^2$.

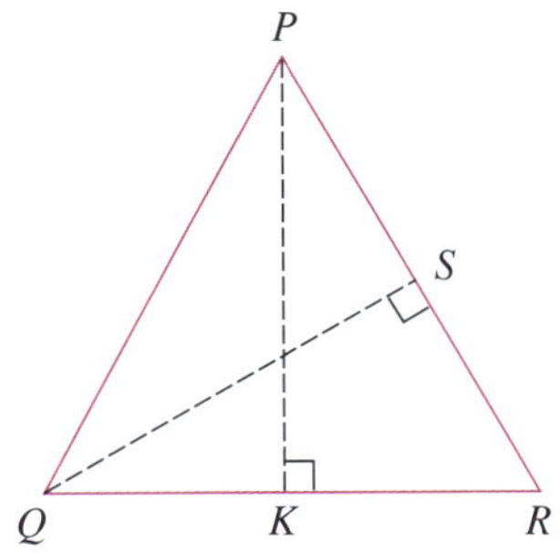

**15.** Find the area of each of the following shaded parts.

**(a)**

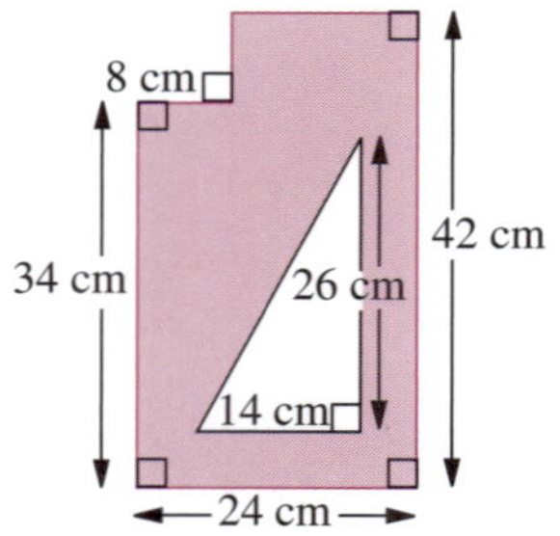


**(b)**

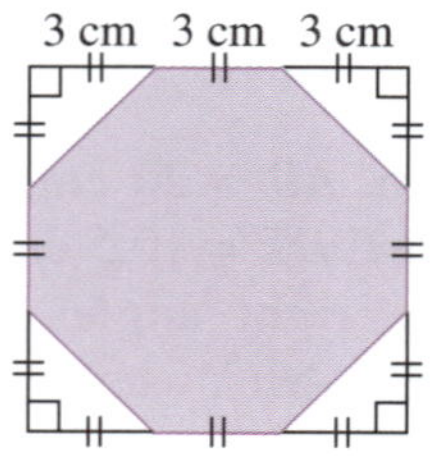


**(c)**

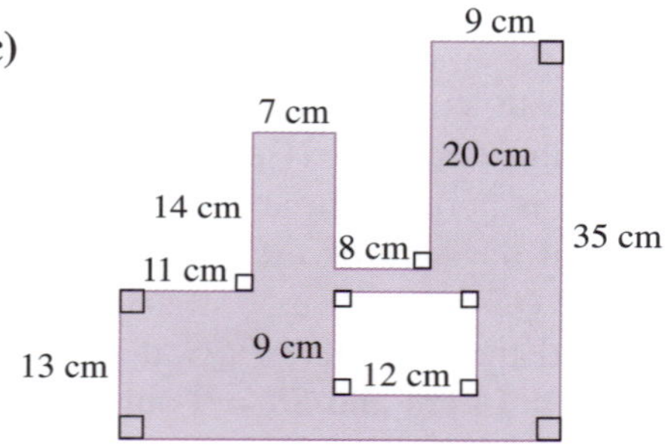


**(d)**

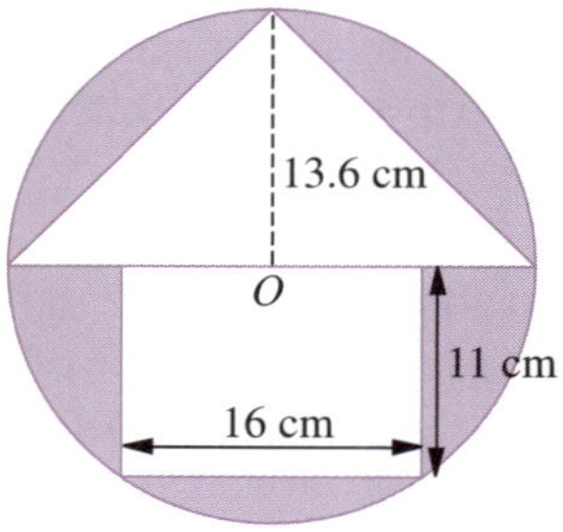


**16.** Find the number of 15-centimetre square tiles required to cover a rectangular floor 5.4 m long and 4.05 m wide.

**17.** As many 8-cm diameter discs as possible are cut from a sheet of rectangular cardboard measuring 170 cm by 90 cm. Find the area of the sheet that is left.

# Area of a Parallelogram

A **parallelogram** is a quadrilateral in which the opposite pairs of sides are equal and parallel.

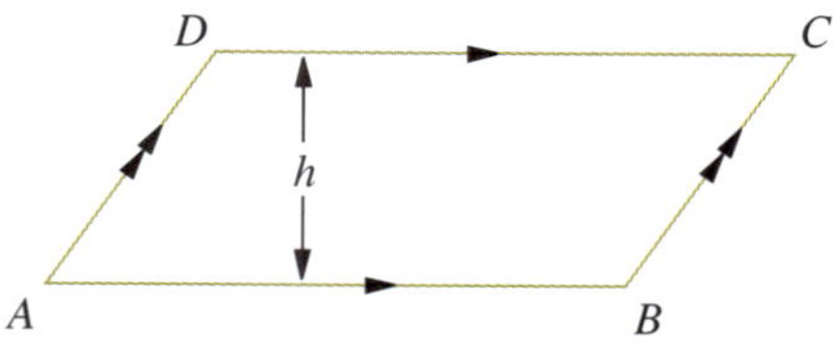


i.e. $AB = DC$, $AD = BC$ and $AB \,//\, DC$, $AD \,//\, BC$.

We can obtain a rectangle from a parallelogram. To do so, draw on a piece of paper a parallelogram *ABCD* as shown in Fig. 8.3. The perpendicular height between *AB* and *DC* is *h*.

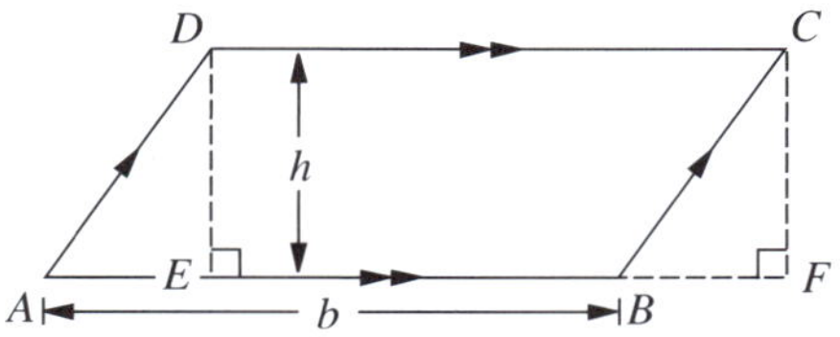


*Fig. 8.3*

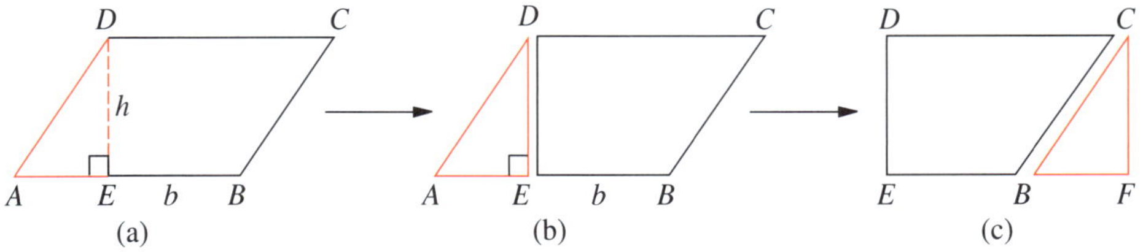


*Fig. 8.4*

Cut off $\triangle AED$ and place it in the position *BFC* (Fig. 8.4(c)). A rectangle *EFCD* is obtained. Do you agree that the parallelogram *ABCD* and the rectangle *EFCD* have the same area?

Area of the parallelogram *ABCD* = Area of the rectangle *EFCD*
$= DC \times DE$

But $DC = AB$ (opposite sides of a parallelogram)

$\therefore$ area of parallelogram *ABCD* $= AB \times DE$
$= b \times h$
= base × height

In general,

**area of a parallelogram = base × height = $b \times h$**

**Just FOR fun**

*Four rectangular cards of identical size are arranged as shown below. You are to move only one card so as to form a square.*

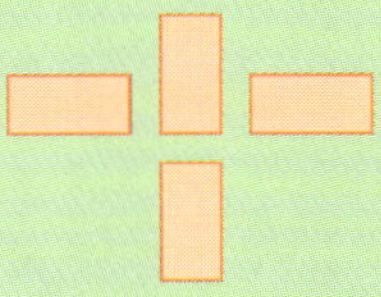

## Explorations

The area of a parallelogram can also be obtained by cutting the parallelogram into two triangles and then using the formula of the area of a triangle.

Work out the formula of the area of a parallelogram based on the above method.

# Area of a Trapezium

A **trapezium** is a quadrilateral with one pair of parallel sides. Fig. 8.5 shows a trapezium *ABCD* in which *AD* is parallel to *BC* with a height of *AH*. The trapezium is divided into $\triangle ABC$ and $\triangle ACD$.

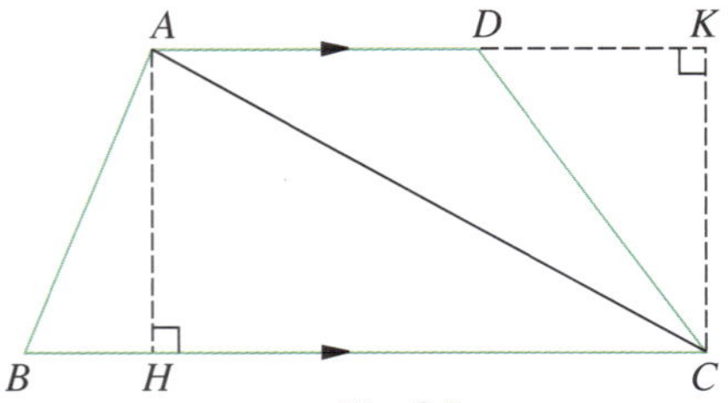


*Fig. 8.5*

*12 toothpicks are arranged as shown below. Remove only 2 toothpicks so as to leave only 2 squares.*

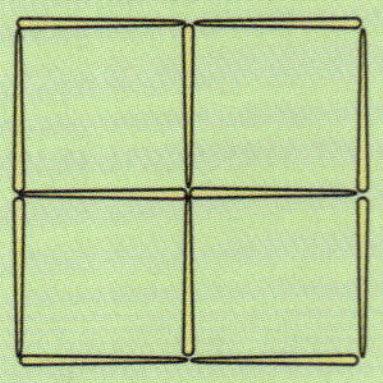

$$\text{Area of trapezium } ABCD = \text{area of } \triangle ABC + \text{area of } \triangle ACD$$

$$= \left(\frac{1}{2} \times BC \times AH\right) + \left(\frac{1}{2} \times AD \times KC\right)$$

Since $AH = KC$,

$$\text{Area of trapezium } ABCD = \frac{1}{2}AH(BC + AD)$$

$$= \frac{1}{2} \times \text{height} \times \text{sum of parallel sides}$$

In general,

$$\textbf{area of trapezium} = \frac{1}{2} \times \textbf{height} \times \textbf{sum of parallel sides}$$

## Explorations

Work in groups.

Try to work out as many possible alternative methods of getting the above formula of the area of a trapezium as possible.

## Example 9

*Find the areas of the following figures:*

(a)

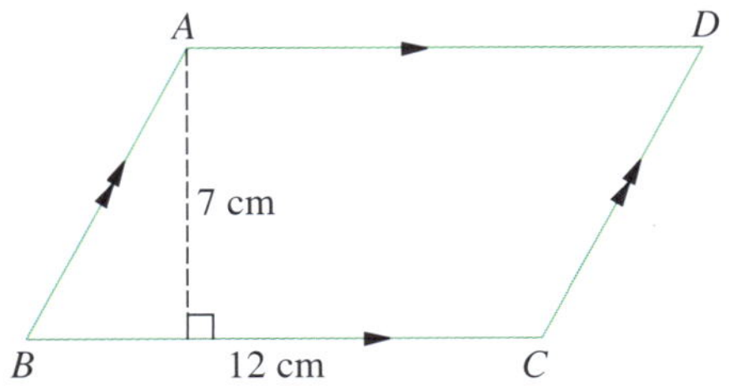


(b)

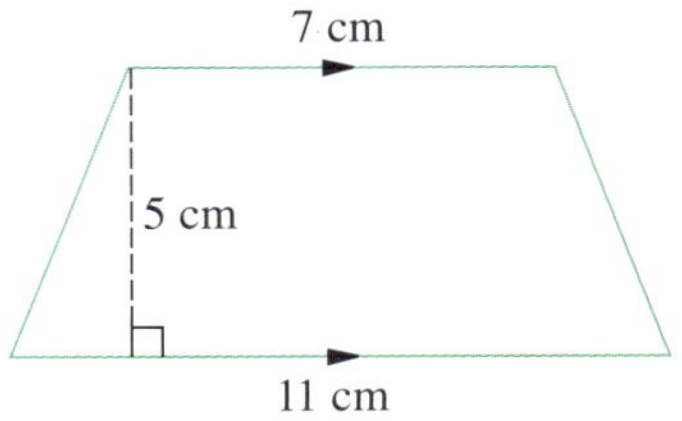


**Solution**

(a) Area of parallelogram $ABCD$ = base × height

$= (12 \times 7)\ \text{cm}^2 = 84\ \text{cm}^2$

(b) Area of trapezium $= \frac{1}{2} \times \text{height} \times \text{sum of parallel sides}$

$= \frac{1}{2} \times 5 \times (7 + 11)\ \text{cm}^2 = 45\ \text{cm}^2$

## Example 10

*Find the value of x in each of the following figures:*

(a)

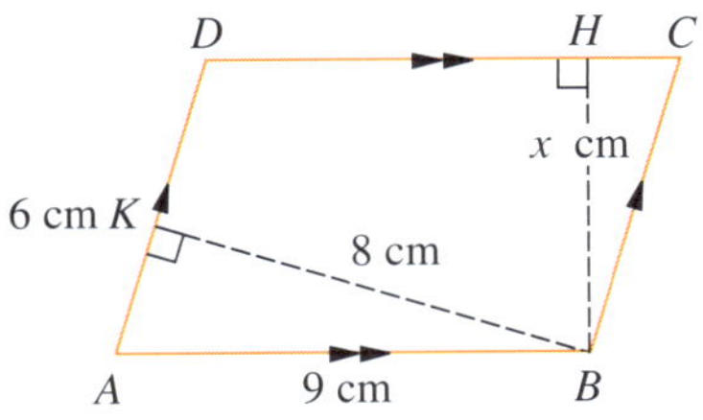


(b)

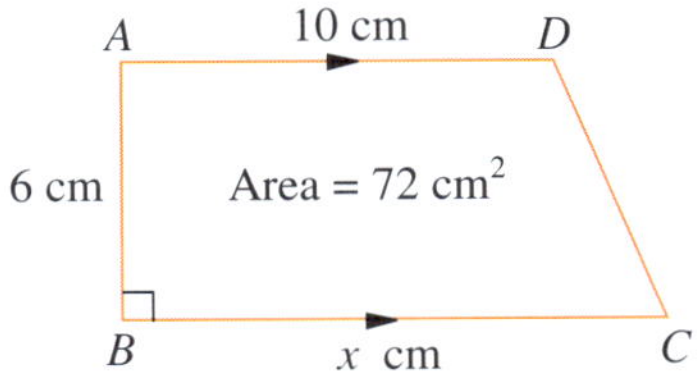


**Solution**

(a) Area of parallelogram $ABCD = AB \times BH$

$= AD \times BK$

$9 \times x = 6 \times 8$

$\therefore \quad x = \frac{6 \times 8}{9} = 5\frac{1}{3}$

(b) Area of trapezium $ABCD = \frac{1}{2} \times AB \times (AD + BC)$

$$72 = \frac{1}{2} \times 6 \times (10 + x)$$
$$24 = 10 + x$$
$$\therefore \quad x = 14$$

## Example 11

*Find the total area of the shaded parts in the diagram.*

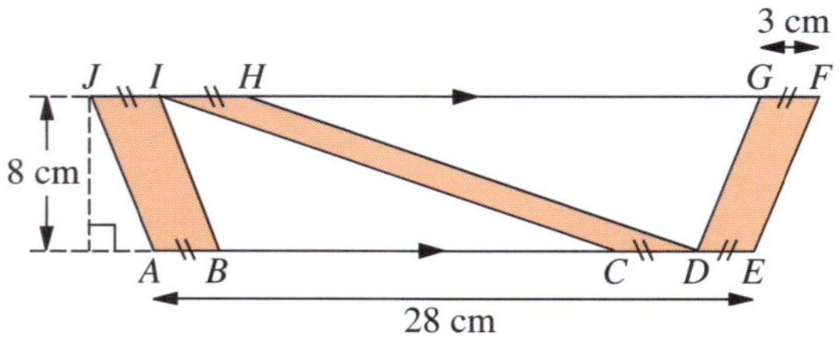


**Solution**

The total area of the shaded parts is made up of 3 parallelograms of the same base length 3 cm and of the same height 8 cm.

$\therefore$ area of shaded region = $3 \times (3 \times 8)$ cm$^2$ = 72 cm$^2$

**Problem Solving Tips**

*Observation of the diagram is essential to solve the problem.*

## Exercise 8b

1. Copy and complete the following table for each parallelogram:

| | (a) | (b) | (c) |
|---|---|---|---|
| *Base* (cm) | 12 | | 7.8 |
| *Height* (cm) | 7 | 6 | |
| *Area* (cm$^2$) | | 42 | 42.9 |

2. Find the areas of the following shaded parts:

(a)

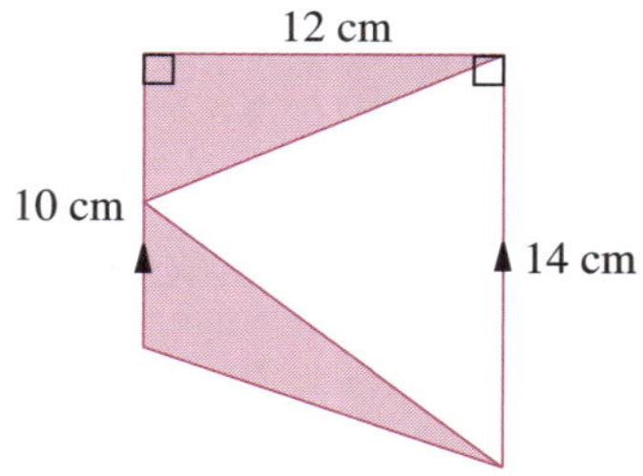

**(b)**

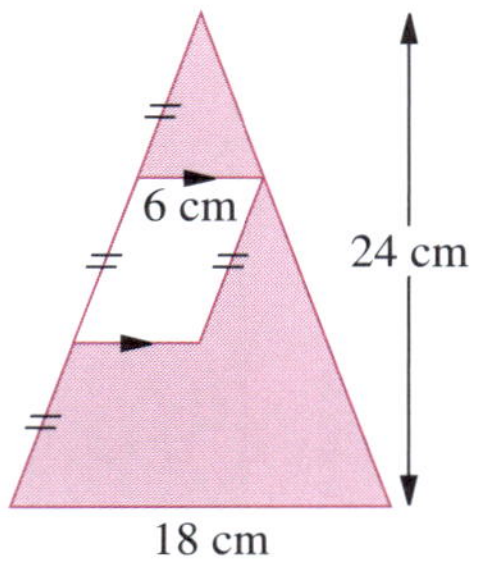


**(c)**

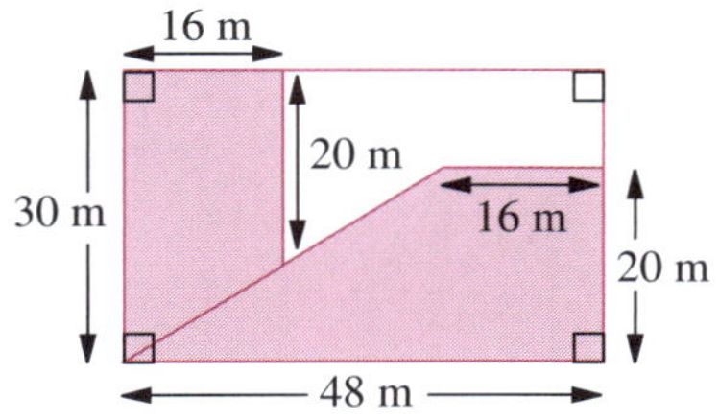


**(d)**

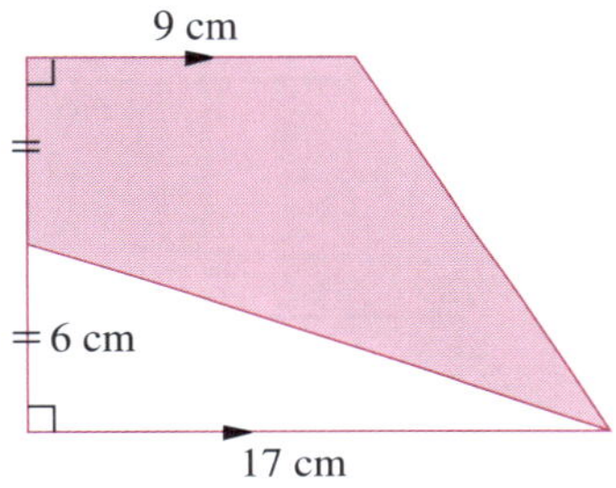


**3.** Copy and complete the following table for each trapezium:

| | (a) | (b) | (c) |
|---|---|---|---|
| *Height* (cm) | 6 | 14 | 8 |
| *Parallel side 1* (cm) | 7 | 8 | |
| *Parallel side 2* (cm) | 11 | | 5 |
| *Area* ($cm^2$) | | 126 | 72 |

**4.** What is the cost of spraying insecticide on a field measuring 2000 m by 3200 m if the cost is \$22 per hectare? (1 ha = 10 000 $m^2$)

**5.** Find the unknowns, marked $x$, in the following figures:

**(a)**

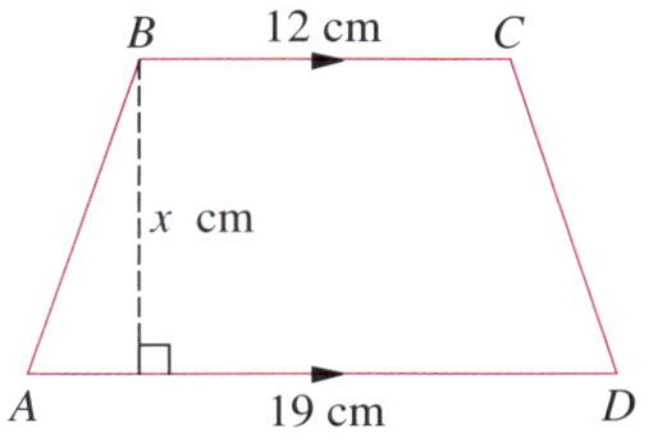


Area of $ABCD = 124\ cm^2$

**(b)**

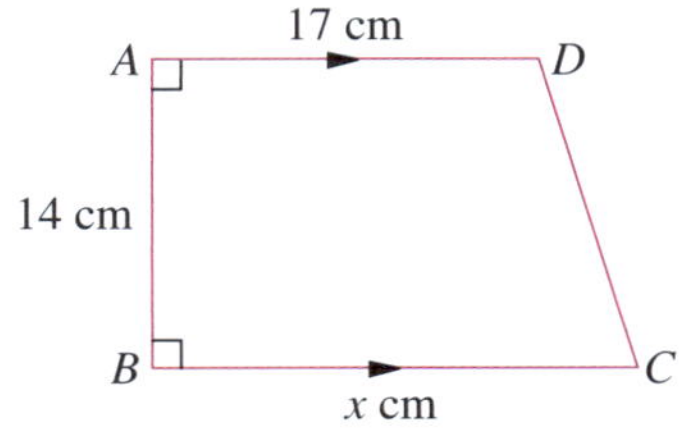


Area of $ABCD = 280\ cm^2$

**(c)**

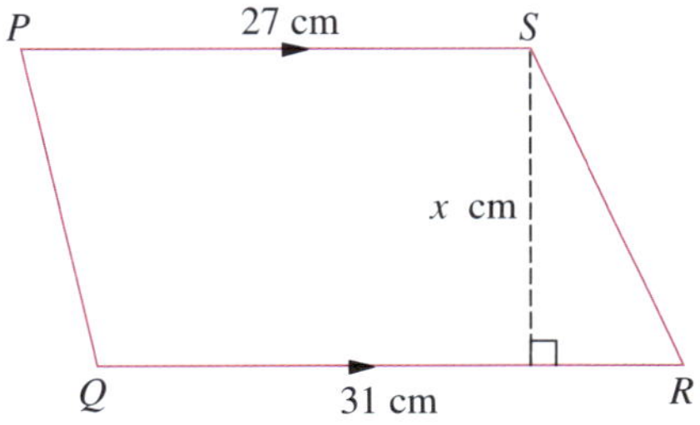


Area of $PQRS = 348\ cm^2$

**(d)**

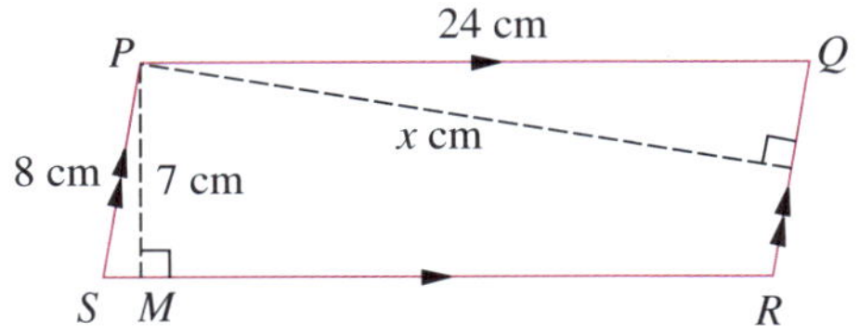

## Example 12

*In the figure, ABCD is a rectangle of length 24 cm and breadth 16 cm. Given that $CQ = PB = \frac{1}{2}PQ$, calculate the area of the trapezium PQRS.*

### Solution

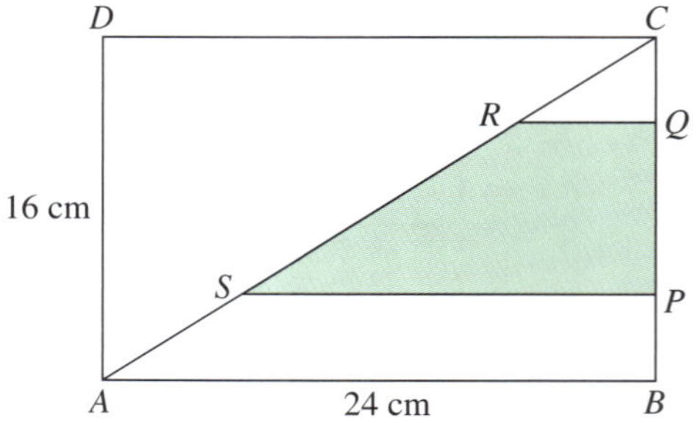


**Strategy 1: Use the formula**

$PQ = 8$ cm, $RQ = \frac{1}{4}AB = 6$ cm and $SP = \frac{3}{4}AB = 18$ cm

Using the formula $\frac{1}{2} \times$ height $\times$ sum of parallel sides

= area of trapezium,

area of $PQRS = \frac{1}{2} \times 8 \times (6 + 18)$

$= 96 \text{ cm}^2$

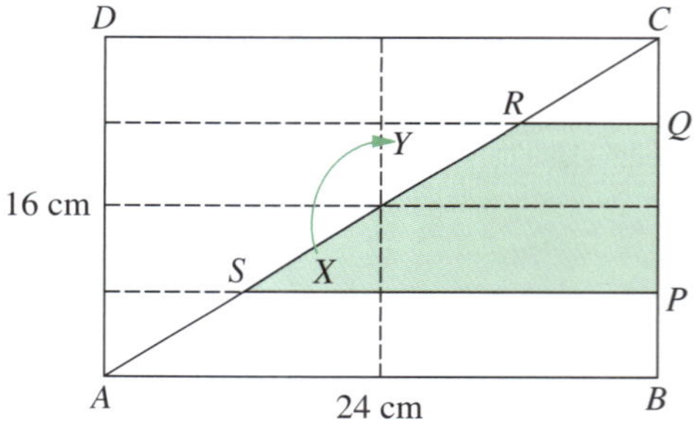


**Strategy 2: Draw a diagram**

Divide the rectangle into 8 equal parts as shown. If we move the shaded triangle $X$ onto $Y$, the total shaded area is equal to $\frac{1}{4}$ of the big rectangle.

$\therefore$ area of $PQRS = \frac{1}{4} \times 16 \times 24$

$= 96 \text{ cm}^2$

## Example 13

*The diagram shows a trapezium ABCD with AD // BC, POD is a diameter of the semi-circle with centre O. Given that AB = 21 cm, BC = 34 cm, AP = 5 cm and PD = 18 cm, calculate the area of the shaded region. Take π = 3.142 and give your answer correct to 2 decimal places.*

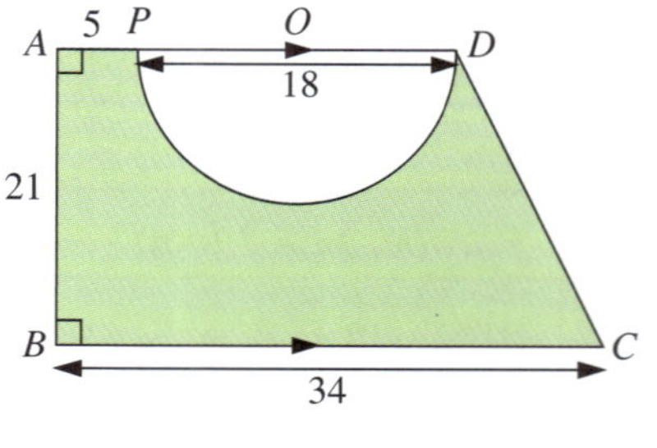


### Solution

Area of shaded region = area of trapezium – area of semicircle

$= \frac{1}{2} \times 21 \times [(5 + 18) + 34] - \frac{1}{2} \times \pi\left[\frac{18}{2}\right]^2$

$= \frac{1}{2} \times 21 \times 57 - \frac{1}{2} \times 3.142 \times 9^2$

$= 471.25 \text{ cm}^2$ (correct to 2 dec. pl.)

*Which area is larger and by how much — a half-km square or a half-square km?*

## Exercise 8c

For this exercise, take $\pi = 3.142$ and give all your answers correct to 2 decimal places where necessary.

1. Find the area of the shaded region in each of the following figures. All units are in cm.

**(a)**

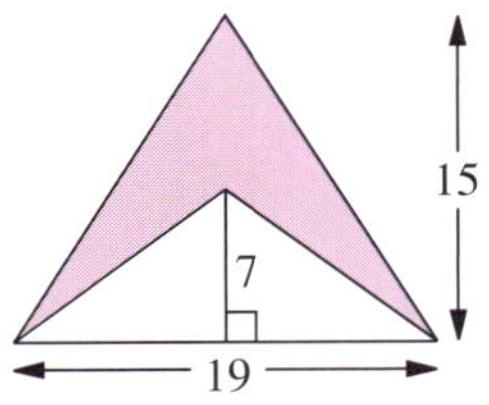


**(b)**

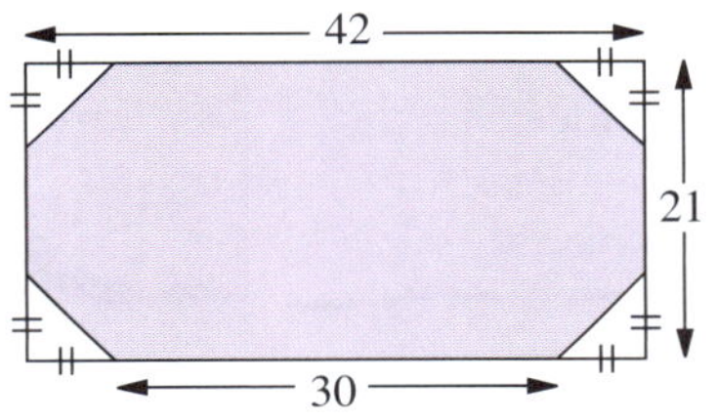


**(c)**

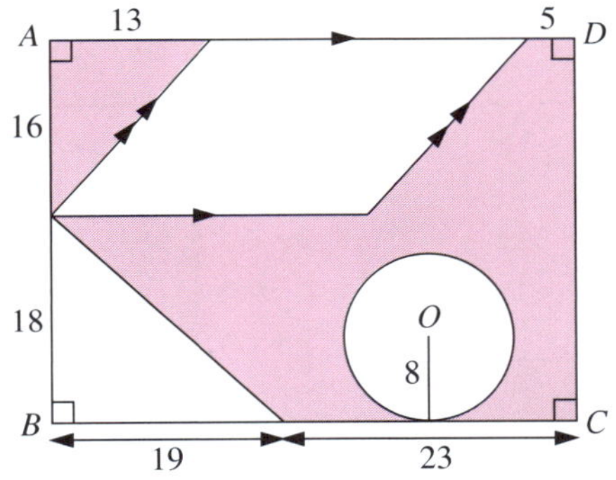


**(d)**

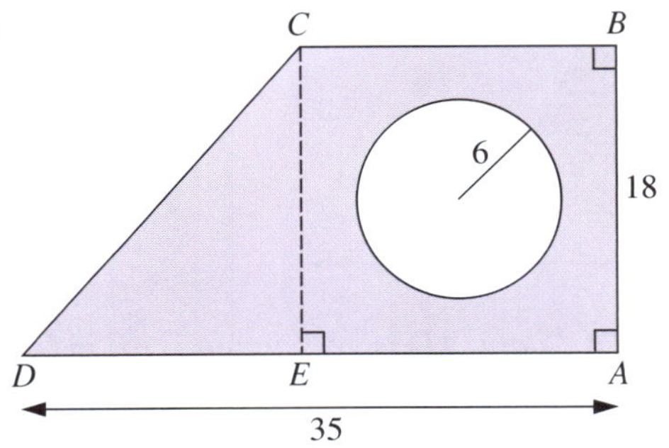


*ABCE* is a square

**(e)**

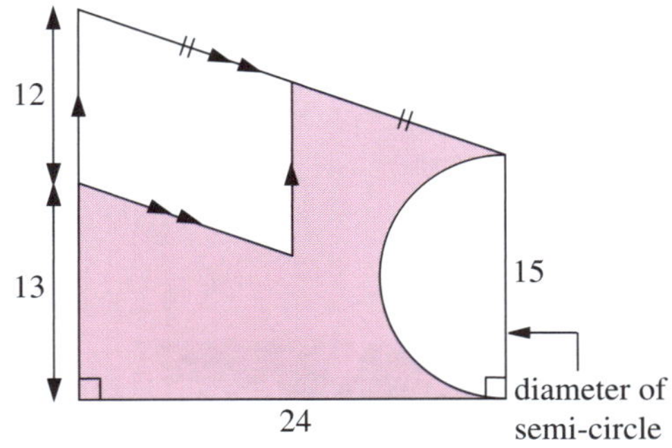


**(f)**

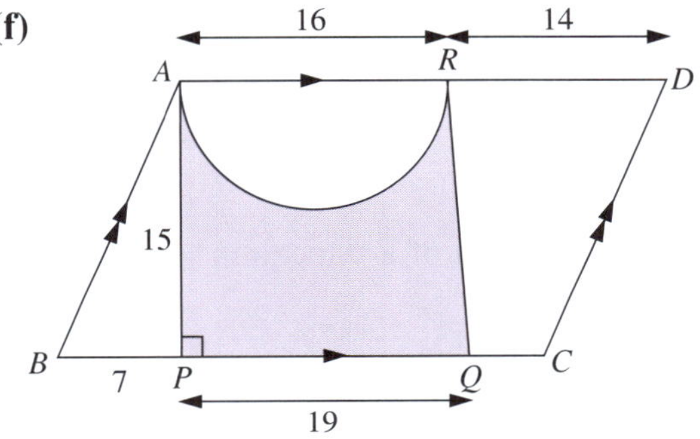


2. In the diagram, *ABCD* is a parallelogram, *BFE* and *CDE* are straight lines. If the area of the parallelogram *ABCD* is 80 cm$^2$, $DE = CD$ and $AF = FD$, find the area of
   **(a)** $\triangle BCE$
   **(b)** $\triangle ABF$
   **(c)** $\triangle EFC$

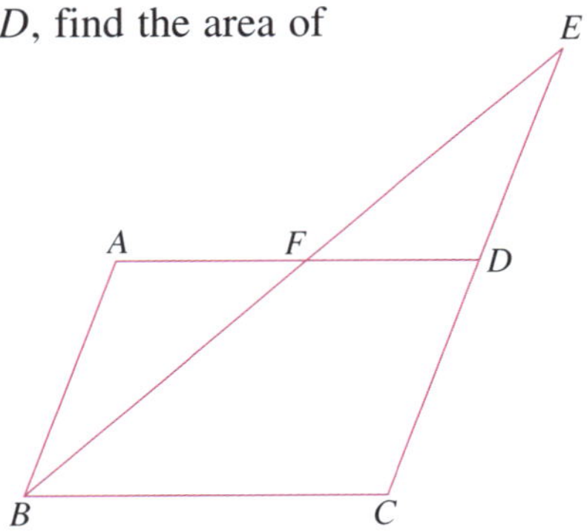


3. Find the difference between the perimeter of a square of area 1 m$^2$ and the circumference of a circle of the same area.

4. Find the speed of a point on the rim of a 24-cm diameter fly-wheel which is turning at 2800 revolutions per minute. Give your answer in metres per second.

**5.** A lorry travels at 50 km/h. Given that the diameter of its wheel is 88 cm, find how many revolutions per minute the wheel is turning. Give your answer to the nearest whole number.

**6.** The figure shows a rectangular cardboard with 4 semicircles being cut off. Find the area of the remaining cardboard in terms of $r$.

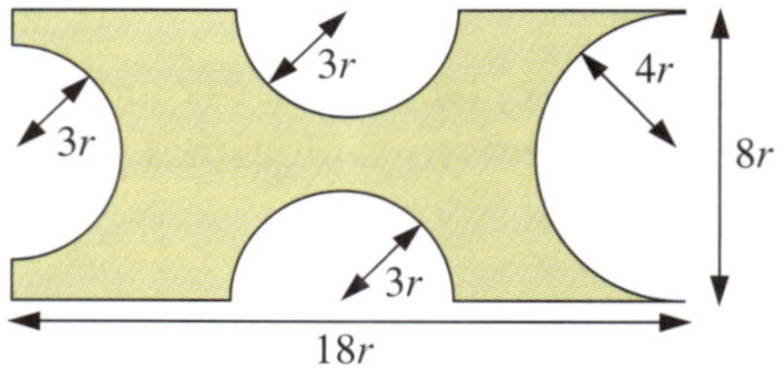


## SUMMARY

1. Area of a parallelogram = base × height

2. Area of a trapezium = $\frac{1}{2}$ × height × sum of parallel sides

## Review Examples 1

### Example 1

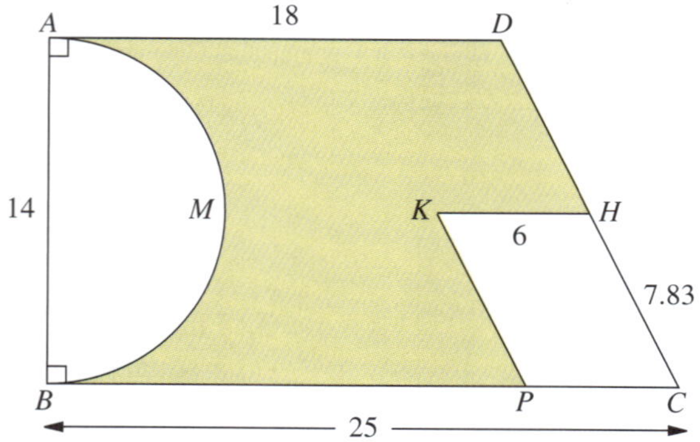


*In the diagram, ABCD is a trapezium with AB = 14 cm, BC = 25 cm, AD = 18 cm and* $D\hat{A}B = C\hat{B}A = 90°$. *AB is the diameter of the semicircle and PCHK is a parallelogram with CH = 7.83 cm, HK = 6 cm and H is the mid-point of CD. Calculate the perimeter and area of the shaded region. (Take* $\pi = 3.142$*)*

**Solution**

The perimeter of the shaded region = $BP + PK + KH + HD + AD + AMB$

$BP = 25 - 6 = 19$ cm, $HD = 7.83$ cm and $AMB = (\pi \times 7)$ cm

height of parallelogram $PCHK = 7$ cm

$$\therefore \quad \text{perimeter} = 19 + 7.83 + 6 + 7.83 + 18 + 7\pi \approx 80.7 \text{ cm}$$

Area of shaded region = area of $ABCD$ – area of semicircle – parallelogram $PCHK$

$$= \frac{1}{2}(14)(18 + 25) - \frac{1}{2} \times \pi \times 7^2 - 6 \times 7$$
$$\approx 182 \text{ cm}^2$$

## Example 2

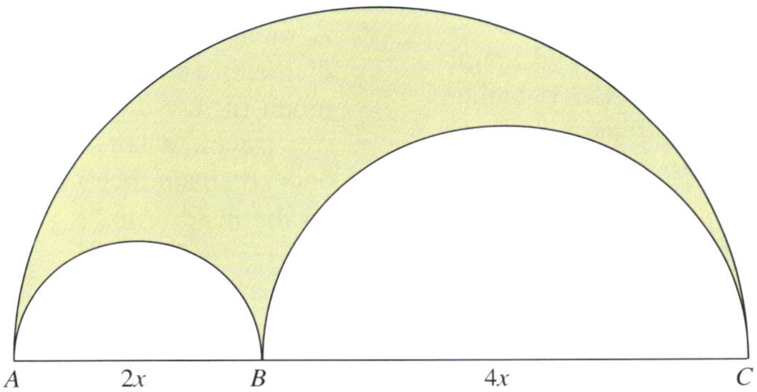


*The diagram shows three semicircles, the diameter of the smallest semicircle is 2x cm, that of the medium semicircle is 4x cm and AC is the diameter of the largest semicircle. Find the perimeter and area of the shaded region in terms of x and π.*

**Solution**

Radius of the largest semicircle = $3x$ cm

$$\text{The perimeter of the shaded region} = \frac{1}{2}(\text{circumference of the three semicircles})$$
$$= \frac{1}{2}(\pi \times 2x + \pi \times 4x + \pi \times 6x)$$
$$= 6\pi x \text{ cm}$$

$$\text{Area of the shaded region} = \text{area of largest semicircle} - \text{area of the 2 smaller semicircles}$$
$$= \frac{1}{2}\pi(3x)^2 - \frac{1}{2}\pi(2x)^2 - \frac{1}{2}\pi(x)^2$$
$$= \frac{9}{2}\pi x^2 - 2\pi x^2 - \frac{1}{2}\pi x^2$$
$$= 2\pi x^2 \text{ cm}^2$$

# Review Questions 1

Take $\pi$ to be $\frac{22}{7}$ for this exercise.

1. $\triangle APQ$ is enclosed within the rectangle $ABCD$ as shown in the figure below. Calculate the area of $\triangle APQ$.

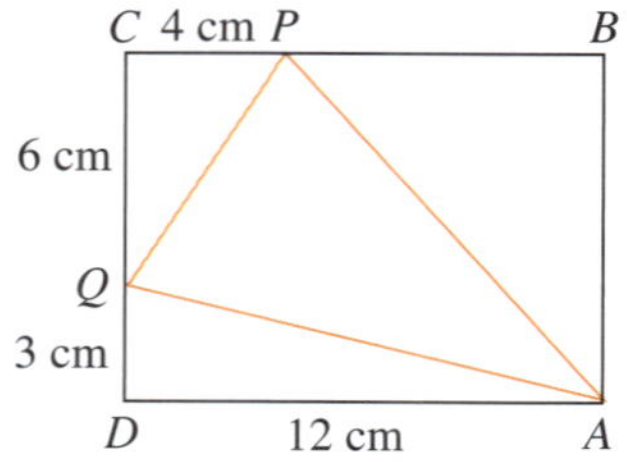


2. In the figure, $ABCD$ is a rectangle of length 8 cm and breadth 6 cm. If $BP = CP$, calculate the area of trapezium $ABPQ$, where $AQC$ is a diagonal of the rectangle.

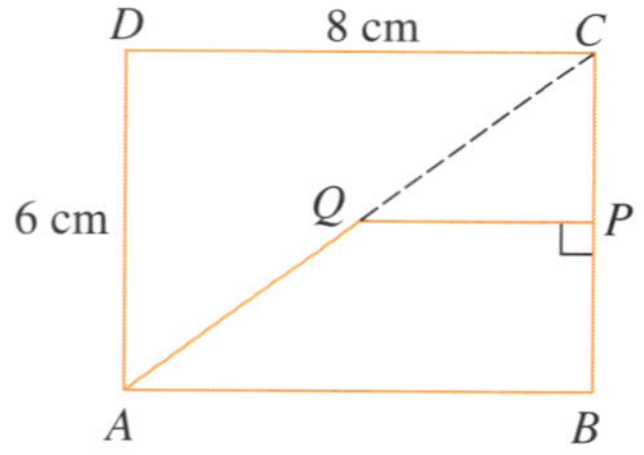


3. A piece of wire 48 cm long is bent to form a rectangle whose length is twice its width. Calculate its area.

4. The length of a rectangle is 4 cm longer than its width and its perimeter is 44 cm. Find the length and area of the rectangle.

5. A rectangular driveway 12 m long and $4\frac{1}{2}$ m wide is to be covered by similar square tiles of side 25 cm each. Find the number of tiles needed to cover the driveway.

6. The area of a trapezium is 36 $cm^2$ and the perpendicular distance between its parallel sides is 6 cm. If the lengths of these parallel sides are $x$ cm and $y$ cm, find the value of $(x + y)$. Given further that $x$ is twice as big as $y$, find the values of $x$ and $y$.

7. A bucket of water is brought up from a well 9.68 m deep by a rope which winds round a drum 22 cm in diameter. How many turns of the handle are required to bring up a bucket from the bottom of the well?

8. A racing track is a circular ring with inner diameter 140 m and track 7 m wide. How much further does a motorist on the outside rim travel, when he goes round the circuit once, than another who goes round the circuit on the inside rim?

9. The diagram shows three semicircles. Calculate the perimeter and area of the shaded region.

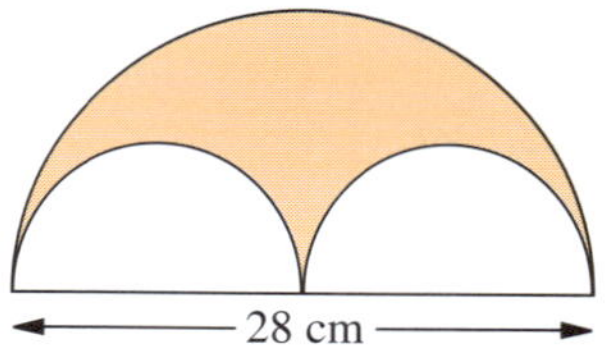


10. Kumar walks round a rectangular field, the length of which is twice its width. He then walks round another rectangular field half as wide but having the same perimeter as the first field. If the difference in area between the two fields is 432 $m^2$, find the length of the second field.

**⋆11.** The diagram shows 4 circles of equal radius touching each other. If the radius of each circle is 12 cm, calculate the area of the shaded region.

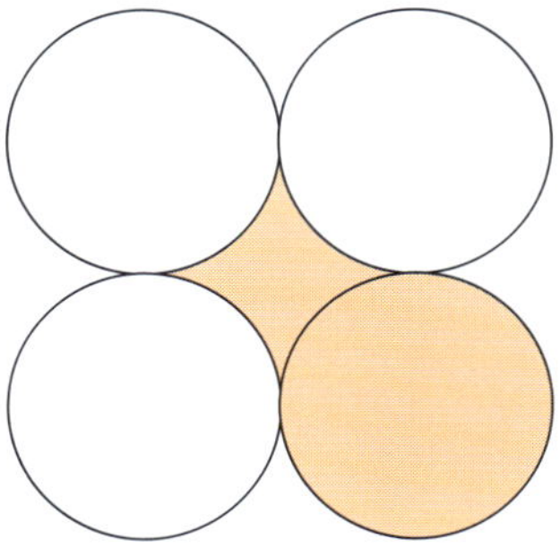

*In this chapter, you will learn how to*

- ***find the volumes and surface areas of cubes, cuboids, prisms and cylinders;***
- ***solve problems involving volumes made up of the above solids.***

CHAPTER

# 9

# Volume and Surface Area

## Introduction

The picture shows a plot of land being cleared for the construction of infrastructure for a new township. The contractor has deliberately left some heaps of soil behind. Do you know that the purpose of this is to estimate the volume of soil taken from the site?

In primary school, you learned about cubes and cuboids and, how to calculate their volumes. We shall have a quick revision of these before proceeding to find the volumes and surface areas of prisms and cylinders.

The following is a summary of what was covered in primary school.

| **Name** | **Figure** | **Volume** | **Total surface area** | **Sample Nets** |
|---|---|---|---|---|
| Cube | 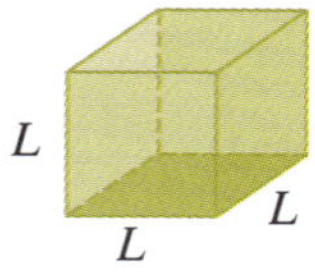 | $L^3$ | $6L^2$ | 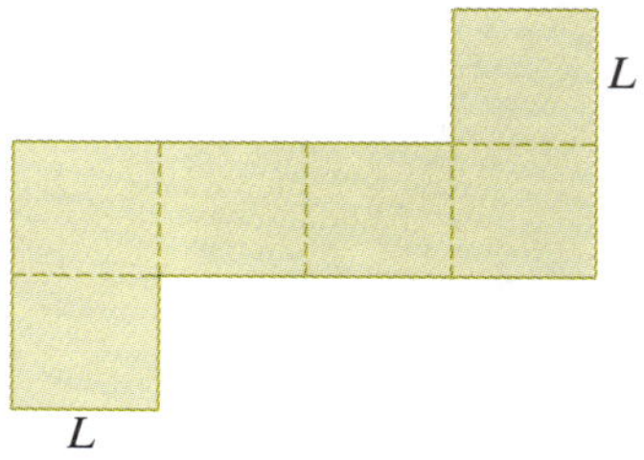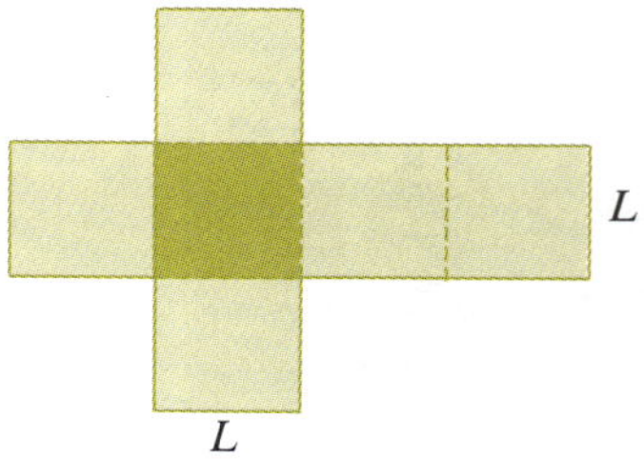 |
| Cuboid | 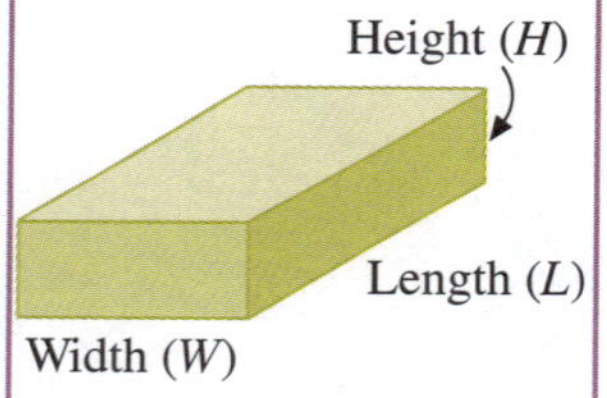 | $L \times W \times H$ | $2(L \times W + L \times H + W \times H)$ | 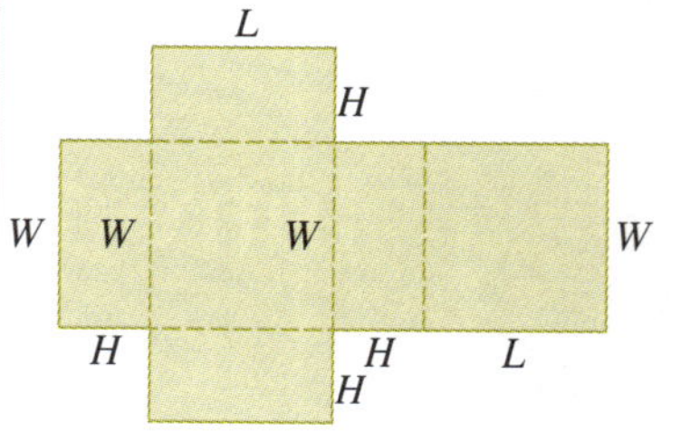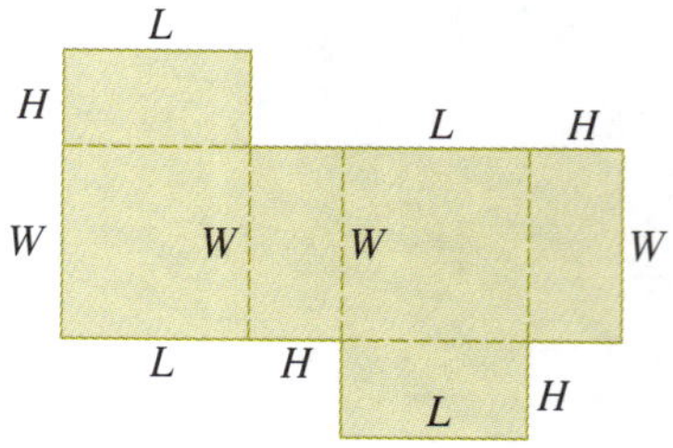 |

# Concept of Volume

Fig. 9.1 shows a piece of wood, a brick and a matchbox.

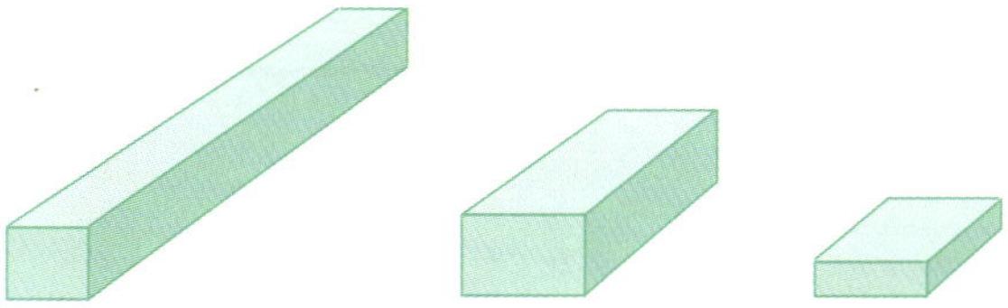


*Fig. 9.1*

Which one of them occupies the least space? Obviously, the matchbox occupies the least space. But which of the other two, the piece of wood or the brick, occupies more space? To answer this question, we first have to make some measurements and then obtain the volume of each object.

The volume of an object is the amount of space it occupies. The object that occupies more space is said to have a greater volume.

As with the case of the area of a plane figure, we compare the volume of an object with a standard unit. A standard unit for volume is a cube with side 1 cm (see Fig. 9.2). We call this 1 cubic centimetre, written as 1 $cm^3$.

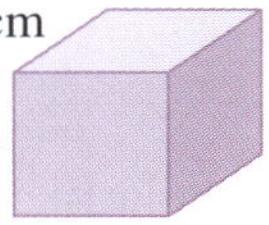


*Fig. 9.2*

Similarly, a cube with side 1 mm has a volume of 1 $mm^3$ and that with side 1 m has a volume of 1 $m^3$.

# Volume of Fluids

The volume of a fluid or liquid is measured using special units. These units are the millilitre (m*l*), the litre (*l*) and the kilolitre (k*l*). Normally, we buy milk and petrol by the litre and we take medicine by the millilitre.

$$1 \text{ m}l = 1 \text{ cm}^3$$
$$1 \text{ litre} = 1000 \text{ m}l = 1000 \text{ cm}^3$$
$$1 \text{ kilolitre} = 1000 \text{ litres} = 1 \text{ m}^3$$

## Explorations

Find out the names and capacities of the largest and the smallest reservoirs in Singapore. Also, find out the average daily consumption of water in Singapore in 2006.

## Example 1

*Express (a) 1 $cm^3$ in $mm^3$ and (b) 1 $m^3$ in $cm^3$.*

(a) Since 1 cm = 10 mm, a cube with side 10 mm has a volume 1 $cm^3$.

i.e., 1 $cm^3$ = (10 × 10 × 10) $mm^3$
= 1000 $mm^3$

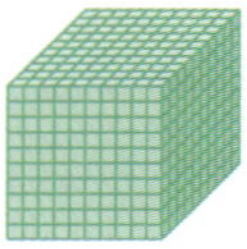

(b) Similarly, since 1 m = 100 cm.

1 $m^3$ = (100 × 100 × 100) $cm^3$
= 1 000 000 $cm^3$

## Example 2

*Express (a) 3 600 000 $mm^3$ in (i) $cm^3$ and (ii) ml;*
*(b) 0.7 $m^3$ in (i) $cm^3$ and (ii) litres.*

### Solution

(a) (i)
$$10 \text{ mm} = 1 \text{ cm}$$
$$1000 \text{ mm}^3 = 1 \text{ cm}^3$$
$$3\,600\,000 \text{ mm}^3 = \frac{3\,600\,000}{1000} \text{ cm}^3 = 3600 \text{ cm}^3$$

(ii)
$$1 \text{ cm}^3 = 1 \text{ m}l$$
$$3600 \text{ cm}^3 = 3600 \text{ m}l$$

(b) (i)
$$1 \text{ m} = 100 \text{ cm}$$
$$1 \text{ m}^3 = 1\,000\,000 \text{ cm}^3$$
$$0.7 \text{ m}^3 = (0.7 \times 1\,000\,000) \text{ cm}^3 = 700\,000 \text{ cm}^3$$

(ii)
$$1000 \text{ cm}^3 = 1 \text{ litre}$$
$$700\,000 \text{ cm}^3 = \frac{700\,000}{1000} \text{ litres} = 700 \text{ litres}$$

## Example 3

*The figure below shows a rectangular cuboid 9 cm long and 7 cm wide. Given that the volume of the cuboid is 315 cm³, find*

*(a) the height of the cuboid;* *(b) its total surface area.*

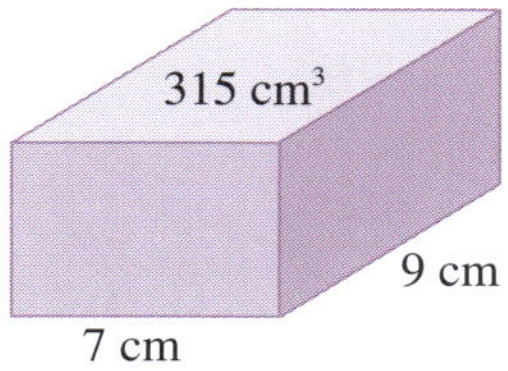


**Solution**

(a) $L = 9, W = 7, H = ?$

Volume of the cuboid $= L \times W \times H$

$$315 = 9 \times 7 \times H$$

$$H = \frac{315}{9 \times 7} = 5$$

$\therefore$ the height of the cuboid is 5 cm.

(b) The surface area of the cuboid $= 2(9 \times 7 + 9 \times 5 + 7 \times 5)\ \text{cm}^2$

$= 286\ \text{cm}^2$

## Explorations

*The diagram below shows the net of a cuboid.*

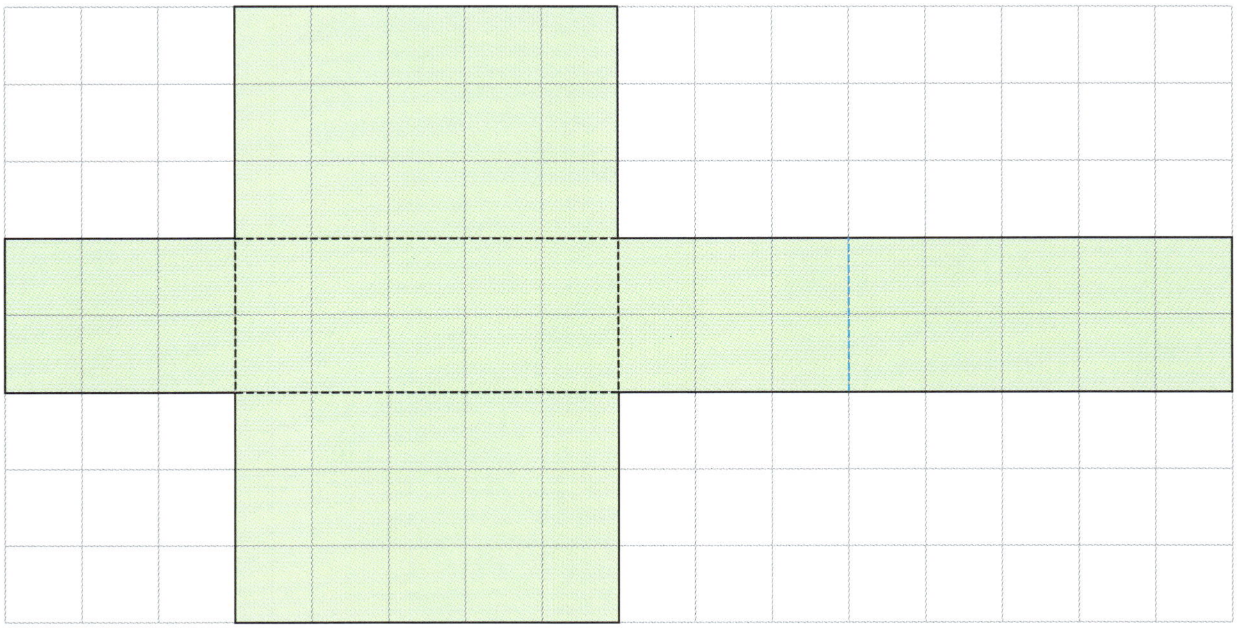

It can be folded to form one of the following cuboids.

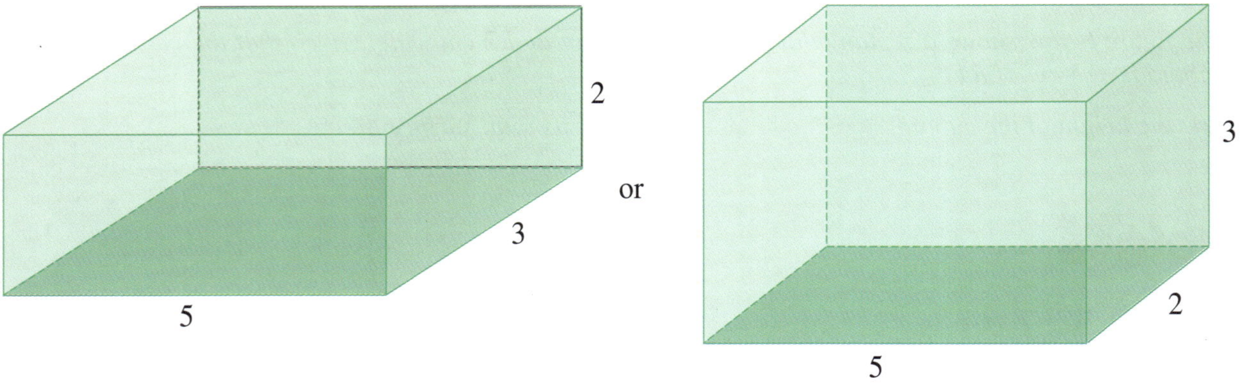


In each case the volume of the cuboid = $(5 \times 3 \times 2)$ units$^3$
$= 30$ units$^3$

The following diagrams show the nets of other cuboids. Sketch each cuboid and find its volume.

**(a)**

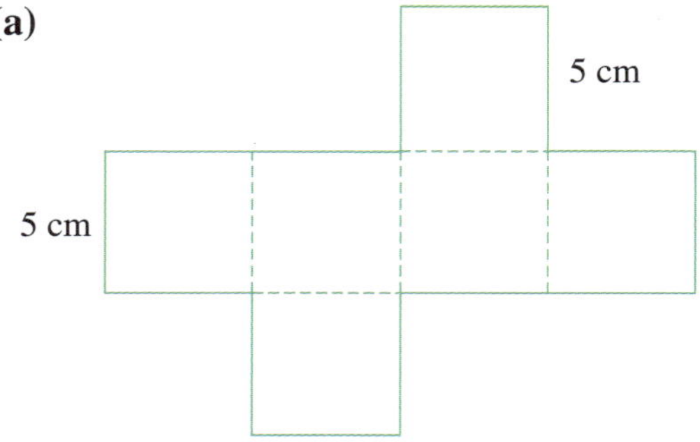


**(b)**

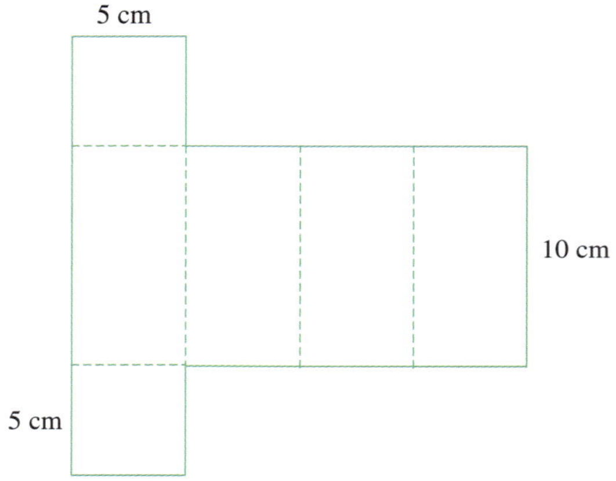


**(c)**

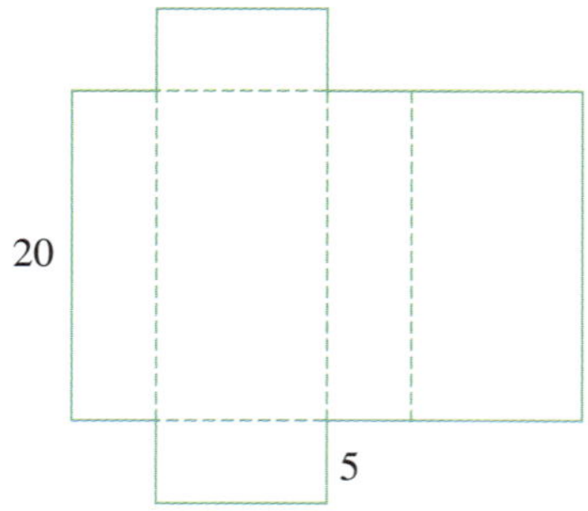


**(d)**

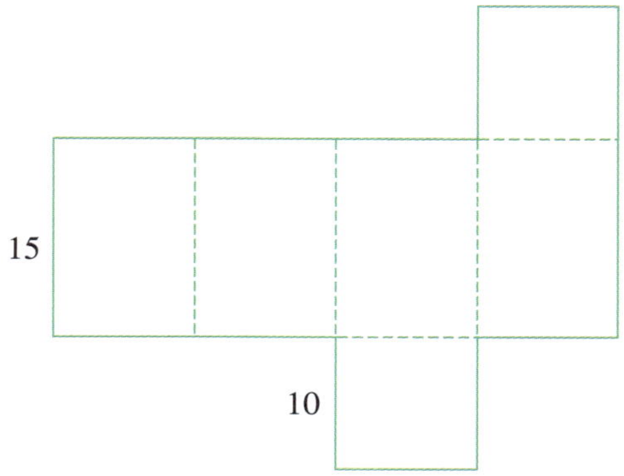

## Example 4

*An open rectangular tank of length 16 cm, width 9 cm and height 13 cm contains water up to a height of 8 cm. Calculate*

*(a) the volume of water in litres;*
*(b) the total surface area of the tank that is in contact with the water.*

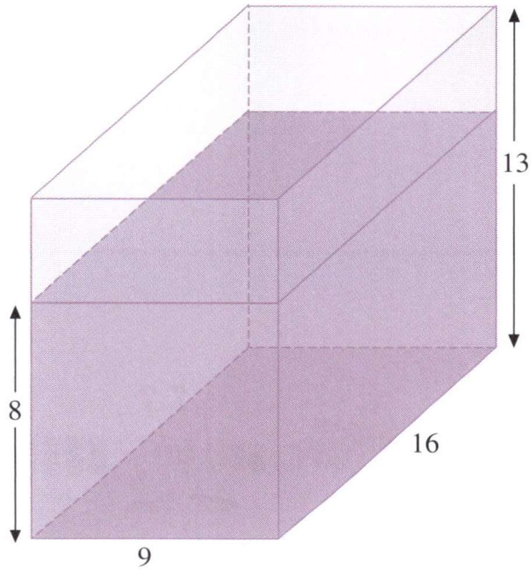


### Solution

**(a)** The volume of water $= (16 \times 9 \times 8)\ \text{cm}^3 = 1152\ \text{cm}^3$
$= 1.152$ litres

**(b)** The surface area of the tank that is in contact with the water includes the 4 walls and the base.

Total surface area $= (9 \times 16) + 2(9 \times 8 + 16 \times 8)\ \text{cm}^2$
$= 544\ \text{cm}^2$

## Explorations

A cube of side 3 cm is painted green on all its 6 faces. It is to be cut into twenty-seven 1-cm cubes. How many cuts do you need to make?

How many small cubes have
(1) no painted faces at all;
(2) 1 face painted green;
(3) 2 faces painted green;
(4) 3 faces painted green;
(5) 4 faces painted green?

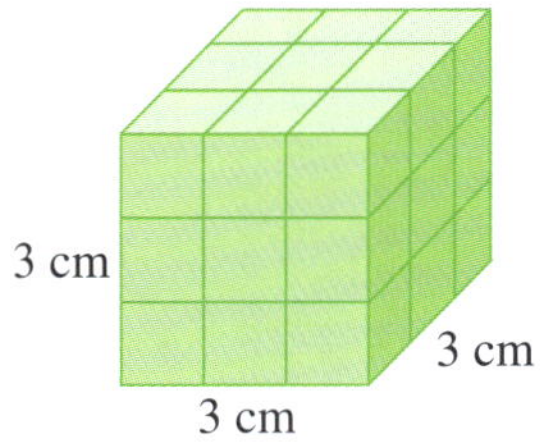

You are now given a 4-cm cube which is also painted green on all its faces. How many cuts do you need to make to reduce it to sixty-four 1-cm cubes?

How many small cubes have
(1) 4 faces painted green;
(2) 3 faces painted green;
(3) 2 faces painted green;
(4) 1 face painted green;
(5) no painted faces at all?

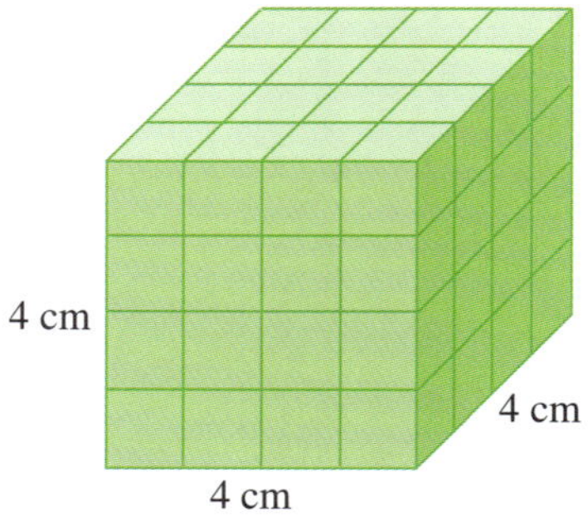


## Exercise 9a

1. Find the volume and surface area of each of the following cuboids:

(a)
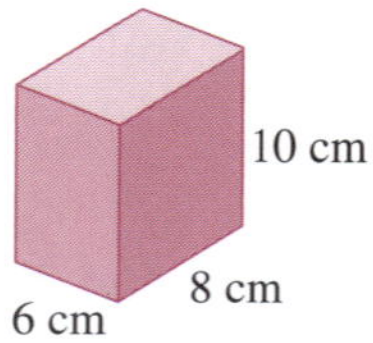


(b)
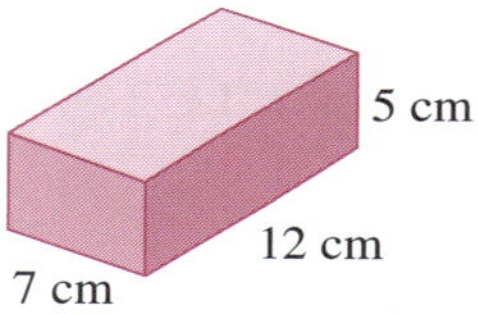


(c)
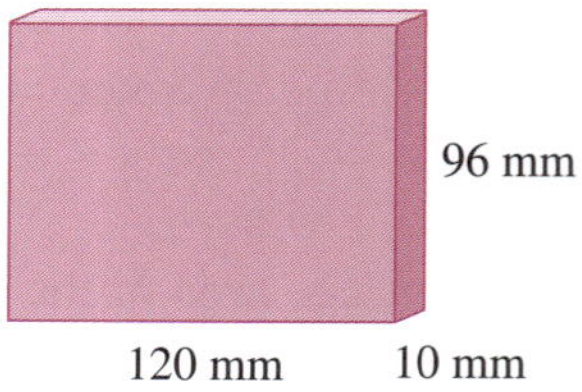


(d)
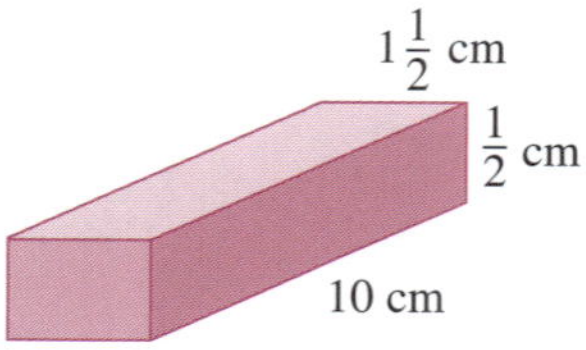


(e)
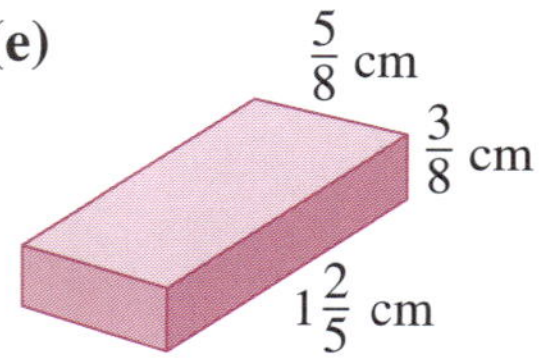


(f)
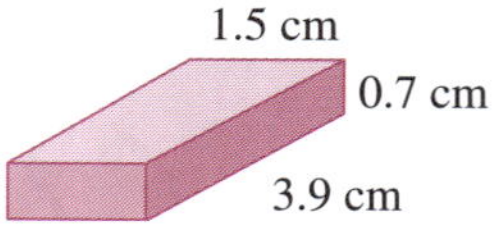

2. Copy and complete the following table for each cuboid:

| | *Length* | *Width* | *Height* | *Volume* | *Surface Area* |
|---|---|---|---|---|---|
| **(a)** | 24 mm | 18 mm | 5 mm | | |
| **(b)** | 5 cm | 3 cm | | 120 $cm^3$ | |
| **(c)** | | 6 cm | $3\frac{1}{2}$ cm | 52.5 $cm^3$ | |
| **(d)** | 12 m | | 6 m | 576 $m^3$ | |
| **(e)** | $2\frac{1}{4}$ cm | 8 cm | | $58\frac{1}{2}$ $cm^3$ | |
| **(f)** | 9 cm | 12 cm | | | 426 $cm^2$ |

3. Find the capacity of each of the following rectangular tanks, giving your answer in litres:
   **(a)** Height = 3.6 m, length = 5.5 m, width = 3.5 m.
   **(b)** Height = 2.7 m, length = 4.75 m, width = 2.6 m.
   **(c)** Height = 0.15 m, length = 0.24 m, width = 0.19 m.
   **(d)** Height = 38 cm, length = 52 cm, width = 18 cm.

4. Find the total surface area of a solid cube of volume 64 $cm^3$.

5. An open water tank with length 20 cm and width 15 cm holds 4.8 litres of water. Calculate the height of the water level in the tank and the total surface area of the cuboid in contact with the water.

6. A rectangular tank measures 4 m long, 2 m wide and 4.8 m high. Initially it is half filled with water. Find the depth of water in the tank after 4000 litres more of water are added to it.

7. A rectangular water tank of length 60 cm and width 40 cm contains water up to a depth of 30 cm. A piece of ice measuring 20 cm by 15 cm by 12 cm is dropped into the tank of water. Calculate the new depth of water when the ice melts completely, assuming its volume decreases by $\frac{1}{10}$.

8. It took two and a half years and 2.85 million $m^3$ of earth to fill the disused Sin Seng quarry at Rifle Range Road.
   **(a)** If each truck can carry a maximum load of 6.25 $m^3$ of earth per trip, how many trips are needed to fill the entire quarry?
   **(b)** If the cost of transport, material and administration for each truck load is $55, how much would it cost to fill the quarry?
   The quarry site now provides an area of approximately 3 hectares for future development. Calculate the cost of one $m^2$ of the land. (1 hectare = 10 000 $m^2$)

9. In November 1998, the government announced in Parliament a $10.5 billion package to help the country overcome the Asian economic crisis. If the $10.5 billion is to be issued in $2 notes, what will be the volume of all the $2 notes, assuming that a $2-note has a length of 13.3 cm, a width of 6.4 cm and a thickness of 0.15 mm. Give your answer in $m^3$.

# Right Prisms

Cut out a large number of identical triangles from a piece of cardboard and pile them up as shown in Fig. 9.3. A solid is formed. This solid is called a triangular prism. The two parallel planes, $PQR$ and $P'Q'R'$, are triangular in shape and the *triangular prism* takes its name from these planes.

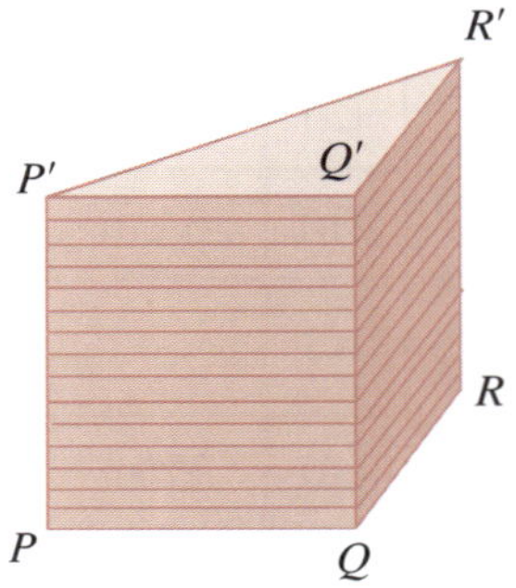


*Fig. 9.3*

A cuboid is a prism with rectangular planes (see Fig. 9.4). Hence, it is called a rectangular prism.

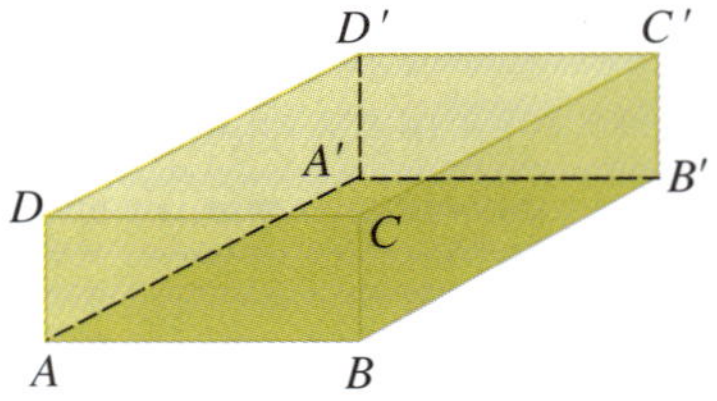


*Fig. 9.4*

Notice that, in Fig. 9.3, the other three surfaces, which are called the lateral surfaces of the triangular prism, are all rectangular, and that $PP'$, $QQ'$ and $RR'$ are all perpendicular to the planes $PQR$ and $P'Q'R'$. Similarly, in Fig. 9.4, the four lateral surfaces of the cuboid are rectangular and $AA'$, $BB'$, $CC'$ and $DD'$ are perpendicular to the planes $ABCD$ and $A'B'C'D'$. These prisms are called **right prisms**.

Fig. 9.5 and Fig. 9.6 show a right pentagonal and a hexagonal prism respectively.

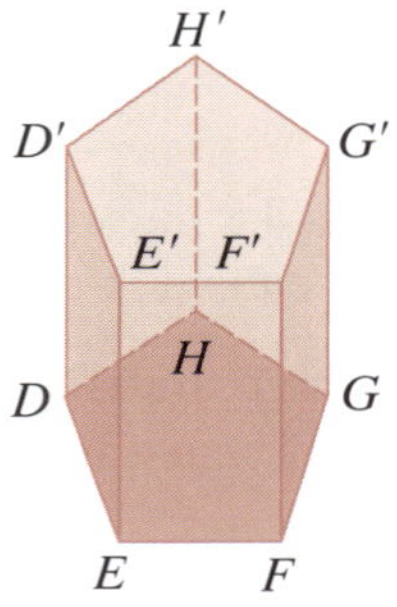


*Fig. 9.5*

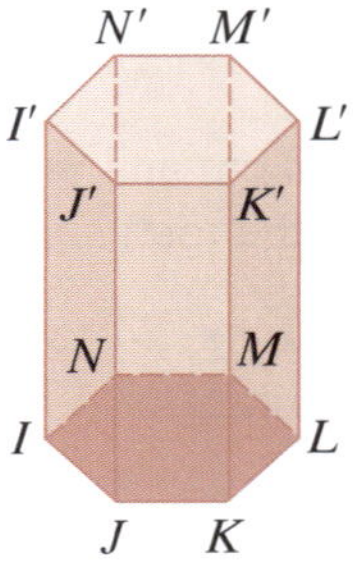


*Fig. 9.6*

In general, a right prism is a solid which has two parallel planes of the same shape and size. Also, its lateral surfaces are perpendicular to its parallel ends.

A right prism has a **uniform cross-section**, i.e., the cross-section of the prism is identical to the two parallel ends (see Fig. 9.7).

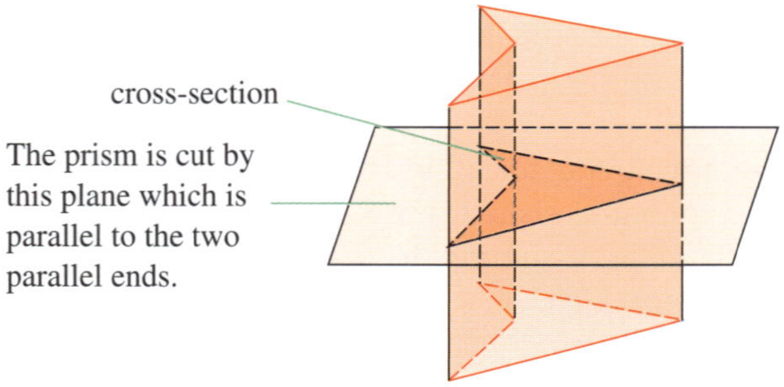


*Fig. 9.7*

# Volume of a Prism

The three prisms shown in Fig. 9.8 are obtained by stacking up a large number of respective identical shapes cut out from cardboard.

*Fig. 9.8*

The volume of the right rectangular prism or cuboid
= area of base × height
= area of an identical cardboard × height of rectangular stack
= area of rectangular cross-section × distance between parallel rectangular ends

## Explorations

(1) Show how we can get:
The volume of a right triangular prism
= area of triangular cross-section × distance between parallel triangular ends

(2) Try to obtain a similar formula as the ones above for the volume of a hexagonal prism.

In general, for a right prism, the volume is given by

**volume = area of cross-section × distance between parallel ends**
**= base area × height**

# Surface Area of a Prism

Fig. 9.9 shows a right prism whose base is a polygon.

Let $A$ denote the surface area of the prism. Suppose the height of the prism is $H$ and the lengths of the sides of the base are $L_1$, $L_2$, $L_3$, $L_4$, $L_5$ and $L_6$.

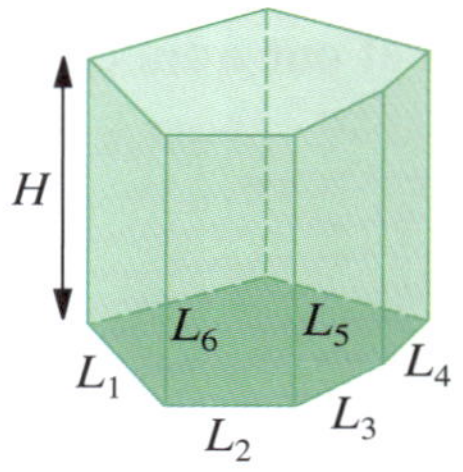


*Fig. 9.9*

A net of the prism is shown on the right.

The dotted lines indicate the folds.

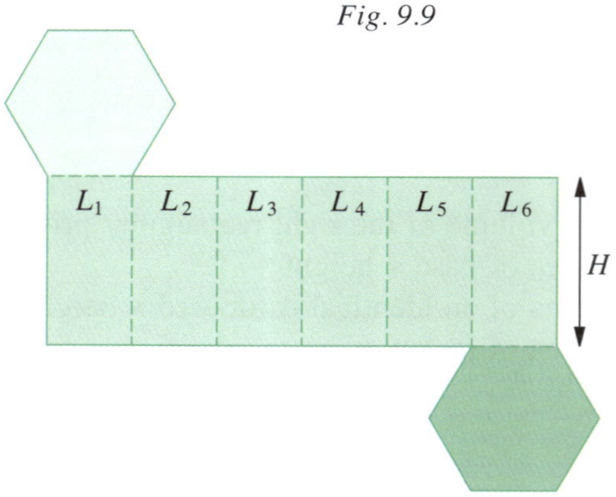


The area, $A$, is given by

$$A = L_1H + L_2H + L_3H + L_4H + L_5H + L_6H + 2(\text{base area})$$
$$= (L_1 + L_2 + L_3 + L_4 + L_5 + L_6)H + 2(\text{base area})$$
$$= \text{perimeter of base} \times \text{height} + 2(\text{base area})$$

In general,

**surface area of a right prism = perimeter of the base × height + 2(base area)**

## Example 5

*Draw a net of the right prism shown on the right and then find its volume and surface area.*

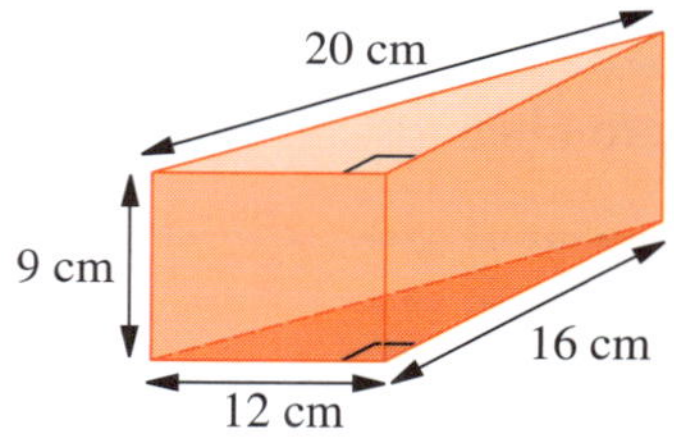


**Solution**

Here is a net of the right prism.

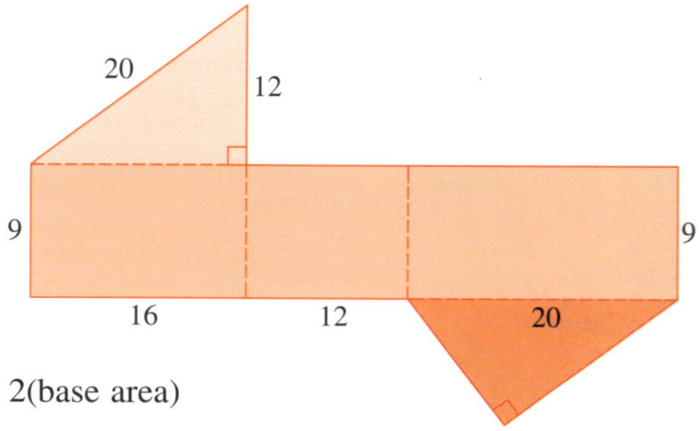


Area of the base $= \left(\frac{1}{2} \times 12 \times 16\right)$ cm$^2$ = 96 cm$^2$

Volume of the solid = area of the base × height
= $(96 \times 9)$ cm$^3$ = 864 cm$^3$

Perimeter of the base = $(12 + 16 + 20)$ cm = 48 cm

Total surface area = perimeter of the base × height + 2(base area)
= $[48 \times 9 + 2(96)]$ cm$^2$ = 624 cm$^2$

## Example 6

*Find the volume and surface area of the right pentagonal prism shown.*

**Solution**

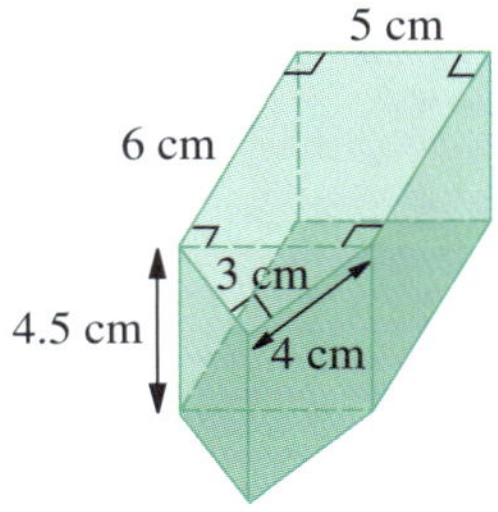


Area of the base = $\left[(6 \times 5) + \left(\frac{1}{2} \times 3 \times 4\right)\right]$ cm$^2$ = 36 cm$^2$

$\therefore$ volume of the prism = $(36 \times 4.5)$ cm$^3$
= 162 cm$^3$

Area of the lateral surfaces = perimeter of the base × height
= $[(6 + 5 + 6 + 4 + 3) \times 4.5]$ cm$^2$
= 108 cm$^2$

$\therefore$ total surface area of the prism = $[108 + 2(36)]$ cm$^2$
= 180 cm$^2$

## Example 7

*Find the volume of wood used in making an open rectangular box 2 cm thick, given that its internal dimensions are 54 cm long, 46 cm wide and 18 cm deep.*

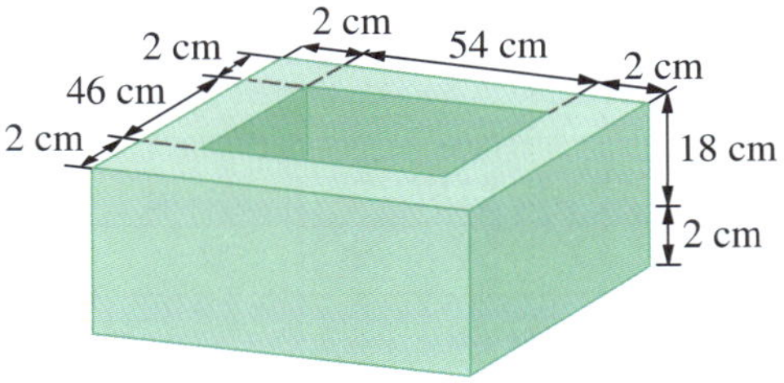


**Solution**

External length = $(54 + 2 + 2)$ cm = 58 cm

External breadth = $(46 + 2 + 2)$ cm = 50 cm

External height = $(18 + 2)$ cm = 20 cm

External volume = $(58 \times 50 \times 20)$ cm$^3$ = 58 000 cm$^3$

Internal volume = $(54 \times 46 \times 18)$ cm$^3$ = 44 712 cm$^3$

$\therefore$ volume of wood used = $(58\ 000 - 44\ 712)$ cm$^3$
= 13 288 cm$^3$

# Exercise 9b

**1.** Find the volumes of the following right prisms.

**(a)**

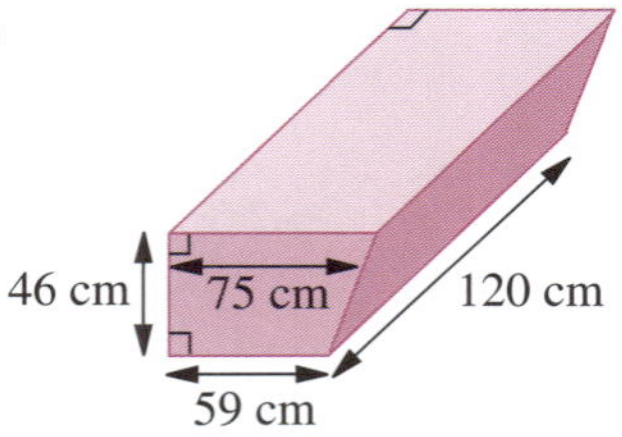


**(b)**

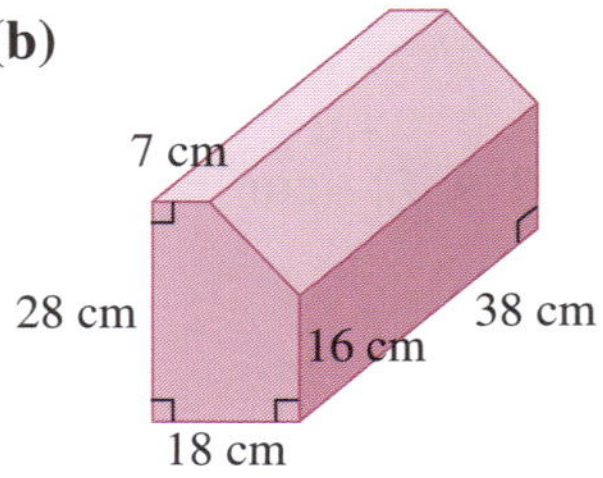


**(c)**

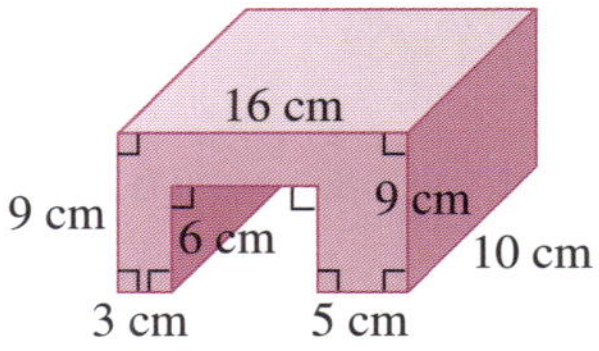


**(d)**

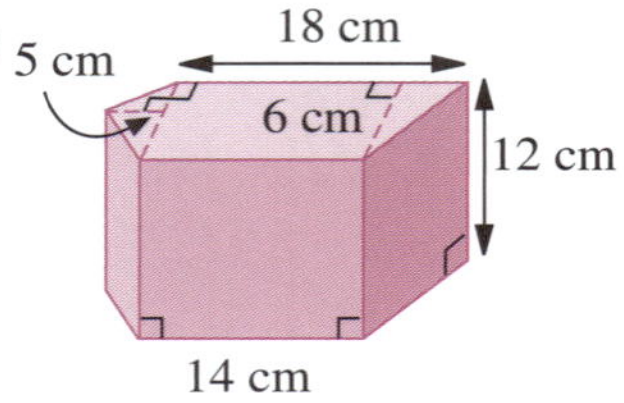


**(e)**

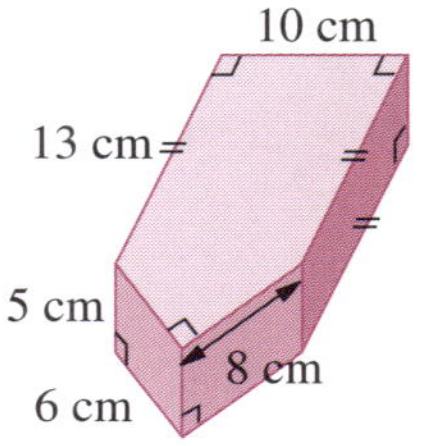


**(f)**

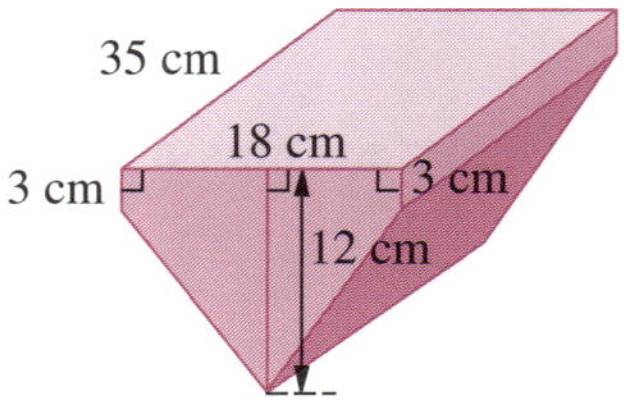


**2.** The figure shows a right prism standing on a horizontal, rectangular base *BCDE*. The triangle *ABC* is a vertical cross-section of the solid prism.

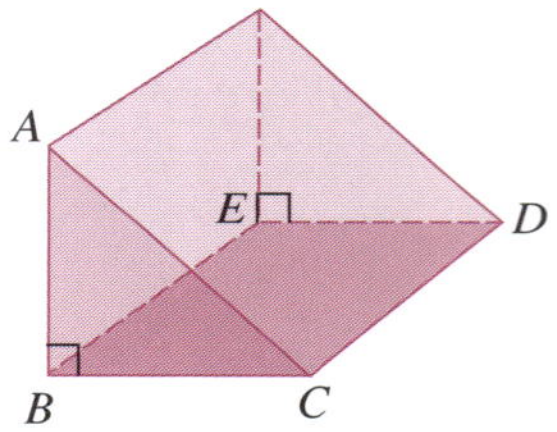


Copy and complete the table below:

| | *AB* | *BC* | *CD* | *Area of* $\triangle ABC$ | *Volume of prism* |
|---|---|---|---|---|---|
| **(a)** | 3 cm | 4 cm | 7 cm | | |
| **(b)** | 9 cm | | 11 cm | 63 $cm^2$ | |
| **(c)** | | 15 cm | 300 cm | | 72 000 $cm^3$ |
| **(d)** | 24.6 cm | 7.8 cm | | | 38 376 $cm^3$ |

**3.** Without taking into consideration the thickness of the walls and roof, find the air space in the hall with the dimensions given in the figure.

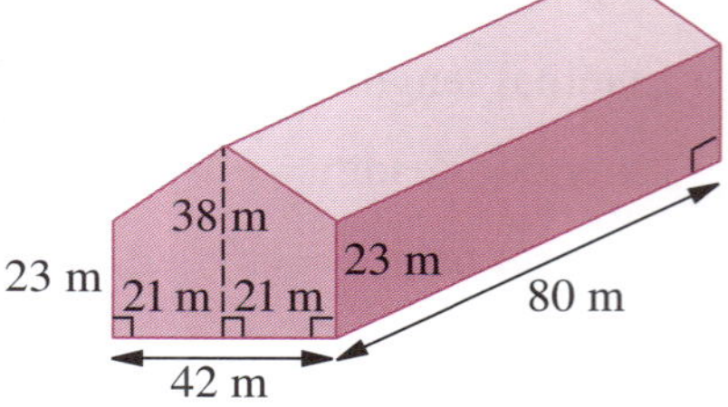


**4.** Find the volume and the surface area of the solid, which is in the shape of a right prism, as shown:

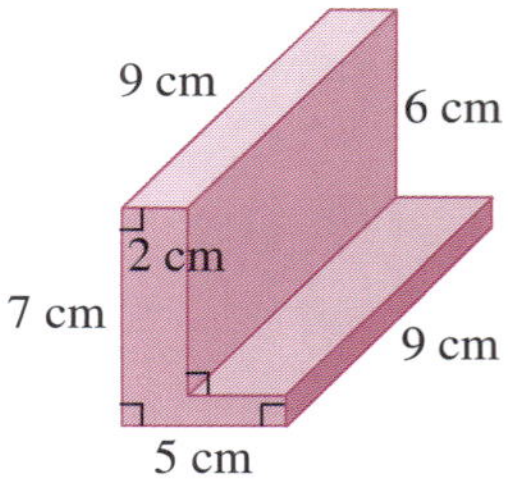

5. A swimming pool is 50 m long and 25 m wide. It is 1 m deep at the shallow end and 8 m deep at the other end. Find the volume of water in the pool when it is full and the total area of the pool which is in contact with the water (refer to figure).

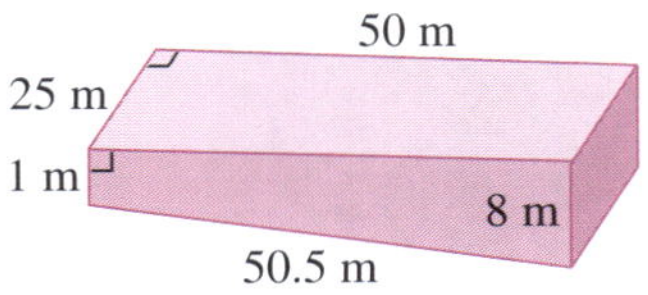


6. A closed box is 135 cm long, 80 cm wide and 60 cm deep internally. It is to be lined on its sides and bottom with cedar veneer of negligible thickness. Find, in square metres, the area of veneer needed.

7. The parallel ends of a right prism, 15 cm long, are isosceles triangles with measurements shown. Find
   **(a)** the volume;
   **(b)** the surface area of the prism.

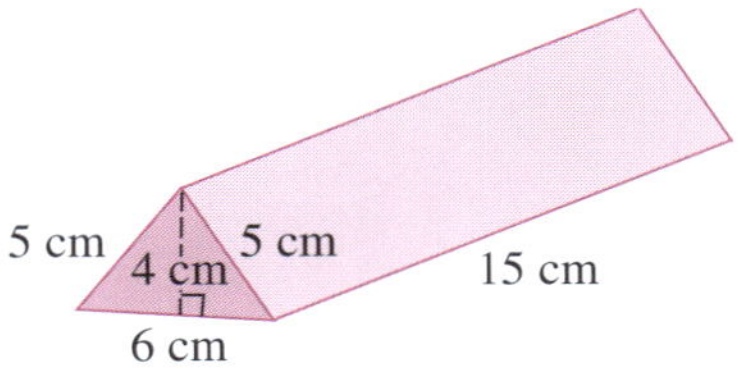


8. The internal dimensions of an open concrete tank are 1.8 m long, 0.8 m wide and 1.2 m high. Find the capacity of the tank in litres. If the concrete is 0.1 m thick, find also, in cubic metres, the volume of concrete used.

9. A trough, in the form of an open rectangular box, is 1.85 m long, 45 cm wide and 28 cm deep externally. If the trough is made of wood 2.5 cm thick, find, in cubic centimetres, the volume of wood required.

10. A tin which is 12 cm long, 9 cm wide and 4 cm deep holds 120 g of tea. If 1 kg of the same tea is packed into a tin which has a 12-cm square base, how tall will the tin have to be?

11. The cross-section of a drain is a rectangle 30 cm wide. If water 3.5 cm deep flows along the drain at a rate of 22 cm per second, how many litres of water will flow through each minute?

## Cylinders

We can form a cylindrical solid by vertically stacking up a pile of 50-cent coins as shown in the figure. This solid is called a right cylinder. Its cross-sectional area is a circle. Steel pipes, oil drums and many tin containers for liquids and canned food are all common examples of cylinders. Can you name other objects which are cylindrical in shape?

a stack of
50-cent coins

*Formally, the cylinder shown here is called a right circular cylinder. In this book, we use 'cylinder' to mean a 'right circular cylinder'.*

## Volume of a Cylinder

For a cylinder, we can find its volume by applying the same method used in finding the volume of a right prism,

i.e., **volume of a cylinder = base area × height**

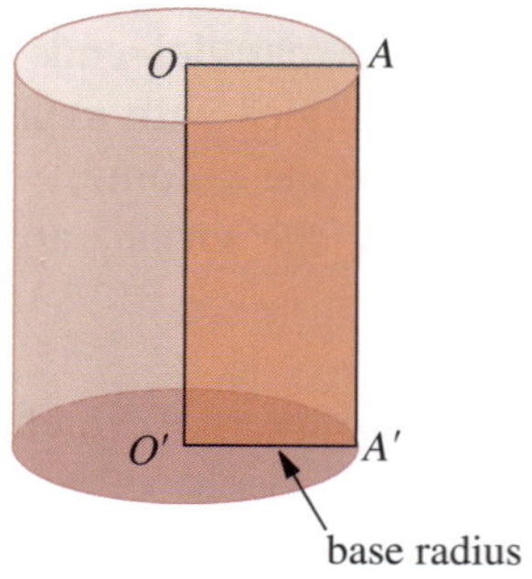


Thus, the volume of a cylinder of base radius $r$ and height $h$ is given by

$$\textbf{volume} = \pi r^2 h$$

## Surface Area of a Cylinder

Fig. 9.10(a) shows two equal circles of radius $r$ and a rectangle of height $h$. It is a net of a cylinder. To form the cylinder shown in Fig. 9.10(b), we roll up the rectangle and bring the two edges $AB$ and $CD$ together. The two equal circles formed will become the top and base circles of the cylinder. Obviously, the length of the rectangle is equal to $2\pi r$, the circumference of each circle.

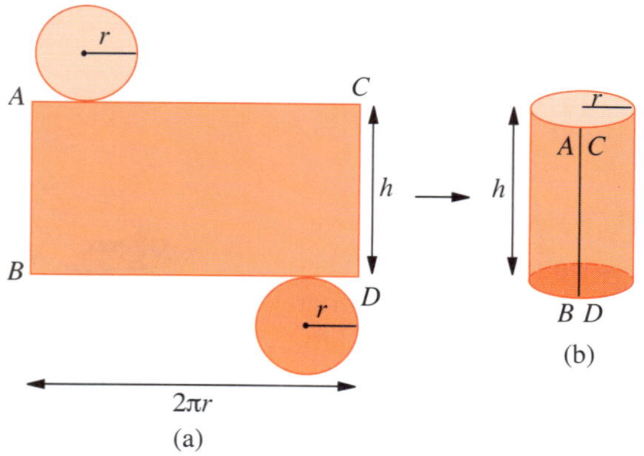


*Fig. 9.10*

$\therefore$ the area of the curved surface of the cylinder
$= \text{area of rectangle } ABCD = 2\pi rh$

**Surface area of a solid cylinder = the area of curved surface + 2 × the area of the base circle**
$$= 2\pi rh + 2\pi r^2 = 2\pi r(h + r)$$

## Example 8

*The diameter of the base of a right circular cylinder is 14 cm and its height is 10 cm. Find the volume and surface area of the solid cylinder. (Take $\pi = 3.142$)*

### Solution

$r = \frac{14}{2} = 7$, $h = 10$

Volume $= \pi r^2 h = (3.142 \times 7^2 \times 10)$ cm$^3$ $= 1540$ cm$^3$

$\therefore$ the volume is 1540 cm$^3$.

Surface area $= 2\pi r(h + r) = [2 \times 3.142 \times 7 \times (10 + 7)]$ cm$^2$ $= 748$ cm$^2$

$\therefore$ the surface area is 748 cm$^2$.

## Example 9

*If water flows through a 56-mm diameter pipe at a rate of 3 m/s, what volume of water, in litres, is discharged per minute? (Take $\pi = 3.142$)*

### Solution

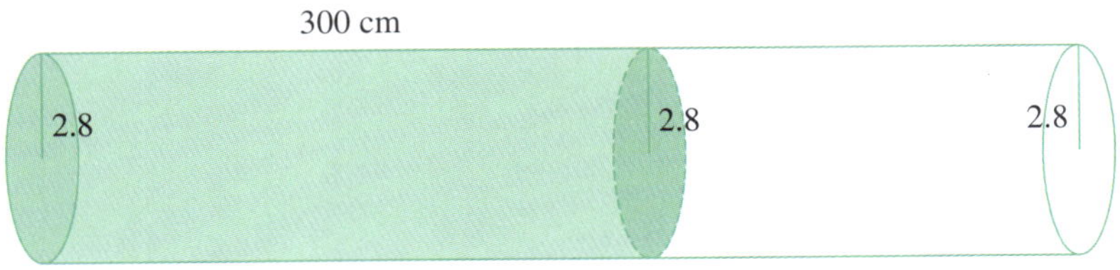


amount of water discharged in 1 second

$r = 28$ mm $= 2.8$ cm, $\quad h = 3$ m $= 300$ cm

Volume of water discharged per second $= \pi r^2 h$

$= (3.142 \times 2.8 \times 2.8 \times 300)$ cm$^3$

$= 7392$ cm$^3$

Volume of water discharged per minute $= (7392 \times 60)$ cm$^3$

$= 443\ 520$ cm$^3$

$= 443.5$ litres (correct to 1 decimal place)

$\therefore$ 443.5 litres of water are discharged per minute.

**Just for fun**

*A man wishes to take 4 litres of water out of a big tank of water. But he has only one 5-litre and one 3-litre jars. How can he do it?*

## Example 10

*The diagram shows a solid made up of a rectangular prism of length 32 cm by 12 cm by 18 cm and half a cylinder with a diameter of 12 cm. Calculate the total surface area and also find the volume of the solid. (Take $\pi = 3.142$ and give your answer correct to 1 decimal place.)*

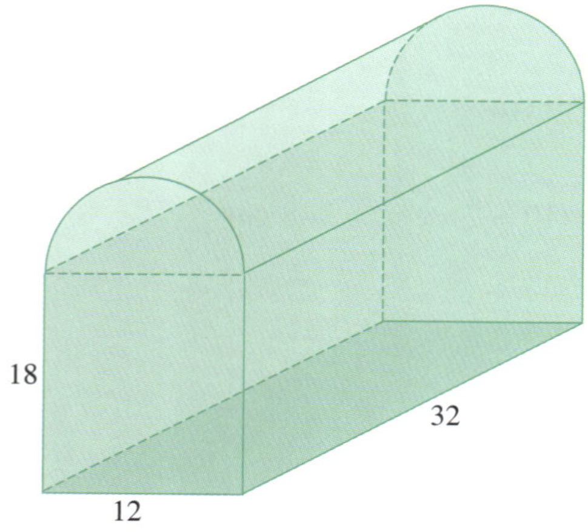


> **Just for fun**
>
> *If the population of the world is $5 \times 10^9$, what is the length of the edge of a cubical box that could hold this many people assuming that the volume of an average person is $5.4 \times 10^{-2}$ $m^3$?*

**Solution**

$$\text{Total surface area} = 2\left(12 \times 18 + \frac{1}{2} \times \pi \times 6^2\right) + (\pi \times 6) \times 32 + 2(32 \times 18) + (12 \times 32)$$

$$= 2683.9 \text{ cm}^2$$

$$\text{Volume of solid} = \left(12 \times 18 + \frac{1}{2} \times \pi \times 6^2\right) \times 32$$

$$= 8720.6 \text{ cm}^3$$

> **Just for fun**
>
> *Which is heavier,*
> *1 kg of iron or 1 kg of feathers?*

## Example 11

*A circular metal sheet 30 cm in diameter and 0.25 cm thick is melted and then recast into a cylindrical bar of diameter 5 cm. Find the length of the bar.*

**Solution**

Volume of circular sheet = $\left(\pi \times 15 \times 15 \times \frac{1}{4}\right)$ cm$^3$

Let the length of the bar be $x$ cm.

Volume of bar = $\left(\pi \times \frac{5}{2} \times \frac{5}{2} \times x\right)$ cm$^3$

Volume of bar = Volume of circular sheet

$$\pi \times \frac{5}{2} \times \frac{5}{2} \times x = \pi \times 15 \times 15 \times \frac{1}{4}$$

$$x = \frac{\pi \times 15 \times 15 \times \frac{1}{4}}{\pi \times \frac{5}{2} \times \frac{5}{2}} = 9$$

$\therefore$ the bar is 9 cm long.

In this example, no numerical value is used for $\pi$.

> **Pay Attention**
> *Do not substitute the value of $\pi$ until it is clear that such a substitution is necessary.*

## Example 12

*The figure on the right shows a section of a steel pipe. Given the internal radius of the pipe is 2.1 cm, the external radius is 2.5 cm and the length of the pipe is 12 cm, find*

*(a) the volume of steel used;*
*(b) its total surface area. (Take $\pi$ = 3.142)*

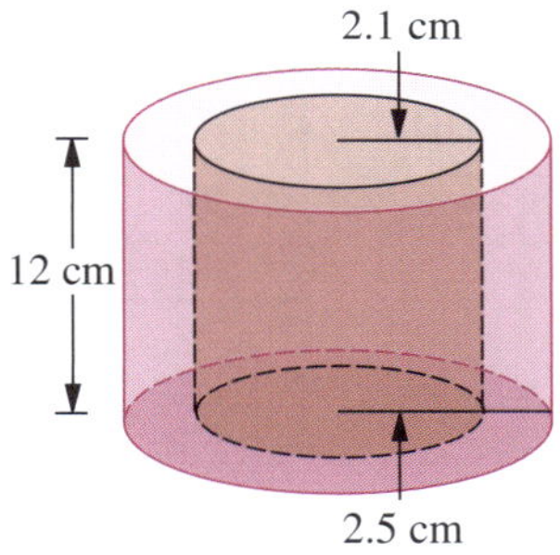


**Solution**

(a) $R = 2.5$, $r = 2.1$, $h = 12$

The cross-section of the pipe is a ring.

The area of the cross-section of the ring is equal to the difference between the areas of the outer circle and the inner circle = $\pi R^2 - \pi r^2$

Area of ring = $[\pi(2.5)^2 - \pi(2.1)^2]$ cm$^2$
= $1.84\pi$ cm$^2$

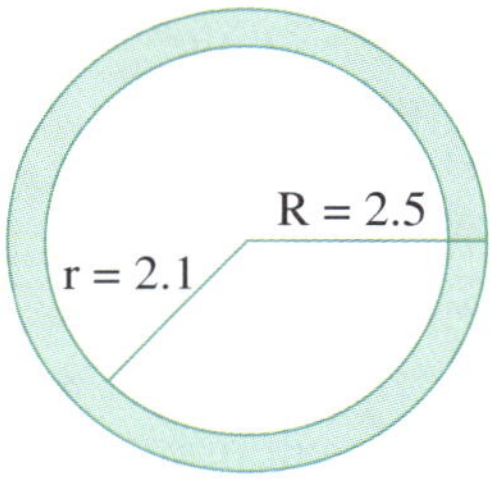


Volume of pipe = $(1.84\pi \times 12)$ cm$^3$
= $(1.84 \times 3.14 \times 12)$ cm$^3$
= 69.3 cm$^3$ (correct to 1 decimal place)

$\therefore$ the volume of steel used = 69.3 cm$^3$.

(b) Total surface area of pipe
= areas of internal and external curved surfaces + area of 2 rings
$= [(2\pi \times 2.1 \times 12) + (2\pi \times 2.5 \times 12) + (2 \times 1.84\pi)]\ \text{cm}^2$
$= (50.4\pi + 60\pi + 3.68\pi)\ \text{cm}^2$
$= 358.2\ \text{cm}^2$ (correct to 1 decimal place)

## Exercise 9c

In this exercise, take $\pi$ to be 3.142 unless otherwise stated.

**1.** Find the volume and total surface area of each of the following cylindrical solids:

**(a)**

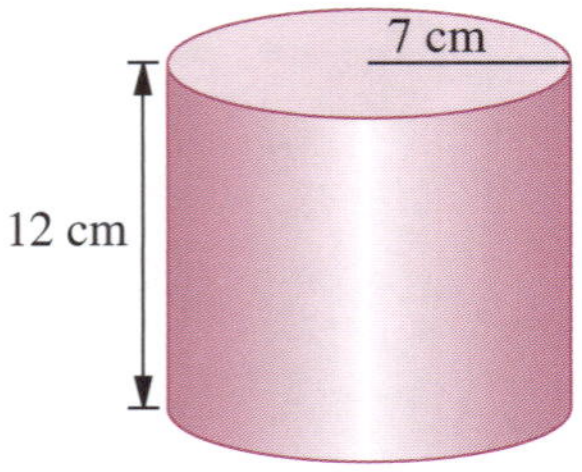


**(b)**

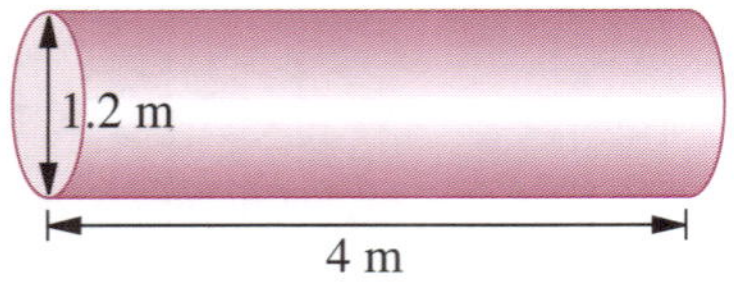


**(c)**

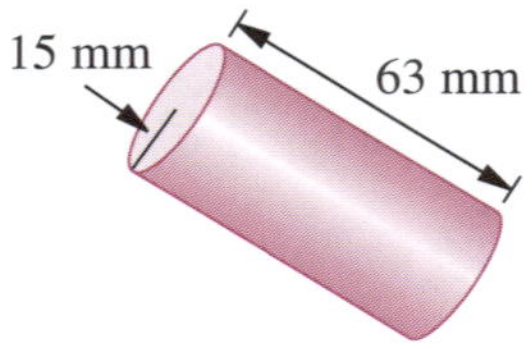


**2.** Find the diameters of the cylinders given the following:
**(a)** volume 704 $\text{cm}^3$, height 14 cm;
**(b)** volume 12 320 $\text{cm}^3$, height 20 cm.

**3.** Find the heights of the cylinders given the following:
**(a)** volume 528 $\text{cm}^3$, diameter 4 cm;
**(b)** volume 1056 $\text{m}^3$, radius 4 m.

**4.** The diagram shows a drinking trough in the shape of a half-cylinder with dimensions as shown. Find its capacity in litres.

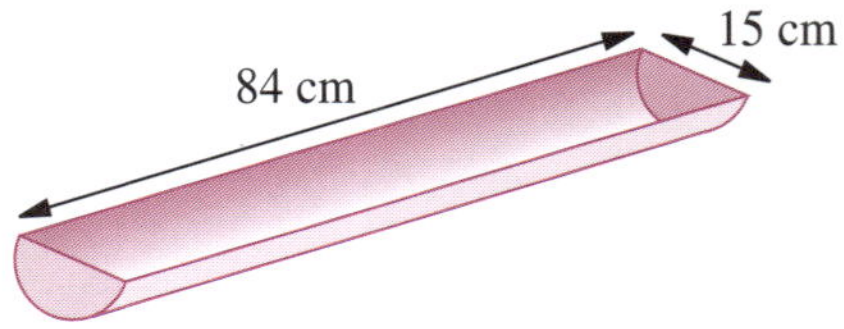


**5.** The diagram shows a metal pipe with an outer diameter of 28 mm and an inner diameter of 20 mm. Its length is 3.5 m. Find the volume, in $\text{cm}^3$, of the metal used in making the pipe.

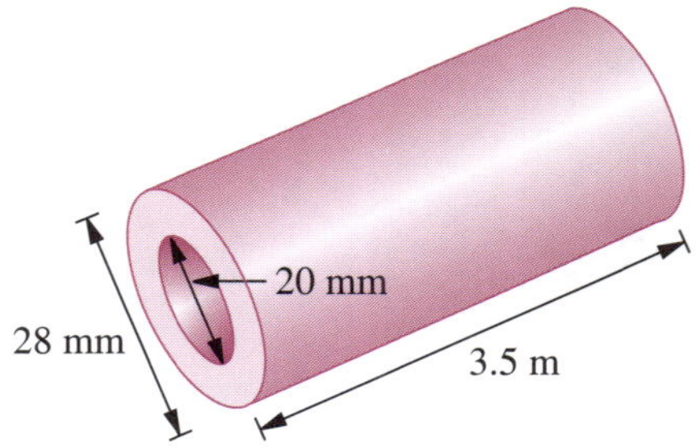


**6.** In a toy factory, 200 wooden solid cylinders 7 cm long and 35 mm in diameter have to be painted. What is the total surface area, in $\text{cm}^2$, that needs to be painted?

**7.** 500 cylindrical cans, without top lids and each 14 cm high with diameter 8 cm, are to be made from a sheet of metal. Find, in $\text{m}^2$, the total area that needs to be painted externally. (Take $\pi = 3.14$)
Correct your answer to one decimal place.

8. A railway tunnel 147 m long is to be bored with a circular cross-section of radius 5 m. What volume of soil has to be excavated? If the soil is to be taken away in wagons of capacity 75 $m^3$ each, how many wagons are needed?

9. A beer cask has a height of 63 cm and a diameter of 50 cm. Find its capacity in litres. How many glasses full of beer can it serve if the capacity of each glass is 0.6 litre?

10. Assuming that a \$1 coin is cylindrical with a diameter of 2.24 cm and a thickness of 2.5 mm, find the volume of the coin, giving your answer in $cm^3$. If the density of the coin is 5.4 $g/cm^3$, find its mass, correct to 2 decimal places.

    The Singapore government announced a \$10.5 billion recovery package in Parliament in 1998 to help the country overcome the Asian economic crisis. If the \$10.5 billion is to be given out in \$1 coins, what is the volume of all the coins? (Give your answer in $m^3$, correct to 1 decimal place.) Also find the mass of all the \$10.5 billion coins, giving your answer in tonnes, correct to the nearest tonnes.

## SUMMARY

1. Volume of a right prism = base area × height.

2. For a cylinder of base radius $r$ and height $h$, curved surface area = $2\pi rh$, total surface area = $2\pi rh + 2\pi r^2$ or $2\pi r(h + r)$ and volume = $\pi r^2 h$.

# Review Examples 9

## Example 1

*150 litres of water is poured into a cylindrical drum of diameter 48 cm. Find the depth of the water in the drum. Take $\pi = 3.142$ and give your answer correct to 2 decimal places.*

### Solution

Let the height of water in the drum be $h$ cm and the radius of the drum is 24 cm.

$$\pi(24)^2 h = 150 \times 1000$$

$$\therefore \quad h = \frac{150\ 000}{\pi(24^2)} = \frac{150\ 000}{3.142 \times 24^2}$$

$$= 82.88 \text{ cm}$$

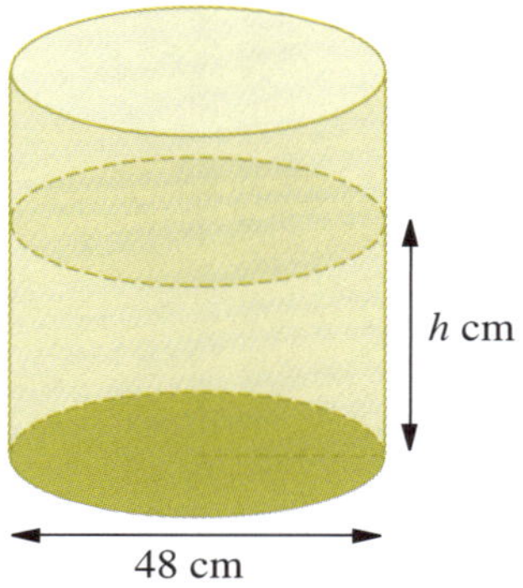


## Example 2

*The figure shows a closed container of uniform cross-section. The cross-section consists of rectangle ADCB and a quadrant DEC of a circle, centre D. Given AB = 14 cm, AD = 9 cm and BH = CG = EF = AI = 20 cm, calculate*

*(a) the area of the cross-section ADECB;*

*(b) the volume of the container;*

*(c) the area of the surface BCEFGH.*

*(Take $\pi = 3.142$)*

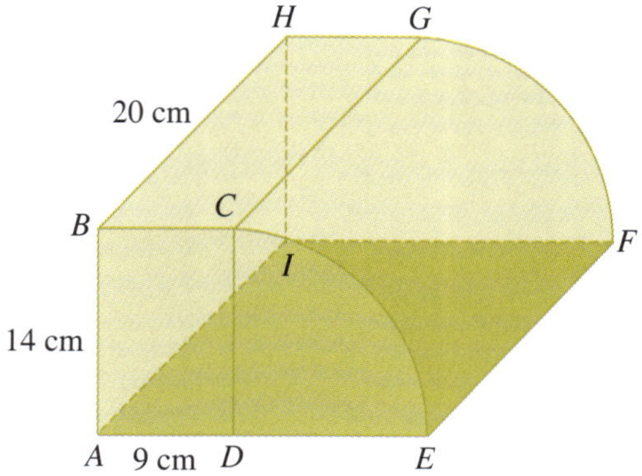

**Solution**

(a) The area of the cross-section $ADECB = 14 \times 9 + \frac{1}{4} \times \pi \times 14^2$

$= 280 \text{ cm}^2$

(b) Volume of the container $= 280 \times 20$

$= 5600 \text{ cm}^3$

(c) The area of the surface $BCEFGH = 9 \times 20 + \frac{1}{4} \times 2 \times \pi \times 14 \times 20$

$= 620 \text{ cm}^2$

## Review Questions 9

Take $\pi$ to be 3.142, where necessary, for the following questions.

★**1. (a)** Find the volume and surface area of each of the following right prisms:

**(i)**

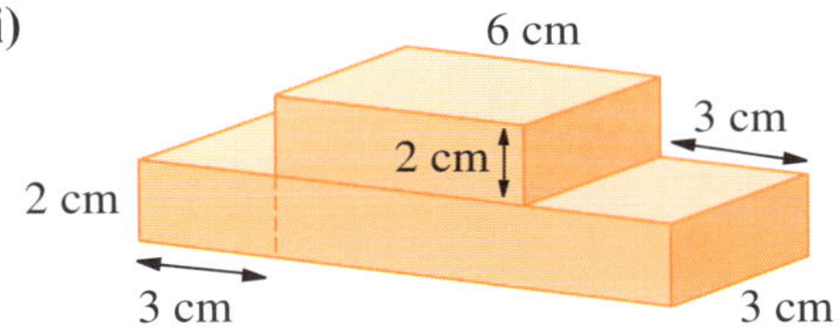


**(ii)**

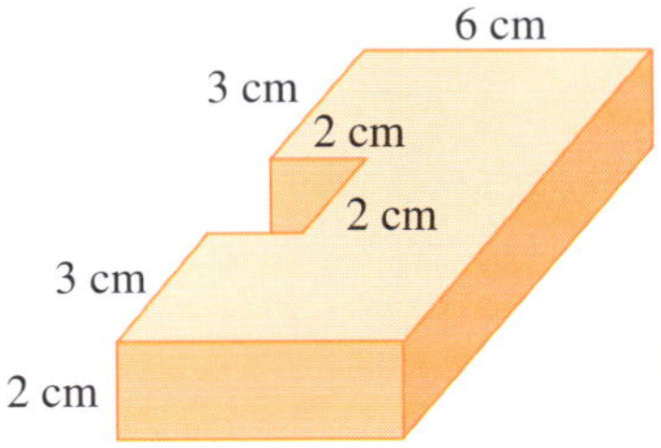


**(b)** Find the volume and surface area of the following solid, which is made up of two cylinders:

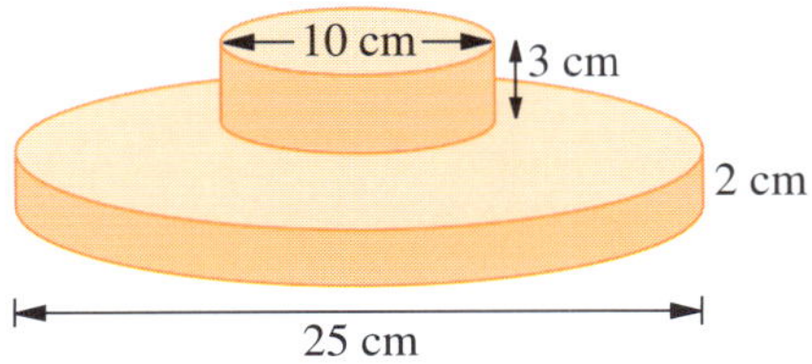


**2. (a)** Calculate the volume and surface area of each of the following right prisms:

**(i)**

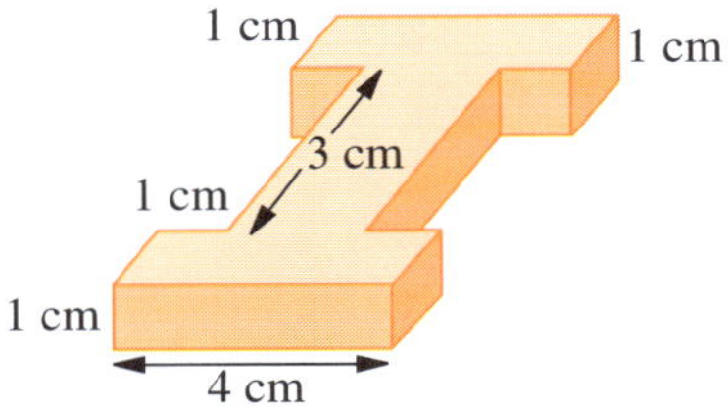

(ii) 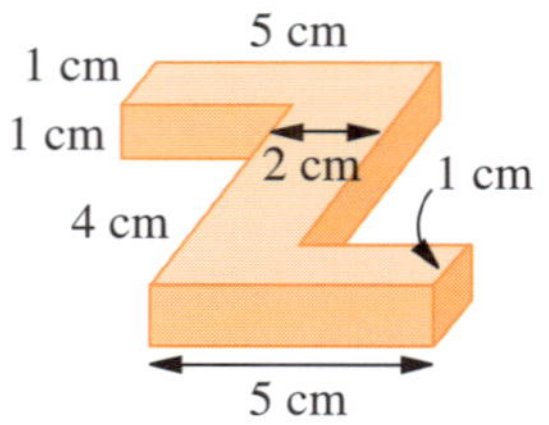


**(b)** Calculate the volume and surface area of the following solid, which is made up of a cylinder and a right prism:

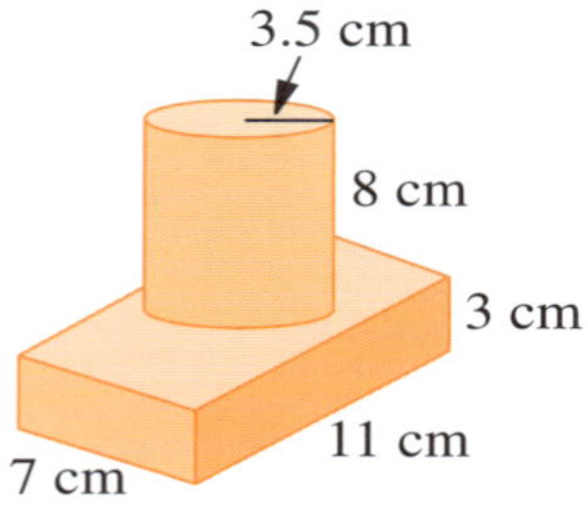


**3.** A brick measures 18 cm by 9 cm by 6 cm. Find the number of bricks that will be needed to build a wall 4.5 m wide, 18 cm thick and 3.6 m high.

**4.** A water tank, 0.8 m long, 0.8 m wide and 2.4 m deep is half-full of water. How many times can a watering-can be filled if its capacity is approximately 12 litres?

**5.** How many matchboxes, each 80 mm by 75 mm by 18 mm, can be packed into a box 72 cm by 60 cm by 45 cm internally?

**6.** The following figure shows a trough 15 m long. Its cross-section is a trapezium. Find the amount of water that the trough can hold in litres.

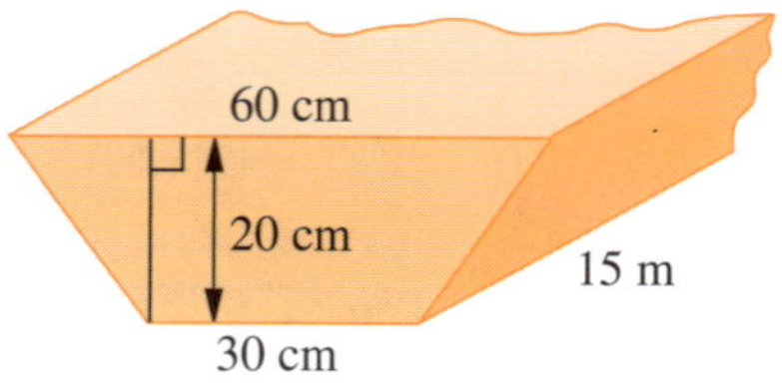


**7.** The figure shows the dimensions of the cross-section of a girder which is 2.5 m long. Find
**(a)** the volume of the girder;
**(b)** the surface area;
**(c)** its weight if the material weighs 7.8 g per $cm^3$.

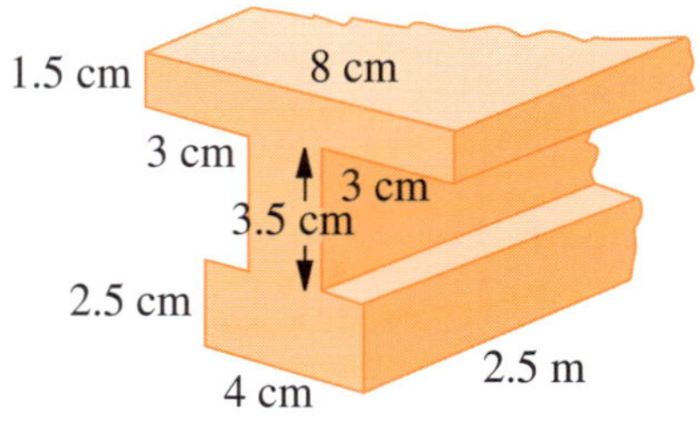


**8.** Find the mass of the water that has fallen onto a flat roof 10.4 m long and 6.5 m wide, when 22 mm of rain is recorded. (The mass of 1 $cm^3$ of water is 1 g.)

**9.** A section of a metal pipe has internal diameter 4.2 cm and external diameter 5.0 cm. If the length of the metal used for the pipe is 8.9 cm, calculate the volume of the metal used for making the pipe. If the metal costs \$8 per kg and 1 $m^3$ of the metal has a mass of 2700 kg, find the cost of the pipe.

**✶10.** A cuboid of dimensions 70 cm by 50 cm by 30 cm has "square holes" measuring 10 cm by 10 cm in the centres of the faces of the cuboid, as shown. The holes cut through the length of each face of the solid. Calculate the volume and the surface area of the solid.

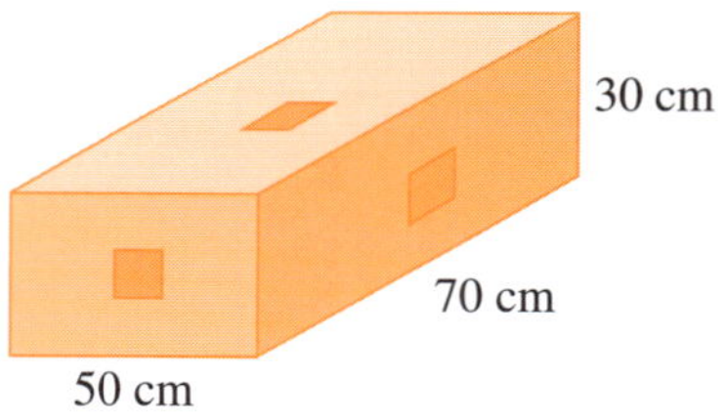


**11.** Ten open cylindrical containers are to be painted on the outside, including the base. Each container has a radius of 30 cm and a height of 28 cm. Given that 150 g of paint is needed to paint an area of 1 $m^2$, find the amount of paint required to paint the ten cylinders. Give your answer in kg.

**12.** A cylindrical barrel 70 cm in diameter and 80 cm in height is filled with water. A leak at the bottom drains away 0.2 litre of water every minute. How long will it take for the water level to drop by 6 cm?

**13.** The Singapore Expo has an exhibition area of 60 000 $m^2$, making it the largest exhibition centre in the region. If the average height of the exhibition centre is 4.85 m, find the volume of the air in the centre. If the density of air is approximately 1.26 $kg/m^3$, find the mass of the air contained in the centre.

## Revision Exercise II No. 1

**1.** Solve the following equations:

**(a)** $3x - (2x - 1) = 7$

**(b)** $8(7 - x) = (3x - 1)$

**(c)** $8 - \frac{3}{4}(x - 4) = \frac{1}{8}(x + 1)$

**(d)** $\frac{x-1}{2} - \frac{x-2}{3} = \frac{x-3}{4}$

**2. (a)** Find the value of $3x^2 + 14x - 7$ when $x = 5$.

**(b)** If $x = 2$ is a solution of the equation $x^2 - 7x + k = 0$, find the value of $k$.

**3. (a)** A father is 36 years old and his son is 6 years old. In how many years' time will the father be twice as old as his son?

**(b)** John is $1\frac{1}{3}$ times as heavy as Mary. If their total mass is 112 kg, find the mass of John.

**4. (a)** Write down the next two terms in the following number sequences:

**(i)** 5, 9, 13, 17, ...

**(ii)** $\frac{1}{2}, \frac{2}{3}, \frac{3}{4}, \frac{4}{5}, \ldots$

**(iii)** $1 \times 3, 2 \times 4, 3 \times 5, 4 \times 6, \ldots$

**(b)** Consider the following pattern:

$$\frac{1}{1 \times 2} = 1 - \frac{1}{2}$$

$$\frac{1}{2 \times 3} = \frac{1}{2} - \frac{1}{3}$$

$$\frac{1}{3 \times 4} = \frac{1}{3} - \frac{1}{4}$$

$$\vdots \qquad \vdots$$

$$\frac{1}{342} = \frac{1}{a} - \frac{1}{b}$$

**(i)** Write down the 8th line in the pattern.

**(ii)** Using the above, find the value of $\frac{1}{50} - \frac{1}{51}$.

**(iii)** Find the value of $a$ and of $b$.

**5. (a)** The product of two numbers is $1\frac{1}{5}$. If one of the numbers is $\frac{1}{5}$, find the sum of the two numbers.

**(b)** Solve the equation

$$\frac{4x+1}{2} = \frac{2x-1}{4} + \frac{x+3}{8}.$$

**6.** Simplify the following algebraic fractions.

**(a)** $2x + \frac{3y - 4x}{3} - \frac{2x - 4y}{5}$

**(b)** $\frac{2x-5}{3} - \frac{2(3x+1)}{5} - 2x$

**7. (a)** The figure shows a semicircle of radius 7 cm being enclosed in a rectangle of sides 14 cm by 12 cm. Calculate the area of the shaded region. (Take $\pi = 3.142$)

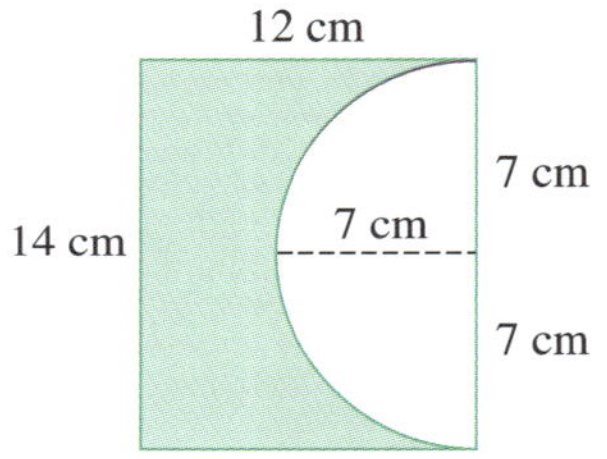


**(b)** It was projected that Singapore will need 1.2 million cubic metres of water daily at the turn of the 20th century. The Singapore Power (SP) planned to invest $900 million to build a desalination plant capable of producing 30 million gallons of water per day. If 1 gallon is equivalent to 4.5 litres,

**(i)** how many $m^3$ of water can the desalination plant produce per day?

**(ii)** how many desalination plants, correct to the nearest whole number, must be built if there are no other sources of water supply?

8. The height of water inside a rectangular cuboid of length 8 cm and width 6 cm is $2\frac{1}{2}$ cm. This water is poured into an empty cylindrical tank of diameter 14 cm. Find the height of water in the cylinder. (Take $\pi = 3.142$)

9. Factorise each of the following completely:
   **(a)** $3x + 9y + 12z$
   **(b)** $2x(5h - k) - 4y(5h - k)$
   **(c)** $2cx + 5dx - 6cy - 15dy$
   **(d)** $6by - 4ay - 3bx + 2ax$

## Revision Exercise II No. 2

1. Solve the following equations:
   **(a)** $\frac{x}{4} + \frac{1}{6} = \frac{x}{2} + \frac{1}{8}$;
   **(b)** $\frac{3x + 2}{4} + \frac{x - 2}{3} = 2 + \frac{x - 5}{2}$.

2. **(a)** I think of a number, halve it and then subtract 1 from it. The result is double the amount obtained by dividing the number by 3 and subtracting 4 from it. Find the number.

   **(b)** Ali earns \$960 a month and Ahmad earns \$1440 a month. How many months must Ali work to earn as much as Ahmad does in 6 months?

3. **(a)** Consider the number pattern:

$$
\begin{aligned}
1^2 - 2 \times 1 &= -1, \\
2^2 - 2 \times 2 &= 0, \\
3^2 - 2 \times 3 &= 3, \\
4^2 - 2 \times 4 &= 8, \\
5^2 - 2 \times 5 &= 15, \\
&\vdots \\
x^2 - 2x &= 63, \\
&\vdots
\end{aligned}
$$

   **(i)** Write down the 6th line in the number pattern.
   **(ii)** Find the value of $x$.

   **(b)** Write down the next three terms in the following number sequences:
   **(i)** 91, 87, 82, 76, ...
   **(ii)** –48, –36, –24, –12, ...

4. **(a)** The volume of a rectangular box is 96 $cm^3$ and its height is 4 cm. What is the area of its base? If the length of the base is 8 cm, find the width.

   **(b)** Find the total surface area and volume of a solid cylindrical block of diameter 28 cm and height 12 cm. (Take $\pi = 3.142$)

5. **(a)** Find the area of the shaded part in the figure below. If the figure represents the cross-section of a solid whose height is 7.5 cm, find the volume.

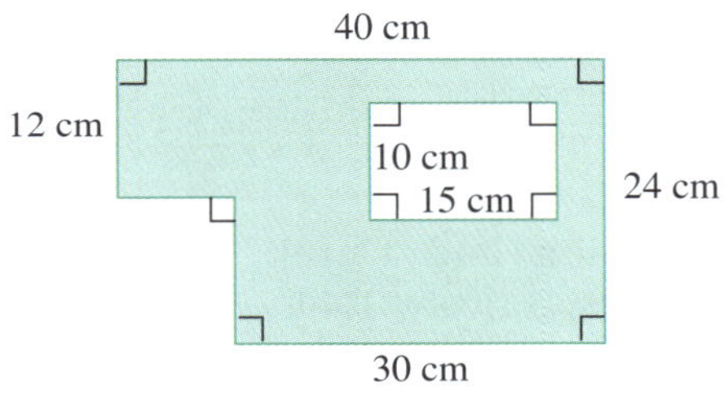


   **(b)** A pipe discharges 48 litres of water per minute into a rectangular tank of length 3.2 m and width 2.5 m. After 30 minutes, find the height of the water in the tank.

6. **(a)** A father is 45 years old and his son is 9 years old. In how many years' time will the father be three times as old as the son?

   **(b)** John has \$3.75 and David has twice as much money as John. How much money do they have altogether?

7. Consider the pattern

$$1 = 1 = 1^2$$
$$1 + 3 = 4 = 2^2$$
$$1 + 3 + 5 = 9 = 3^2$$
$$1 + 3 + 5 + 7 = 16 = 4^2$$
$$\vdots \quad \vdots$$

(i) Write down the tenth line in the pattern.
(ii) Find the value of $1 + 3 + 5 + ... + 29$.
(iii) Given that $1 + 3 + 5 + ... + (2x - 1) = 13^2$, find the value of $x$.

8. Solve $\frac{5 - x}{3} = 1 + \frac{1 - x}{4}$,

9. (a) The sum of two numbers is 35; $\frac{1}{2}$ of the smaller number is equal to $\frac{1}{3}$ of the greater number. Find the numbers.

(b) Simplify $\frac{a - 1}{5} + \frac{2a - 3}{4}$.

(c) John and Derek had \$24 and \$50 respectively. They spent the same amount of money on some books. How much money did each boy spend if Derek had 3 times as much money as John after buying the books?

10. Factorise each of the following completely:
(a) $3a(3x - y) - 4b(y - 3x)$
(b) $6ax + 12by - 9ay - 8bx$
(c) $15hx + 3hy - 5xk - ky$

## Revision Exercise II No. 3

1. Solve the following equations:
(a) $\frac{1}{3}x + 15 = 2x$
(b) $3(1.5x - 0.2) = 0.5x$
(c) $\frac{x + 1}{5} + \frac{x - 1}{5} = 6$
(d) $\frac{3x - 5}{6} + 2 = \frac{5x - 4}{5}$

2. (a) A bookshelf can hold 45 books, each 6.3 cm thick. How many books can it hold if each book is 2.1 cm thick?

(b) A mother is now three times as old as her daughter. If the sum of their ages five years ago was fifty, how old is the mother now?

3. (a) Write down the next two terms in each of the following number sequences:
(i) $-9, -6, -3, 0, ...$
(ii) 33, 44, 55, 66, ...
(iii) 13, 26, 39, 52, ...

(b) Identify a rule for each of the following number sequences, and then complete it.
(i) 3, 5, 8, 12, 17, 23, ____, ____
(ii) $\frac{2}{3}, 1, 1\frac{2}{3}, 2\frac{2}{3}, 4, 5\frac{2}{3}$, ____, ____

4. (a) Solve the equation $\frac{5x}{6} - \frac{2}{3} = \frac{x}{2} + \frac{1}{6}$.

(b) Find the value of $a$ from the formula $A = \frac{1}{2}h(a + b)$ if $A = 117$, $h = 13$ and $b = 11$.

5. Copy and complete the following statements with '<' or '>':
(a) 2 ____ 21 (b) $-1$ ____ 2
(c) $-1$ ____ $-2$

6. (a) Given that $H = \frac{V(P - R)}{500}$, find the value of $P$ when $V = 80$, $H = 3.2$ and $R = 65$.

(b) Solve $\frac{x}{3} - \frac{3x - 5}{2} = 6$.

7. (a) Ah Beng has \$38 and Ah Lian has \$20. If Ah Beng gives some money to Ah Lian, Ah Beng will then have one-third of what Ah Lian has. How much does Ah Beng give to Ah Lian?

(b) Mr Li's age is four times his son's. Six years ago, Mr Li was ten times as old as his son. How old are they now?

**8. (a)** A rectangular tank $1\frac{1}{2}$ m long and 88 cm wide contains water to a depth of 65 cm. The water is transferred to an empty rectangular tank 2 m long and 1 m wide. Find the depth of the water in centimetres.

**(b)** Find the total surface area and volume of a solid cylinder of diameter 14.6 cm and height 16.8 cm. If 1 $cm^3$ of the solid weighs 3.8 g, calculate its mass.
(Take $\pi = 3.142$)

**9. (a)** The figure shows a rectangle with a semicircle on one of its sides. Calculate the perimeter and the area of the figure.
(Take $\pi = 3.142$)

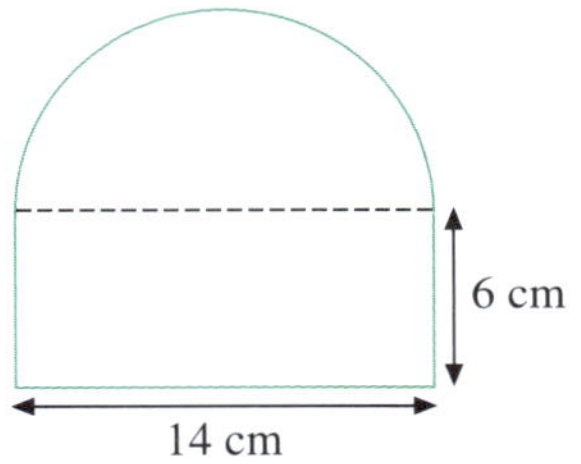


**(b)** Find the volume of a solid of uniform cross-section with an area of 6.5 $cm^2$ and height, 24 cm.

**(c)** The perimeter of a rectangle is 24 cm and its length is twice its breadth. Find the area of the rectangle.

**(d)** A swimming pool is 50 m long and 20 m wide and its depth varies uniformly from 1 m to 3.5 m. Find the volume of the pool in cubic metres.

**10. (a)** Solve the following inequalities:

**(i)** $2x < 16$ **(ii)** $\frac{3}{4}x > 9$

**(iii)** $\frac{1}{2}x > -4$

**(b)** Factorise each of the following:

**(i)** $3x + 18xh - 12xk$

**(ii)** $2h(5x - y) - k(y - 5x)$

**(iii)** $5ky + 3xh - 15xk - hy$

## Revision Exercise II No. 4

**1. (a)** A student paid \$5.40 for 30 pencils. Some pencils cost 10 cents each and others cost 20 cents each. How many pencils of the 10-cents type did he buy?

**(b)** A boy cycles $12x$ km in 5 hours. How far can he cycle in $3y$ hours at the same average speed?

**2.** Find $x$ when

**(a)** $9x = 15$; **(b)** $\frac{3}{4}x = 6$;

**(c)** $18 = 32 - x$; **(d)** $\frac{1}{2}x + 6 = 8$;

**(e)** $6x - 3x = 31 - 25$;

**(f)** $1.2x = x + 1$.

**3.** Peter is 24 years younger than his father. In 5 years' time, his father will be 3 times as old as Peter.

**(a)** How old is Peter now?

**(b)** How old will Peter's father be in 25 years' time?

**4. (a)** Solve $2 - \frac{1}{2}(x - 2) = \frac{3}{4}(x + 3) - \frac{1}{4}(x - 5)$.

**(b)** A man is 3 times as old as his son. Five years ago, he was 4 times as old as his son. Find their present ages.

**(c)** The product of two numbers is $2\frac{2}{3}$. If one number is $\frac{3}{5}$, find their sum.

**5.** The sum of three consecutive odd numbers is 237. Find the largest of the three numbers.

**6. (a)** The height of water in a cylinder of radius 12 cm is 28 cm. This water is transferred into a rectangular cuboid of length 24 cm, width 18 cm and height 40 cm. Find the height of water level in the rectangular cuboid. (Take $\pi = 3.142$)

**(b)** Find the area of the cross-section of the metal plate shown in the figure. If its thickness is 1.5 cm, find its volume.

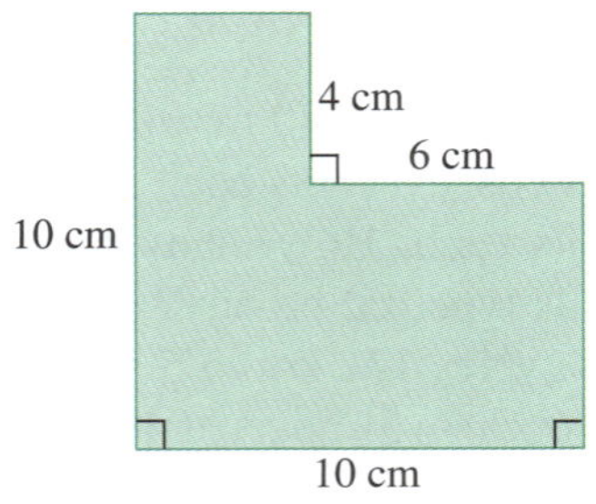


7. A rectangular tank measured internally is 2.5 m long, 1.5 m wide and 0.8 m high. How many litres of water are needed to fill the tank? The water in the tank is transferred into small cylindrical tanks each of diameter 70 cm and height 40 cm. Find the number of complete cylindrical tanks that can be filled. (Take $\pi = 3.142$)

8. **(a)** Consider the number pattern

$$1^2 - 0^2 = 1$$
$$2^2 - 1^2 = 3$$
$$3^2 - 2^2 = 5$$
$$4^2 - 3^2 = 7$$
$$\vdots$$
$$x^2 - y^2 = 157$$

**(i)** Write down the tenth line of the pattern.
**(ii)** Find the value of $99^2 - 98^2$.
**(iii)** Find the value of $x$ and $y$ in $x^2 - y^2 = 157$.

**(b)** Simplify $\sqrt[3]{64x^6y^9} - \frac{1}{2}[7 - (5 - 6x)]$.

9. **(a)** If the value of $3x^3 + 2x^2 + xy + 7$ is equal to 40 when $x = -2$, find the value of $y$.

**(b)** $\frac{4}{7}$ of the passengers in an MRT train are men, $\frac{1}{3}$ of them are women and the rest are children. If there are 42 children, find
**(i)** the total number of people in the train;
**(ii)** how many more men than women there are in the MRT train.

10. **(a)** Solve the following inequalities:
**(i)** $\frac{2}{3}x < 22$
**(ii)** $4y \geq 44$
**(iii)** $\frac{2}{5}x > \frac{-4}{25}$

**(b)** Factorise each of the following:
**(i)** $7x - 14xh + 21khx$
**(ii)** $3x(2a - 3b) + y(3b - 2a)$
**(iii)** $12ax + 10by - 8bx - 15ay$

## Revision Exercise II No. 5

1. **(a)** Solve $\frac{6x}{5} - \frac{x-1}{4} = x$.

**(b)** Solve the equation:
$3(m^2 - 2m) = 7 - m^2 + m(4m - 8)$.

2. Find $x$ when
**(a)** $7x = 42$; **(b)** $6x = 16$;
**(c)** $2x + 19 = 41$; **(d)** $3x = 7 + 8$;
**(e)** $6x - 32 = 2x$; **(f)** $3.4x = x - 3$.

3. Simplify
**(a)** $7 \times 4a$; **(b)** $8a + 2a \div 4$.

4. Write down the next three terms in the following number sequences:
**(a)** 2, 5, 10, 17, ...
**(b)** 10, 8, 6, 4, ...
**(c)** 1, 4, 5, 9, 14, ...
**(d)** $1, \frac{1}{2}, \frac{1}{4}, \frac{1}{8}, \ldots$

5. The cost of printing name cards is given by the equation $y = k + \frac{3000}{x}$ where $y$ cents is the cost per card, $x$ is the number of cards printed and $k$ is a constant. Given that $y = 10$ when $x = 600$, find the value of $k$.
**(a)** Calculate the cost per card if 750 cards are printed.
**(b)** How many cards will be printed if the cost per card is to be 7.5 cents?

**6. (a)** The sum of three consecutive odd numbers is 141, find the largest of the three numbers.

**(b)** Five teachers took a group of students for a movie. Each adult ticket cost \$7.20 and students' tickets were sold at half price. If the total cost for the group was \$212.40, calculate the number of children in the group.

**7.** Solve the equation $x - 2 = \frac{x - 4}{3}$.

**8. (a)** If the area of a parallelogram is 32.8 $cm^2$ and the height between the two parallel sides is 16.4 cm, calculate the length of the base.

**(b)** Mrs Li has 2 m of cake frill. She uses it for trimming a square cake of side 20 cm and the circumference of a round cake of diameter 28 cm. How much of the cake frill is left over? (Take $\pi = 3.142$)

**9. (a)** The result of adding 90 to a number is the same as multiplying that number by 6. Find the number.

**(b)** John is 4 years older than David and Joe is 2 years younger than David. If the sum of their ages is 41, how old will John be in 8 years' time?

**(c)** If $x = \frac{5y}{z - 4y}$, find the value of $y$ when $x = 3$ and $z = 2$.

**10. (a)** Solve the following inequalities:

**(i)** $\frac{1}{2}x < 2\frac{1}{2}$

**(ii)** $2\frac{1}{2}x \geq 12\frac{1}{2}$

**(b)** Factorise each of the following:

**(i)** $x(2h - k) + 3y(k - 2h)$

**(ii)** $-2ay + 3by + 2ax - 3bx$

*In this chapter, you will learn how to*

- ***find the ratio of two or more quantities;***
- ***recognize and use common measures of rate;***
- ***solve problems involving ratios;***
- ***solve problems involving speed and average speed.***

CHAPTER

# 10

# Ratio, Rate and Speed

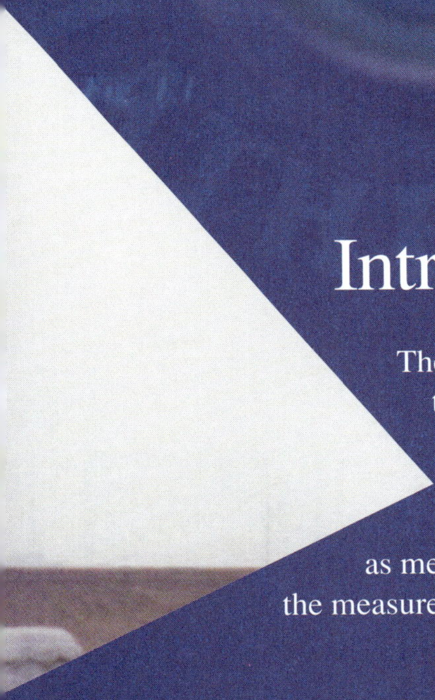

## Introduction

The picture shows an ancient sundial used in China for measuring the time of the day.

The measurement of time is essential to calculate the speed of a moving object. It is also needed in many other situations such as measuring the rate of one's heart beat, the rate of change of force in the measure of power and the rate at which your body digests food.

# Ratio

In primary school, you learned about ratios and solved problems involving ratios. We shall have a quick recapitulation of these.

A **ratio** is a comparison of like quantities measured in like units or quantities of the same kind.

The ratio of $a$ to $b$, where $a$ and $b$ represent two quantities and $b$ is not zero, is written as $a : b$ or as a fraction $\frac{a}{b}$.

A ratio is said to be in its simplest form when $a$ and $b$ are integers with no common factor (other than 1).

Thinking Time

If the ratio of the length of Fatimah's hair to that of Fandi's hair is 3 : 1, can we make the following conclusions?

1. Fatimah's hair is 3 times as long as Fandi's hair;
2. Fatimah's hair is very long;
3. Fatimah is 3 times as old as Fandi;
4. Fatimah is 3 times as tall as Fandi.

# Equivalent Ratios

We know that $\frac{30}{15} = \frac{2}{1}$, thus 30 : 15 = 2 : 1. We say that 30 : 15 and 2 : 1 are **equivalent ratios**.

The **order** in which the ratio is expressed is important. *For example,* in a group of 45 pupils where there are 15 girls and 30 boys, the boy-girl ratio is 30 : 15 = 2 : 1, or $\frac{2}{1}$, while the girl-boy ratio is 15 : 30 = 1 : 2, or $\frac{1}{2}$.

Notes

*Expressing a ratio in its simplest form is the same as reducing it to its lowest terms. We usually express a ratio in its simplest form.*

## Example 1

*Find the ratio of*

*(a) 700 g to 1 kg,* *(b)* $\frac{2}{3}$ *to* $\frac{5}{8}$.

## Solution

(a) 700 g and 1 kg are of different units and thus we have to express them in the same units first. It is easier to express 1 kg as 1000 g.

$\therefore$ the ratio of 700 g to 1 kg is 700 : 1000 or 7 : 10.

(b) **Method 1**

Convert both fractions into the same denominator.

$$\frac{2}{3} : \frac{5}{8} = \frac{16}{24} : \frac{15}{24} = 16 : 15$$

**Method 2**

Multiplying both fractions by the LCM (i.e. 24)

$$\frac{2}{3} : \frac{5}{8} = 16 : 15$$

Ratios can be used to compare more than two quantities. For example, three men, $A$, $B$ and $C$, share the profit of a business. They receive \$4000, \$3000 and \$1000 respectively. The ratio of their share of the profit is then 4000 : 3000 : 1000 or 4 : 3 : 1.

## Example 2

*If* $x : y = 11 : 8$ *and* $y : z = 6 : 7$, *find*

*(a)* $x : z$, *(b)* $x : y : z$.

## Solution

(a) Since $\frac{x}{y} = \frac{11}{8}$ and $\frac{y}{z} = \frac{6}{7}$

$$\frac{x}{z} = \frac{x}{y} \times \frac{y}{z} = \frac{11}{\cancel{8}_4} \times \frac{\cancel{6}^3}{7} = \frac{33}{28}$$

$\therefore$ $x : z = 33 : 28$

(b) Now $\frac{x}{y} = \frac{11}{8} = \frac{33}{24}$ and $\frac{y}{z} = \frac{6}{7} = \frac{24}{28}$

We have $x : y : z = 33 : 24 : 28$.

# Exercise 10a

1. Express each of the following ratios in its simplest form:
   (a) $\frac{5}{6} : \frac{7}{10}$   (b) $\frac{13}{40} : \frac{3}{8}$
   (c) 3.6 : 4.5   (d) 0.4 : 20
   (e) $1\frac{1}{2} : 2$   (f) 32 : 40 : 24
   (g) $1\frac{1}{3} : \frac{2}{3} : \frac{1}{6}$   (h) 1.2 : 2 : 2.8
   (i) $1.4 : \frac{2}{5} : \frac{1}{3}$   (j) $6\frac{2}{5} : 9.6 : 16$

2. Express each of the following as a ratio of the first quantity to the second, in its lowest term, **(i)** in the form $a : b$, **(ii)** as a fraction:
   (a) 25 cents, 80 cents
   (b) 210°, 360°
   (c) 250 cm, 1 m
   (d) 80 cents, $1.20
   (e) 1 kg 250 g, 3 kg
   (f) 3 min 30 s, 1 h

3. Given that $a : b = 7 : 9$ and $\frac{b}{c} = \frac{6}{13}$, find
   (a) $a : c$   (b) $a : b : c$
   (c) $\frac{2a}{5c}$

4. In a carpark, the ratio of red cars to green is 5 : 6, while that of green cars to blue is 3 : 10. Find the ratio of red cars to green cars to blue cars in its simplest form.

5. A school has an enrolment of 630 Chinese students, 120 foreign students and 60 Malay students. Find the ratio of foreign students to Chinese students to Malay students.

6. A man earns $1200 and spends $450 per month. Find the ratio of
   (a) his income to his expenditure and
   (b) his savings to his income.

7. Three people, *A*, *B* and *C*, share $416 among themselves. *A* receives $169 and *B* receives $156. Find the ratio in which the sum of money is shared.

8. The interior angles of a quadrilateral are 40°, 60°, 120° and 140°. Find the ratio of these angles according to the order given.

9. The sides of two squares are 4 cm and 6 cm. Find the ratio of
   (a) their areas and   (b) their perimeters.

10. The table below shows how 117 people travel to work.

| *Taxi* | *MRT Train* | *Bus* | *Car* |
|---|---|---|---|
| 9 | 21 | 72 | 15 |

Find the ratio of people who use taxi to MRT train to bus and to car.

# Increase and Decrease in Ratio

If the number of teachers in a school is increased from 45 to 55, then the ratio of the number of present staff to the number of previous staff = 55 : 45 = 11 : 9.

$$\frac{\text{number of present staff}}{\text{number of previous staff}} = \frac{55}{45} = \frac{11}{9}$$

We say that the number of teachers has been **increased in the ratio** 11 : 9, or $\frac{11}{9}$.

**Notes**

*$\frac{11}{9}$ is an improper fraction and is also a rational number*

A newspaper agent orders 84 copies of newspapers everyday. During the holidays, he decreases his order to 63 copies. The ratio of the number of copies ordered during the holidays to the usual number of copies ordered = 63 : 84 = 3 : 4, or $\frac{3}{4}$.

We say that the number of copies ordered per day has been **decreased in the ratio** 3 : 4, or $\frac{3}{4}$, during the holidays.

**Notes**

*$\frac{3}{4}$ is a proper fraction and is also a rational number.*

## Example 3

*Increase \$20 in the ratio 6 : 5; what is the result?*

**Solution**

The new value is \$20 × $\frac{6}{5}$ = \$24.

## Example 4

*In what ratio must 40 $m^3$ be decreased to become 24 $m^3$?*

**Solution**

The required ratio = new value : old value
= 24 $m^3$ : 40 $m^3$
= 24 : 40 = 3 : 5

## Exercise 10b

1. Find the result of increasing or decreasing the quantities in the given ratios:
   **(a)** 40 kg, 5 : 8
   **(b)** 56 m, 8 : 7
   **(c)** 35 hectares, 2.5 : 1
   **(d)** 2.5 $cm^2$, 2 : 5

2. In what ratio must 35 be increased to become 49?

3. In what ratio must 144 kg be decreased to become 108 kg?

4. A photograph measuring 5.5 cm by 9 cm is enlarged in the ratio 7 : 5. Find the dimensions of the enlarged photograph.

5. The cost of mutton has increased in the ratio 9 : 7. If the original price was $5.60 per kg, what is the new price?

6. Due to import duty, the price of a car increases in the ratio 11 : 8. What is the new price of a car which originally cost $25 600?

## Example 5

*Suppose a sum of money is divided among 3 people, John, Mary and Peter, in the ratio 2 : 3 : 7.*
*This means that John's share : Mary's share : Peter's share = 2 : 3 : 7.*
*If the sum of money is $192, how much will each of them receive?*

### Solution

**Strategy 1: Use ratio**

Consider the sum of money divided into 2 + 3 + 7 = 12 equal parts.
John, Mary and Peter will get 2, 3 and 7 parts respectively.

John's share : $192 = 2 : 12

$\therefore$ John's share = $\frac{2}{12} \times \$192 = \$32$

Similarly,

Mary' share = $\frac{3}{12} \times \$192 = \$48$

and Peter's share = $\frac{7}{12} \times \$192 = \$112$

Check: John's share + Mary's share + Peter's share
= $(32 + 48 + 112) = $192

**Notes**

*9 hens can lay 9 eggs in 9 days. How many eggs can 3 hens lay in 3 days?*

**Strategy 2: Use an equation**

Let John's share be $\$(2x)$.

$\therefore$ Mary's share = $\$(3x)$ and Peter's share = $\$(7x)$

$$
\begin{aligned}
\$(2x + 3x + 7x) &= \$192 \\
\text{i.e.,}\quad 12x &= 192 \\
x &= 16
\end{aligned}
$$

$$
\begin{aligned}
\therefore\quad \text{John's share} &= \$(2 \times 16) = \$32 \\
\text{Mary's share} &= \$(3 \times 16) = \$48 \\
\text{and}\quad \text{Peter's share} &= \$(7 \times 16) = \$112
\end{aligned}
$$

## Example 6

*Two chemicals, A and B, are combined in the ratio 5 : 11 by volume to form another solution X. A second solution Y is formed by combining chemicals A, B and C in the ratio 2 : 3 : 15 by volume. Calculate*

*(a) the volume of each chemical in 400 litres of solution X;*
*(b) the volumes of chemicals A and C in 400 litres of solution Y;*
*(c) the ratio of the three chemicals in a solution Z formed by mixing equal volumes of solutions X and Y;*
*(d) the volume of chemical C in 400 litres of solution Z.*

**Solution**

(a) Volume of chemical $A$ in 400 $l$ of solution $X = \left(\frac{5}{16} \times 400\right) l$
$= 125\ l$

Volume of chemical $B$ in 400 $l$ of solution $X = (400 - 125)\ l$
$= 275\ l$

(b) Volume of chemical $A$ in 400 $l$ of solution $Y = \left(\frac{2}{20} \times 400\right) l$
$= 40\ l$

Volume of chemical $C$ in 400 $l$ of solution $Y = \left(\frac{15}{20} \times 400\right) l$
$= 300\ l$

(c) Volume of chemical $B$ in 400 $l$ of solution $Y = (400 - 40 - 300)$
$= 60\ l$

In 800 $l$ of solution $Z$, the volumes of chemicals $A$, $B$ and $C$ are $(125 + 40) = 165\ l$, $(275 + 60) = 335\ l$ and $300\ l$ respectively.

$\therefore$ the ratio of chemicals $A$, $B$ and $C$ in solution $Z$ is 165 : 335 : 300, i.e., 33 : 67 : 60.

(d) $33 + 67 + 60 = 160$

$\therefore$ volume of chemical $C$ in 400 $l$ of solution $Z = \left(\frac{60}{160} \times 400\right)$
$= 150\ l.$

## Exercise 10c

**1.** Divide 110 g of salt in the ratio 5 : 8 : 9; what is the result?

**2.** Three partners of a firm divide the profit of \$6720 among themselves in the ratio 2 : 3 : 7. What is the biggest share of the profit?

**3.** A sum of money is divided among three persons, *X*, *Y* and *Z*, in the ratio 10 : 7 : 5. If *Y* gets \$14 more than *Z*, how much will *X* get and what is the total sum of money?

**4.** Three families, *P*, *Q* and *R*, share 480 kg of rice. *Q* receives twice as much rice as *P*, and *R* receives half as much rice as *Q*. How much rice does each family get?

**5.** Tea at \$2.40 per kg is mixed with tea at \$3.20 per kg in the ratio 1 : 3.
**(a)** Calculate the weight of each type of tea in 40 kg of the mixture.
**(b)** Calculate the price per kg of the mixture.

✶ **6.** A shopowner blends three types of coffee, *A*, *B* and *C*, in the ratio 3 : 5 : 7.
**(a)** Calculate the weight of each type of coffee in 45 kg of the blended mixture.
**(b)** Given that type *A* coffee costs \$7 per kg, type *B* coffee costs \$10 per kg and type *C* coffee costs \$13 per kg, calculate the cost per kg of the blended mixture.

✶ **7.** Two types of coffee, *P* and *Q*, are blended in the ratio 3 : 13 by weight to form a standard blend *X*. A second standard blend, *Y*, is formed by blending type *P* coffee, type *Q* coffee and type *R* coffee in the ratio 1 : 7 : 12 by weight. Calculate
**(a)** the weight of each type of coffee in 800 g of *X*;
**(b)** the weights of type *P* coffee and type *R* coffee in 800 g of *Y*;
**(c)** the ratio of the three types of coffee in a third standard blend *Z*, formed by mixing equal weights of *X* and *Y*;
**(d)** the weight of type *P* coffee in 800 g of *Z*.

# Rate

Let us consider the following questions:

A : If one dozen eggs cost \$1.80, what is the cost of 30 eggs?

B : A car travels 570 km on 60 litres of petrol. If the car travels 190 km, what will the petrol consumption be in litres?

C : A boy works 5 hours and is paid \$12.50. How much will he be paid if he works 12 hours?

> **Just for fun**
>
> *If it takes four minutes to boil one egg, how long will it take to boil three eggs?*

To answer each of the above questions, we must first find

A : The cost of 1 egg = $\$\frac{1.80}{12}$ (1.80 ← dollars, 12 ← eggs) or $\frac{180}{12}$ (180 ← cents, 12 ← eggs)

B : The petrol consumption for 1 km = $\frac{60}{570}$ (60 ← litres, 570 ← km)

C : The pay for 1 hour = $\$\frac{12.50}{5}$ (12.50 ← dollars, 5 ← hours)

Each of the above results is different from a ratio in that it involves two quantities of different kinds. Each of them is called a **rate**. It is a comparison of quantities of different kinds and thus it is necessary to state the units used.

A : The rate = $\$\frac{1.80}{12}$ = \$0.15 per egg

$\therefore$ the cost of 30 eggs = \$0.15 × 30 = \$4.50.

B : The rate = $\frac{60}{570} = \frac{2}{19}$ litre per km

$\therefore$ if the car travels 190 km, the petrol consumption $= \frac{2}{19} \times 190$

$= 20$ litres.

C : The rate = $\$\frac{12.50}{5}$ = \$2.50 per hour

$\therefore$ he will be paid \$2.50 × 12 = \$30.00 for working 12 hours.

> **Notes**
>
> *We normally use the word "per" or the symbol "/" to denote a rate. Thus we have \$2.50 per hour or \$2.50/hour.*

## Example 7

*How far can a car travel on 15 litres of petrol if it can travel 91 km on 7 litres of petrol? How much does the owner of the car spend on petrol, which costs \$1.10 per litre, when he travels 260 km?*

**Solution**

Distance travelled on 1 litre of petrol = $\frac{91}{7}$ km = 13 km.

$\therefore$ distance travelled on 15 litres of petrol = $(13 \times 15)$ km = 195 km.

Consumption of petrol per km = $\frac{7}{91}$ litre = $\frac{1}{13}$ litre.

Consumption of petrol for 260 km = $\left(\frac{1}{13} \times 260\right)$ litres = 20 litres.

$\therefore$ the owner spends \$1.10 × 20 = \$22 on petrol when he travels 260 km.

**Problem Solving Tips**

*In Example 8, we have to answer the following questions before we solve the actual problem.*

*1. How far can a car travel on 1 litre of petrol?*

*2. How much petrol is needed to travel 1 km?*

## Exercise 10d

1. Copy and complete the following:
   (a) If a typist types 900 words in 1 hour, her rate of typing is _____ words per minute.
   (b) If a man pays \$600 rent for 3 months, the rental rate is _____ dollars per month.
   (c) If \$14.06 is charged for 74 units of electricity, the rate is _____ cents per unit.
   (d) A machine is used to stamp bottle caps. If 150 bottle caps can be stamped in 30 seconds, the rate is _____ caps per second.

2. A man earns \$250 in a five-day week. What is his pay for 3 days?

3. A car uses 40 litres of petrol to travel 340 km. How far can it travel if it has only 32 litres of petrol?

4. A machine stamps 720 bottle caps in 2 minutes. How many bottle caps can it stamp in 40 seconds?

5. A wire 22 cm long has a mass of 374 g. What is the mass of 13 cm of this wire?

6. A shopkeeper buys 72 articles for \$82.80. How much will he have to pay if he buys 150 such articles?

7. 40 cm of a certain type of piping cost \$2.00. What is the cost of 1 km of such piping?

8. The cost of a long-distance call lasting 4 minutes and 20 seconds was \$23.40. At this rate, what was the cost of a call lasting 6 minutes 30 seconds?

9. 250 $cm^3$ of a liquid weighs 125 g. Find the weight of 1000 $cm^3$ of the liquid.

10. 200 g of fertilizer is required for a land area of 8 $m^2$. At this rate,
    (a) how many grams of fertilizer are needed for a land area of 1 $m^2$?
    (b) how many grams of fertilizer are required for a land area of 14 $m^2$?
    (c) for what land area will 450 g of fertilizer be sufficient?

*11. A cook uses fifteen 2-kg bottles of cooking-oil over a 4-week period. If he decides to buy 5-kg tins of oil instead, how many tins of cooking oil will he use over a 10-week period if the rate of using it remains unchanged?

# Average Rate

We have learnt average speed in primary school. Indeed average speed is one of the most popular example in average rate. The two examples below illustrate the basic concept of average rate.

A. David ate 5 hamburgers in 10 minutes. His rate of eating is $\frac{1}{2}$ of a hamburger per minute.

B. Mrs Lim's car travelled 800 km on 64 litres of petrol. Its rate of petrol consumption is 12.5 km/litre.

In the two examples above, we are assuming the rate given is constant throughout the given duration of time (as in A) and throughout the journey (as in B).

In real life, the rate may not be constant and the value calculated is indeed the average rate for a given duration of time (or for a journey).

### Thinking Time

Do you think that David will take 2 minutes to eat the first burger? Will he take longer to eat the fifth burger?

Do you think that Mrs Lim's car can travel 12.5 km on one litre of petrol when she drives around town as well as when she drives on expressway?

### Explorations

(A) You are to work in pairs.
Student A is to jump up and down on the spot 10 times. Immediately after the jump student B will take student A's pulse for 20 seconds. After a rest of 1 minute, student B will then take student A's pulse and record the result in the table below.

Student B will then do the jumping on the spot while student A will take his pulse as above.

| After | Pulse count in 20 seconds | Pulse count per minute |
|---|---|---|
| (a) 20 seconds | | $X =$ |
| (b) 1 minute | | $Y =$ |

Is there a difference between the values of $X$ and $Y$?
Can you suggest some reasons?

(B) Work in groups of 4
- Brainstorm for at least 10 situations where average rate is used in everyday life situations.
- Brainstorm for at least 5 situations where constant rate occurred in everyday life situations.
- Present your result to the class.

## Exercise 10e

1. Jason took 135 shots using a digital camera in 15 minutes. How many shots did Jason take in (i) 1 second? (ii) 1 minute?

2. The canteen staff find that they can butter 65 sandwiches with a 500 g packet of butter which costs \$3.85.
   **(a)** How many sandwiches can they butter with 1 g of butter?
   **(b)** If they make 403 sandwiches each day, what is the cost of the butter per day?

3. Jane used 2.8 m of material to make a dress. She bought the material at \$8.55 for 0.6 m.
   **(a)** How much was the material per metre?
   **(b)** What was the cost of the material for Jane's dress?

4. Bill Gates obtained a total score of 1980 points in three computer games. His average score per game on these games was 24 points below his previous game average. What was his average score before these three games?

5. Ali punted the ball 4 times in a football game. His longest kick was 45 m and shortest kick was 27 m. His average rate was 38 m/punt. If the other two punts were equal in length, find the distance that each travelled.

6. A full bathtub will empty at an average rate of 10 minutes. It will fill up to full at an average rate of 8 minutes if the plug is in. How long will it take to fill up to full if the plug is removed and the tap is turned on?

## Time

To record the time of the day, we can either use the 12-hour clock or the 24-hour clock. In the 12-hour clock, morning (from midnight to just before noon) is denoted by a.m.; afternoon, evening and night are denoted by p.m. In the 24-hour clock, four digits are used to indicate time. The first two digits denote hours and the last two denote minutes.

**For your information**

*a.m. stands for "ante meridiem" (Latin word) meaning "before midday". p.m. stands for "post meridiem" meaning "after midday".*

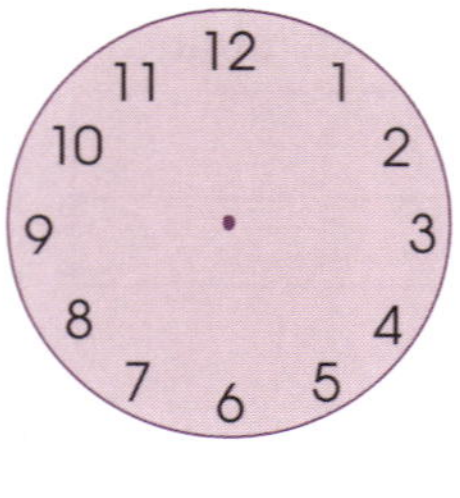


12-hour clock

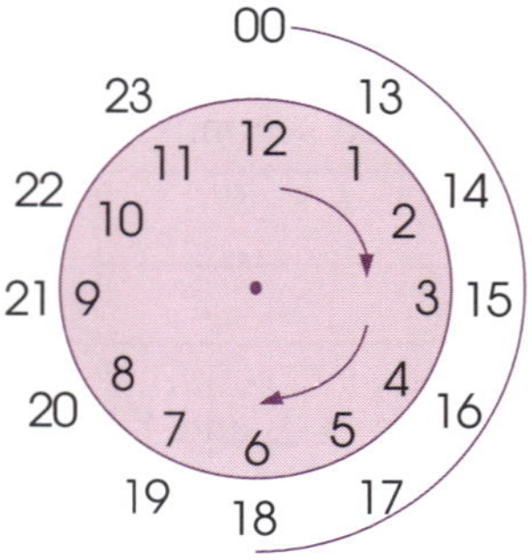


24-hour clock

The table below shows some examples.

| *Time* | *12-hour clock* | *24-hour clock* |
|---|---|---|
| 2 o'clock early morning | 2.00 a.m. | 02 00 |
| 5 to 11 in the morning | 10.55 a.m. | 10 55 |
| Noon | 12.00 p.m. | 12 00 |
| Half past 12 early afternoon | 12.30 p.m. | 12 30 |
| Quarter to 3 in the afternoon | 2.45 p.m. | 14 45 |
| 5 past 8 in the evening | 8.05 p.m. | 20 05 |
| One minute to midnight | 11.59 p.m. | 23 59 |
| Midnight | 12.00 a.m. | 00 00 |
| One minute past midnight | 12.01 a.m. | 00 01 |

## Example 8

*A car leaves Singapore at 21 15 on Wednesday and arrives in Kuala Lumpur $5\frac{1}{2}$ hours later. At what time and day does the car arrive in Kuala Lumpur?*

**Just for fun**

*A cashier of a bank is given one million one-cent coins to count. How long will she take if she can count five coins in one second?*

**Solution**

$5\frac{1}{2}$h = 5h 30 min

$$21\ 15 \xrightarrow{+5\text{ h}} 26\ 15\ (02\ 15) \xrightarrow{+30\text{ min}} 02\ 45$$

∴ the car arrives in Kuala Lumpur at 02 45 or 2.45 a.m. on Thursday.

## Example 9

*A train left Singapore at 07 35 and arrived in Seremban at 13 05 on the same day. How long was the train journey?*

**Solution**

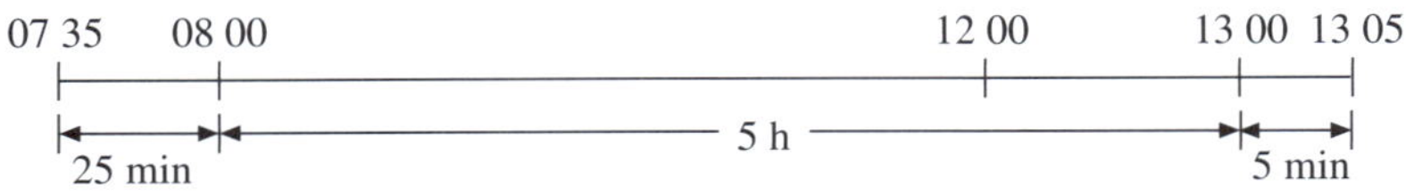


25 min + 5 min = 30 min

∴ the train journey was 5 h 30 min long.

**Notes**

*One hour is converted to 60 minutes.*

## Exercise 10f

1. Convert the following times to 24-hour clock notation:
   (a) 8.00 a.m. (b) 9.42 p.m.
   (c) midnight (d) 2.42 a.m.

2. Convert the following times to 12-hour clock notation:
   (a) 03 30 (b) 23 12
   (c) 19 15 (d) 24 00

3. Copy and complete the following table:

| | *Departure time* | *Journey time* | *Arrival time* |
|---|---|---|---|
| (a) | 02 40 | 55 mintues | |
| (b) | 22 35 | 8 hours | |
| (c) | 15 45 | | 17 50 |
| (d) | 09 48 | | 22 16 |
| (e) | 20 35 (Tue) | | 07 15 (Wed) |
| (f) | | $1\frac{1}{4}$ hours | 23 50 |

4. A train left a station at 8.35 a.m. and arrived at its destination at 3.12 p.m. How long did the journey take?

5. An overnight train left at 21 55 on a journey that took 9 h 18 min. Find the time at which it arrived at its destination.

6. A car arrived at a town at 15 06 after travelling for $4\frac{1}{4}$ hours. Find the time the car started its journey.

7. According to a timetable, a coach was due to leave a station at 22 55 and arrive at its destination at 06 05 the next day. How long would the journey take? If the train actually arrived 35 minutes early, at what time did it arrive?

*8. Shown below is the schedule of the arrival and departure times of a long-distance express overnight coach.

| *Destination* | *Arrival* | *Departure* |
|---|---|---|
| Singapore | — | 21 30 |
| Johor Baru | 22 15 | 22 30 |
| Seremban | 02 25 | 02 30 |
| Kuala Lumpur | 03 50 | 04 20 |
| Ipoh | 07 50 | 08 00 |
| Taiping | 09 20 | 09 30 |
| Butterworth | 10 45 | — |

Find the time taken for the coach to travel from
(a) Singapore to Seremban;
(b) Johor Baru to Ipoh;
(c) Seremban to Taiping;
(d) Kuala Lumpur to Butterworth;
(e) Singapore to Butterworth.

# Speed and Average Speed

Cyclist $A$ and cyclist $B$ travel 90 km, in a race, in 5 hours and $4\frac{1}{2}$ hours respectively. Which cyclist travels faster?

Cyclist $B$ travels faster since he takes less time to complete the race.

We can also find the **speed** at which each cyclist travels to find out who travels faster.

**Speed** is the rate of distance travelled per unit time.

$$\text{Cyclist } A\text{'s speed} = \frac{90 \text{ km}}{5 \text{ h}} = 18 \text{ km/h}$$

$$\text{Cyclist } B\text{'s speed} = \frac{90 \text{ km}}{4\frac{1}{2} \text{ h}} = 20 \text{ km/h}$$

Cyclist $B$ travels faster since he travels at a greater speed.

When calculating the speed of each cyclist, we assume that one travels at the same speed all the time. This is considered the **uniform speed** of each cyclist.

Thinking Time

In real life, do you think the cyclist is able to cycle at the same speed all the time?
What about when he is cycling up a slope or going down a slope?

In real life, cyclist $A$ is unlikely to cycle at 18 km/h throughout the race. What we have calculated is the average speed. This means that on average, he travels 18 km every hour.

The average speed can be obtained by using the formula:

$$\textbf{Average speed} = \frac{\textbf{Total distance travelled}}{\textbf{Total time taken}}$$

Notes

*If a cyclist is equipped with a speedometer, which gives his speed at a particular instant, the reading of the speedometer will change from time to time.*

# Problems Involving Speed and Average Speed

## Example 10

*A car travelled 510 km in 6 hours. Find the average speed of the car for the whole journey.*

**Solution**

Distance travelled = 510 km

Time taken = 6 hours

$$\text{Average speed} = \frac{\text{Distance travelled}}{\text{Time taken}} = \frac{510 \text{ km}}{6 \text{ h}} = 85 \text{ km/h}$$

> **For your information**
>
> *The highest speed limit for cars on Singapore roads is 90 km/h. How many demerit points will a motorist be given if he is caught speeding on the expressway at*
> *(a) 100 km/h;*
> *(b) 120 km/h'*
> *(c) 160 km/h?*

## Example 11

*A cyclist travels at an average speed of 18 km/h.*

*(a) Express his average speed in m/s.*
*(b) Find the distance he travels in 3 hours.*
*(c) Find how far he travels in 25 seconds.*

**Solution**

(a) 18 km = 18 × 1000 m, 1 h = (60 × 60) s

$$\therefore \quad 18 \text{ km/h} = \frac{18 \text{ km}}{1 \text{ h}} = \frac{(18 \times 1000) \text{ m}}{(60 \times 60) \text{ s}} = 5 \text{ m/s}$$

(b) In 3 hours, the cyclist travels (18 × 3) km = 54 km

(18: Average speed in km/h; 3: Time taken in hours)

(c) In 25 seconds, he travels (5 × 25) m = 125 m

(5: Average speed in m/s; 25: Time taken in seconds)

In general,

**Distance travelled = Average speed × Time taken**

> **Just for fun**
>
> *A sports car leaves Singapore for Kuala Lumpur at the same time as a bus, which leaves Kuala Lumpur for Singapore. They travel along the same road, the sports car at 110 km/h and the bus at 55 km/h. Which vehicle is further away from Singapore when they meet?*

## Example 12

*A train travels at an average speed of 15 m/s.*

*(a) Express its average speed in km/h.*
*(b) Find the time taken by the train to travel 750 m.*
*(c) If the train sets off from Station A at 8.00 a.m., find the arrival time of the train at Station B which is 36 km away.*

> **For your information**
> 1. *The cheetah is the fastest land animal. It can acquire a speed of 110 km/h in a matter of seconds.*
> 2. *The men's world record for the 100 m sprint is approximately 36.7 km/h.*
> 3. *The speed of sound is about 34 times faster than the speed of the fastest 100 m human sprinter.*

### Solution

(a) $15 \text{ m} = \dfrac{15}{1000} \text{ km}, \; 1 \text{ h} = 3600 \text{ s}$

In 1 second, the train travels $\dfrac{15}{1000}$ km.

In 1 hour, the train travels $\left(\dfrac{15}{1000} \times 3600\right)$ km = 54 km.

$\therefore$ its average speed is 54 km/h.

(b) $15 \text{ km/s} = \dfrac{15 \text{ m}}{1 \text{ s}}$

In 1 second, the train travels 15 m.

The time taken to travel 1 m is $\dfrac{1}{15}$ s.

$\therefore$ the time taken by the train to travel 750 m $= \dfrac{1}{15} \times 750$ s

$= \dfrac{750}{15}$ s (750: Distance travelled in m. 15: Average speed in m/s.)

$= 50$ s

(c) Similarly in 1 hour, the train travels 54 km.

The time taken to travel 1 km is $\dfrac{1}{54}$ h.

$\therefore$ the time taken by the train to travel 36 km $= \dfrac{1}{54} \times 36$ h

$= \dfrac{36}{54}$ h (36: Distance travel in km. 54: Average speed in km/h.)

$= \dfrac{2}{3}$ h

$= \dfrac{2}{3} \times 60 \text{ min} = 40 \text{ min}$

$\therefore$ the arrival time of the train at Station B is 8.40 a.m.

In general,

$$\textbf{Time taken} = \frac{\textbf{Distance travelled}}{\textbf{Average speed}}$$

## Example 13

*Two brothers, Peter and Paul, are vegetable sellers. One morning, they left home together at 5 o'clock for the market, each pushing a cart full of vegetables. Peter travelled at a constant speed of 100 m/min and Paul a constant speed of 60 m/min. After arriving at the market, Peter spent 5 minutes unloading the vegetables and immediately returned along the same route at the same speed to help Paul. If Peter met Paul at 5 minutes to 6 o'clock, what is the distance between their house and the market?*

### Solution

The diagram drawn below shows Peter's path and Paul's path between their house and the market.

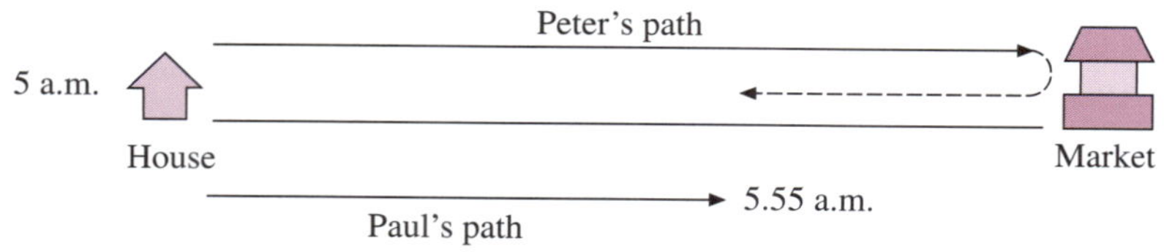


**Problem Solving Tips**

*Here we use the strategy of using a diagram or a model.*

Peter travelled 100 × 50 = 5000 m in 50 min.

Paul travelled 60 × 55 = 3300 m in 55 min.

They travelled a total distance of 5000 + 3300 = 8300 m.

From the diagram, the distance between their house and the market is half of the total distance travelled.

**Your Attention**

*Peter spent 5 minutes unloading.*

$\therefore$ the distance between their house and the market = $\frac{1}{2} \times 8300 = 4150$ m.

## Exercise 10g

**1.** Copy and complete the following. The first one has been done for you.

| | *Distance travelled* | *Time taken* | *Average speed* |
|---|---|---|---|
| **(a)** | 180 km | $1\frac{1}{2}$ h | 120 km/h |
| **(b)** | 200 m | 25 s | |
| **(c)** | 40 m | 1 min | |
| **(d)** | | $5\frac{1}{2}$ h | 80 km/h |
| **(e)** | | $\frac{1}{3}$ min | 25 m/s |
| **(f)** | 100 m | | 20 m/s |

**2.** Express the following in m/s:
**(a)** 18 km/h **(b)** 72 km/h
**(c)** 90 km/h

**3.** Express the following in km/h:
**(a)** 10 m/s **(b)** 35 m/s
**(c)** $\frac{1}{2}$ km/s

**4.** How long will a man take to run, once, round a circular track of radius 28 m at an average speed of 8 m/s? $\left(\text{Take } \pi = \frac{22}{7}\right)$

**5.** A cyclist begins on a 24-km journey at 09 23. When will he complete his journey if he travels at an average speed of 16 km/h?

**6.** A train leaves Town $X$ at 12 57 and arrives at Town $Y$ 45 minutes later.
**(a)** At what time does the train arrive in Town $Y$?
**(b)** What is the average speed of the train, in km/h, if the distance between the two towns is 84 km?

**7.** A car travels at an average speed of 24 km/h. Find, in metres, the distance travelled by the car in 12 seconds.

**8.** A car travelled on a $B$ class road for 20 minutes at an average speed of 57 km/h. It then travelled a distance of 55 km in 30 minutes on an expressway. Find
**(a)** the distance the car travelled on the $B$ class road;
**(b)** the average speed, in km/h, of the car when it travelled on the expressway.

**9.** A man cycles for two hours at an average speed of 16 km/h and then walks for 3 hours at an average speed of 6 km/h. Find his average speed for the whole journey.

**10.** A train travels 68 km at an average speed of 51 km/h. It then travels another 20 km at an average speed of 40 km/h before reaching its destination. Calculate the average speed for the whole journey.

⋆**11.** Two points, $X$ and $Y$, are 120 m apart. $M$ is the mid-point of $X$ and $Y$. An object travels from $X$ to $M$ in 12 seconds and then from $M$ to $Y$ at an average speed of 15 m/s. Calculate
**(a)** the average speed of the object from $X$ to $M$;
**(b)** the time taken to travel from $M$ to $Y$;
**(c)** the average speed for the whole journey from $X$ to $Y$.

⋆**12.** Three points, $L$, $M$ and $N$, lie on a straight line with $LN$ = 160 m. An object travels from $L$ to $M$ at an average speed of 10 m/s in 6 seconds and then from $M$ to $N$ at an average speed of 25 m/s. Calculate
**(a)** the distance from $L$ to $M$;
**(b)** the time taken to travel from $M$ to $N$;
**(c)** the average speed for the whole journey from $L$ to $N$.

⋆**13.** A passenger train travelled at a speed of 72 km/h. A man on the passenger train observed a goods train travelling at a speed of 54 km/h in the opposite direction. If the goods train passed him in 8 seconds, find the length of the goods train. (*Hint:* model the problem and draw a diagram).

⋆**14.** Steven left Town $A$ and walked towards Town $B$ at a speed of 100 m/min. At the same time, Jason and Melvin started from Town $B$ and walked towards Town $A$ at a speed of 80 m/min and 75 m/min respectively. If Steven met Melvin six minutes after passing Jason, find the distance between Town $A$ and Town $B$.

## Summary

1. A **ratio** is expressed as a fraction of the first quantity over the second. To find the ratio of two quantities, we must first express them in the same units. A ratio has no unit.

2. A **rate** expresses a relationship involving two quantities of different kinds.

3. Average speed of a moving object is given by the formula:

$$\text{Average speed} = \frac{\text{Total distance travelled}}{\text{Total time taken}}$$

4. Conversion of Units:

$$x \text{ km/h} = \frac{(x \times 1000)\text{ m}}{60 \text{ min}} = \frac{1000x}{60} \text{ m/min.}$$

$$y \text{ km/h} = \frac{(y \times 1000)\text{ m}}{60 \times 60 \text{ s}} = \frac{1000y}{3600} \text{ m/s.}$$

## Review Examples 10

### Example 1

*The ratio of the number of apples to oranges in stall A and stall B are 2 : 5 and 5 : 9 respectively. If the total number of fruits in stall A is twice that in stall B,*
*(a) find the ratio of the number of apples in stall A to the numbers of apples in stall B;*
*(b) find the number of oranges left in stall A after 84 oranges are sold from stall A and the ratio of apples to oranges became 3 : 4 in stall A.*

#### Solution

We shall use algebraic method to solve this problem.

(a) Let $x$ be the number of fruits in stall B, then $2x$ will be the number of fruits in stall $A$.

$$\therefore \quad \frac{\text{the number of apples in stall } A}{\text{the number of apples in stall } B} = \frac{\frac{2}{7} \times 2x}{\frac{5}{14} \times x} = \frac{8}{5}$$

(b) Let the number of apples and oranges in stall $A$ be $2k$ and $5k$ respectively, where $k$ is a constant.

we have
$$\frac{2k}{5k-84} = \frac{3}{4}$$
$$8k = 15k - 252$$
$$k = 36$$

$\therefore$ the number of oranges left in stall $A = 5k - 84$
$= 5 \times 36 - 84$
$= 96$

## Example 2

*Cheetah is the fastest land animal, capable of attaining a top speed of 110 km/h. The men's world record for the 100 m sprint is 9.77 seconds in 2005. Find*

*(a) how many times the cheetah is as fast as the fastest 100 m human sprinter;*

*(b) the time taken by the fastest 100 m human sprinter to run one round along the perimeter of a basketball court with the dimensions of 28.65 m by 15.24 m.*

**Solution**

9.77 s to cover 100 m

$\therefore$ 1 second the man can cover $\frac{100}{9.77}$ m

$\therefore$ In 1 hour he can cover $\frac{100}{9.77} \times 60 \times 60$ m $= 36.85$ km

i.e. the speed of the man is 36.85 km/h

$$110 \div 36.85$$
$$= 2.99 \approx 3$$

$\therefore$ The cheetah is about 3 times as fast as the fastest 100 m human sprinter.

(b) Time to cover 100 m = 9.77 s

Perimeter of the basketball court $= (28.65 + 15.24) \times 2$
$= 87.78$ m

$\therefore$ time required $= 87.78 \div 100 \times 9.77$
$= 8.59$ s (correct to 3 sig. fig.)

# Review Questions 10

1. The bill for domestic power in a home is reduced from \$120 to \$90 per month. By what ratio has the bill decreased?

2. A poster measuring 150 cm by 180 cm is enlarged in the ratio 8 : 5. Find the length and breadth of the enlarged poster.

3. A manufacturing firm plans to increase its output in the ratio 2.25 : 1 next year. Its present output is 148 000 articles. How many articles does it hope to produce next year?

4. Goat's milk contains 27 g of protein, 30 g of fat and 36 g of carbohydrate. Find the ratio of protein to fat to carbohydrate in goat's milk.

5. Pupils in a class are told to choose one out of three sports options: tennis, basketball and badminton. Given that the pupils choose the options in the ratio 4 : 2 : 3 and that 20 choose tennis, find
   **(a)** the number of pupils in the class;
   **(b)** the number of pupils who choose badminton.

6. A student took 18 minutes to walk from home to school.
   **(a)** Given that he arrived in school at 07 05, find the time at which he left home.
   **(b)** Given also that he walked to school at an average speed of 4 km/h, calculate how far he had to walk.

7. A car took 2 hours and 15 minutes to travel 198 km. If it arrived at its destination at 12 06, find
   **(a)** the time it started its journey;
   **(b)** the average speed of the car, giving your answer in kilometres per hour.

8. The cost of 1 kg of fish is \$4.12.
   **(a)** Find the cost of $4\frac{1}{2}$ kg of fish.
   **(b)** How many kilograms of such fish can be bought for \$20.60?

9. **(a)** Given that a typist can type 575 words in 25 minutes, how long will she take to type
   **(i)** 3680 words?
   **(ii)** 8855 words?
   **(b)** On another occasion, the typist started to type a report at 10 35 and finished it at 11 28.
   **(i)** How long did she take to type the report?
   **(ii)** Assuming she made no mistakes and typed non-stop at her usual speed, find the number of words in the report.

10. A rope is cut into three pieces in the ratio 1 : 3 : 5. Given that the length of the longest piece is 35 m, find
    **(a)** the length of the original rope;
    **(b)** the length of the shortest piece of rope.

11. A car travelled 100 km with half the distance at 40 km/h and the other half at 80 km/h. Find the average speed of the car for the whole journey.

12. **(a)** A photography competition offers \$2100 in prize money. Given that the prize money is divided among the first, second and third prize winners in the ratio 7 : 5 : 2, find the amount each prize winner gets.
    **(b)** It was decided later to introduce a fourth and fifth prize. The ratio of the first, second, third, fourth and fifth prize is then adjusted to 7 : 5 : 2 : 1 : 1. Given that the prize money is increased to \$2800, find how much each prize winner gets.

**13.** A lorry leaves a factory on a journey of 195 km at 08 45, travelling at an average speed of 52 km/h.

**(a)** Find the time at which the lorry arrives at its destination.

**(b)** On the return journey, the lorry leaves at 14 55 and arrives at the factory at 18 15. Calculate the time taken and the average speed of the lorry on the return journey.

**14.** A man parked his car in a carpark at 08 30 and retrieved it at 15 45 on the same day.

**(a)** How long did he park his car in the carpark?

**(b)** If the parking charges are $1.50 for the first hour and 80 cents for each subsequent half hour or part thereof, how much must he pay for parking his car there?

✶**15.** Peter cycles to visit his grandmother and then returns home by the same route. He always cycles at 4 km/h when going uphill, 12 km/h when going downhill, and 6 km/h when on level ground. If his total cycling time is 2 hours and 20 minutes, what is the total distance he cycles in km?

✶**16.** Two candles of the same height are lit at the same time. Each candle burns at a constant rate and the first candle takes 5 hours while the second candle takes 4 hours to burn completely. Find the time, in hours, taken for the height of the first candle to be four times that of the second candle.

✶**17.** A container is filled with 56 litres of pineapple juice. 8 litres of pineapple juice are extracted and the container is refilled with mango juice. The content of the container is thoroughly mixed and 8 litres of the mixture are then extracted and the container is again refilled with mango juice. What is the ratio of mango juice to pineapple juice in the final mixture?

✶**18.** Twelve men take 6 hours to finish a piece of work. After the 12 men have worked for 1 hour, the contractor decides to call in 8 more men so that the work can be completed earlier. How many more hours would 20 men take to complete the remaining work?

✶**19.** Ahmad and Kumar together can paint a house in 12 days. Kumar and Chong Beng together can complete the same job in 15 days while Ahmad and Chong Beng together will take 20 days to paint the same house. How many days will it take Ahmad, Kumar and Chong Beng to complete the job together?

*In this chapter, you will learn how to*

- *express one quantity as a percentage of another;*
- *compare two quantities by percentage;*
- *manipulate percentages and solve problems involving percentages;*
- *solve problems on personal and household finance and simple financial transactions.*

CHAPTER

11

# Arithmetical Problems, Percentages

## Introduction

Singaporeans consumed about 1240 million eggs in 1999. Assuming that the population of Singapore was 3.85 million then, how many eggs on average did each Singaporean consume in 1999?

In 2005, Singaporeans consumed approximately 1150 million eggs and the population was 4.08 million. Find the percentage change in the average consumption of eggs per Singaporean from 1999 to 2005.

# Percentages, Fractions and Decimals

## Percentages

We often see phrases like "Up to 75% off all items", "90% Housing Loan with Low Interest Rates" and "Fantastic Savings: 10 – 40% Genuine Discounts" in advertisements. 75%, 90%, 10% and 40% are examples of percentages.

Suppose there are 400 pupils in a school. During one afternoon, 120 pupils remain in school to participate in co-curricular activities. We say that 120 out of 400 pupils participate in CCA that afternoon. The fraction of pupils who participate in CCA is equal to $\frac{120}{400}$, which can also be written as $\frac{30}{100}$. We say that 30 **per cent** (or percent) of the pupils participate in CCA. The expression "per cent" simply means **for every hundred**, or **out of every hundred**. Thus, a percentage is a fraction whose denominator is **100**. We use the symbol % to represent **per cent**.

A percentage can be greater than 100%. For example, Mary had \$100 in her savings box. After one year, she saved \$150 more, the percentage of the increase in her savings box is $\frac{150}{100} \times 100\% = 150\%$.

FOR YOUR Information

*The word per cent originated from the Latin word "per centum" meaning "per hundred".*

Copy and complete the following table:

| Percentage | Fraction | Decimal |
|---|---|---|
| 65% | $\frac{65}{100} = \frac{13}{20}$ | $\frac{65}{100} = 0.65$ |
| 1.36% | | |
| | | 0.27 |
| | $\frac{3}{5}$ | |
| $13\frac{1}{2}\%$ | | |
| | | 0.095 |
| | | 1.12 |
| | $\frac{3}{4}$ | |
| 134.5% | | |

# Expressing One Quantity as a Percentage of Another

In a secondary school, 56 out of 70 teachers are female. What percentage of the teachers are female? What percentage of them are male?

We know that the fraction of female teachers in the school is $\frac{56}{70}$. Changing this fraction to percentage, we have $\frac{56}{70} \times 100\% = 80\%$. Hence 80% of the teachers in the school are female. The percentage of male teachers in the school = 100% – 80% = 20%.

In general, to express one quantity, $a$, as a percentage of another quantity, $b$, we

1. **write $a$ as a fraction of $b$,**
2. **multiply the fraction $\frac{a}{b}$ by 100% to convert it to a percentage.**

## Example 1

*A pupil scored 108 out of 150 marks in Mathematics and 112 out of 160 marks in English. Find the percentage mark for each subject.*

**Just for fun**

*Ninety-nine boys and one girl are in a lecture theatre. How many boys must leave the theatre so that the percentage of boys becomes 98%?*

**Solution**

The pupil's percentage mark for Mathematics $= \frac{108}{150} \times 100\%$
$= 72\%$

The pupil's percentage mark for English $= \frac{112}{160} \times 100\%$
$= 70\%$

# Finding the Percentage of a Number

**For your information**

*About 70% of the earth's surface is covered by water.*

If you are told that 75% of the pupils in a class of 40 passed a Mathematics test, how many of them failed?

The number of pupils who passed the test = 75% of the 40 pupils

$$= \frac{75}{100} \times 40$$
$$= 30$$

$\therefore$ number of pupils who failed the test = 40 – 30 = 10

**Thinking Time**

Is there another way of solving the question?

## Exercise 11a

1. During the economic downturn, a company retrenched 24 out of its 400 employees. What percentage of employees was retrenched?

2. In a survey, 120 pupils were asked which learning site they would like to visit as part of the National Education learning journey.

   24 said National Heritage Board
   36 said Battle Box
   54 said Singapore Discovery Centre

   The rest said they would like to visit the People's Association.
   Write these results as percentages.

3. Calculate the following:

   **(a)** $15\frac{1}{2}\%$ of \$640
   **(b)** 6.5% of 5000 people
   **(c)** 125% of 50 cm
   **(d)** 30.6% of 300 $l$

4. In a certain constituency, there are 8500 voters and on election day, 15% of them failed to vote. Calculate the number of people who voted.

5. A company finds that $4\frac{1}{4}\%$ of the tyres made are defective. The company made 28 000 tyres. How many tyres were defective?

6. **(a)** Soya beans on average contain 39.5% protein. A bushel of soya beans weighs 120 kg. How many kilograms of proteins are contained in a bushel?
   **(b)** A 100-acre field yields 50 bushels per acre. How many kilograms of protein does the field yield?

# Comparing Two Quantities by Percentages

The following conversation was heard during dinner time in the Lee's family.

Eng Hua: Dad, I lost 6 kg out of 60 kg in the past 3 months.

Ming Hua: Dad, I lost 5.5 kg out of 50 kg in the past 3 months.

Mrs Lee: Does it mean that Eng Hua has lost more weight in percentage on the weight-losing exercise?

Percentage lost by Eng Hua $= \frac{6}{60} \times 100\% = 10\%$.

Percentage lost by Ming Hua $= \frac{5.5}{50} \times 100\% = 11\%$.

Obviously, Ming Hua has lost more weight in percentage compared to Eng Hua.

## Example 2

*Teck Ann is a taxi driver who earns \$1300 per month while Ming Onn is a manager who earns \$6800 per month. In 2006, Teck Ann donated a total of \$1000 to charity while Ming Onn donated a total of \$4500 to charity. Who has donated a higher percentage of their annual income to charity?*

### Solution

Total income of Teck Ann in 2006 = \$1300 × 12 = \$15 600

$$\begin{aligned}\text{\% of income donated} &= \frac{1000}{15\,600} \times 100\% \\ &= 6.41\% \quad \text{(correct to 3 sig. fig.)}\end{aligned}$$

Total income of Ming Onn in 2006 = \$6800 × 12 = \$81 600

$$\begin{aligned}\text{\% of income donated} &= \frac{4500}{81\,600} \times 100\% \\ &= 5.51\%\end{aligned}$$

$\therefore$ Teck Ann had donated a higher percentage of his income to charity.

**Thinking Time**

Jean surveyed 20% of a group of students and 80% of another group of students.

In conclusion, Jean has done a survey on $\frac{20\% + 80\%}{2} = 50\%$ of the total number of students.

Do you think the conclusion is correct?

Explain your reason clearly.

## Percentages Greater than 100%

### Example 3

*The cost of hiring a car for 2 days in 2006 was \$264 which was 20% more than what it was in 2001. What was the cost of hiring a car for 2 days in 2001?*

**Solution**

"20% more than what it was in 2001" means that the hiring cost in 2006 was 120% of the hiring cost in 2001,

i.e. $\text{cost in 2001} \times \frac{120}{100} = \$264$

$\therefore \quad \text{cost in 2001} = \$264 \times \frac{100}{120} = \$220$

### Example 4

*The value of an antique vase appreciates each year by 15% of its value at the beginning of the year. If the vase costs \$156 000 in January 2006, what is its value at the end of 2007?*

**Solution**

**Method 1**

At the end of 2006, the value of the vase $= \$156\,000 \times \frac{100 + 15}{100}$
$= \$179\,400$

At the end of 2007, the value of the vase $= \$179\,400 \times \frac{115}{100}$
$= \$206\,310$

**Method 2**

Value of the vase at the end of 2007 $= \$156\ 000 \times \left(\frac{115}{100}\right)^2$

$= \$206\ 310$

## Exercise 11b

1. The value of an antique art piece appreciates each year by 12% of its value at the beginning of the year. If this art piece costs \$22 800 in January, what is its value at the end of 3 years? Correct your answer to the nearest 100 dollars.

2. In Mulu Village, there were 30 thousand people and it was estimated that about 4000 people attended a party to celebrate the new year. In Koka Village, there were 25 thousand people and 2800 people attended a celebrating party in the same year. Which village has the higher percentage of people who attended the party?

3. A science competition is marked out of 60. James obtains 43 marks, Roy 46 marks and Ben 49 marks. The examination board decided that those who score 80% and above will get a gold award, those who score 70% to 79% inclusive will get a silver award. What award does each boy get?

4. Billy reads 150 pages of a novel on Friday and 40% of the remaining pages on Sunday. If there are 600 pages in the novel, what percentage of the book remains to be read?

5. Nana sold her tea at 65¢ per gram in 2006, which was 25% more than what it was in 2004. What was the price of the tea in 2004?

6. A man's salary is \$1850 per month. Out of this, 20% goes to his rent, \$690 goes to food and \$940 on other expenses. What percentage has he over spent?

7. After spending 128% of his income, a man has to borrow \$756 from his friend to make up the difference. Find his income.

8. For each of two successive years, bean prices rose by 10%. At the beginning of the first year, 1 kg of bean cost \$20. At the end of the second year, how many grams of bean, to the nearest ten grams could be bought for \$50?

★9. A litre of orange fruit drink contains 22% orange juice. How many millilitres of orange juice must be added to produce a mixture containing 50% orange juice?

★10. Which would you rather have: 12% of \$330 or 330% of \$12? Explain why by showing detailed calculation.

★11. When negotiating for more pocket money with his father, Alan is given a choice: either an increase of 30% this year followed by a 20% increase next year, or an increase of 20% this year followed by a 30% increase next year. Which would Alan choose and why?

# Increasing/Decreasing a Quantity by a Given Percentage

The change in the value of an item can be expressed as a percentage increase or decrease in the original value. An **increase** of, say, 5% in the salary of a worker means that for every $100 in the **original** salary, there is an increase of $5, i.e., each $100 in the original salary becomes $105 in the **new** salary. Suppose the original salary for the worker is $1600, how much will his new salary be after a 5% increase?

$$\text{New salary} : \text{Original salary} = 105 : 100$$

$$\frac{\text{New salary}}{\text{Original salary}} = \frac{105}{100}$$

$$\text{New salary} = \frac{105}{100} \times \text{Original salary}$$

$$= \frac{105}{100} \times \$1600 = \$1680$$

**Notes**

*1. Original salary increases in the ratio 105 : 100 and the new salary is 105% of the original salary.*

*2. Original salary decreases in the ratio 95 : 100 and the new salary is 95% of the original salary.*

On the other hand, a **decrease** of, say, 5% in the salary means that for every $100 in the **original** salary, there is a decrease of $5, i.e., each $100 in the original salary becomes $95 in the **new** salary. In the above case,

$$\text{New salary} : \text{Original salary} = 95 : 100$$

$$\frac{\text{New salary}}{\text{Original salary}} = \frac{95}{100}$$

$$\text{New salary} = \frac{95}{100} \times \text{Original salary}$$

$$= \frac{95}{100} \times \$1600 = \$1520$$

## Example 5

*The workers of an electronic company were given an increase amounting to 8% in their monthly salaries.*

*(a) If Peter earned $1800 per month originally, find his monthly salary after the increment.*
*(b) If Paul's monthly salary after the increment is $1728, find his original monthly salary.*

**Solution**

(a) Peter's monthly salary after the increment is 100% + 8% = 108% of his original monthly salary.

His monthly salary after the increment = 108% of $1800
= $1800 × 1.08 = $1944

or $= \$1800 \times \frac{108}{100} = \$1944$

Alternatively, the increment = 8% of $1800 = $144

and thus Peter's new monthly salary = $1800 + $144 = $1944

(b) **Method 1: Use ratio**

Paul's salary after the increment : Paul's original salary = 108 : 100

Paul's original salary = $\$1728 \times \dfrac{100}{108} = \$1600$

**Method 2: Use an equation**

Let Paul's original salary be $\$x$.

His monthly salary after the increment = 108% of $\$x = \$\dfrac{108}{100}x$.

$$\therefore \quad \frac{108}{100}x = 1728$$

$$x = 1728 \times \frac{100}{108} = 1600.$$

$\therefore$ Paul's original salary is \$1600.

## Example 6

*After 6% of a bill has been deducted, \$282 remains to be paid. How much was the original bill?*

**Solution**

**Strategy 1: Use ratio**

Following a discount of 6%, 94% of the bill remains to be paid.

Original bill : \$282 = 100 : 94

$\therefore$ original bill = $\$282 \times \dfrac{100}{94} = \$300$

**Strategy 2: Use an equation**

Let his original bill be $\$x$.

The deduction = $\$(x - 282)$

$\therefore$ 6% of $\$x = \$(x - 282)$

i.e., $\dfrac{6}{100}x = x - 282$

or $\dfrac{94}{100}x = 282$

$$x = 282 \times \frac{100}{94} = 300$$

$\therefore$ the original bill is \$300.

## Example 7

*The cost of a television set is raised from \$600 to \$624. Find the percentage increase.*

**Solution**

The increase = \$624 – \$600 = \$24

$$\text{The percentage increase} = \frac{\text{Increase}}{\text{Original cost}} \times 100\%$$

$$= \frac{\$24}{\$600} \times 100\% = 4\%$$

## Example 8

*The cost of a piece of furniture is calculated as follows:*

*Wood \$300; Paint \$200; Wages \$200*

*If the costs of the wood and paint are increased by 12% and 7% respectively, while the wages are decreased by 10%, find the percentage increase or decrease in the cost of the furniture.*

**Solution**

| | *Original cost (\$)* | *Percentage change* | *New cost (\$)* |
|---|---|---|---|
| Wood | 300 | +12% | $300 \times \frac{112}{100} = 336$ |
| Paint | 200 | +7% | $200 \times \frac{107}{100} = 214$ |
| Wages | 200 | –10% | $200 \times \frac{90}{100} = 180$ |
| Furniture | 700 | | 730 |

The percentage increase in the cost of the furniture

$$= \frac{\text{Increase}}{\text{Original cost}} \times 100\% = \frac{730-700}{700} \times 100\% = \frac{30}{700} \times 100\% = 4\frac{2}{7}\%$$

## Explorations

Answer the following questions and compare the results.

1. In 2006, a man was given a pay increase of 10%. In 2007, he suffered a pay cut of 10%. Was he better off or worse off as compared to 2005?
2. In 2006, a man was given a pay cut of 10%. In 2007, he was given a pay increase of 10%. Was he better or worse off as compared to 2005?

## Exercise 11c

1. Increase 28 by 125%; what is the result?

2. Decrease 216 by $37\frac{1}{2}$%; what is the result?

3. The result of a number, when increased by 15%, is 161. Find this number.

4. The result of a number, when decreased by 20%, is 192. Find this number.

5. A flat costs 36% more today than when it was built. If the original cost of the flat was \$90 000, find its price today.

6. If 10% is deducted from a bill, \$58.50 remains to be paid. How much is the bill?

7. A property company sold 20% more houses in 2007 than it did in 2006. If the company sold 426 houses in 2007, calculate how many houses it sold in 2006.

8. The height of a tree was 4.8 m. After one year, the height of the tree was increased by 12.5%. Find its new height.

9. A machinist is hired at \$7.20 per hour.
   **(a)** On a particular day, he clocked in at 8.00 a.m. and clocked out at 3.45 p.m. Find his wage for that day.
   **(b)** At the end of a one-year probation period, his wage will increase by $27\frac{1}{2}$%. If the machinist successfully completes the apprenticeship,
   **(i)** what will his pay be per hour?
   **(ii)** what will his pay be per week, if he starts with a forty-hour week?

10. The production cost of a printer is as follows: overheads — \$80; wages — \$120; raw materials — \$100.

    If the cost of overheads increases by 20%, wages by 15% and raw materials by 11%, find the percentage increase in the production cost of the printer.

11. A new car costs \$120 000. After 1 year, its value decreases by 20%. For the second year, its value decreases a further 10%. What is the value of the car after 2 years?

12. James is 8% taller than John, and Wilson is 10% shorter than John. By what percentage is James taller than Wilson?

13. In 2006, a train carried 8% more passengers than in 2005. In 2007, it carried 8% more passengers than in 2006. What was the total percentage increase in the number of passengers from 2005 to 2007?

14. During 2006, a swimming lesson at a particular school lasted 45 minutes.
    **(a)** In 2007, the lesson time will be 50 minutes. Find the increase in the lesson time from 2006 to 2007 as a percentage of the lesson time in 2006.
    **(b)** The lesson time in 2007 will be 25% more than it was in 2005. Find how long the lesson lasted in 2005.
    **(c)** At present, James takes 240 seconds to swim 100 metres. In a year's time, he would have **improved** his swimming **time** by 25%. Calculate the time he will take to swim 100 metres in a year's time.
    **(d)** Robert, at present, also takes 240 seconds to swim 100 metres. In a year's time, he would have **increased** his swimming **speed** by 25%. Calculate the time he will take to swim 100 metres in a year's time.
    **(e)** Who will be the faster swimmer, James or Robert, in a year's time?

# Discount

Very often, retailers cannot sell merchandise with minor defects, overstocked items, discontinued models and so on at the retail selling prices. To clear the merchandise in stock, the retailers usually sell the items at a lower price, called the sale price. The difference between the original selling price or the **marked price** and the cheaper price or the **sale price** is called the **discount**.

**Discount = Marked price – Sale price**

Notes

***A discount***
*A reduction in the price of an article or commodity.*

Discount is often expressed as a percentage of the original price.

## Example 9

*A watch priced at \$160 is sold for \$140. Find the percentage discount.*

Solution

Discount = Marked price – Selling price
= \$160 – \$140 = \$20

$\therefore$ percentage discount = $\frac{20}{160} \times 100\% = 12.5\%$

## Example 10

*The charges for printing photographs from a digital camera in a shop are as shown in the table.*

*Mr Lee wants 30 photographs printed.*

*(a) How much does it cost him?*
*(b) How much will it cost him if a 15% discount is given?*

| *Charges* | |
|---|---|
| The first 20 prints | \$12.00 |
| The next 20 prints or less | \$0.50 per print |

Solution

(a) $12 + (30 - 20) \times 0.50 = 17$
$\therefore$ it costs him \$17.

(b) Discounted price : original price = 85 : 100

$\therefore$ discounted price = $\$17 \times \frac{85}{100} = \$14.45$. It will cost him \$14.45.

Alternatively, discount = 15% of the original price = \$17 × 15% = \$2.55
$\therefore$ it will cost him \$17 – \$2.55 = \$14.45.

## Example 11

*A ladies' bag selling for \$175 was marked down by 20% in a special promotion. Later, the sale price was marked further down by $17\frac{1}{2}$%. When the bag was still not sold, it was marked further down to a price 60% off the original selling price.*

*(a) What were the first two sale prices of the bag?*
*(b) What was the final selling price of the bag?*

### Solution

(a) The first sale price of the bag = (100 – 20)% of \$175 = \$175 × $\frac{80}{100}$ = \$140.

The second sale price of the bag = $\left(100 - 17\frac{1}{2}\right)$% of \$140 = 140 × $\frac{165}{2}$ × $\frac{1}{100}$ = \$115.50

(b) The final selling price of the bag = (100 – 60)% of \$175 = \$175 × $\frac{40}{100}$ = \$70.

## Commission

A **commission** is the payment an agent gets for selling or buying something on behalf of another person. It is usually given as a percentage of the cost price or the selling price.

## Example 12

*A flat was sold for \$220 000 by an agent who received a commission of 1.5%. How much commission did he receive?*

### Solution

1.5% of \$220 000 = \$220 000 × $\frac{1.5}{100}$
= \$3300

∴ the agent received \$3300 as commission.

# Value-added Tax and GST

In some countries, when you buy an article, you have to pay a certain amount of tax known as the value-added tax in addition to the price of the article. The tax payable is usually given as a certain percentage of the selling price. In Singapore, a goods and services tax (GST) of 3% is imposed on goods bought and services rendered, which came into effect on 1 April 1994. It was later revised to 5% in 2004.

## Example 13

*An article has a value-added tax of 15% imposed on it. If the marked price of the article is $64, calculate the total amount a man has to pay if he wants to buy it.*

**Solution**

Value-added tax payable = 15% of \$64 = $\$64 \times \frac{15}{100} = \$9.60$

$\therefore$ the total amount the man has to pay = \$64 + \$9.60 = \$73.60.

## Example 14

*An ink jet printer is advertised as $630 inclusive of a 5% GST. What is the original price of the printer?*

**Solution**

Original price : \$630 = 100 : 105

$\therefore$ original price = $\$630 \times \frac{100}{105} = \$600$

**Thinking Time**

Mr Ong and his family patronized a restaurant owned by his former pupil. The pupil gave Mr Ong a 20% discount. However, Mr Ong still had to pay a 10% service charge. The pupil gave Mr Ong two choices:
(a) claim the discount before paying the service charge,
(b) pay the service charge and then claim the discount.
Which of the two options is better for Mr Ong?

## Exercise 11d

1. Find the percentage discount of the following, given the marked prices and the sale prices:

| | Marked Price | Sale Price |
|---|---|---|
| **(a)** | \$100 | \$88 |
| **(b)** | \$580 | \$464 |

2. A supermarket gives a 10% discount on the marked prices of all items during a sale. Find for the following marked prices
   **(i)** the amount of discount given and
   **(ii)** the sale price:
   **(a)** \$45 **(b)** \$110

3. Find the marked prices of the following, given the amount of discount:
   **(a)** A 7% discount which is \$49.
   **(b)** A $33\frac{1}{3}\%$ discount which is \$270.

4. Find the marked prices of the following, given the percentage discount and the sale prices:
   **(a)** 12% discount, \$77
   **(b)** 25% discount, \$123

5. At a sale, the price of a washing machine was reduced by 12% to \$440. What was the original price of the washing machine?

6. A shopkeeper marks the price of an article at \$80. Find the selling price of the article if
   **(a)** he gives a discount of 10%,
   **(b)** he gives two successive discounts of 5% each.

7. During a sale a shopkeeper reduced the prices of all his goods by 15%. Calculate the original selling price of a calculator, which was sold for \$23.80 during the sale.

8. John bought an air-conditioner priced at \$800, but was given a discount of 12.5%. Calculate the price he paid.

9. The price of a camcorder originally marked at \$2000 was reduced by 20% during a special promotion. The price was then reduced by an additional 30% to clear the stock. What was the sale price for each reduction? Will the final price be the same if the camcorder was reduced by 30% first and followed by 20%?

10. The cost of a colour film for 36 photographs is usually \$3.60. Additionally, the cost of processing and printing the 36 photographs is usually \$12.60.
    **(a)** Show that the **total** cost of each photograph is usually 45 cents.
    **(b)** A shop keeps the price of the film at \$3.60, and offers a reduction of 20% off the cost of processing and printing. Calculate the total cost of each photograph at this shop.
    **(c)** Calculate the percentage reduction in the total cost, compared to the usual cost, of a photograph obtained from this shop.

11. A property agent charges a commission of 5% on the first \$10 000 and $2\frac{1}{4}\%$ on the remainder of the selling price. Find the amount of commission he will receive if he sells a piece of property for \$46 000.

12. Mr. Goh's monthly income consists of \$500 plus a commission of 4% on all his sales. Find his total sales for a particular month if his income was \$1220 that month.

13. A tour guide earns commission by bringing tourists to patronise a certain handicraft shop. Given that the commission he receives is 3% of the total sales, calculate his commission on a particular day when the 12 tourists he brought to the shop spent an average of \$250 each.

14. Three salesmen went to 400 households to sell a certain brand of rice cooker costing \$60 each. 15% of the households bought a rice cooker each from them. If the three salesmen sold the rice cookers in the ratio 3 : 4 : 5, find the commission each salesman received if the commission is 4% of the total sales.

15. Calculate the total amount of money a man has to pay for an article marked at \$85 with a 14% value-added tax imposed on it.

16. How much does a customer have to pay for an article costing \$240 with a 5% GST imposed on it?

17. An article is advertised for \$262.50 inclusive of a 5% GST. Find the original price.

## SUMMARY

1. A percentage is a fraction whose denominator is 100 and we use % to represent percent. A percentage can be converted to a fraction by dividing it by 100.

2. Discount = Original Selling price – Sale price

## Review Examples 1

### Example 1

*A shopkeeper sold two computer games for \$48 each.*

*He made a 25% profit on one computer game and a loss of 20% on the other. What was his net gain or loss on the sale of the two computer games?*

**Solution**

Suppose the cost price of the first computer game is \$$x$.

Then $x + 25\%$ of $x = 48$

That is $125\%$ of $x = 48$

So $\qquad x = 48 \div 125\%$
$\qquad\quad = 38.40$

The profit of selling the first computer game is \$(48 – 38.40) = \$9.60

Suppose the cost price of the second computer game is \$$y$.

Then $\qquad y - 20\% \text{ of } y = 48$

That is $\qquad 80\% \text{ of } y = 48$
$\qquad\qquad y = 48 \div 80\% = 60$

The loss of selling the second computer game is \$(60 – 48) = \$12.

So the net loss of selling the two computer game is \$(12 – 9.60) = \$2.40.

## Example 2

*The diagram shows a cuboid with a square base of side 10 cm and a height y cm. Given that the length of the base increases by 10% while the height decreases by 10%. Find the percentage difference in the volume of the cuboid.*

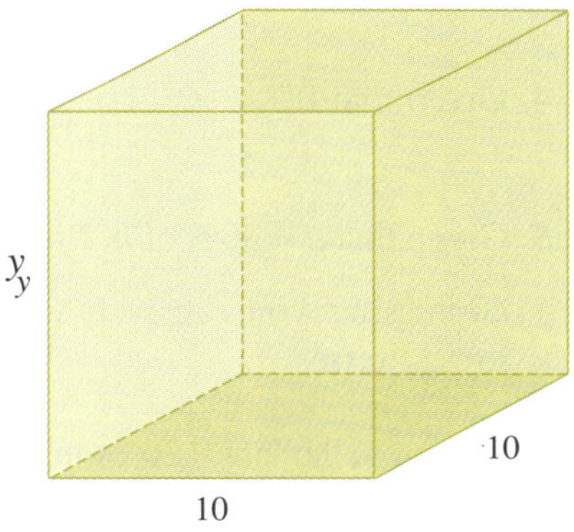


**Solution**

Original volume of the cuboid = $10^2y = 100y$

The side of the new base = $\frac{110}{100} \times 10 = 11$

The new height = $\frac{90}{100}y = 0.9y$

$\therefore$ new volume = $(11)^2 \times 0.9y = 108.9y$.

Increase in volume = $108.9y - 100y$
$= 8.9y$

$\therefore$ Percentage difference = $\frac{8.9y}{100y} \times 100\%$
$= 8.9\%$

# Review Questions 1

1. A school is given a 65% discount for buying textbooks in bulk. How much does the school have to pay for 100 books if each book costs \$10 before the discount?

2. Yiwei receives \$28 per week in pocket money. If she decides to save 20% of it, how much will she
   **(a)** save and
   **(b)** spend in a year?
   Assume that there are 52 weeks in a year.

3. Arrange 1.74, $1\frac{2}{3}$, 1.56, 173% and $1\frac{3}{4}$ in descending order.

4. Given that $0.7 : \frac{7}{8} = 7\% : x$, find $x$.

5. Given that $a = 30\%$ of $b$, find the value of $\frac{a}{4b}$, expressing your answer as a fraction in its lowest terms.

✶ 6. A fruit seller has 120 oranges. Given that he has 20% more apples than oranges and 40% less oranges than pears, find the number of apples and the number of pears the fruit seller has.

7. Mr Chen decided to buy a new car costing \$60 000. He paid for his new car in cash and was given a discount. Given that he paid \$57 000 for his new car, calculate the percentage discount he received.

✶ 8. **(a)** A retailer bought a compact disc from a manufacturer for \$20. In addition to that, he paid a 15% value-added tax. If he sold the disc to a customer for \$26, calculate the cash profit he made.
   **(b)** The manufacturer later increased the price of the compact disc by 20%. At the same time, the value-added tax was increased to 25%.
   **(i)** If the retailer made the same cash profit as before, calculate the price a customer had to pay for a disc.
   **(ii)** Find the cash profit the retailer made on each disc if he sold each one at a price which was 30% more than the cost price.

✶ 9. A jacket costs \$300. During a sale, a discount of 15% is given for cash payment only. John is paying by cash and a \$100 voucher. The sales girl calculates the payment in the following way: John's cash payment is $\$(300 - 100) \times 85\% = \$170$.
   If John pays the whole amount by cash, would it be cheaper? What is the difference?

✶ 10. John accepted a reduction of 15% in his salary when his company was not doing well. Now his company's financial position has improved and his boss wants to restore his original salary. By what percentage must his reduced salary be increased?

✶ 11. The radius of a cylinder is increased by 15% and its height is decreased by 20%. Find the percentage change in the volume of the cylinder.

✶ 12. Mr Sim instructed his assistant to place an order for 5 pairs of leather shoes and a number of pairs of canvas shoes. A pair of leather shoes costs two and a half times as much as a pair of canvas shoes. His assistant made a mistake in the order and the number of pairs of the two types of shoes had been interchanged. This increased the bill by $33\frac{1}{3}\%$.
   Find the ratio of the number of pairs of canvas shoes to the number of pairs of leather shoes in the order that Mr Sim instructed his assistant to place.

∗**13.** Kelvin read 60 pages of a book on the first day. This was 20% more than the number of pages he read on the second day. Given that he read $\frac{1}{6}$ of the book on the second day, find the number of pages in the book.

**14.** With a \$1000 purchase of a sofa set, Mrs Dora is offered 3 successive discounts of 10%, 20% and 25% in any order she wishes. Which order of discount will benefit her the most? Explain why.

*In this chapter, you will learn how to*

- *plot coordinate points on a graph;*
- *plot a set of ordered pairs satisfying a function;*
- *understand the gradient of a straight line.*

CHAPTER

# 12

# Functions and Graphs

## Introduction

An explorer found a map of an island that hinted that there are hidden treasures at four locations on the island. He and his team landed at the jetty and he was wondering how to lead his team to the various locations. We can locate the various locations with the help of the concept of coordinate geometry.

Can you help the explorer to locate these points after the lesson?

# Rectangular Coordinates in Two Dimensions

Have you ever noticed how a seat in a cinema is identified? How about the position of a chess piece on the chessboard and the location of a place in a sectional map in the Singapore Street Directory? Can you use the same idea to identify a desk in a classroom where the desks are arranged neatly in rows and columns?

At the beginning of a new school year, some teachers will prepare a seating arrangement plan of a class (as shown below) in order to know the students quickly. Each box represents the position of a pupil.

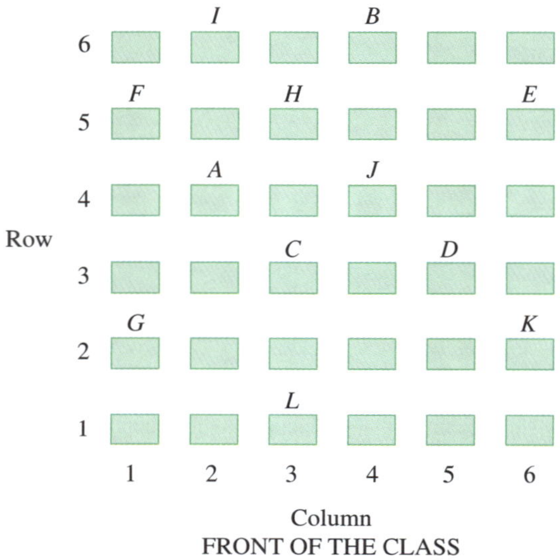


*Fig. 12.1*

To identify each student more easily, the teacher can connect each student with the column and the row in which he or she is sitting. Thus, student $A$ sits in column 2, row 4, while student $B$ sits in column 4, row 6 and so on.

The teacher can write a pair of numbers against the name of a student in the class list as follows:

| | | |
|---|---|---|
| $A(2, 4)$ | $B(4, 6)$ | $C(3, 3)$ |
| $D(5, 3)$ | $E(6, 5)$ and so on. | |

From the pair of numbers (2, 4), we know that student $A$ is in column 2 and in row 4. The pair of numbers (4, 6) tells us that student $B$ is in column 4 and in row 6.

Can you write down the pairs of numbers corresponding to students $F$, $G$, $H$, $I$, $J$, $K$ and $L$?

**FOR YOUR INFORMATION**

*Rene Descartes (1596–1650), the great French mathematician, developed the idea of combining algebra with geometry. This resulted in the formation of a new field of mathematics called analytical geometry or Cartesian geometry in honour of him.*

**Thinking Time**

1. Is the order in which the two numbers are written important? That is, do (5, 3) and (3, 5) indicate the same position?

2. The pairs of numbers (2, 4), (4, 6), (3, 3), (5, 3), (6, 5) and so on are examples of **ordered pairs**. Do you know why they are called ordered pairs?

Figure 12.2 displays the same classroom plan on graph paper with horizontal lines and vertical lines drawn through the centres of the boxes showing the positions of the students. The horizontal lines and vertical lines are numbered as shown.

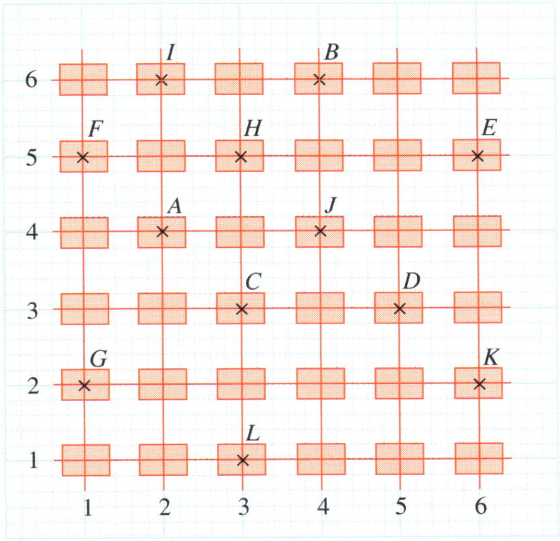


*Fig. 12.2*

Thus, the first number in each ordered pair used to locate a student refers to the horizontal scale while the second number refers to the vertical scale. To simplify it even further, we can now use a point (indicated by crosses) to replace the boxes to show the students' positions. This gives us an idea of locating a point in a plan.

# The Rectangular or Cartesian Plane

In Chapter 3, we discussed how to locate a point representing a real number on a number line. We shall now discuss how to use an ordered pair of real numbers ($a$, $b$) to determine a point in a plane.

**Just for fun**

*In chess, the knight is normally represented by a horse's head. A knight moves to its new position by moving two steps (two squares) horizontally (either left or right) and one step vertically (either upwards or downwards) or two steps vertically and one step horizontally as shown in the diagram.*

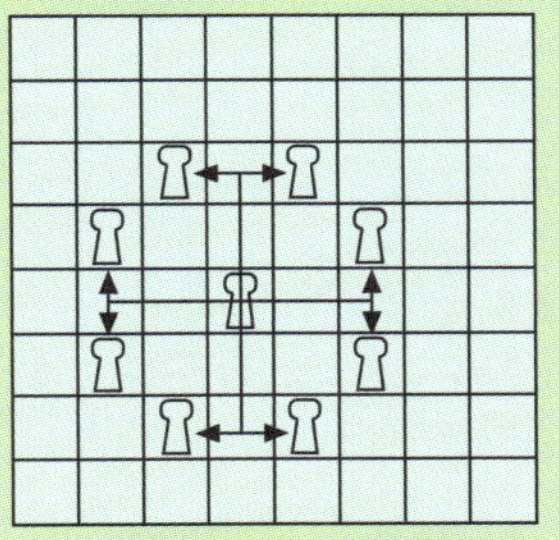

*(a) If a knight starts from position 1 in the diagram below, what is the minimum number of moves required for it to reach the shaded square?*

*(b) Find the minimum number of moves required in each case if the knight starts from 2, 3, 4, 5, 6, 7 or 8.*

*(c) If the knight starts from position 1, is it possible for it to visit all the 64 squares once only? (Use a pencil to cross out the squares it has visited.)*

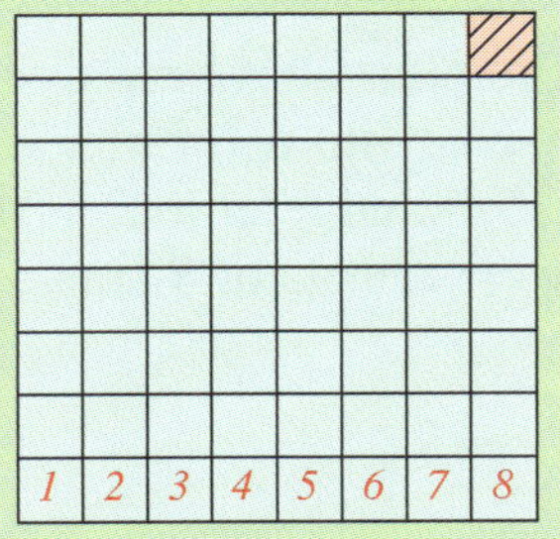

Fig. 12.3 below shows a **rectangular** or **Cartesian** plane consisting of two number lines $Ox$ and $Oy$ intersecting at right angles at the point $O$.

The point $O$ is known as the **origin**. $Ox$ and $Oy$ are called the **coordinate axes** or the **horizontal axis** and **vertical axis** respectively, or more simply as the ***x*-axis** and ***y*-axis**.

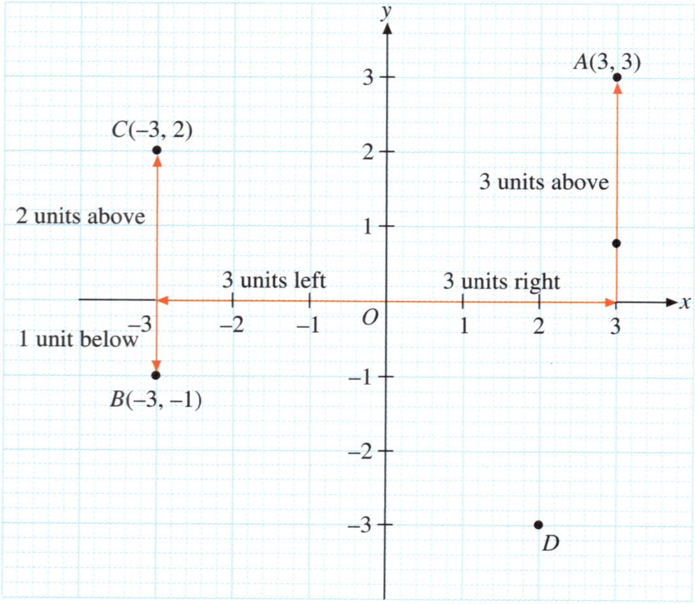


*Fig. 12.3*

## Coordinates of a Point

The position of any point in the plane can be determined by its distance from each of the axes. In Fig. 12.3, point $A$ is 3 units to the **right** of the $y$-axis and 3 units **above** the $x$-axis; its position is described by the ordered pair (3, 3). The ordered pair (–3, –1) determines point $B$ since it is 3 units to the **left** of the $y$-axis and 1 unit **below** the $x$-axis. Point $C$ is 3 units to the **left** of the $y$-axis and 2 units **above** the $x$-axis and thus it is represented by the ordered pair (–3, 2).

Can you name the ordered pair that determines point $D$? What ordered pair represents the origin $O$?

From the above discussion, each point $P$ in the plane is located by an ordered pair $(a, b)$. We call $a$ the ***x*-coordinate** (or **abscissa**) of $P$, and $b$ the ***y*-coordinate** (or **ordinate**). We say that $P$ has **coordinates** $(a, b)$ and refer to $P$ as the point $(a, b)$ or $P(a, b)$. Thus, $A$ has coordinates (3, 3). The $x$-coordinate and $y$-coordinate of $B$ are –3 and –1 respectively. $C$ is the point (–3, 2).

## Exercise 12a

1. Write down the coordinates of the points in the diagram below:

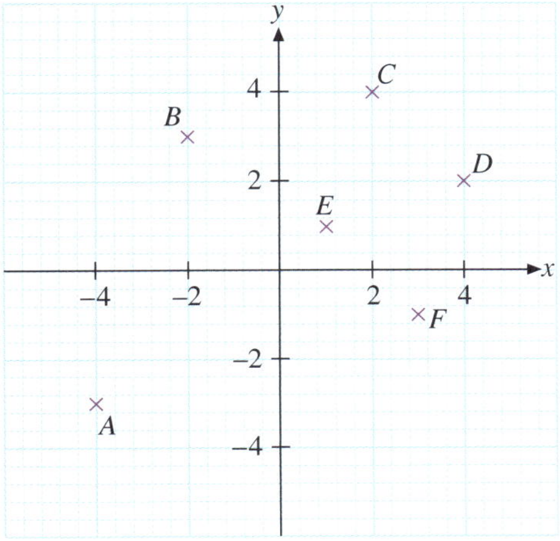


2. Write down the coordinates of the points shown in the diagram below:

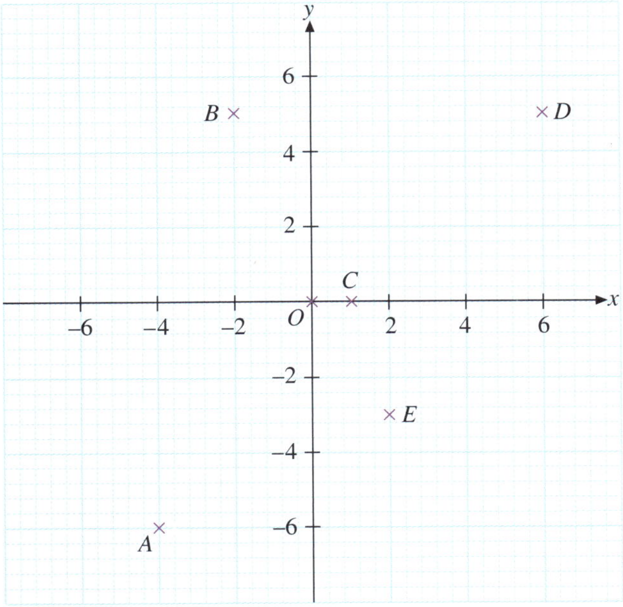


3. Plot the following points on a piece of graph paper:
$A(2, 5)$, $B(1, 2)$, $C(-3, -1)$, $D(6, -2)$, $E(3, -2)$, $F(-1, 2)$.

4. Plot the following points on a piece of graph paper:
$W(-4, 0)$, $X(3, 6)$, $Y(0, 4)$, $Z(5, -5)$.

5. Plot each set of the given points on graph paper. Join the points in order with straight lines and identify the geometrical shapes obtained.
   **(a)** (6, 4), (−6, 4), (−6, −4), (6, −4).
   **(b)** (0, 5), (−6, 0), (0, −5), (6, 0).
   **(c)** (0, 0), (0, 8), (5, 4).
   **(d)** (1, 0), (0, 3), (−1, 4), (−5, −2).
   **(e)** (4, 0), (0, 5), (−4, 0)
   **(f)** (5, 2), (−1, 3), (−1, −3), (5, −2).
   **(g)** (0, −3), (6, −3), (4, 2), (−1, 2)

6. The vertices of a right-angled triangle are $A(1, 0)$, $B(7, 0)$ and $C(1, 8)$. Plot the points $A$, $B$ and $C$. Hence, find the area of $\triangle ABC$.

7. Plot the following points on a piece of graph paper:

   (3, −5), (2, −3), (1, −1), (0, 1), (−1, 3), (−2, 5), (−3, 7).

   Do you notice that the points lie in a special pattern? Describe the pattern.

8. **(a)** Plot the following points on a piece of graph paper:
   (−3, −1), (−2, 0), (−1, 1), (0, 2), (1, 3), (2, 4), (3, 5).
   **(b)** Draw a line through all the points.
   **(c)** What is the shape obtained?

9. **(a)** Plot the following points on a piece of graph paper:

   (−3, 5), (−2, 0), (−1, −3), (0, −4), (1, −3), (2, 0), (3, 5).

   **(b)** Join all the points with a smooth curve.

# The Idea of Functions

There are numerous relationships involving two variables in our world.

Let us look at some examples involving the numerical relationship between two variables.

(a) The total cost of 20 tickets to a particular musical at the Esplanade by the Bay, at \$75 per ticket is \$1500. In this example, the total cost is the first variable and the second variable is the number of tickets.

Let $y$ be the first variable and $x$ be the second variable. Then, we have $y = 75x$. We say that $y$ is a function of $x$.

(b) The perimeter of a square is equal to 4 times its side. In this case, the perimeter is the first variable and the second variable is the length of the sides of the square.

Let $y$ be the first variable and $x$ be the second variable. Then, we have $y = 4x$ as the function relating $x$ and $y$.

(c) The monthly pay of a salesman is the sum of a basic pay of \$450 plus a commission of \$20 for every unit of goods sold. The monthly pay is the first variable and the number of units of goods sold is the second variable in this case.

Let $y$ be the first variable and $x$ be the second variable, then we have the function $y = 450 + 20x$.

**FOR YOUR Information**

*Sometimes, we also use the notation $f(x) = 75x$, $f(x) = 4x$ and $f(x) = 450 + 20x$ to represent the functions.*

For a function, each input produces only one output. For example, for $y = 75x$, given $x = 1$, we get only one value of $y$, that is 75.

## Example 1

*In the function $y = 4x$, calculate the value of $y$ when*
*(a) $x = 2$,* *(b) $x = 5$.*

**Solution**

(a) When $x = 2$, $y = 4(2) = 8$
(b) When $x = 5$, $y = 4(5) = 20$

## Example 2

*(a) In the function $y = 20 - 4x$, calculate the value of y when*
*(i) $x = 3$,* *(ii) $x = 8$.*

*(b) In function $y = 3x - 2$, calculate the value of y when*
*(i) $x = 6$,* *(ii) $x = -7$.*

### Solution

(a) (i) When $x = 3$, $y = 20 - 4(3) = 8$
(ii) When $x = 8$, $y = 20 - 4(8) = -12$

(b) (i) When $x = 6$, $y = 3(6) - 2 = 16$
(ii) When $x = -7$, $y = 3(-7) - 2 = -23$

## Ordered Pairs Satisfying a Function

Consider the function $y = 2x$, we shall look at all pairs of numbers $x$ and $y$ that satisfy the function $y = 2x$.

When $x = 1$, $y = 2(1) = 2$
$x = 2$, $y = 2(2) = 4$
$x = 3$, $y = 2(3) = 6$
$x = 4$, $y = 2(4) = 8$ etc

Thus the pair of numbers $x = 1$, $y = 2$ satisfy the function $y = 2x$. Similarly, $x = 3$, $y = 6$ also satisfy the function $y = 2x$ etc. Obviously, we cannot list all the values of $x$ and $y$ that satisfy the function $y = 2x$. Can we find a way to display the function $y = 2x$?

The table below shows five pairs of values of $x$ and $y$ that satisfy the function $y = 2x$.

| $x$ | 1 | 2 | 3 | 4 | 5 |
|---|---|---|---|---|---|
| $y = 2x$ | 2 | 4 | 6 | 8 | 10 |

The five pairs of numbers $(x, y)$ in the table above are plotted as points in the Cartesian plane as shown in Fig. 12.4 below. The scale for both the $x$- and $y$-axes is 1 cm to 1 unit.

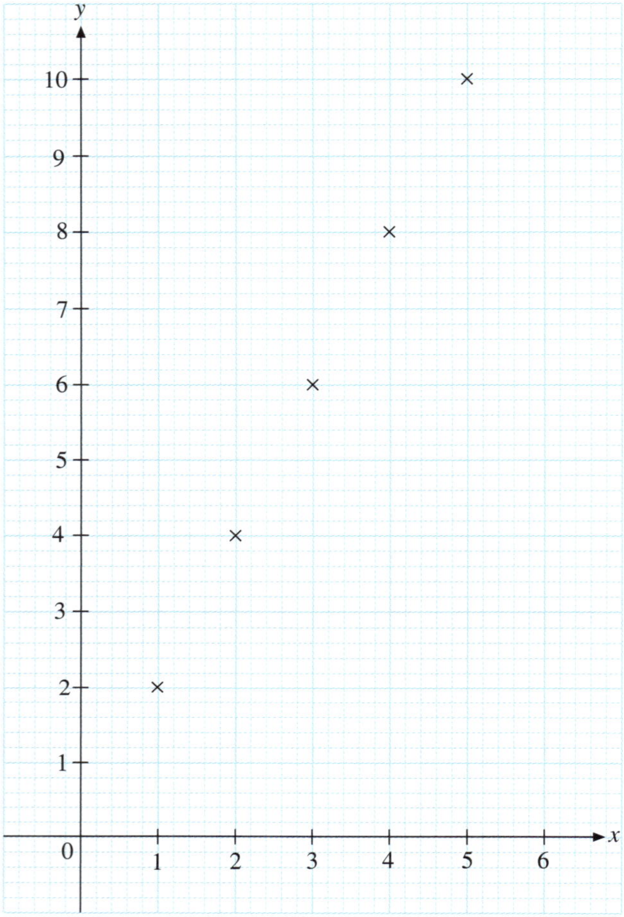


*Fig 12.4*

Do you notice any pattern formed by the points? How does the value of $y$ changes as the value of $x$ increases by 1? Do all the values of $y$ change accordingly as the values of $x$ increase?

In the following graph, Fig. 12.5, a straight line is drawn to pass through the five points.

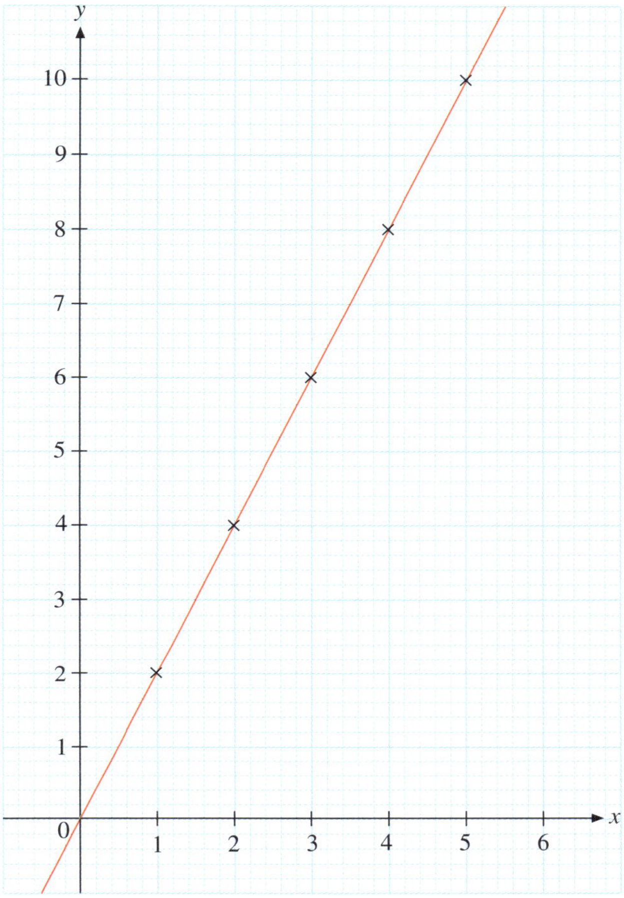


*Fig 12.5*

The line is said to be a graph of the linear function $y = 2x$. Do you agree that every pair of numbers $(x, y)$ satisfying $y = 2x$ appear as a point somewhere on the line and that every point on the line has coordinates that satisfy the function $y = 2x$? The graph of the function is the set of all points whose coordinates satisfy the function.

Let us consider another function $y = 2x + 1$, we shall look at some pairs of numbers $x$ and $y$ that satisfy the function $y = 2x + 1$.

When $x = -2$, $y = 2(-2) + 1 = -3$
$x = -1$, $y = 2(-1) + 1 = -1$
$x = 0$, $y = 2(0) + 1 = 1$
$x = 1$, $y = 2(1) + 1 = 3$
$x = 2$, $y = 2(2) + 1 = 5$

Thus the pairs of numbers $x = -2$, $y = -3$, and $x = -1$, $y = -1$ etc satisfy the function $y = 2x + 1$.

The table below shows five pairs of values of $x$ and $y$ that satisfy the function $y = 2x + 1$.

| $x$ | $-2$ | $-1$ | 0 | 1 | 2 |
|---|---|---|---|---|---|
| $y = 2x + 1$ | $-3$ | $-1$ | 1 | 3 | 5 |

The five pairs of numbers $(x, y)$ appearing in the table on the previous page are plotted as points in the Cartesian plane as shown in Fig. 12.6 below where we use 1 cm to represent both the $x$ and $y$-axes.

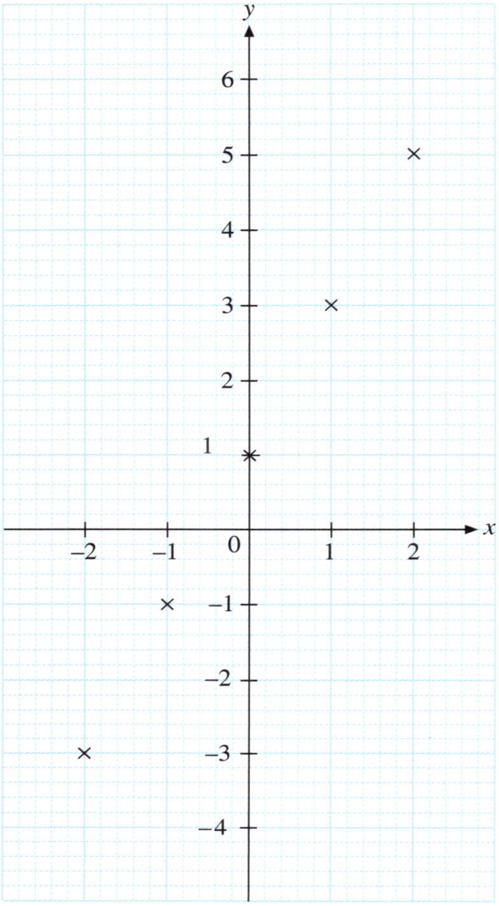


*Fig 12.6*

Do you notice any pattern formed by the points? How does the value of $y$ change as the value of $x$ increases by 1? Do all the values of $y$ change accordingly as the values of $x$ increase?

In the following graph, Fig. 12.7, a straight line is drawn to pass through the five points.

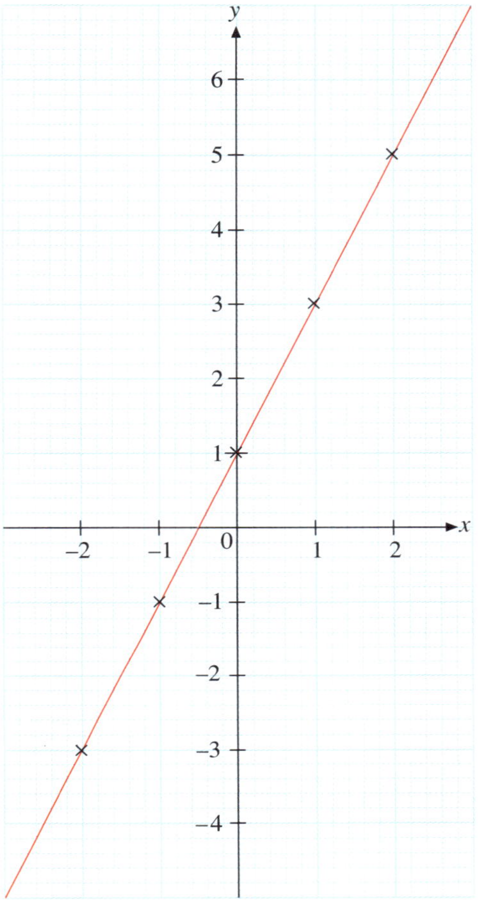


*Fig 12.7*

The line is the graph of the linear function $y = 2x + 1$.

In the previous two examples, we had plotted five points and a line is then drawn to join the five points. What is the minimum number of points we need to draw a linear function? How do we ensure that we have calculated the coordinates of the points satisfying the functions accurately? How many points should we plot to draw a line accurately?

# Explorations

1. (a) On a piece of graph paper and using a scale of 1 cm to represent 1 unit on the $x$-axis, draw the $x$-axis for values of $x$ from –8 to 8.
   (b) Using a scale of 1 cm to represent 2 units on the $y$-axis, draw the $y$-axis for values of $y$ from –10 to 16.
   (c) Select any three pairs of values of $(x, y)$ satisfying the function $y = 2x + 2$. Display these three pairs of values in the form of a table.

| $x$ | | | |
|---|---|---|---|
| $y = 2x + 2$ | | | |

   (d) Plot the three points on the graph paper and join them using a straight line.
   (e) Repeat steps (c) and (d) for the following functions.
   (i) $y = 2x + 8$ (ii) $y = 2x - 3$
   (iii) $y = 2x - 6$ (iv) $y = 2x - 8$
   Do you notice any similarity in all the lines you have plotted?

2. Repeat the steps of (1) above for the following functions on a fresh sheet of graph paper
   (a) $y = 3x + 7$ (b) $y = 3x + 5$
   (c) $y = 3x - 3$ (d) $y = 3x - 6$
   What do you notice about the similarity in the lines you have drawn?

3. Repeat the steps of (1) above for the following functions on another piece of graph paper.
   (a) $y = -2x + 5$ (b) $y = -2x + 3$
   (c) $y = -2x$ (d) $y = -2x - 4$
   (e) $y = -2x - 7$
   Is there any similarity in the lines you have drawn?

4. Repeat the steps of (1) above for the following functions on a new piece of graph paper.
   (a) $y = -4x + 8$ (b) $y = -4x + 2$
   (c) $y = -4x$ (d) $y = -4x - 3$
   (e) $y = -4x - 6$
   Do you notice that a linear function has the form $y = mx + c$ where $m$ and $c$ are constants?

# Gradient of a Straight Line

Have you ever climbed up a mountain and noticed that some paths are rather steep while others are more gentle? The steeper the path, the harder it is to climb.

We call the measure of the steepness **gradient**. It is a measure of the ratio of the vertical change to the horizontal change.

Explorations

Consider the linear functions that we have drawn so far. Some of these are produced below, Fig. 12.8(a) to Fig. 12.8(d).

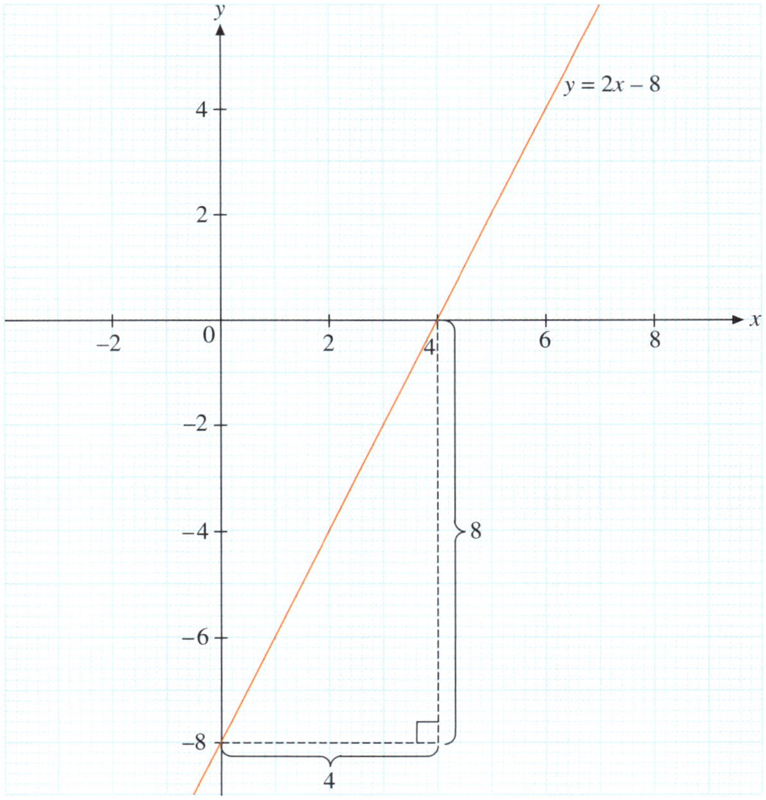


*Fig 12.8a*

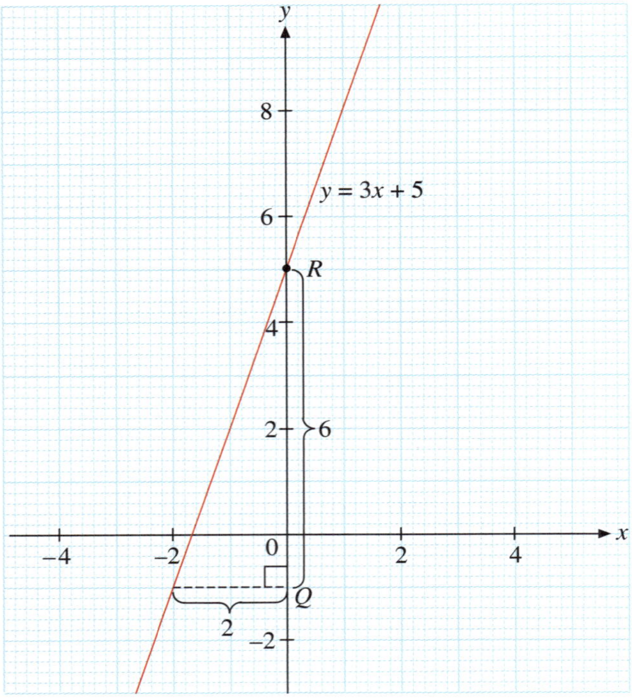


*Fig 12.8b*

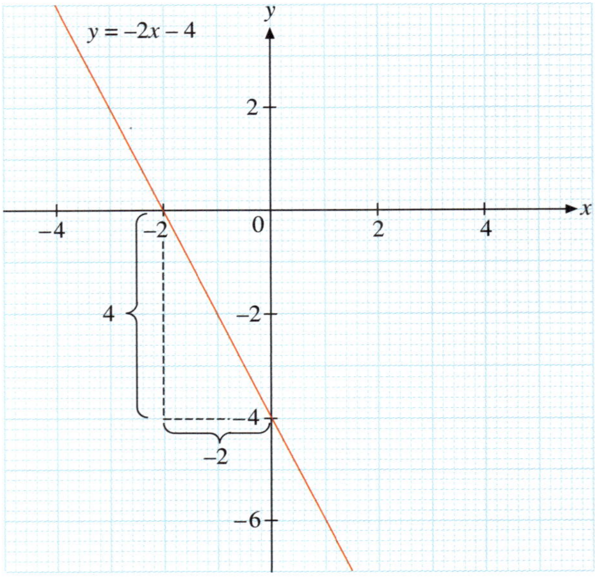


*Fig 12.8c*

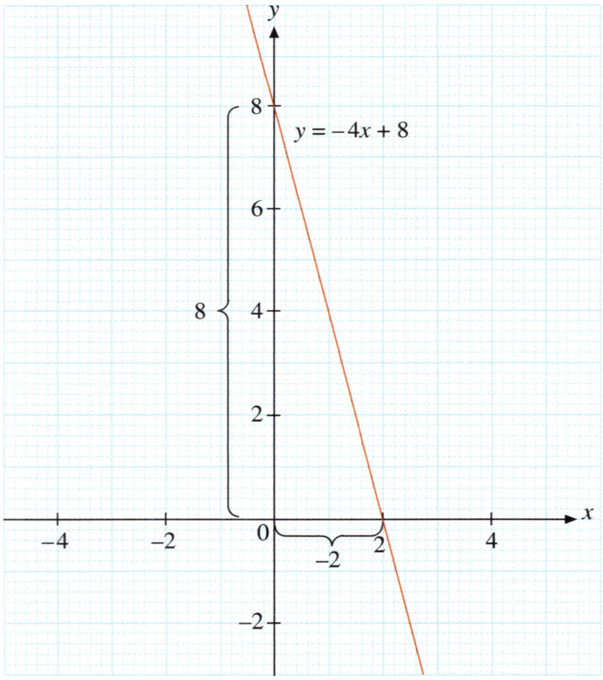


*Fig 12.8d*

In fig. 12.8(a), the horizontal change ($AB$) is equal to 4 units and the vertical change ($BC$) is 8 units. The ratio of $\frac{\text{Vertical change}}{\text{Horizontal change}}$ is thus equal to $\frac{8}{4} = 2$.

In Fig. 12.8(b), the horizontal change ($PQ$) is equal to 2 units and the vertical change ($QR$) is 6 units.

Hence, the ratio of $\dfrac{\text{Vertical change}}{\text{Horizontal change}} = 3$.

In Fig. 12.8(c) and Fig. 12.8(d), the ratios of $\dfrac{\text{Vertical change}}{\text{Horizontal change}}$ are $-2$ and $-4$ respectively.

Complete the table shown below.

| Line | Horizontal change | Vertical change | Gradient = $\dfrac{\text{Vertical change}}{\text{Horizontal change}}$ | Diagram |
|---|---|---|---|---|
| 1. $y = 2x - 8$ | 4 | 8 | $\dfrac{8}{4} = 2$ | 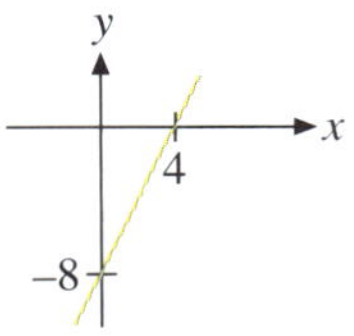 |
| 2. $y = 3x + 5$ | | | | |
| 3. $y = -2x - 4$ | $-2$ | 4 | $\dfrac{4}{-2} = -2$ | 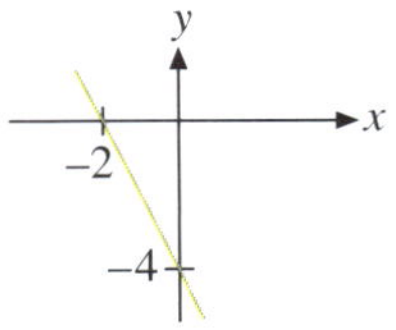 |
| 4. $y = -4x + 8$ | | | | |

Do you notice how the gradient affects the orientation of the straight line?

For the line $y = 2x - 8$, the gradient is a positive 2, the line rise from left to right  .

For the line $y = -2x - 4$, the gradient is a negative 2, the line falls from left to right  .

*In real life, we can liken the ratio of the vertical change to the horizontal change to the steepness of the slope of a hill, if you are to climb the slopes represented by the lines $y = 2x - 8$ and $y = 3x + 5$, which line is steeper?*

## Example 3

*Draw the graph of each of the following linear functions*

*(a)* $y = 4x + 6$ *(b)* $y = -2x + 5$

*and hence find the gradient of each function.*

**Solution**

(a) $y = 4x + 6$

When $x = 0$, $y = 4(0) + 6 = 6$

When $x = 2$, $y = 4(2) + 6 = 14$

The graph of $y = 4x + 6$ is shown on the right.

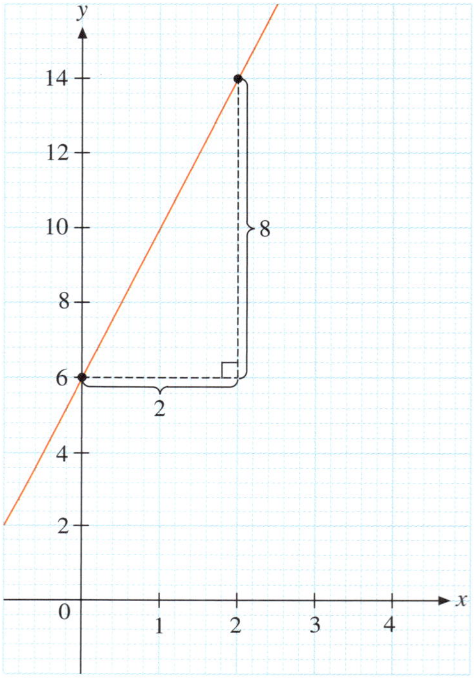


From the graph, the vertical change is 8 units and the horizontal change is 2 units.

$\therefore$ the gradient of the line $= \frac{8}{2} = 4$.

(b) $y = -2x + 5$

When $x = 0$, $y = -2(0) + 5 = 5$

When $x = 3$, $y = -2(3) + 5 = -1$

The graph of $y = -2x + 5$ is shown on the right

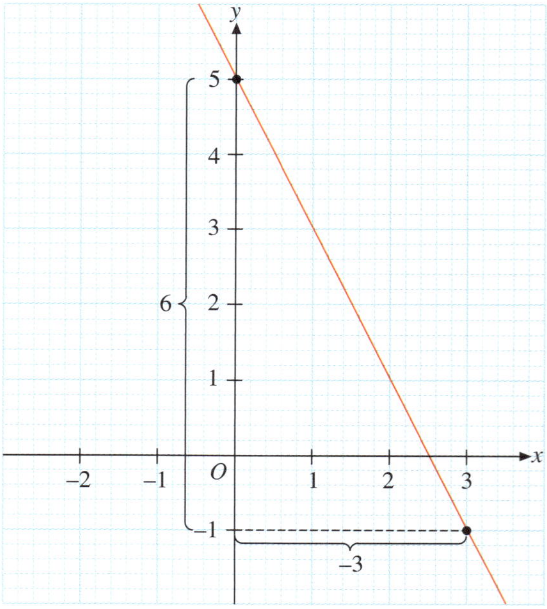


From the graph, the vertical change is 6 units and the horizontal change is –3 units.

$\therefore$ the gradient of the line $= \frac{6}{-3} = -2$

## Exercise 12b

**1.** Given a function $y = 4x + 5$, find the values of $y$ when

**(a)** $x = -2$, **(b)** $x = 3$,

**(c)** $x = 7$.

**2.** Given a function $y = 25 - 3x$, find the values of $y$ when

**(a)** $x = -5$, **(b)** $x = 0$,

**(c)** $x = 15$.

**3.** Given a function $y = \frac{2}{3}x + \frac{1}{3}$, find the values of $y$ when

**(a)** $x = -3$, **(b)** $x = 1\frac{1}{2}$,

**(c)** $x = 3\frac{1}{4}$.

**4.** Given a function $y = 20 - 6x$, find the values of

**(a)** $x = -5$, **(b)** $x = 1$,

**(c)** $x = 9$.

**5.** Draw the graph of each of the following linear functions.

**(a)** $y = 3x + 4$ **(b)** $y = 2x - 7$

**(c)** $y = -x + 5$ **(d)** $y = -4x - 2$

**6.** Find the gradient of each of the following linear functions.

**(a)**

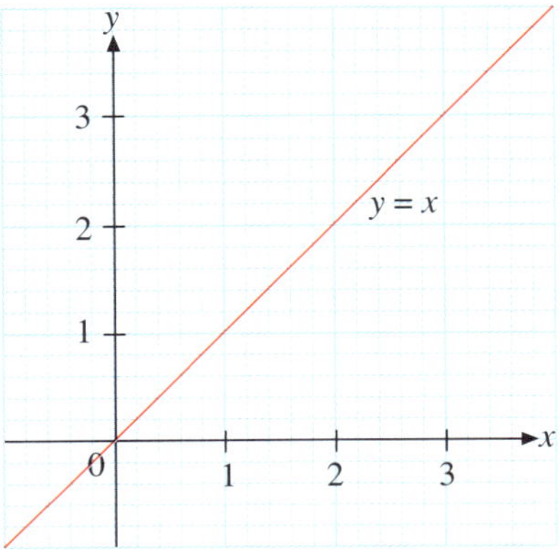


**(b)**

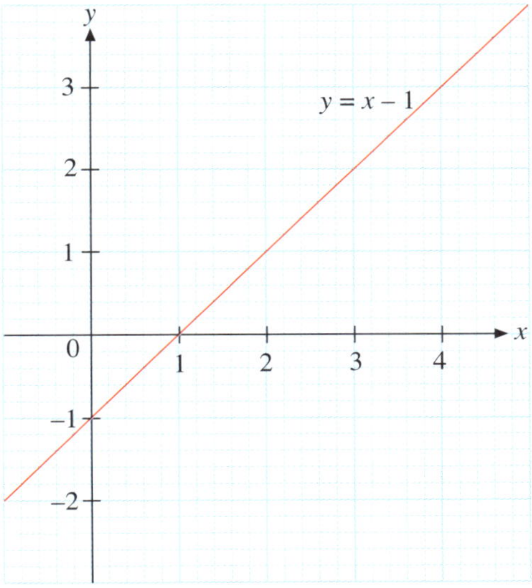


**(c)**

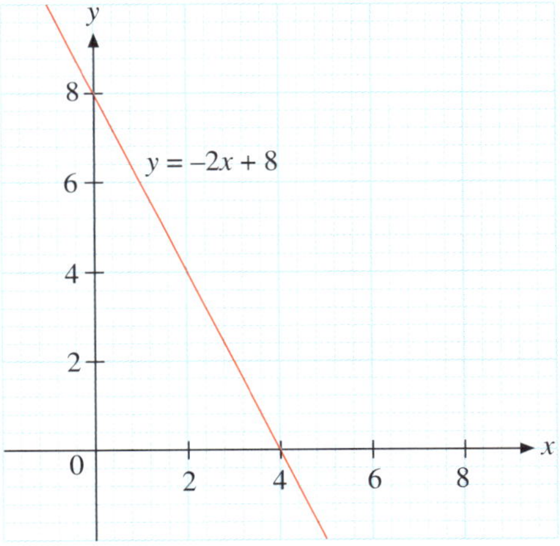

**(d)**

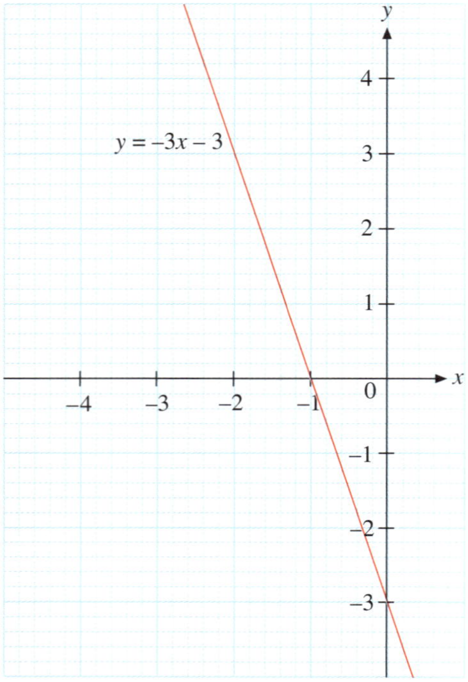


**7.** Find the gradient of each of the following linear functions.

**(a)**

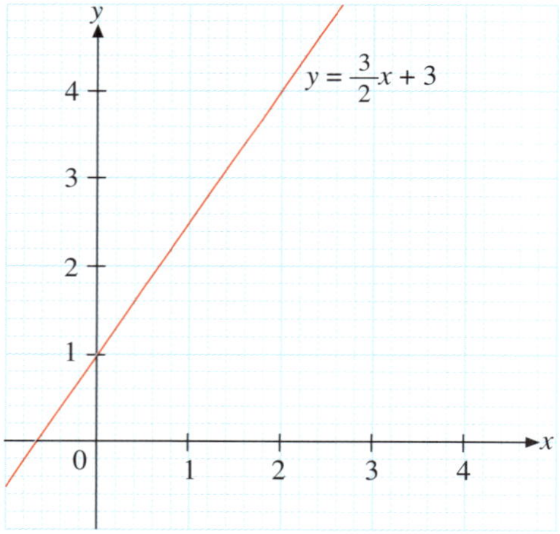


**(b)**

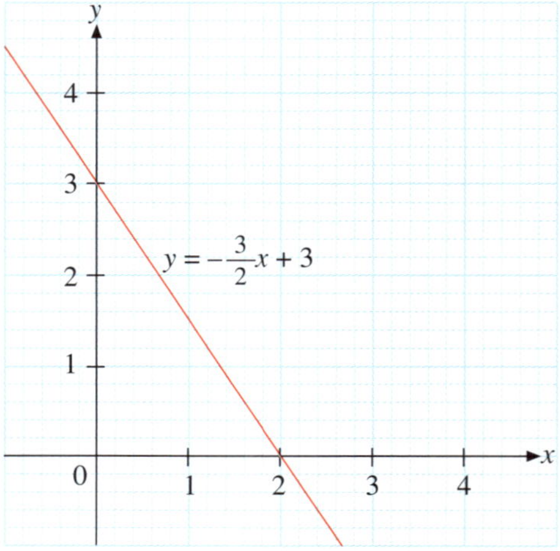


**(c)**

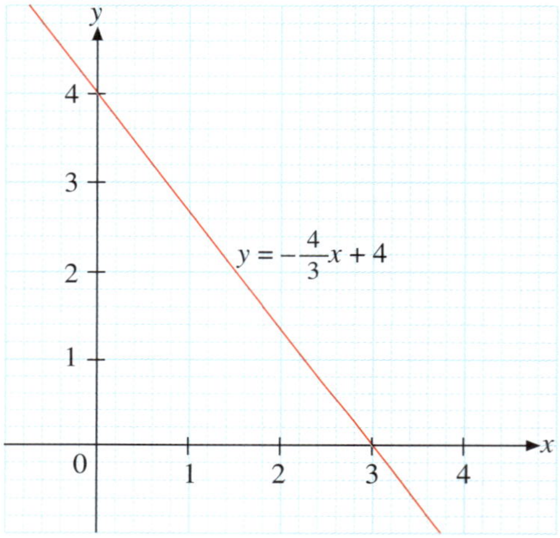


**(d)**

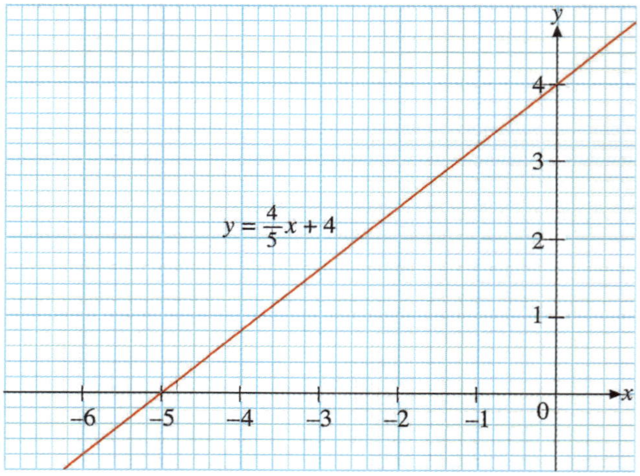

# SUMMARY

1. A Cartesian plane consists of two axes, the **x-axis** and the **y-axis**, intersecting at right angles at the **origin** $O(0, 0)$.

2. An ordered pair $(a, b)$ locates a point $P$ in the Cartesian plane, $a$ being the **x-coordinate** and $b$ the **y-coordinate** of $P$.

3. A function is a relationship that expresses a dependent variable in terms of an independent variable and each value of the independent variable will produce a unique value for the dependent variable.

4. The gradient of a straight line is a measure of the ratio of the vertical change to the horizontal change.

# Review Examples 12

## Example 1

*Joanne's father gives her \$100 as pocket money at the beginning of each month. If she spends an average of \$3 per day,*

*(a) find the amount of money that Joanne has after*

*(i) 2 days, (ii) 5 days, (iii) 10 days.*

*(b) Set up a table of values displaying the 3 pairs of values linking the number of days that has passed (x days) and the amount of money that Joanne has left (\$y) as shown below.*

| *x days* | 2 | 5 | 10 |
|---|---|---|---|
| *\$y* | | | |

*(c) Plot the pairs of values of (x, y) on a piece of graph paper.*

*(d) Express y as a function of x.*

**Solution**

(a) (i) The amount of money that Joanne has left after 2 days = \$100 – 3 × 2 = \$94

(ii) The amount of money that Joanne has left after 5 days = \$100 – 3 × 5 = \$85

(iii) The amount of money that Joanne has left after 10 days = \$100 – 3 × 10 = \$70

(b)

| $x$ | 2 | 5 | 10 |
|---|---|---|---|
| $y$ | 94 | 85 | 70 |

(c)

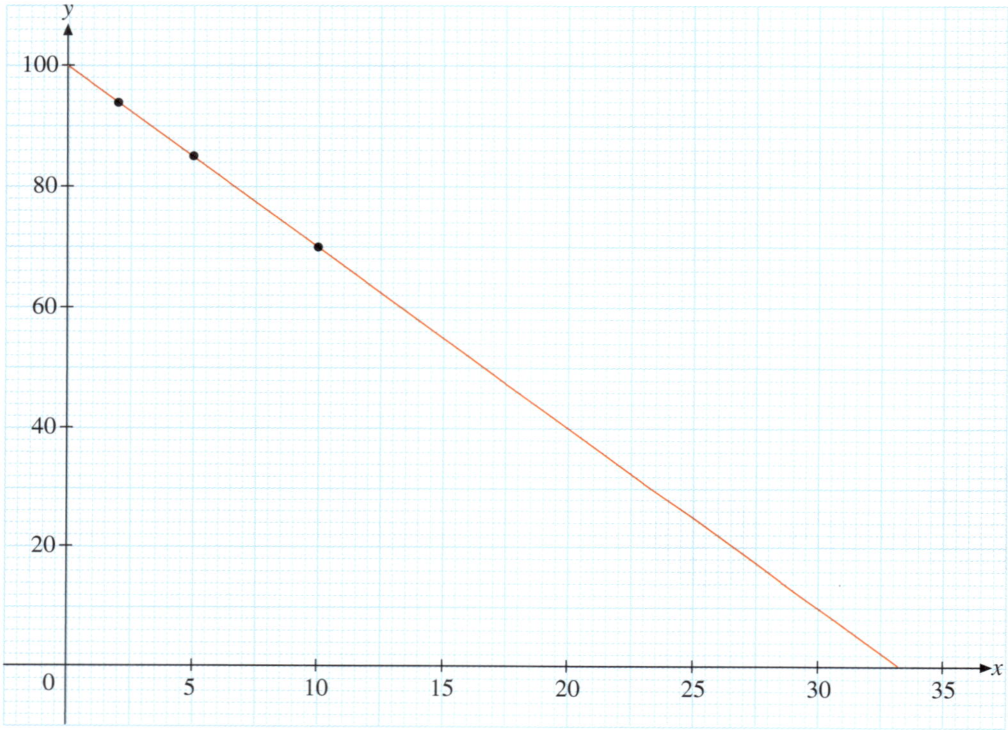


(d) $y = 100 - 3 \times x$

i.e. $y = 100 - 3x$

**Thinking Time**

(1) What is the value of $y$ when

(a) $x = -5$, (b) $x = 35$?

(2) Can you attach a meaning to the above context? Give your reasons clearly.

# Review Questions 12

1. Plot the following points on a piece of graph paper. Join the points in order and give the name of the figure formed.
   **(a)** $P(-2, 2)$, $Q(-2, 6)$, $R(4, 6)$, $S(4, 2)$
   **(b)** $A(2, -2)$, $B(6, 2)$, $C(2, 6)$, $D(-2, 2)$
   **(c)** $C(2, -4)$, $D(8, 4)$, $E(6, 8)$, $F(-2, 4)$
   **(d)** $W(0, 7)$, $X(2, 7)$, $Y(2, 5)$, $Z(-4, 1)$

2. In the function $y = 250 - 20x$, calculate the value of $y$ when
   **(a)** $x = 5$, **(b)** $x = -10$,
   **(c)** $x = 70$.

3. Given a function $y = 4x - 1\frac{1}{2}$, calculate the value of $y$ when
   **(a)** $x = -\frac{1}{2}$, **(b)** $x = 2\frac{1}{2}$,
   **(c)** $x = 12$.

4. Find any three points $(x, y)$ that satisfy the linear function $y = 2\frac{1}{2}x + 3$. Display these points in a table.

| $x$ | | | |
|---|---|---|---|
| $y = 2\frac{1}{2}x + 3$ | | | |

   Plot the graph of $y = 2\frac{1}{2}x + 3$ and henc, find the gradient.

5. Draw the graph of each of the following linear functions and hence, find the gradient.
   **(a)** $y = 5x + 7$
   **(b)** $y = -7x - 4$
   **(c)** $y = -\frac{1}{2}x + 3$
   **(d)** $y = \frac{2}{3}x - 2$

6. The fuel tank of Mr Ong's car is 45 litres. He fills up the fuel tank in Singapore before heading to Penang. The fuel consumption of his car is approximately 10.8 km/litre, find the amount of fuel left (in litres) in the fuel tank after he has travelled
   **(a)** 48 km, **(b)** 245 km.
   Express the amount of fuel ($y$ litres) that was left in the fuel tank after he has travelled $x$ km. If $x = -25$, what is the value of $y$? What will the value of $y$ be when $x = 558$? Can you explain the significance of the values you get from the above?

*In this chapter, you will learn how to*

- *collect, classify and tabulate data;*
- *read and interpret tables and statistical diagrams;*
- *construct bar graphs, pie charts, pictograms, line graphs and histograms with equal intervals;*
- *construct a group frequency table and a histogram to represent the group frequency table.*

CHAPTER

# 13

# Statistics

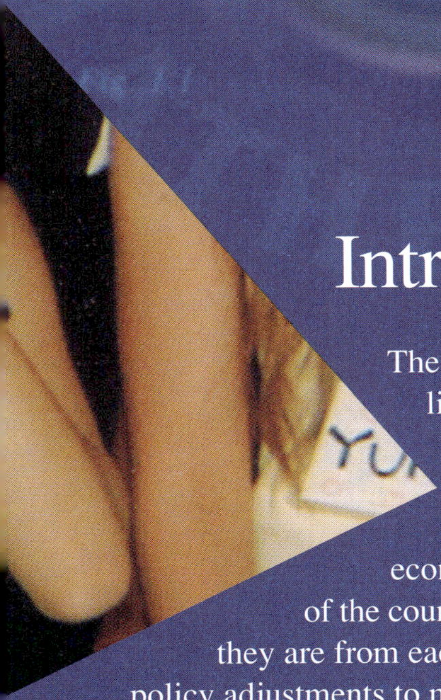

## Introduction

The government collects numerical information about all aspects of life in a country. They collect data such as the number of births in each year, the average age of newly wedded couples in a particular year, the number of pupils who successfully completed their secondary, post secondary and tertiary education each year, the economic growth for each quarter of the year, the unemployment rate of the country in each quarter, the number of tourist arrivals and the countries they are from each month. The information is crucial to the goverment in making policy adjustments to meet the needs and the trends of the time. For example, from the number of babies born each year the government will be able to project the number of primary 1 places required in the schools in Singapore 7 years down the line and to provide these places and train enough teachers to meet the needs at that time. From the trend of tourist arrivals, the government will be able to work with the private sector to build or upgrade the hotels to meet the needs of the tourists from the various countries and regions.

The picture shows a group of Japanese tourists in Singapore.

# Introduction to Numerical Data

We live in a world of figures. Consider the following statements:

(1) Deaths resulting from smoking were about 18 500 in a city in 2005.

(2) Singapore has a labour force of 2.342 million and an unemployment rate of 3.4% in 2004.

(3) The literacy rate among the local residents aged 15 and above was about 96% in 2004.

(4) The number of tourists arriving in Singapore reached a monthly record high of 876 000 in July 2005.

(5) Singaporeans donated a total of $43.2 million to the National Council of Social Services (NCSS) in 2004.

(6) There were about 380 000 private cars on Singapore roads in 2003. This worked out to be about 11.1 cars per 100 residents.

We can continue to give figures like those mentioned above. Have you ever wondered how such figures are collected, summarised and presented? There are many ways of obtaining data, organising them and presenting them, for example, by means of pictures, tables, charts and graphs. We shall look at some of the basic ways of collecting data and presenting them in simple charts, tables, and graphs.

**For Your Information**

*Statistics is a science of collecting, organising, interpreting and analysing data.*

***Misleading data***

*Statistical data can be misleading. Therefore, it is often useful to consider some of the following points before we draw our conclusions from given statistical information:*

1. *the source and authority of the statistical information,*
2. *the size of the sample involved,*
3. *deliberate omissions of information,*
4. *any bias involved.*

# Collection, Organisation and Interpretation of Data

We collect and analyse data to answer questions as well as solve problems.

Consider the following example:

Suppose Mr Wang of ABC Boys' School wants to know the type of sports his 40 students enjoy.

**1.** To **collect** the information, he gives each student a survey form containing the names of several sports. Each student is to select only one favourite sport.

2. When the counting is done, he **organises** the results (information or data collected) in the form of a table as shown below.

| *Favourite sport* | Basketball | Badminton | Hockey | Soccer | Table-tennis | **Total** |
|---|---|---|---|---|---|---|
| *Number of students* | 10 | 7 | 5 | 12 | 6 | 40 |

3. He **presents** the information found in the table using a **diagram**.

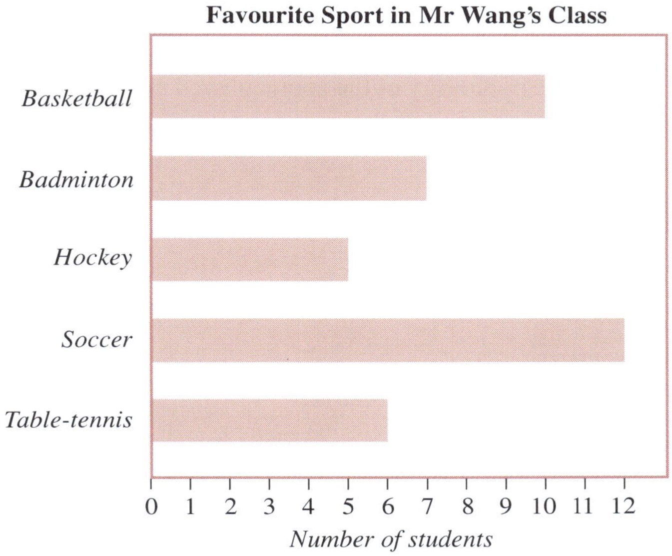


4. He **interprets** the information and draws the following conclusions:
   **(a)** The most popular sport among his students is soccer.
   **(b)** More than half of the class prefer either soccer or basketball.
   **(c)** Hockey is the least popular sport.

The example above shows a very simple way of collecting and organising data. In practice, data can be collected in many ways depending on the amount of data available and the purpose of the survey.

In Step 4, the aim of collecting and organising data is to draw conclusions. However, when interpreting and analysing the data obtained, we are likely to be misled by the data.

Are the following statements true?

1. Students who smoke perform worse academically than non-smokers. Thus, smoking will make one stupid.
2. It is more dangerous to drive today than it was in 1950 because statistics have shown that there are more road accidents today than there were in 1950.

3. Of all car accidents, 20% are caused by people who drink alcohol and drive, whereas 80% are caused by those who are sober. Therefore, there is no cause for alarm when people drink alcohol and drive.

## Collection of Data Using a Questionnaire

Suppose the principal of ABC Boys' School has to plan the annual sports budget. To make a better decision, he wants to know the relative popularity of the sports played in the school. He asks Mr Wang to conduct a survey by giving each student a copy of the questionnaire shown below:

**QUESTIONNAIRE**
**on**
**Favourite Sports at ABC Boys' School**

Which is your favourite sport?

Check the box next to the sport you like best.

Choose only ONE sport.

- Badminton ☐
- Soccer ☐
- Basketball ☐
- Table-tennis ☐
- Hockey ☐

Mr Wang records the results of the survey in a table as shown below.

| *Favourite Sport* | *Tally* | *No. of students* |
|---|---|---|
| Badminton | 𝍸 𝍸 𝍸 𝍸 𝍸 𝍸 𝍸 𝍸 𝍸 𝍸 𝍸 𝍸 𝍸 𝍸 𝍸 𝍸 𝍸 𝍸 𝍸 𝍸 𝍸 𝍸 𝍸 𝍸 𝍸 𝍸 𝍸 / | 136 |
| Basketball | 𝍸 𝍸 𝍸 𝍸 𝍸 𝍸 𝍸 𝍸 𝍸 𝍸 𝍸 𝍸 𝍸 𝍸 𝍸 𝍸 𝍸 𝍸 𝍸 𝍸 𝍸 𝍸 𝍸 𝍸 𝍸 𝍸 𝍸 𝍸 𝍸 𝍸 𝍸 𝍸 𝍸 𝍸 𝍸 𝍸 | 180 |
| Hockey | 𝍸 𝍸 𝍸 𝍸 𝍸 𝍸 𝍸 𝍸 𝍸 𝍸 𝍸 𝍸 𝍸 𝍸 𝍸 𝍸 //// | 84 |
| Soccer | 𝍸 𝍸 𝍸 𝍸 𝍸 𝍸 𝍸 𝍸 𝍸 𝍸 𝍸 𝍸 𝍸 𝍸 𝍸 𝍸 𝍸 𝍸 𝍸 𝍸 𝍸 𝍸 𝍸 𝍸 𝍸 𝍸 𝍸 𝍸 𝍸 𝍸 𝍸 𝍸 𝍸 𝍸 𝍸 𝍸 𝍸 𝍸 𝍸 𝍸 𝍸 𝍸 // | 212 |
| Table-tennis | 𝍸 𝍸 𝍸 𝍸 𝍸 𝍸 𝍸 𝍸 𝍸 𝍸 𝍸 𝍸 𝍸 𝍸 𝍸 𝍸 𝍸 𝍸 𝍸 𝍸 𝍸 /// | 108 |
| | **Total** | 720 |

He does this in an organised manner by using the method of **tallying**. A tally is put in the corresponding space in the table, matching the checked box in the questionnaire. Tallies are grouped in fives (卌) with the fifth tally crossing the first four for counting convenience.

From the table in the previous page, can you tell the relative popularity of the various types of sports in the school? You might have difficulty understanding and interpreting the information displayed in the table. You will probably be able to make better comparisons of the data if you present the information in a diagram as shown in the following sections.

## Pictograms

Mr Wang can draw different diagrams to display the results of the survey, so as to give a clearer picture of the relative popularity of the various sports. One such diagram is a **pictogram**, shown in Fig. 13.1 below.

Notes

*A **pictogram** uses pictures to represent statistics.*

**Favourite Sports played at ABC Boys' School**

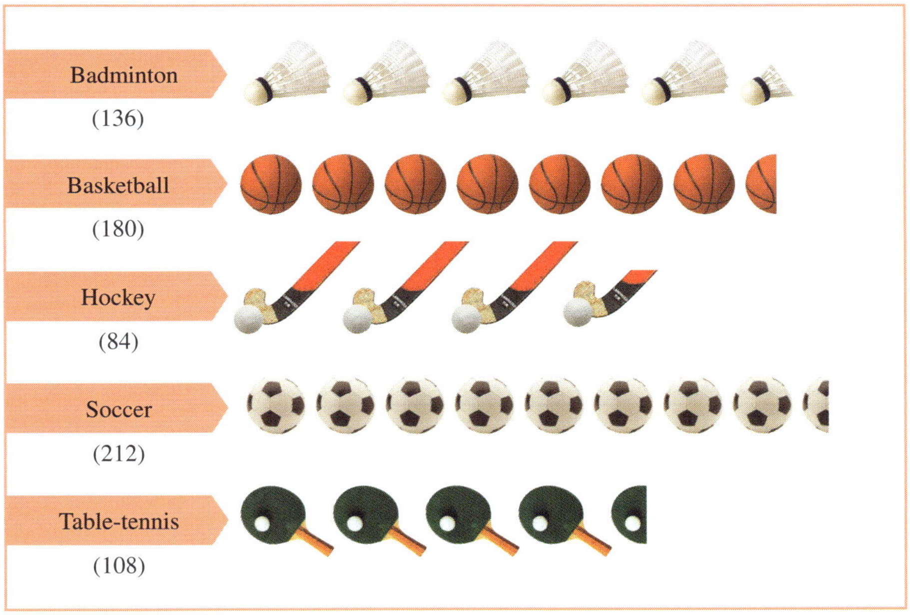


*Each figure represents 25 pupils.*

*Fig. 13.1*

Each figure in the pictogram above represents 25 pupils and a fraction of a figure means a corresponding fraction of 25. It should be noted that this is not a very accurate method of showing the **exact** number of pupils. It merely gives us a quick comparison of the relative popularity of the sports played in the school.

## Example 1

*The table shows the profits, after taxation, of a company from 2001 to 2006.*

*(a) What was the profit in*
*(i) 2004?*
*(ii) 2006?*

*(b) In which year was the profit the smallest? By how much had the profit decreased that year from the previous year?*

**Profits of Company**

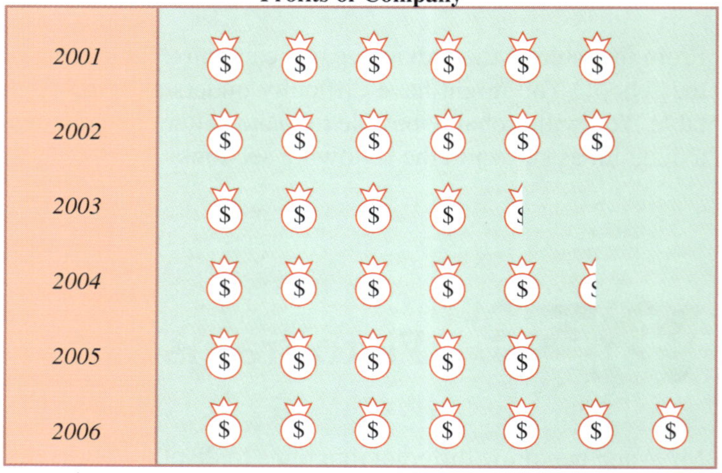


*Each $ represents 1 million dollars.*

### Solution

(a) (i) The profit in 2004 was about $5\frac{1}{2}$ million dollars.

(ii) In 2006, it was 7 million dollars.

(b) In 2003, the profit was the lowest and it had decreased from 2002 by about $1\frac{1}{2}$ million dollars.

## Bar Graphs

Mr Wang may also represent the information in a bar graph as shown in Fig. 13.2.

**Favourite Sports at ABC Boys' School**

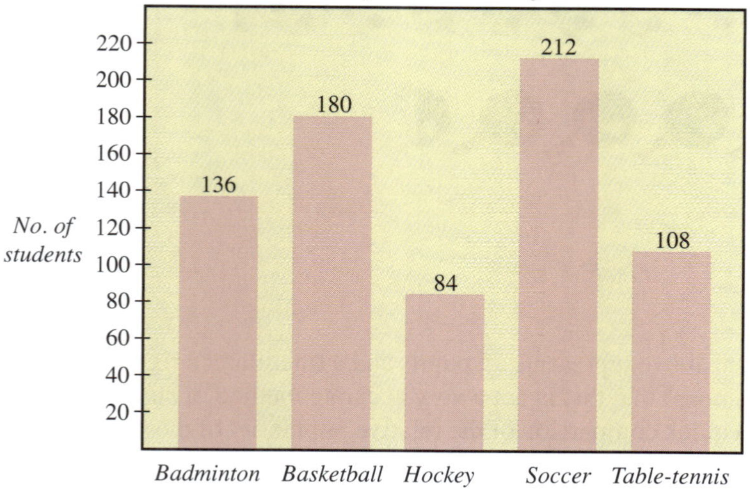


*Fig. 13.2*

The vertical axis shows the number of students and the different categories of favourite sports are labelled along the horizontal axis. A bar is constructed representing each category, with the length of the bar proportional to the number of students in that category. The bars should be of the same width. Note that there is a space between each category on the horizontal axis to distinguish clearly between the categories.

One can see at a glance that soccer is the most popular sport as it has the longest bar.

Can you tell which are the second, third and fourth most popular sport? Which is the least popular sport?

## Example 2

*A company owns five electrical shops. The bar graph shows the number of television sets sold in the five shops in November and December in a certain year.*

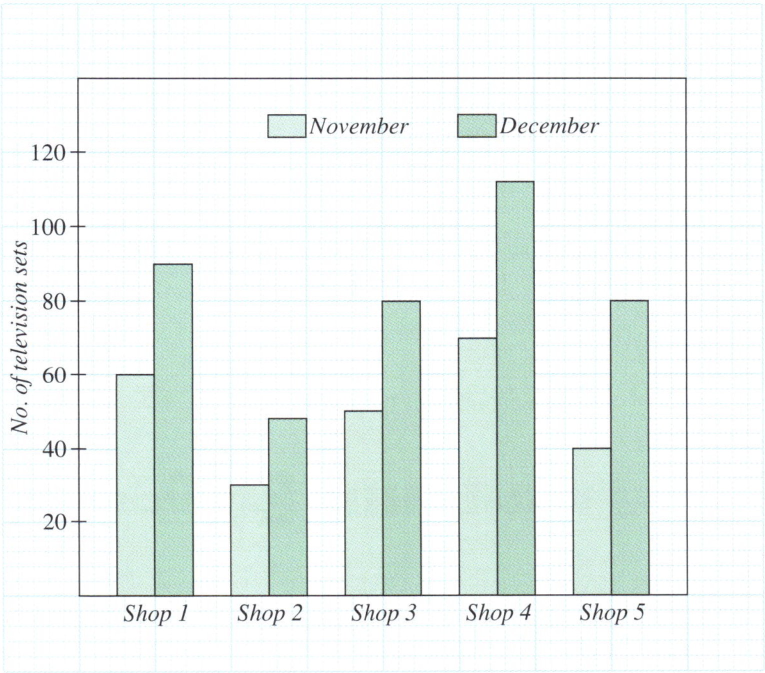


*Study the graph above and answer the following questions:*

*(a) Find the total number of television sets sold in the five shops in*
*(i) November;* *(ii) December.*

*(b) Express the total number of television sets sold in December as a percentage of the total number of television sets sold in November and December.*

*(c) (i) Which shop enjoyed the greatest increase in the number of television sets sold?*
*(ii) Express the increase as a percentage of the number of television sets sold in November.*

### Solution

(a) (i) The total number of television sets sold in the five shops in November
$= 60 + 30 + 50 + 70 + 40$
$= 250$

(ii) The total number of television sets sold in the five shops in December
$= 90 + 48 + 80 + 112 + 80$
$= 410$

(b) Percentage of the total number of television sets sold in December
$= \dfrac{410}{250 + 410} \times 100\%$
$= 62\%$ (correct to the nearest whole number)

(c) (i) Shop 4 enjoyed the greatest increase in the number of television sets sold.

(ii) Percentage increase in sales $= \dfrac{112 - 70}{70} \times 100\%$
$= 60\%$

**Thinking Time**

Note that the increase in sales in Shop 5 is 40. However, its percentage increase in sales
$= \dfrac{80 - 40}{40} \times 100\%$
$= 100\%$

Did Shop 4 or Shop 5 perform better?

## Exercise 13a

**1.** The following pictogram shows the average weekly pocket money that students from each of the five secondary two classes receive.

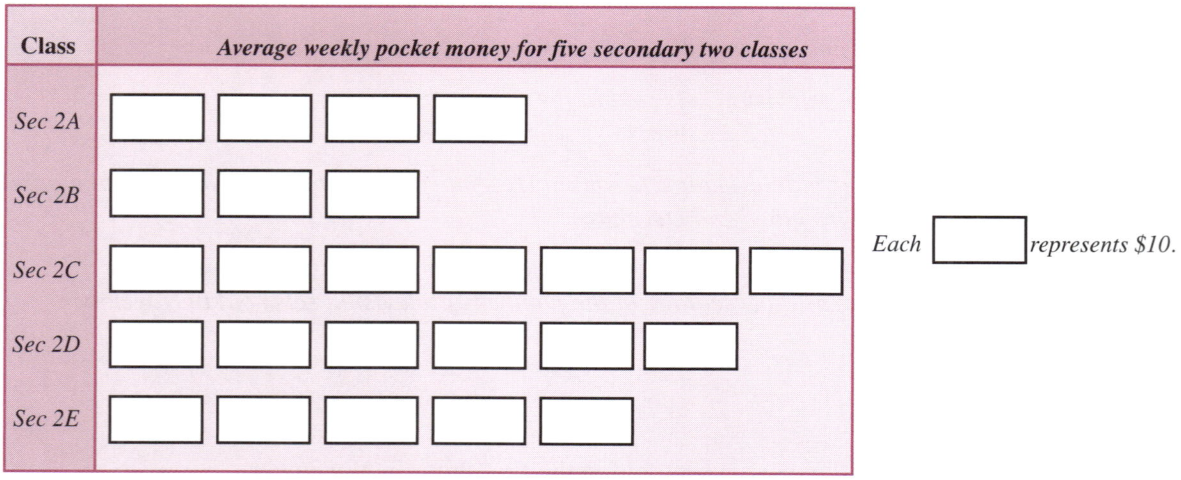


**(a)** What is the ratio of Sec 2D's average weekly pocket money to Sec 2B's average weekly pocket money?
**(b)** Express Sec 2A's average weekly pocket money as a percentage of Sec 2D's average weekly pocket money.

2. The bar graph illustrates the results of a survey carried out in the shops of a certain housing estate. Calculate
(a) the total number of workers;
(b) the percentage of shops hiring 3 or more workers.

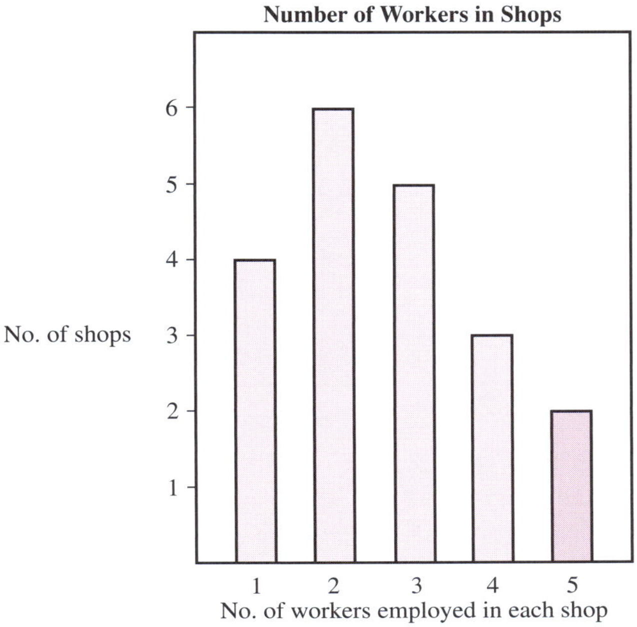


3. The graph below shows the number of candidates who sat for an examination and the number who were successful.

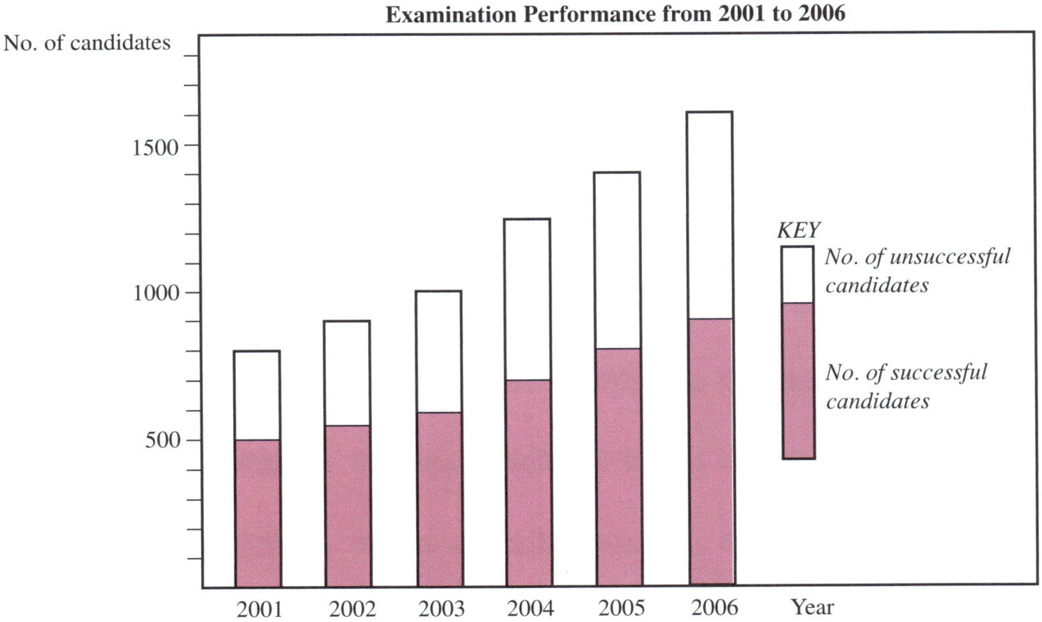


(a) How many candidates sat for the examination in 2003?
(b) How many passed in 2006?
(c) What fraction of the number who took the examination in 2005 passed?

**4.** The graph below shows the product sales of ABC Company in the first four months of 2006.

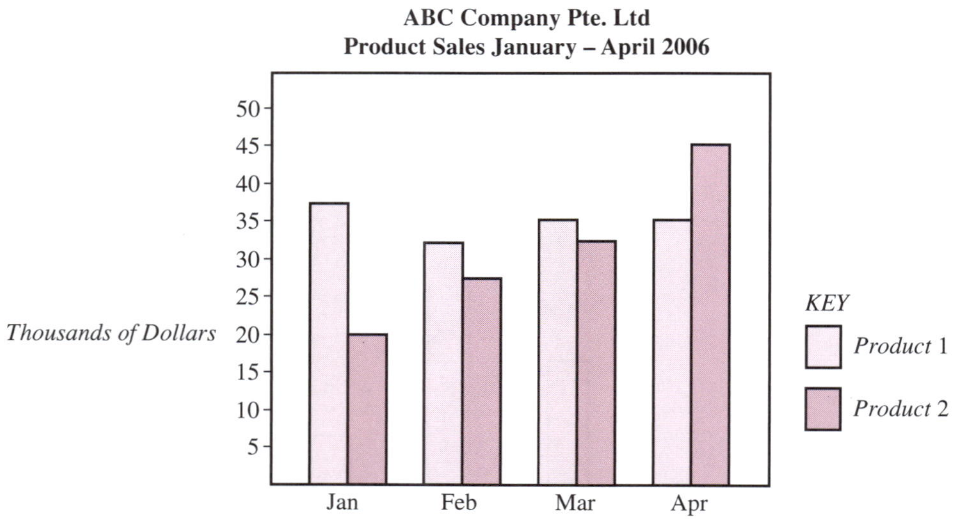


**(a)** Which product has almost the same amount of sales over the four-month period?
**(b)** Which month showed the greatest increase in the sales of product 2?
**(c)** How much were the sales of product 1 in January?
**(d)** How much were the sales of product 2 in April?
**(e)** What were the total sales of product 1?
**(f)** What were the total sales of product 2?
**(g)** What fraction of the total product sales of the company did the sales of product 1 make up during the first four months?

**5.** The pictogram below illustrates the number of cars registered each year from 2001 to 2005 in a certain city.

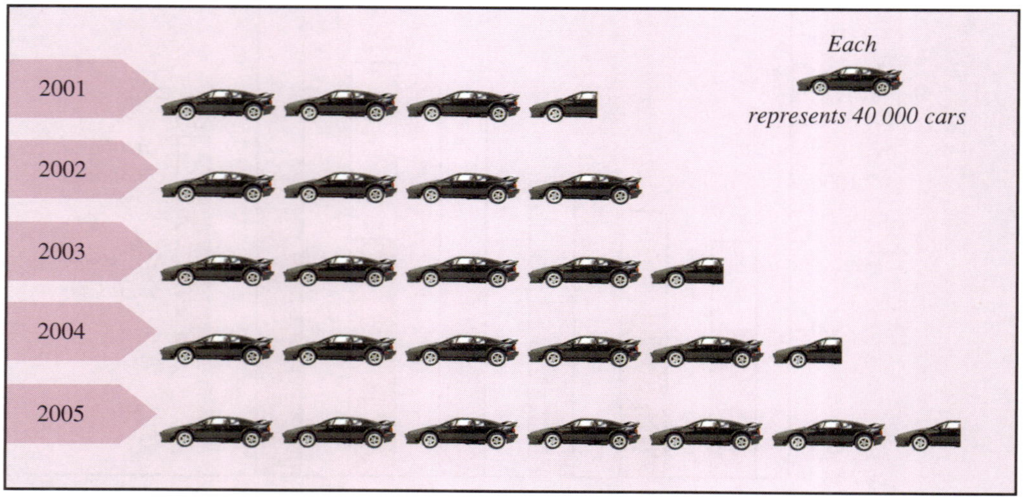


**(a)** In which year was the greatest number of cars registered? Estimate the number of cars registered that year.
**(b)** Estimate the number of cars registered in each of the other years.

(c) If, in 2003, the registration fee for each car was \$500, estimate the total amount the Registry of Vehicles collected for that year.

(d) Give an estimate of the percentage increase in the number of cars registered from 2004 to 2005.

**6.** The circulation of a certain newspaper from 2001 to 2005 is given as follows:

| *Year* | 2001 | 2002 | 2003 | 2004 | 2005 |
|---|---|---|---|---|---|
| *No. of copies (in thousands)* | 250 | 275 | 290 | 315 | 280 |

Use a bar graph to illustrate this information.

**7.** In a certain city, the average daily number of traffic accidents in a week is given as follows:

| *Day* | Sun | Mon | Tue | Wed | Thur | Fri | Sat |
|---|---|---|---|---|---|---|---|
| *No. of accidents* | 70 | 40 | 30 | 35 | 50 | 55 | 80 |

Illustrate this information with a bar graph.

**8.** The table below shows the number of students who play squash, tennis and badminton. Illustrate the data using a pictogram.

| *Sport* | Squash | Tennis | Badminton |
|---|---|---|---|
| *No. of students* | 40 | 60 | 50 |

**(a)** Find the ratio of the number of squash players to the number of badminton players.

**(b)** Calculate the percentage of pupils who play badminton.

## Collection of Data Through Observation

**(a)** Carry out a traffic survey by watching and noting the types of vehicles which pass by your school during a 15-minute period.

**(b)** Copy and complete the survey sheet below.

Name : ____________________ Date : ____________________

Location of survey : ____________________ Time : From ____________________

To ____________________

| *Types of vehicles* | *Tally* | *No. of vehicles* |
|---|---|---|
| Bus | | |
| Taxi | | |
| Private car | | |
| Motorcycle | | |
| Lorry | | |
| Pick-up or Van | | |
| | **Total** | |

**Note:** You may change the list of vehicles to suit your situation. Students in the class should work in small groups and choose different times and locations to conduct the survey.

**(c)** Display the information you collected in the form of a pictogram or a bar chart. Remember to give a short title for your diagram.

**(d)** With the help of your diagram, interpret the information and draw some conclusions.

## Pie Charts

If the principal of ABC Boys' School wants to know what fraction of the school chose a particular sport, then a **pie chart**, as shown in Fig. 13.3 would be useful to him. The pie chart displays clearly the fraction of the school choosing a particular type of sport.

Fig. 13.3 is constructed by dividing a circle into different sectors. Each sector corresponds to the percentage of a category of students who like a particular sport. The angle of each sector is proportional to the number of students the sector represents.

**Notes**

*A **pie chart** represents relative quantities by areas of sectors of a circle.*

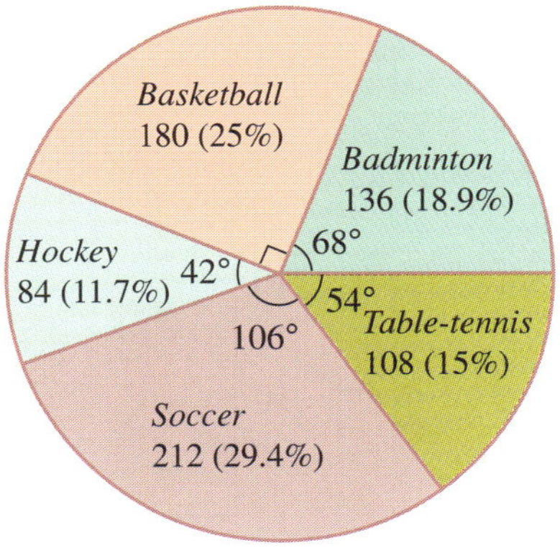


*Fig. 13.3*

The angle of each sector can be found easily, e.g. if 212 out of 720 students prefer soccer, then the angle of the sector for soccer is $\frac{212}{720}$ of $360° = \frac{212}{720} \times 360° = 106°$.

The table below shows the percentage of each category of students and the angle of the sector for each category.

*These days, we usually use software programmes on the computer to help us construct bar graphs, pie charts, etc.*

| *Favourite Sport* | *No. of students* | *Percentage* | *Angle of sector* |
|---|---|---|---|
| Badminton | 136 | $\frac{136}{720} \times 100\% = 18.9\%$ | $\frac{136}{720} \times 360° = 68°$ |
| Basketball | 180 | $\frac{180}{720} \times 100\% = 25\%$ | $\frac{180}{720} \times 360° = 90°$ |
| Hockey | 84 | $\frac{84}{720} \times 100\% = 11.7\%$ | $\frac{84}{720} \times 360° = 42°$ |
| Soccer | 212 | $\frac{212}{720} \times 100\% = 29.4\%$ | $\frac{212}{720} \times 360° = 106°$ |
| Table-tennis | 108 | $\frac{108}{720} \times 100\% = 15\%$ | $\frac{108}{720} \times 360° = 54°$ |

Although pie charts are particularly useful where the proportions of a whole are more important than the actual numerical values, they have some disadvantages. For instance, it is difficult to interpret a pie chart representing data which involves too many categories. In constructing pie charts, long calculations are often needed and actual measurements of angles using protractors are necessary to produce an accurate diagram. Of course, nowadays software can help to produce an accurate pie-chart very quickly and easily.

## Example 3

*The bar graph illustrates the results of a survey conducted to find the number of people in each car in a random survey of 120 cars at a traffic junction.*

*Calculate*

*(a) the angle in a pie chart of the sector which represents cars with 2 people;*
*(b) the total number of people in these 120 cars;*
*(c) the percentage of cars with 4 or more people.*

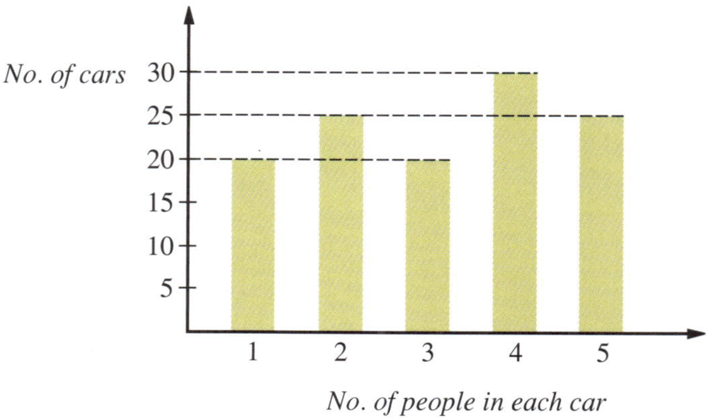


### Solution

(a) No. of cars with 2 people = 25

$\therefore$ the angle of the sector in a pie chart representing this is $\frac{25}{120} \times 360° = 75°$

(b) Total no. of people in these 120 cars $= (1 \times 20) + (2 \times 25) + (3 \times 20) + (4 \times 30) + (5 \times 25)$
$= 375$

(c) No. of cars with 4 or more people = 30 + 25 = 55

$\therefore$ the percentage of cars with 4 or more people $= \frac{55}{120} \times 100\% = 45.8\%$

## Example 4

*The pie chart shows the nutritional composition of a fast-food product.*

*(a) Calculate the value of x.*
*(b) What is the percentage of fat in the fast-food product?*
*(c) Given that one such fast-food product contains 120 grams of carbohydrates, calculate the total mass of the fast-food product.*

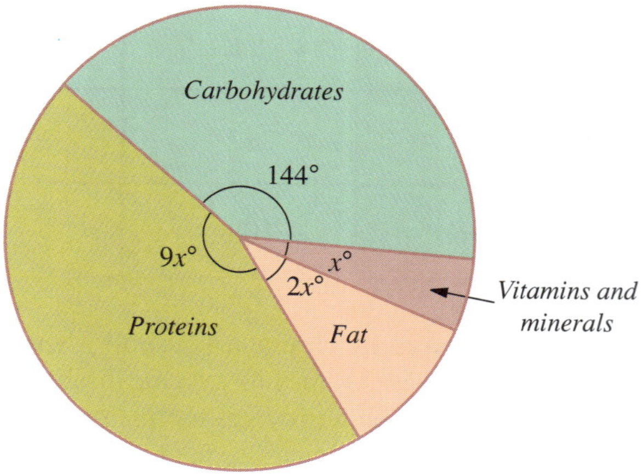


### Solution

(a) $144° + 9x° + 2x° + x° = 360°$
$12x = 216$
$\therefore \quad x = 18$

(b) The angle representing fat in the product $= 2 \times 18°$
$= 36°$

$\therefore$ the percentage of fat in the fast-food product $= \dfrac{36°}{360°} \times 100\%$
$= 10\%$

(c) The angle representing carbohydrates is 144° and this constitutes 120 g.

$\therefore$ the total mass of the product $= \left(\dfrac{360}{144} \times 120\right)$ g
$= 300$ g

## Collection of Data Through Interviews

### Explorations

**(a)** Conduct interviews with students in your class to find out their daily means of transport to school.
**(b)** Decide on the number of students you want to interview.
**(c)** Think of where and when to interview the students, e.g. in the canteen during recess time, or in the morning before students go to their classes for lessons. Do not forget to thank the students you have interviewed for their participation.
**(d)** Design a form for recording the results.
**(e)** Use a diagram, preferably a pie chart, to display your data.
**(f)** Interpret the information from your diagram and draw your conclusions.

### Exercise 13b

**1.** The main products of a manufacturing company are corn oil (50%), margarine (30%), peanut oil (15%) and others (5%). Display the above data using a pie chart.

**2.** 72 students of a certain school were asked to indicate a place of interest in Singapore they would like to visit during their holidays. The table below shows the results.

| *Place of interest* | Bird Park | Chinese Garden | Science Centre | Sentosa Island | Zoo |
|---|---|---|---|---|---|
| *No. of students* | 9 | 5 | 19 | 24 | 15 |

Use a bar graph to illustrate the results.

**3.** 100 people took part in a survey on their favourite fruit. The results were tabulated as follows:

| *Fruit* | Banana | Durian | Mangosteen | Papaya | Rambutan |
|---|---|---|---|---|---|
| *No. of people* | 30 | 25 | 10 | 15 | 20 |

**(a)** Illustrate the results using a
 **(i)** bar graph; **(ii)** pie chart.
**(b)** Which diagram shows more clearly that one quarter of the people like durian most?
**(c)** Which diagram shows more clearly that banana is more popular than durian?

**4.** The pie chart below shows the health care expenditure for the year 2005.

**Health Care 2005 — Where the money goes**
(Total $720 million)

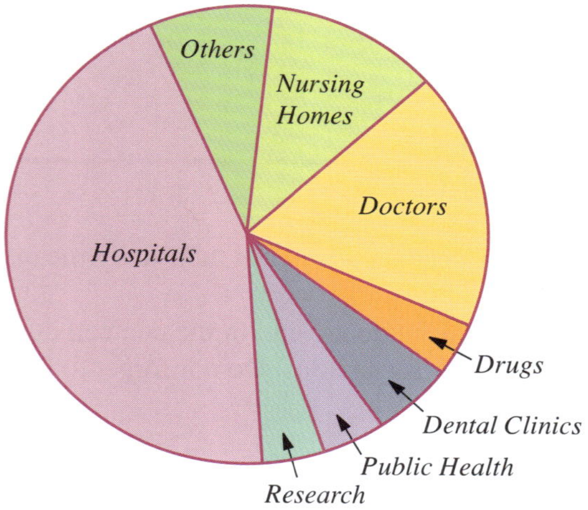


**(a)** Measure the angle of each sector using a protractor and calculate the amount of money allocated to each category.
**(b)** Express the expenditure of each category as a percentage of the total expenditure on Health Care.

**5.** The pie chart shows the number of pupils and teachers in a certain school.
**(a)** Calculate the value of $x$.
**(b)** If there are 45 teachers in the school, how many
 **(i)** boys are there in the school?
 **(ii)** girls are there in the school?

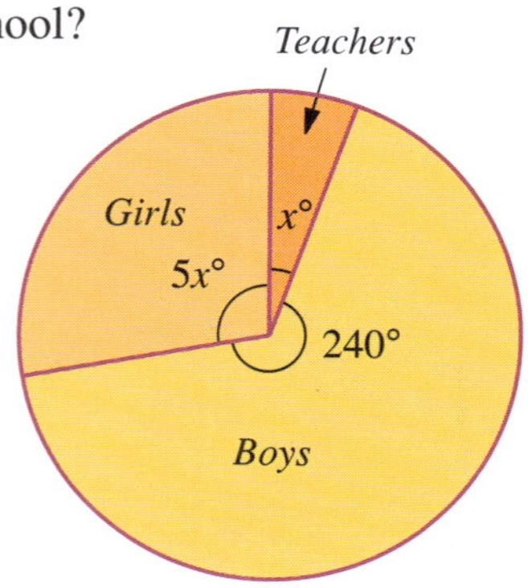

6. The bar graph shows the number of cars of different colours sold in one year in a certain city.
   **(a)** Which colour is the most popular?
   **(b)** Illustrate the information using a clearly labelled pie chart.

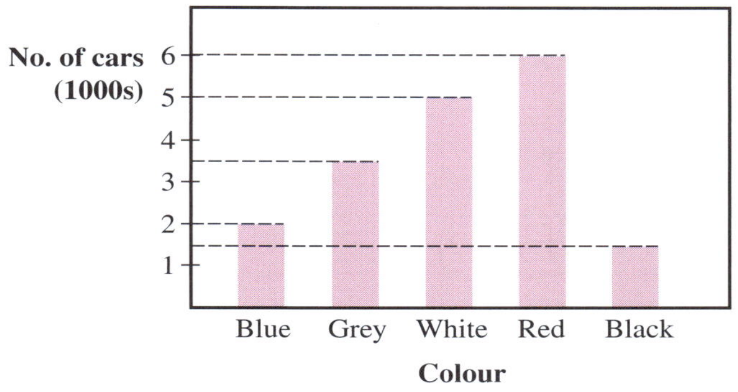


7. Pupils in a class were asked to indicate which one of the four ice-cream flavours — vanilla, chocolate, yam or mango — they preferred. The following pie chart shows the results.
   **(a)** If one-quarter of the class preferred yam flavour, state the angle in the yam sector.
   **(b)** Calculate the angle in the vanilla sector.
   **(c)** If 5 students indicated a preference for mango flavour, calculate the number of students in the class.
   **(d)** Calculate the percentage of students in the class who preferred vanilla flavour.

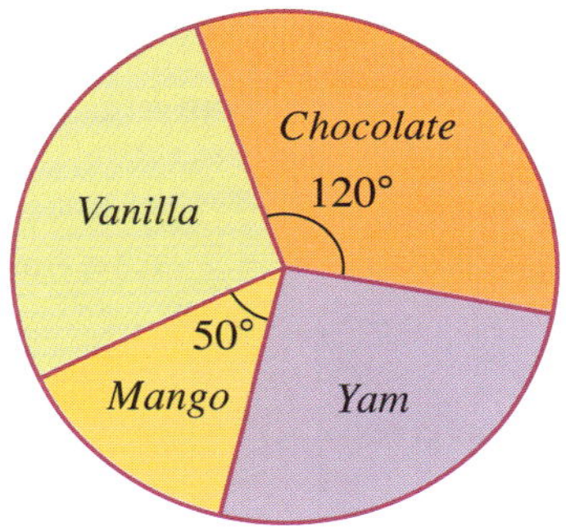


8. The daily output of two products, $X$ and $Y$, in a factory are 6 tonnes and 14 tonnes respectively. If the output is represented by a pie chart, calculate the angle of the sector representing the output of product $Y$.

9. A factory produces three products, $A$, $B$ and $C$, in the ratio of $1 : x : 5$. When the output is illustrated by a pie chart, the angle of the sector representing the output of $C$ is 120°. Find $x$.

10. The pie chart shows the sales of a publishing company.
   **(a)** What percentage of the total sales does each of the following make up?
   **(i)** newspapers **(ii)** magazines
   **(b)** Given that books make up $17\frac{1}{2}\%$ of the total sales, find $x$.

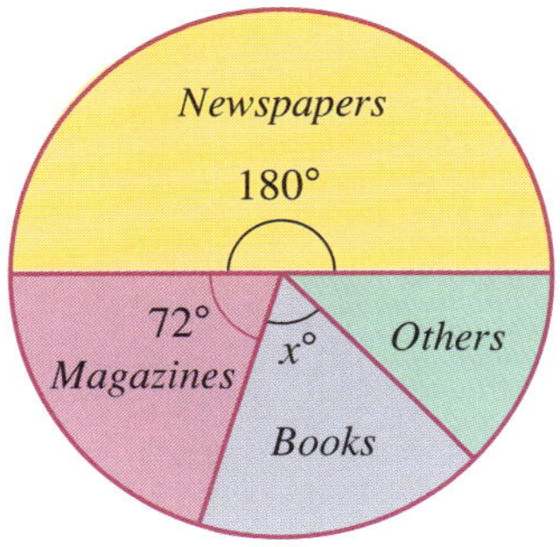

# Line Graphs

The table below shows the recurrent expenditure of an Asean government on university education in thousand dollars.

| *Year* | 87/88 | 88/89 | 89/90 | 90/91 | 91/92 | 92/93 |
|---|---|---|---|---|---|---|
| *Expenditure ('000)* | 293 443 | 259 014 | 288 571 | 351 476 | 841 154 | 412 524 |

| *Year* | 93/94 | 94/95 | 95/96 | 96/97 | 97/98 |
|---|---|---|---|---|---|
| *Expenditure ('000)* | 431 554 | 502 999 | 520 289 | 546 120 | 669 004 |

By plotting the points corresponding to the data and then joining the points by line segments, we obtain the line graph as shown in Fig. 13.4.

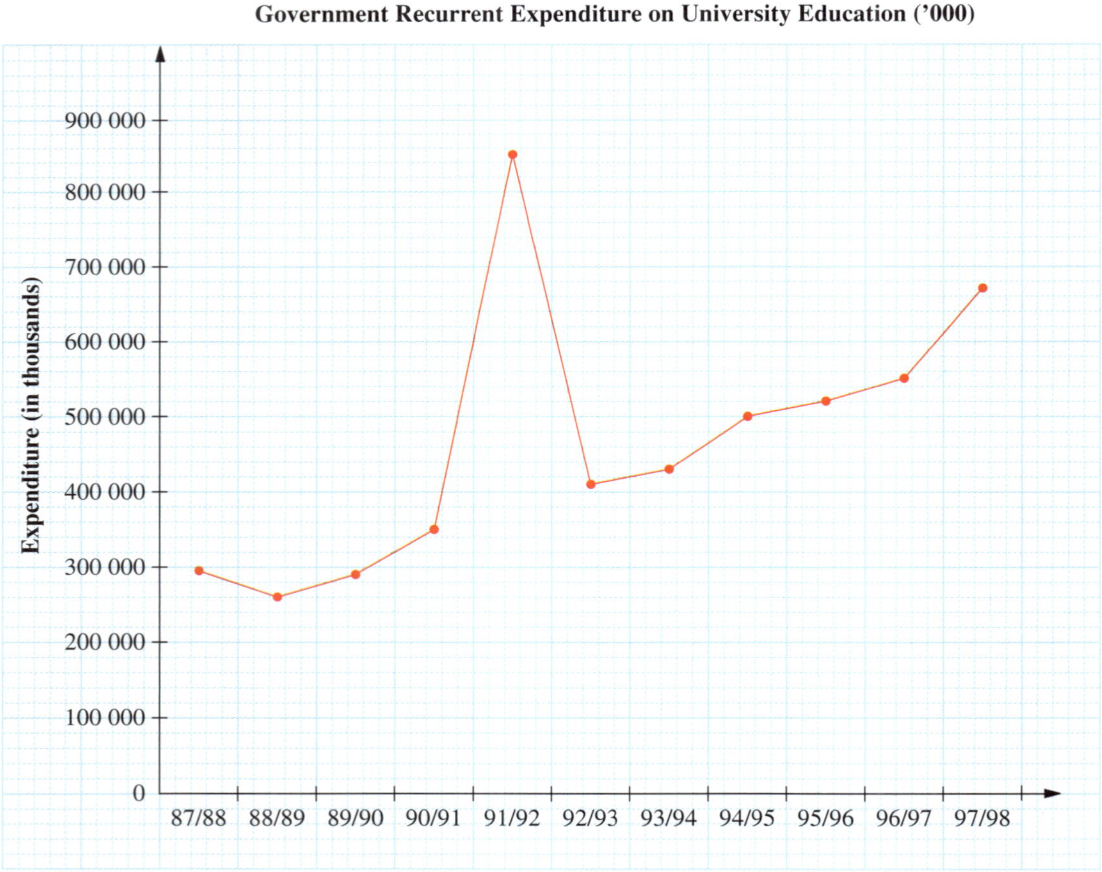


*Fig. 13.4*

A line graph is a suitable graph to construct when we wish to show a rising or a falling trend in a set of data over a period of time. The graph in Fig. 13.4 shows that the government recurrent expenditure on university education reached its lowest point in the 88/89 fiscal year, after which it increased gradually for the next two years. The expenditure hit the highest point in the fiscal year 91/92 following a big jump from the previous year. The expenditure then took a dip in the year 92/93 after which it rose gradually for the next five years.

**Notes**

*Although adjacent points are joined by a line segment, the intermediate values, other than the values recorded, have no meaning.*

## Example 5

*The following graph shows the temperatures of a child who developed fever. His temperature was taken every two hours starting from 2 p.m.*

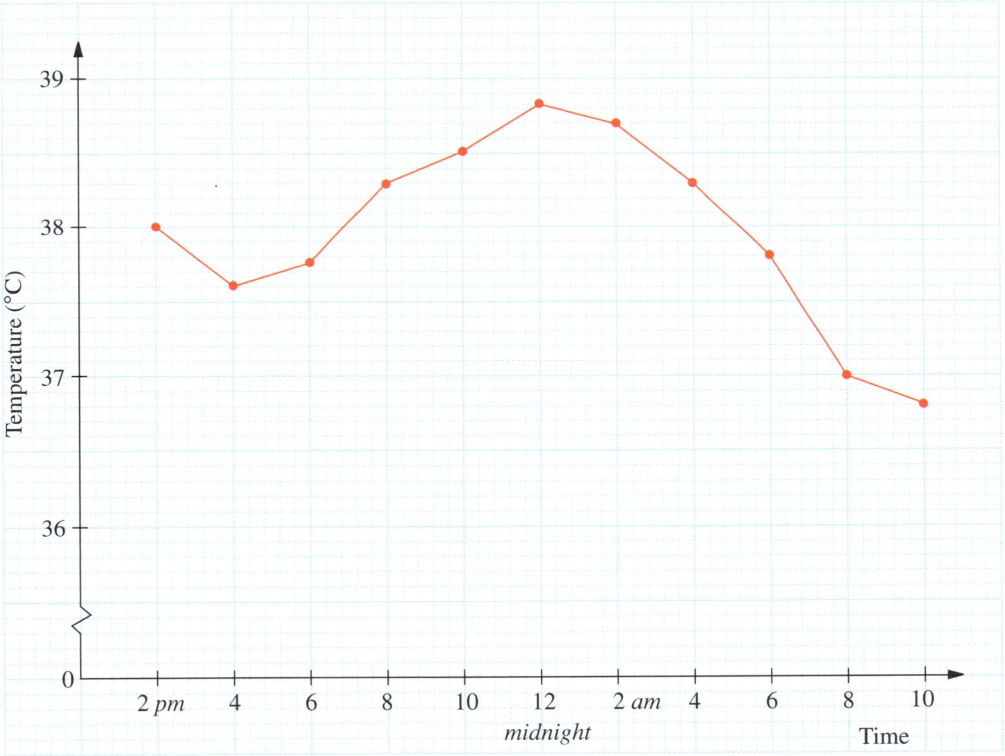


*(a) At what time was his temperature the highest on the graph?*
*(b) What was his temperature at 4 p.m., 9 p.m. and 3 a.m.?*
*(c) When was the child's temperature 37 °C, 37.4 °C and 38 °C?*

**Solution**

(a) His temperature was the highest at 12 midnight.

(b) His temperature at 4 p.m. was 37.6°C. His temperatures at 9 p.m. and 3 a.m. can be estimated using intermediate values because it is reasonable to assume that the temperature changes gradually within a relatively short period of 2 hours.

From the graph,
his temperature at 9 p.m. was about 38.4°C;
his temperature at 3 a.m. was about 38.5°C.

(c) The child had a temperature of 37°C at 8 a.m., 37.4°C at 7 a.m., 38°C at 2 p.m., at about 6.50 p.m. and at about 5.10 a.m.

## Exercise 13c

**1.** The line graph below shows the monthly tin output of a tin mine from July to December in 2005. Study the graph and estimate
**(a)** the largest monthly output;
**(b)** the smallest monthly output;
**(c)** the total output for the six months.

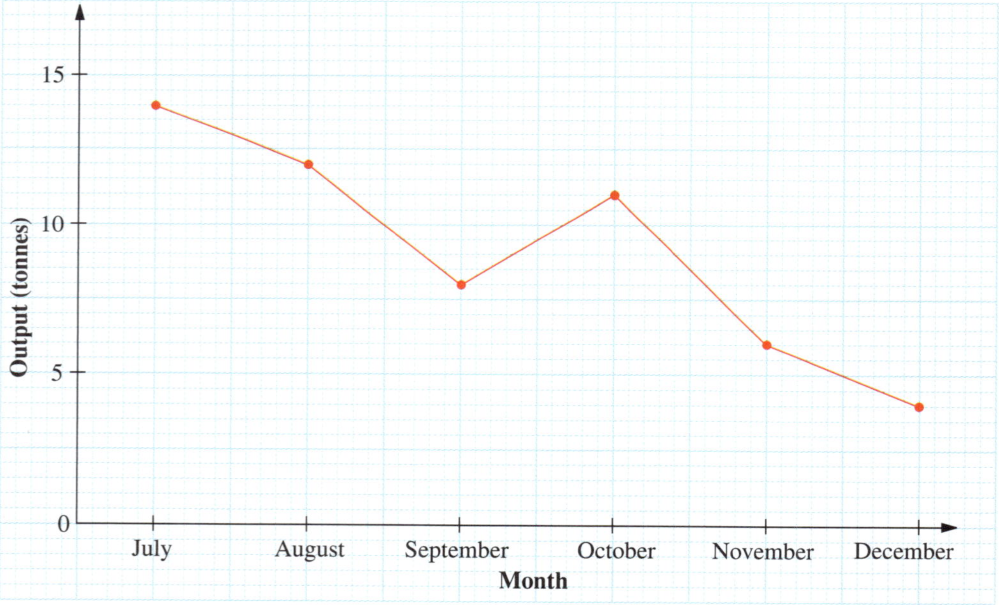


**2.** The temperature of a patient, taken every 3 hours, was recorded as shown in the table below:

| *Time* | 3 p.m. | 6 p.m. | 9 p.m. | 12 midnight | 3 a.m. | 6 a.m. | 9 a.m. |
|---|---|---|---|---|---|---|---|
| *Temperature (°C)* | 39 | 39 | 39.5 | 37.5 | 39 | 38 | 37 |

**(a)** Display the data using a line graph.
**(b)** From the graph, estimate the patient's temperatures at 5 p.m. and at 1 a.m.

**3.** The line graph below shows the population of a town from 1990 to 2000.

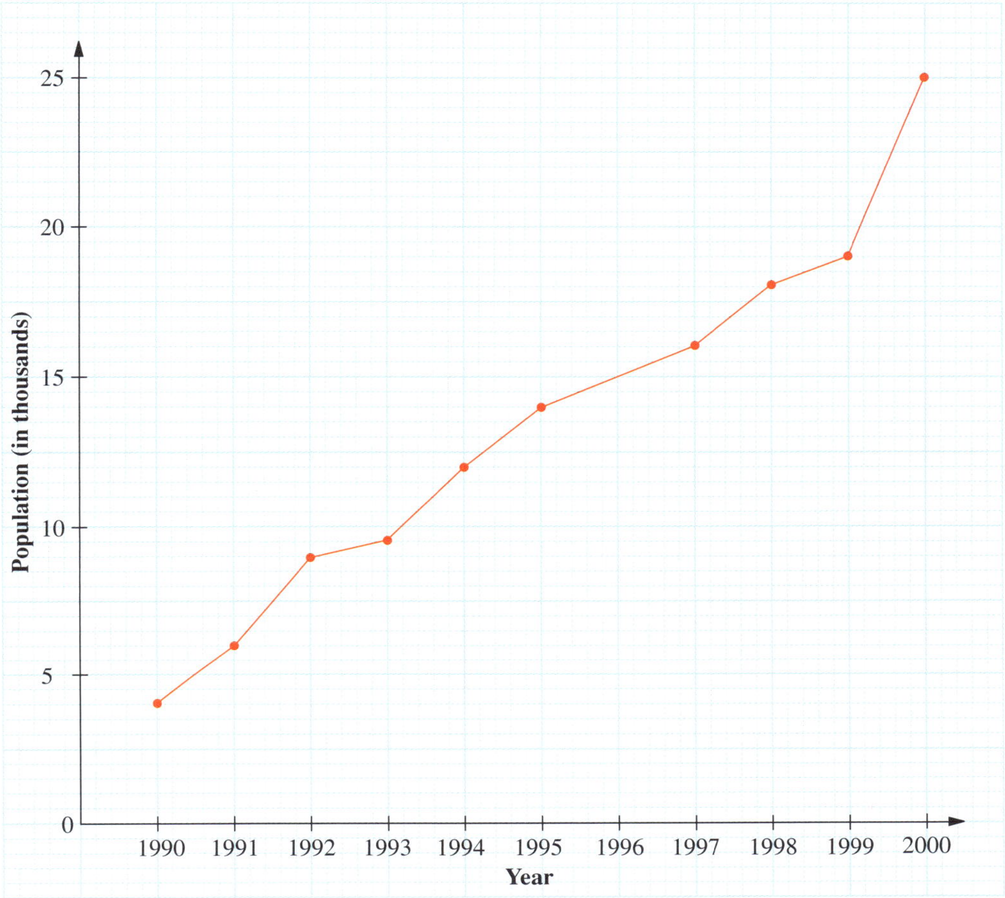


**(a)** Between which two years did the town have the greatest number increase in population?
**(b)** Find the percentage increase in the population from 1997 to 2000.

# Frequency Tables and Histograms

Consider the marks scored by 40 students in a Science test marked out of a total of 10.

| | | | | | | | | | |
|---|---|---|---|---|---|---|---|---|---|
| 8 | 6 | 4 | 3 | 5 | 5 | 2 | 9 | 2 | 7 |
| 9 | 3 | 3 | 7 | 7 | 5 | 8 | 3 | 7 | 3 |
| 4 | 8 | 7 | 8 | 2 | 4 | 6 | 2 | 4 | 1 |
| 7 | 7 | 6 | 2 | 6 | 4 | 4 | 6 | 10 | 6 |

The teacher would like to know the overall performance of the students after the test to get feedback on the students' progress. He may be interested in:

**(a)** the average score
**(b)** the middle score
**(c)** the most common score
**(d)** the highest score
**(e)** the lowest score
**(f)** the range of scores
**(g)** whether the test was too difficult or too easy.

He can find what he wants from the list of marks. But he will have some difficulty in doing so because the marks are not arranged in some convenient order. Hence, as discussed earlier, a set of unsummarised data, which we call raw data, has to be simplified and then arranged in an orderly fashion first, so that we will be able to understand and use them later.

We notice from the list of marks that some marks appear more than once. We arrange the marks according to the number of times each mark appears on the list.

| *Mark* | *Tally* | *Frequency* |
|---|---|---|
| 0 | | 0 |
| 1 | / | 1 |
| 2 | ~~////~~ | 5 |
| 3 | ~~////~~ | 5 |
| 4 | ~~////~~ / | 6 |
| 5 | /// | 3 |
| 6 | ~~////~~ / | 6 |
| 7 | ~~////~~ // | 7 |
| 8 | //// | 4 |
| 9 | // | 2 |
| 10 | / | 1 |
| **Total frequency** | | 40 |

The marks are arranged in order of magnitude. We go through the list of marks and keep a tally as shown in the table above. The number of times each mark appears is called its **frequency**. The table which gives the frequency of each score is called a **frequency table**. It shows the frequency distribution of the marks obtained by the 40 pupils in the test.

A diagram can be constructed to illustrate the information given in the frequency table. Such a diagram is called a **histogram**. The histogram is easier to understand than the frequency table.

Fig. 13.5 below shows the histogram representing the frequency of the marks obtained in the Science test taken by 40 students.

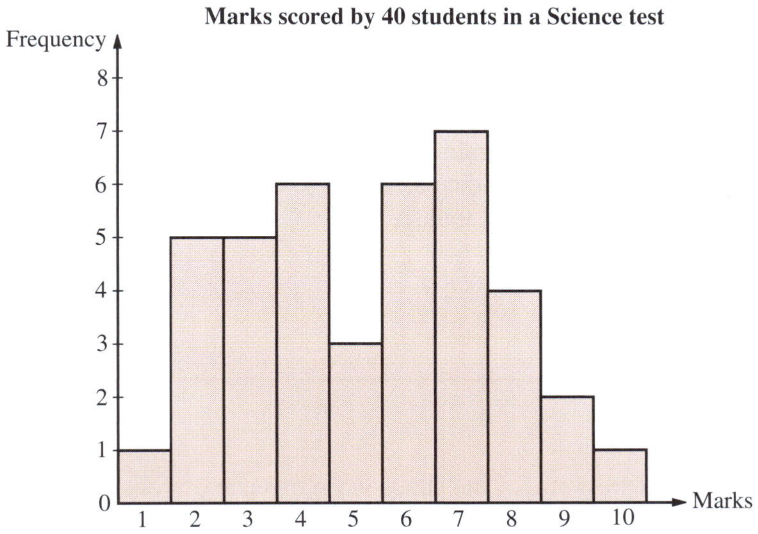


*Fig. 13.5*

*Which letters of the alphabet are used most frequently in written material?*

*Choose an article from a newspaper. Making use of tally marks, record the frequency for each letter by going through the first 600 successive letters in the article. Tabulate the results in a frequency table and construct a histogram to illustrate your findings.*

A histogram is actually a vertical bar graph with no space in between the bars. However, the areas of the bars or the rectangles, and not the heights, are proportional to the numbers they represent. In our example, the bases of the rectangles are equal. So the height of each rectangle will be proportional to the frequency.

Using the histogram in Fig. 13.5, try asking and answering questions like:

(1) What was the most common score?

(2) What was the middle score, that is, the score with half of the pupils scoring less than it?

(3) What was the highest score?

(4) What was the lowest score?

(5) What was the range, that is, the difference between the highest and the lowest scores?

## Collection of Data by Measuring

This method of collecting data is useful in controlling the quality of products from a production line. For example, in a factory producing electrical light bulbs, the quality control process may involve measuring the lifespan of a certain number of bulbs selected at random from the production line. The results of the actual calculations involving this number of bulbs are then analysed to determine whether the electrical light bulbs produced overall are up to the acceptable standard.

### Explorations

**(a)** Measure the lengths, correct to the nearest cm, of the shoes worn by each of your classmates.
**(b)** Record the results and arrange them in the form of a frequency table.
**(c)** Display your information using a histogram.
**(d)** Comment on your results.

### Exercise 13d

**1.** In a spelling test, the number of mistakes incurred by each of the 30 pupils in a primary one class is given below:

| | | | | | | | | | |
|---|---|---|---|---|---|---|---|---|---|
| 3 | 4 | 6 | 0 | 2 | 2 | 4 | 3 | 5 | 3 |
| 4 | 2 | 2 | 3 | 1 | 5 | 3 | 0 | 4 | 5 |
| 4 | 3 | 4 | 0 | 3 | 2 | 6 | 3 | 1 | 0 |

**(a)** Construct a frequency table for the number of spelling mistakes.
**(b)** Draw a histogram to illustrate the results.
**(c)** What is the most common number of mistakes?
**(d)** What is the highest number of mistakes?

**2.** The teachers of a certain school were asked to indicate the average number of hours they spend on marking students' assignments each day. The following set of data was obtained.

| | | | | | | | | |
|---|---|---|---|---|---|---|---|---|
| 6 | 4 | 3 | 1 | 2 | 2 | 3 | 1 | 4 |
| 1 | 2 | 5 | 3 | 4 | 5 | 2 | 2 | 3 |
| 3 | 1 | 2 | 2 | 3 | 1 | 4 | 2 | |

**(a)** Construct a frequency table and draw a histogram illustrating the results.
**(b)** How many teachers responded to the survey?
**(c)** What is the longest number of hours spent?
**(d)** What is the most common number of hours spent?

★**3.** 100 crates of oranges imported from Country *A* were inspected and the number of rotten oranges recorded is shown below:

| *Rotten oranges* | 0 | 1 | 2 | 3 | 4 | 5 | 6 | 7 | 8 | 9 |
|---|---|---|---|---|---|---|---|---|---|---|
| *No. of crates* | 4 | 9 | 12 | 28 | 22 | 15 | 5 | 2 | 2 | 1 |

Another 100 crates of oranges imported from Country *B* were also inspected and the number of rotten oranges recorded was:

| *Rotten oranges* | 0 | 1 | 2 | 3 | 4 | 5 | 6 | 7 | 8 |
|---|---|---|---|---|---|---|---|---|---|
| *No. of crates* | 51 | 30 | 8 | 4 | 1 | 2 | 2 | 1 | 1 |

**(a)** Draw a histogram to represent each frequency table.
**(b)** What was the largest number of rotten oranges for each exporter?
**(c)** Find the total number of rotten oranges for each exporter.

**4.** A company claims that it produces 50 matches per box. Fifty boxes were selected from a large number of boxes manufactured by the company and their contents counted, giving the following results:

| | | | | | | | | | |
|---|---|---|---|---|---|---|---|---|---|
| 53 | 50 | 48 | 52 | 52 | 50 | 49 | 51 | 50 | 52 |
| 49 | 49 | 50 | 51 | 48 | 51 | 50 | 52 | 52 | 49 |
| 49 | 51 | 51 | 50 | 50 | 51 | 48 | 50 | 52 | 49 |
| 50 | 50 | 51 | 52 | 49 | 53 | 50 | 52 | 49 | 52 |
| 51 | 49 | 50 | 53 | 52 | 50 | 49 | 50 | 48 | 50 |

**(a)** Construct a frequency distribution table for the results.
**(b)** Draw a histogram to represent the data.
**(c)** What percentage of the boxes contain exactly 50 matches?

**5.** The number of siblings that each employee in a group of 40 has is shown below:

| | | | | | | | | | | | | | | | | | | | |
|---|---|---|---|---|---|---|---|---|---|---|---|---|---|---|---|---|---|---|---|
| 2 | 2 | 3 | 5 | 3 | 2 | 1 | 4 | 2 | 1 | 1 | 0 | 3 | 2 | 1 | 1 | 2 | 3 | 1 | 4 |
| 3 | 2 | 3 | 1 | 0 | 1 | 2 | 1 | 2 | 3 | 1 | 1 | 2 | 2 | 1 | 3 | 5 | 1 | 1 | 5 |

**(a)** Construct a frequency distribution table for the results.
**(b)** Represent the results with a histogram.

**6.** The table below shows the number of fire incidents per day in a city over a period of 60 days.

| *Number of fire incidents* | 0 | 1 | 2 | 3 | 4 | 5 | 6 | 7 |
|---|---|---|---|---|---|---|---|---|
| *Number of days* | 16 | 12 | 11 | 10 | 6 | 2 | 1 | 2 |

**(a)** Represent the data using a histogram.
**(b)** Express the number of days when there were no fire incidents as a percentage of the total number of days.

# Collection of Data by Using Electronic Means and the Internet

## Explorations

Have you ever taken part in any survey conducted by the various search engines such as Yahoo! Singapore and Google? Other websites hosted by newspapers and television stations also often ask their readers and viewers to vote to express their opinions on certain current topics. Some schools also ask their students to provide their feedback electronically on certain issues. Even the Singapore Statistics department conducts their survey using electronic means. The use of modern electronic tools greatly cut down the many hours of manual work that one has to put in, compared with the use of traditional means. With the use of spreadsheets, information gathered by electronic means can be transmitted and summarised almost instantaneously. Spreadsheets can be used to tabulate the results and present them in the form of charts and diagrams.

**Work in a group of five.**

You are to design a simple survey with the following guidelines and to post the survey on your school's website so that all your classmates may take part.

1. You can choose any topic and formulate the questions to achieve what you set to find out on any aspect of school life.
2. The questionnaire should contain no more than 5 questions.
3. The questions should be in the form of multiple-choice type.
4. Use a spreadsheet to tabulate the data collected, and summarise the findings in the form of tables and charts.
5. You are to present your findings to the class at the end of the assignment.

You can refer to the rubric provided in Chapter 8 to help you write the report and make presentation to your class.

# Grouped Frequency Table

The following set of raw data shows the lengths, in millimetres, measured to the nearest mm, of 40 leaves taken from plants of a certain species.

| | | | | | | | | | |
|---|---|---|---|---|---|---|---|---|---|
| 40 | 54 | 25 | 50 | 58 | 45 | 47 | 49 | 30 | 28 |
| 52 | 31 | 52 | 41 | 47 | 44 | 46 | 39 | 41 | 59 |
| 49 | 38 | 43 | 48 | 43 | 43 | 40 | 51 | 40 | 56 |
| 31 | 53 | 44 | 37 | 35 | 37 | 33 | 38 | 46 | 36 |

**For Your Information**

*The study of statistics began when an Englishman, John Graunt (1620–1674), collected and studied the death records in various cities of Britain and was fascinated by the patterns he found in the whole population even though people died randomly.*

Data represented in this form can be confusing and, it would be very difficult to interpret or analyse the data meaningfully. To better understand the data and to get as much information as we can out of it, we need to arrange the data in an *orderly* way. The number of different measurements that appear in the data makes it impractical to organise the data in the same way as we did in Example 1.

One way is to group this kind of data into a number of **class intervals**. If possible, we want all the intervals to have *equal* size.

Grouping a set of data into class intervals requires the following steps:

(1) Glance through the data to pick out the largest and the smallest observations. In our example above,

the *largest* observation = 59
the *smallest* observation = 25

(2) Find the difference between the largest and the smallest observations. In our example,

the *difference* = 59 – 25 = 34.

(3) Decide on the number of class intervals. The amount of data available and the difference between the largest and smallest observations will help us make the choice. In our example, a 10-mm interval size will give us 3 class intervals as $\frac{34}{10} = 3.4 \approx 3$.

A 5-mm interval size will give us 7 class intervals as $\frac{34}{5} \approx 7$.

A 5-mm interval size is more appropriate for our data. As a working guide, we should choose 5 to 20 class intervals depending on the amount of data available.

(4) Choose the first class interval such that the smallest observation is included. In our example, we will take marks greater than or equal to 25 but less than 30 as the first class interval i.e. $25 \le x < 30$ where $x$ is the length of the leaves. The other class intervals are thus $30 \le x < 35$, $35 \le x < 40$, $40 \le x < 45$, $45 \le x < 50$, $50 \le x < 55$ and $55 \le x < 60$. The last class interval $55 \le x < 60$ must include the largest observation.

(5) Read through the list of observations, making a stroke in the tally column against the class intervals to which each belongs. The table below shows the frequency of the length of the 40 leaves.

| *Lengths (mm)* | *Tally* | *Frequency* |
|---|---|---|
| $25 \le x < 30$ | // | 2 |
| $30 \le x < 35$ | //// | 4 |
| $35 \le x < 40$ | ~~////~~ // | 7 |
| $40 \le x < 45$ | ~~////~~ ~~////~~ | 10 |
| $45 \le x < 50$ | ~~////~~ /// | 8 |
| $50 \le x < 55$ | ~~////~~ / | 6 |
| $55 \le x < 60$ | /// | 3 |
| | Total frequency= 40 | |

The diagram below shows the histogram representing the distribution of the length of the 40 leaves.

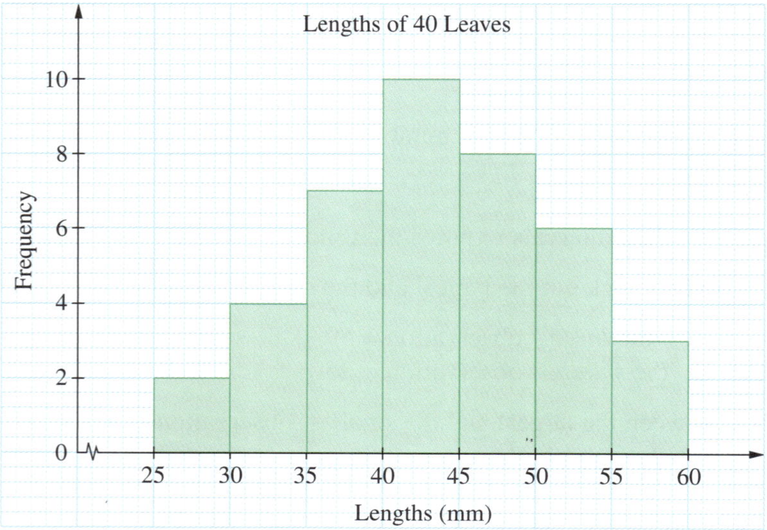


## Exercise 13e

**1.** The lengths, in mm, of 48 rubber tree leaves are given below.

| | | | | | | | | | | | |
|---|---|---|---|---|---|---|---|---|---|---|---|
| 137 | 152 | 127 | 134 | 147 | 141 | 157 | 132 | 153 | 166 | 147 | 136 |
| 146 | 142 | 162 | 169 | 149 | 135 | 166 | 148 | 157 | 141 | 146 | 147 |
| 163 | 133 | 148 | 150 | 136 | 127 | 162 | 152 | 143 | 138 | 142 | 153 |
| 145 | 154 | 144 | 126 | 139 | 126 | 158 | 147 | 136 | 144 | 159 | 161 |

Copy and complete the following table:

| *Lengths* ($x$ mm) | *Tally* | *Frequency* |
|---|---|---|
| $125 < x \leq 130$ | | |
| $130 < x \leq 135$ | | |
| $135 < x \leq 140$ | | |
| $140 < x \leq 145$ | | |
| $145 < x \leq 150$ | | |
| $150 < x \leq 155$ | | |
| $155 < x \leq 160$ | | |
| $160 < x \leq 165$ | | |
| $165 < x \leq 170$ | | |

Draw a histogram to illustrate the distribution.

2. The waiting time, $x$ minutes, for 52 patients at a hospital pharmacy are as follows:

| | | | | | | | | | | | | |
|---|---|---|---|---|---|---|---|---|---|---|---|---|
| 12 | 17 | 25 | 34 | 46 | 18 | 11 | 8 | 13 | 15 | 27 | 44 | 36 |
| 14 | 22 | 45 | 48 | 35 | 52 | 23 | 37 | 16 | 42 | 17 | 36 | 58 |
| 51 | 27 | 19 | 23 | 29 | 32 | 31 | 43 | 16 | 19 | 27 | 29 | 33 |
| 44 | 31 | 16 | 9 | 10 | 14 | 16 | 39 | 29 | 35 | 41 | 31 | 20 |

**(a)** Construct a frequency table using class intervals $5 < x \le 10$, $10 < x \le 15$, $15 < x \le 20$, $20 < x \le 25$, $25 < x \le 30$, and so on.

**(b)** Draw a histogram for the distribution.

3. The following frequency table shows the amount of money spent by 100 smokers per day during a random survey done in 2005.

| *Amount spent per day (\$x)* | *Frequency* |
|---|---|
| $0 < x \le 4$ | 4 |
| $4 < x \le 8$ | 6 |
| $8 < x \le 12$ | 16 |
| $12 < x \le 16$ | 18 |
| $16 < x \le 20$ | 24 |
| $20 < x \le 24$ | 16 |
| $24 < x \le 28$ | 8 |
| $28 < x \le 32$ | 5 |
| $32 < x \le 36$ | 3 |

Draw a histogram to represent this information.

4. The mass, in kg, of 70 girls were recorded as shown in the table below.

| *Mass, x* kg | Frequency |
|---|---|
| $35 \le x < 40$ | 6 |
| $40 \le x < 45$ | 7 |
| $45 \le x < 50$ | 18 |
| $50 \le x < 55$ | 19 |
| $55 \le x < 60$ | 10 |
| $60 \le x < 65$ | 5 |
| $65 \le x < 70$ | 3 |
| $70 \le x < 75$ | 2 |

Draw a histogram to represent this information.

**5.** The histogram shows the life span (in days) of insects in a colony.

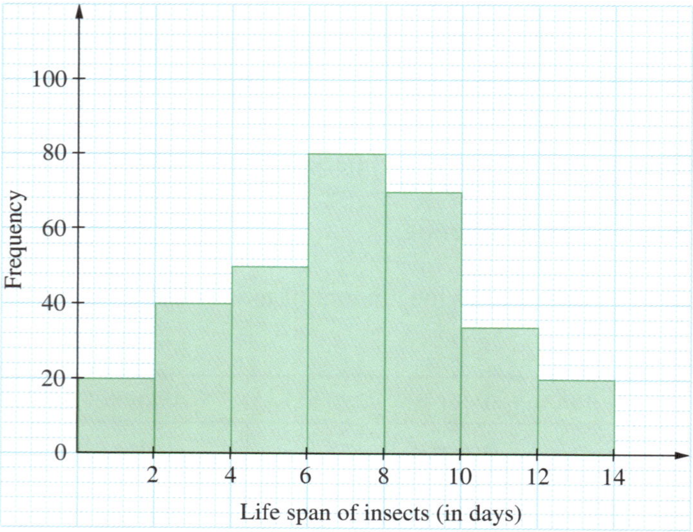


**(a)** Find the total number of insects in the colony.
**(b)** How many insects have a life span of less than 6 days?
**(c)** How many insects have a life span of more than or equal to 10 days?

## SUMMARY

**1.** **(a)** Numerical data can be obtained in many ways.
**(b)** The data collected can be summarised in a systematic way by tabulation.
**(c)** The tabulated data is usually presented in a graphical form.
**(d)** Some common statistical graphs are **pictograms**, **bar graphs**, **pie charts** and **line graphs**.

**2.** **(a)** A set of data, or raw data, can be arranged in an orderly way in the form of a **frequency table**.
**(b)** A frequency table can be represented graphically by a **histogram**.
**(c)** A **histogram** is a vertical bar graph with no space in **between** the bars.
**(d)** The **area** of each bar is proportional to the frequency it represents.

**3.** Suitable choice of statistical graphs depends on the type of data that is collected. In general,
**(a)** a pictogram is most suitable when the data needs to be presented in a lively and interesting manner while the accuracy is not of utmost importance;
**(b)** a bar graph and a histogram are useful for comparing the data clearly;
**(c)** a pie chart is suitable when we need to compare the proportion of a whole rather than the actual numerical values.
**(d)** a line graph is most suitable to show data that changes with time.

## Example 1

*Each pupil in a class was asked how many books they had read in the last month. The results are shown in the following bar graph.*

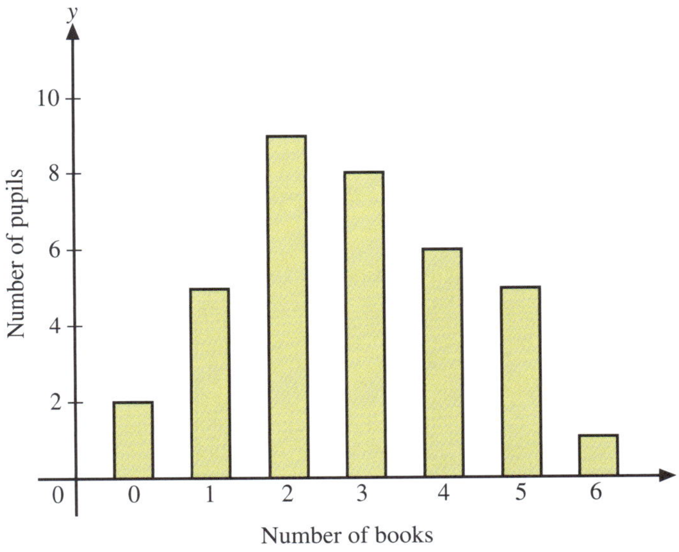


*Find*

*(a) the total number of books read by the pupils,*

*(b) the proportion of pupils who read more than 4 books, giving your answer as a fraction,*

*(c) the angle of the sector in a pie-chart that represents the number of pupils who read less than or equal to 2 books.*

**Solution**

(a) The total number of books read by the pupils

$= 2 \times 0 + 5 \times 1 + 9 \times 2 + 8 \times 3 + 6 \times 4 + 5 \times 5 + 1 \times 6$

$= 102$

(b) Number of pupils who read more than 4 books = 5 + 1 = 6
Total number of pupils = 2 + 5 + 9 + 8 + 6 + 5 + 1 = 36

$\therefore$ proportion of pupils who read more than 4 books = $\frac{6}{36} = \frac{1}{6}$

(c) Number of pupils who read less than or equal to 2 books = 2 + 5 + 9 = 16

$\therefore$ angle in a pie-chart representing this = $\frac{16}{36} \times 360°$
= 160°

## Example 2

*To attract investors to buy into a listed Real Estate Investment Trust, a company put up an advertisment showing the present distributable yield in cents per unit and the projected distributable yield after investors buy into the company.*

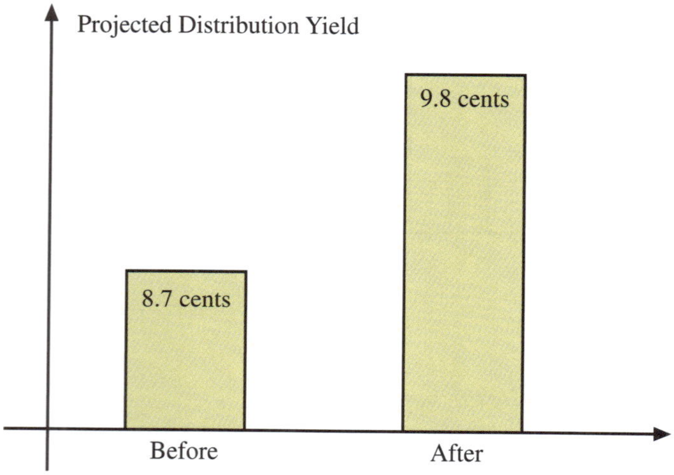


*Briefly explain the impact of the bar graph to a reader.*

### Solution

The height of the bar representing the yield before investors buy into the company is about half the height of that after investors buy into the company. This gives the impression that the yield will increase by 100% after investors buy into the company. In actual fact, the increase is less than 15%. This is created as a result of not using the proper scale and not starting with 0 on the vertical axis.

# Review Questions 13

1. The bar graph below illustrates the monthly expenditure of a family.

   Draw a pie chart to represent this information, marking out the size of the angle in each sector clearly.

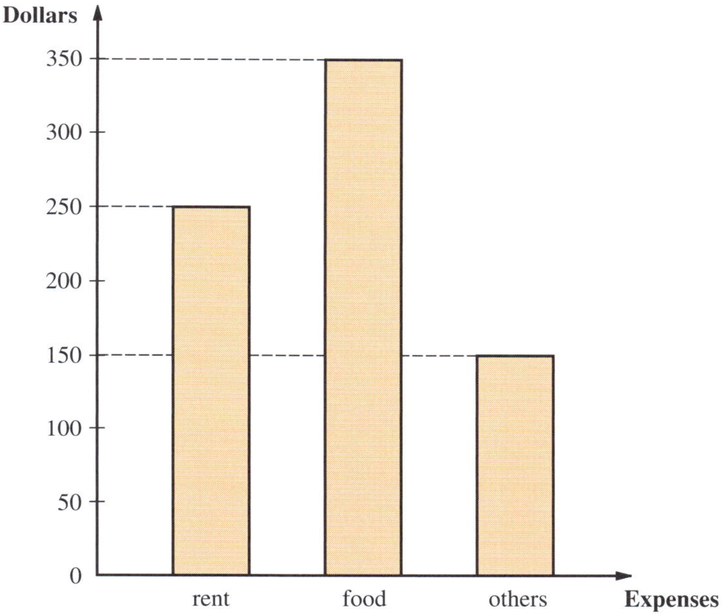


2. For the period of 2004 – 2005, the gross income of a company was \$63 million. The profit of the company before tax was \$8 million after the costs of running the company were deducted from the gross income.
   **(a)** Find the running costs for the period 2004 – 2005.
   **(b)** The running costs are represented in the following pie chart.
      **(i)** What percentage of the running costs belonged to capital charges?
      **(ii)** What is the measurement of the angle used to represent materials in the pie chart?
      **(iii)** Find the actual amount of money spent on fuel.

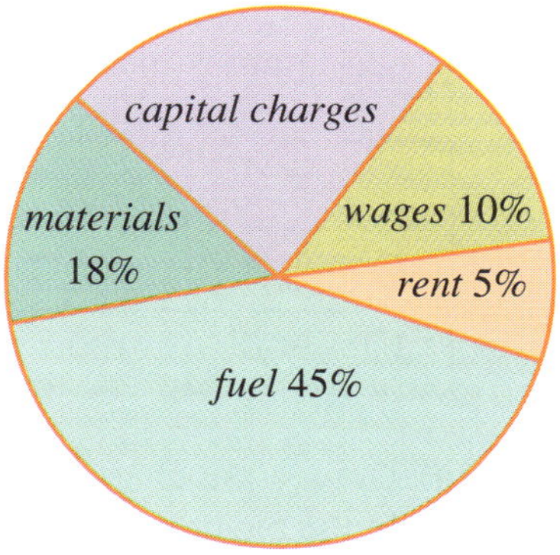


3. The table shows the number of goals scored by each team in a soccer tournament.
   **(a)** How many teams scored more than two goals?
   **(b)** Draw a histogram to represent the data.

| *No. of goals* | 0 | 1 | 2 | 3 | 4 | 5 |
|---|---|---|---|---|---|---|
| *No. of teams* | 15 | 19 | 8 | 7 | 1 | 0 |

★ **4.** During a one-month period, the number of sick leave days of 100 workers in a factory was recorded as shown in the table below:

| *No. of sick leave days* | 0 | 1 | 2 | 3 | 4 |
|---|---|---|---|---|---|
| *No. of workers* | 45 | 32 | 14 | 6 | 3 |

**(a)** Represent the data using a histogram.
**(b)** What is the most common number of days of sick leave?

**5.** The table below shows the number of hours 60 workers work per week.

| *Number of hours* | 37 | 38 | 39 | 40 | 41 | 42 | 43 | 44 | 45 |
|---|---|---|---|---|---|---|---|---|---|
| *Frequency* | 3 | 4 | 9 | 18 | 10 | 6 | 3 | 5 | 2 |

**(a)** Draw a histogram to represent the data.
**(b)** What percentage of workers work at least 40 hours per week?

**6.** The table below shows the amount of vitamin C, in milligrams, found in 102 oranges.

| *Amount of vitamin C* (mg) | 26 | 27 | 28 | 29 | 30 | 31 | 32 | 33 | 34 |
|---|---|---|---|---|---|---|---|---|---|
| *Number of oranges* | 10 | 7 | 16 | 20 | 24 | 12 | 8 | 2 | 3 |

Draw a histogram to represent the above distribution.

**7.** The following table shows the amounts, in dollars, spent on food, rent, clothing, fuel and other items by two families in a week.

| | *Food* | *Rent* | *Clothing* | *Fuel* | *Others* | **Total** |
|---|---|---|---|---|---|---|
| Family $A$ | 160 | 56 | 48 | 32 | 24 | 320 |
| Family $B$ | 180 | $x$ | 63 | 45 | 42 | $y$ |

**(a)** Draw a pie chart of radius 4 cm to represent the amounts spent by Family $A$.
**(b)** In the pie chart for Family $B$, the angle of the sector representing the amount spent on rent is the same as that for Family $A$. Calculate the values of $x$ and $y$.

**8.** The table shows the ages in years of 1500 members of a country club.

| *Age* ($x$ years) | *Frequency* |
|---|---|
| $10 \leq x < 20$ | 25 |
| $20 \leq x < 30$ | 135 |
| $30 \leq x < 40$ | 585 |
| $40 \leq x < 50$ | 350 |
| $50 \leq x < 60$ | 235 |
| $60 \leq x < 70$ | 125 |
| $70 \leq x < 80$ | 35 |
| $80 \leq x < 90$ | 10 |

Draw a histogram to represent the above information.

## Revision Exercise III No. 1

1. A man walks at an average rate of 1.25 m/s. Find the time he takes to walk 3.75 km.

2. Find the ratio of the weight of a Japanese car weighing 4200 kg to that of a German car weighing 7200 kg.

3. What is the difference in total wages between employing 18 men at a wage of \$380 each and 33 women at a wage of \$208 each?

4. A sum of money is distributed among 3 boys, $A$, $B$ and $C$, in the ratio 2 : 4 : 14. If $B$ gets \$1.20 more than $A$, how much money does $C$ get?

5. Each student in a group of 240 was asked to choose his/her favourite subject from Mathematics, English and Geography. The result is represented on the pie chart given below.

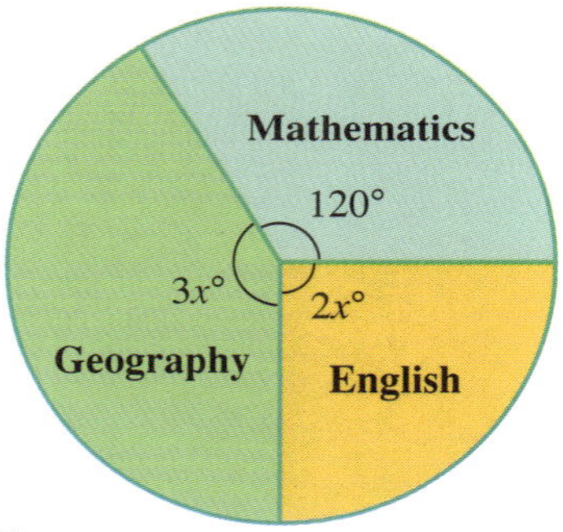


Calculate
(a) the value of $x$;
(b) the number of pupils who chose English;
(c) the fraction of pupils who chose Geography.

6. Consider the number pattern below:

$$1 = 1$$
$$1 + 2 + 1 = 4$$
$$1 + 2 + 3 + 2 + 1 = 9$$
$$1 + 2 + 3 + 4 + 3 + 2 + 1 = 16$$
$$\vdots \quad \vdots$$

(a) Write down the 5th line of the sequence.
(b) Find the value of $1 + 2 + 3 + ... + 8 + 9 + 8 + 7 + ... + 3 + 2 + 1$.
(c) If $1 + 2 + 3 + ... + (x - 1) + x + (x - 1) + ... + 3 + 2 + 1 = 169$, find the value of $x$.

7. (a) A solid copper cylinder, 8 cm long and 4 cm in diameter, is melted and recast into a length of wire 2 mm in diameter. How long is the wire?
(b) A lawn roller is 80 cm wide with a diameter of 42 cm. Find the area it covers in 40 revolutions. Give your answer in square metres and take $\pi$ to be $\frac{22}{7}$.

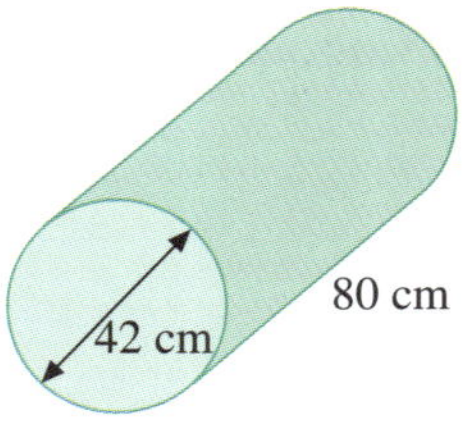


8. An open cylindrical tank with diameter 28 cm contains water to a depth of 30 cm. Find the volume of the water inside the tank, giving your answer in litres. Find also the total surface area of the tank that is in contact with the water.

9. A rectangular field has sides $5x$ metres by $4x$ metres. Calculate the value of $x$ in each of the following cases:
(a) The perimeter of the field is 450 metres.
(b) The area of the field is 2000 square metres.

## Revision Exercise III No. 2

1. (a) How many books costing \$8.40 each can be bought with \$200? How much money will then be left?
(b) A car uses $27\frac{1}{2}$ litres of petrol for a journey of 220 km. How much petrol will it need to cover a distance of 680 km?

2. (a) Express 0.035 as a fraction in its lowest terms.

(b) Express $1\frac{9}{40}$ as a decimal.

(c) Express 2.004 56 correct to two decimal places.

(d) Express 0.025 47 correct to two significant figures.

3. (a) Find the LCM of 25, 80 and 120.
(b) Simplify $3(5x - 7) - 2(3x - 5y) + 2(3y - 7x) - (x - 2y)$.

4. Solve the equation $x - \frac{x-1}{4} = \frac{x+2}{5}$.

5. (a) The area of a trapezium is 1500 $cm^2$. Find the height between the parallel sides if their lengths are 14 cm and 36 cm.

(b) How many litres of water are required to fill a rectangular storage tank 12 m long, 6 m wide and 2 m deep?
(1 litre = 1000 $cm^3$)

6. Consider the following number pattern:

$$2 \times 2 = 1 \times 3 + 1$$
$$3 \times 3 = 2 \times 4 + 1$$
$$4 \times 4 = 3 \times 5 + 1$$
$$5 \times 5 = 4 \times 6 + 1$$
$$\vdots \qquad \vdots$$
$$n \times n = 13 \times 15 + 1$$

(a) Write down the 10th line in the pattern.
(b) Find the value of $n$.
(c) Use the above pattern to calculate the value of $199^2$.

7. The students of a Secondary 1 class were asked to name their favourite subject. The bar graph illustrates the result. Find
(a) the total number of students in the class;
(b) the fraction of students who like Maths;
(c) the fraction of students who like either History or Geography.

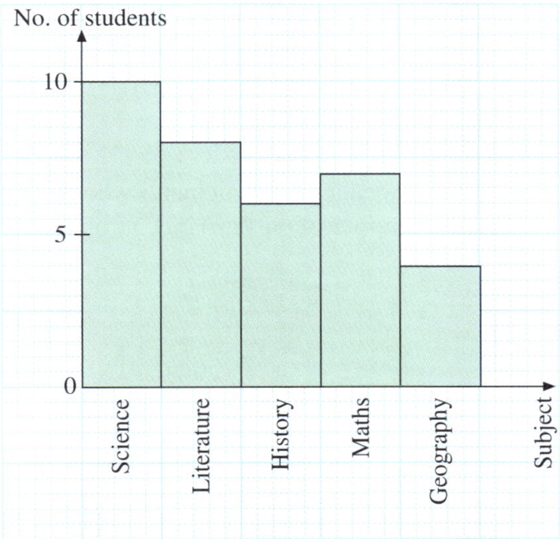


8. (a) A solid concrete cylinder of diameter 70 cm is 120 cm high. What is its volume in cubic metres?

(b) At a sale, a sofa set selling at a discount of 25% fetches \$3570. Find the original price of the sofa set.

(c) A trader buys $x$ pencils at \$$y$ each and sells them at 3 for \$$z$. Find an expression for his total profit.

9. A bicycle wheel has a diameter of 70 cm. Find
(a) the circumference of the wheel;
(b) the distance the bicycle travels after the wheel has made 400 revolutions.

$\left(\text{Take } \pi = 3\frac{1}{7}\right)$

## Revision Exercise III No. 3

1. Two men, $A$ and $B$, can paint a house in 8 days. $A$ alone can paint it in 12 days. How long will $B$ take to paint the house by himself?

2. 784 marbles are shared among $X$, $Y$ and $Z$ in the ratio 3 : 5 : 8. How many marbles does each get?

3. The pie chart below shows the number of people taking part in the big walk organized by the Singapore Sports Council. If 1656 girls took part in the walk, find

**(a)** $x$;

**(b)** the percentage of participants who were men;

**(c)** the number of participants who were boys.

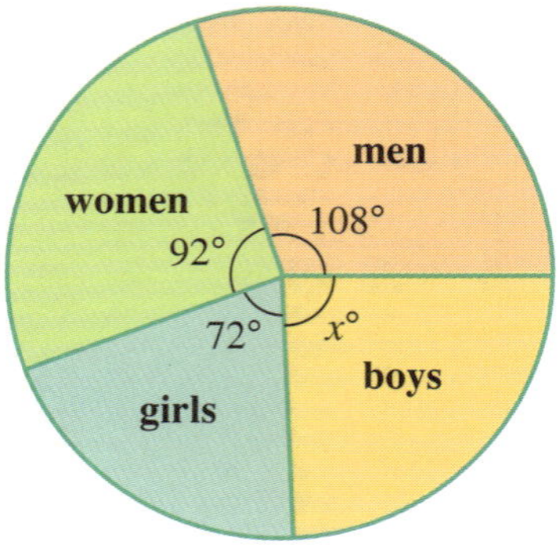


4. In the figure, $ABC$ is a semicircle and $ACD$ is a triangle. Find the area of the whole figure with the dimensions given.

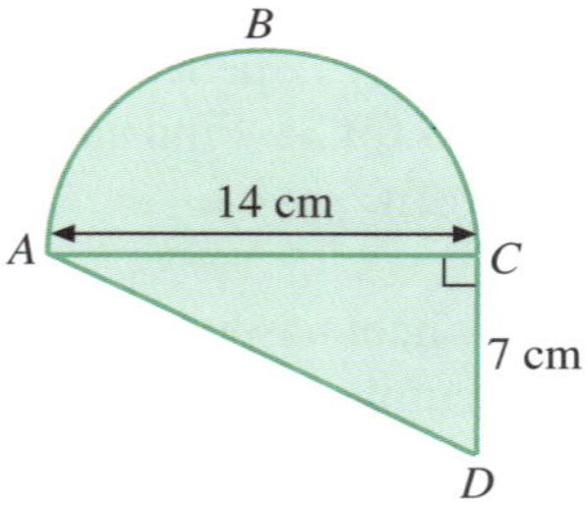


5. **(a)** \$$2x$ is shared among 3 persons, $A$, $B$ and $C$, in the ratio 3 : 2 : 5. If $B$ gets \$45, find the value of $x$.

**(b)** The cost of material, labour and administration for an advertising campaign is in the ratio 8 : 5 : 2. If the total cost of the campaign is \$3525, find the cost of labour.

6. A school field measures 120 m by 60 m. A plan of the field is drawn to a scale of 1 : 500. Find the area of the field on the plan in $cm^2$.

7. The inner dimensions of an open wooden rectangular box are 10 cm by 8 cm for the base and 5 cm for the height. If the thickness of the wood is 1 cm, calculate the volume of the material used in making the box. Find the mass of the box if the density of the wood is 0.7 $g/cm^3$.

8. **(a)** The pie chart below shows the sales of a shop dealing with telecommunications in a week.

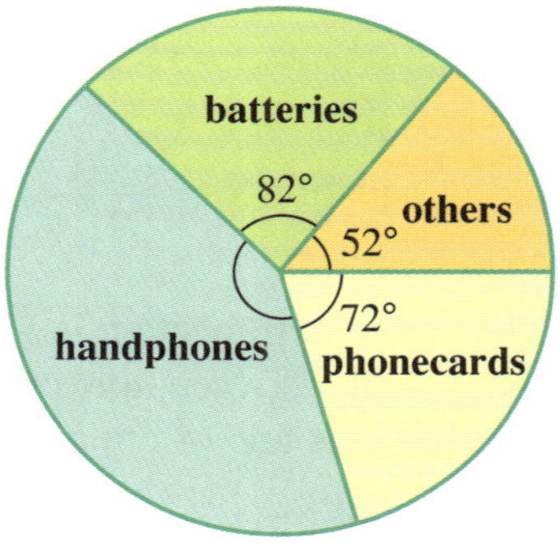


**(i)** What percentage of the sales were from phonecards?

**(ii)** If the total sales of the week amounted to \$72 000, calculate the amount of sales for handphones.

**(b)** 12 men can renovate a house in 15 days if they work 7 hours a day. If 14 men were asked to renovate a similar house in 10 days, how many hours a day must the men work?

9. Express 48 as the product of its prime factors. Hence or otherwise find the smallest possible integer value of $x$ for which $48x$ is a perfect square.

10. Write down all the integer values of $x$ for which $-2 < x \leq 4\frac{1}{4}$.

## Revision Exercise III No. 4

**1. (a)** A man buys 5 kg of beef at \$12.50 per kg. In addition, for every kilogram of beef purchased, he has to pay a consumption tax of 6% on the selling price. Calculate the total amount of money that he has to pay.

**(b)** A man bought 400 dozen pencils at \$0.80 a dozen. He sold half of them at \$1.05 a dozen and the rest at \$0.75 a dozen. Find his profit.

**2. (a)** There are 31 sweets to be shared among 3 boys, $A$, $B$ and $C$. $A$ has 2 sweets more than $B$ and $B$ receives 1 more than $C$. How many sweets does each receive?

**(b)** The driver of a car 3 m long took 5 seconds to drive through a tunnel 44 m long. Find the average speed of the car in km per hour.

**3.** Each pupil in a class of 40 was asked to state the length of time he/she spends on private tuition in a week. The results are shown in the following table.

| *(t hours)* *Length of time spent* | *No. of pupils* |
|---|---|
| $t = 0$ | 8 |
| $0 < t \leq 2$ | 4 |
| $2 < t \leq 4$ | 12 |
| $4 < t \leq 6$ | 10 |
| $6 < t \leq 8$ | 6 |

Illustrate the above information by using a pie chart. What is the angle representing pupils who have 2 to 4 hours of private tuition per week?

**4. (a)** Express 0.086 479
**(i)** correct to 2 decimal places;
**(ii)** correct to 2 significant figures.

**(b)** Express 99 225 in prime factors and hence, find the value of $\sqrt{99\ 225}$.

**5.** Solve the following equations:
**(a)** $5(3 - x) = 7(2x - 5)$

**(b)** $\dfrac{x}{3} - \dfrac{2x - 5}{5} = \dfrac{3}{15}$

**6.** Two towns, $A$ and $B$, are 180 km apart. Motorist $P$ sets off from $A$ for $B$ at 12 20 and travels at a speed of 50 km/h. Motorist $Q$ sets off from $A$ for $B$ at 12 45 at a speed of 60 km/h. Who will arrive at $B$ first and at what time?

**7.** The histogram shows the number of absentees of a class during the first six months in a school in Singapore.

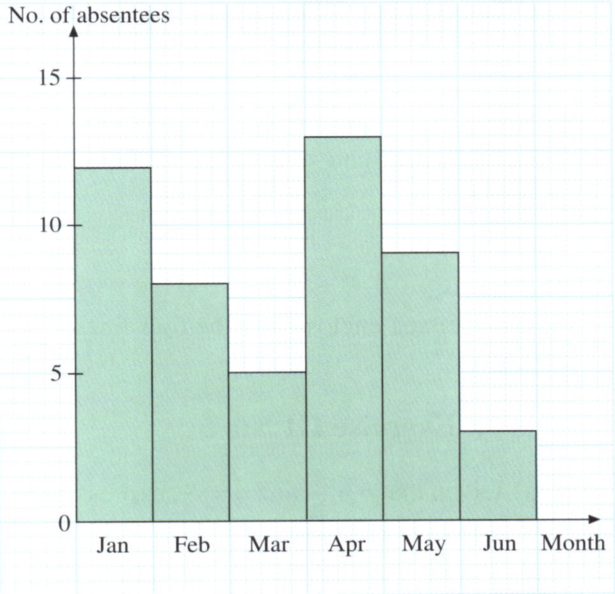


**(a)** Find the total number of absentees.
**(b)** Suggest a simple reason why there were fewer absentees in the months of March and June.

**8. (a)** If $2 : x = 5 : 9$, find $x$.

**(b)** If $p = \dfrac{3}{4}$ and $q = 1\dfrac{5}{7}$, express $\dfrac{p}{q}$ as a percentage.

**(c)** Find two consecutive odd numbers such that the greater number added to 3 times the smaller number gives a total of 86.

**9.** A rectangular hole measuring 2 cm by 3 cm is cut out from a solid cyclinder of radius 12 cm and height 14 cm as shown in the diagram.

**(a)** Find the volume of the remaining solid.

**(b)** If the density of the material used to make the solid is 1.5 g/cm$^3$, find the mass of the remaining solid.

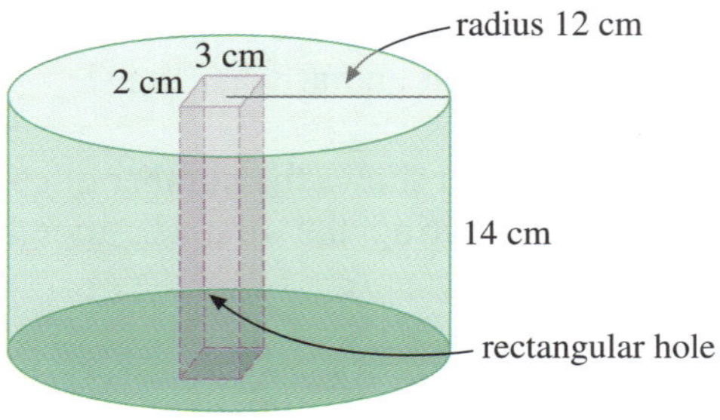


**10.** On the same piece of graph paper and using the same axes and scales, draw the graphs of

**(a)** $y = x + 2$,

**(b)** $y = x - 3$,

**(c)** $y = 2$,

**(d)** $y = -3$.

Find the area enclosed by the four lines.

## Revision Exercise III No. 5

**1.** **(a)** Given that $p = \frac{1}{4}$ and $q = \frac{2}{7}$, express $\frac{p}{q}$ as a percentage.

**(b)** A school has 750 pupils and 32% of them wear spectacles. How many pupils do not wear spectacles?

**(c)** The average speed of a car is 108 km/h. How many metres can the car travel in 3 minutes?

**2.** **(a)** A shopkeeper buys a flower vase for \$60 and prices it at \$85. Find the cash price a customer has to pay if the shopkeeper gives him a 5% discount when he pays cash.

**(b)** *A*, *B* and *C* share a sum of money in the ratio 3 : 4 : 9. If *B* has \$2.80 more than *A*, how much is *C*'s share?

**3.** Meiling's mother is 8 times as old as Meiling. In 10 years' time, Meiling's mother will be only 3 times as old as Meiling. Find their present ages.

**4.** The road distance between two towns, *A* and *B*, is 550 km. A car leaves *A* for *B* at an average speed of 72 km/h and a lorry leaves *B* for *A*, travelling along the same road as the car, at an average speed of 38 km/h at the same time. How long will it take before the two vehicles meet?

**5.** **(a)** A motorist takes $5\frac{3}{4}$ hours to travel from *P* to *Q* at *V* km/h. If he increases his speed by 5 km/h, the time taken will be reduced by half an hour. Find *V*.

**(b)** The sum of 3 consecutive even numbers is 108. Find the numbers.

**6.** The histogram displays the results of a survey on the number of children the residents of a newly completed condominium have.

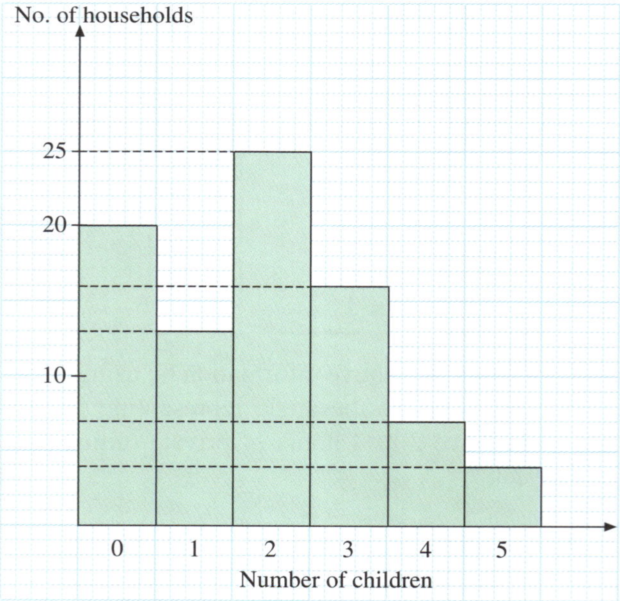


Find

**(a)** the number of households;

**(b)** the total number of children in the survey;

**(c)** the percentage of households having exactly 2 children.

**7.** The results of a survey of the mode of transport used by the students of a school is illustrated by the pie chart. If the number of students travelling by car is 160, how many students travel by MRT?

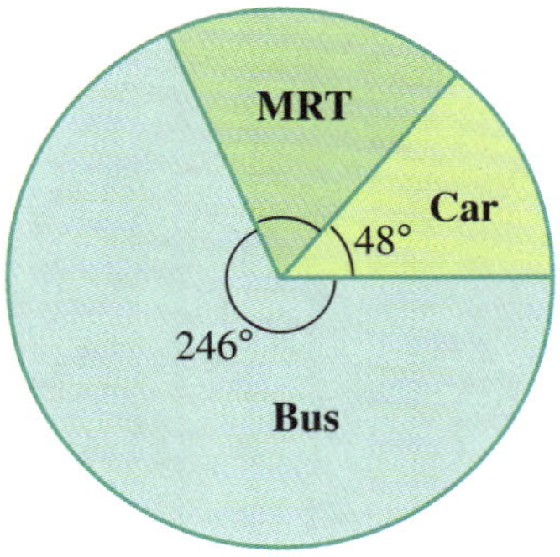


**8. (a)** The perimeter of a square is 36 cm. Find its area.

**(b)** The area of a triangle $ABC$ is 36 $cm^2$ and $AB = 18$ cm. Find the perpendicular height from $C$ to $AB$.

**9.** A farmer uses $\frac{1}{3}$ of his land for rubber, $\frac{3}{8}$ for palm oil, $\frac{1}{6}$ for durians and the remaining 23 hectares for cocoa. Find the total area of the land.

**10.** Draw the graphs of each of the following equations on the same piece of graph paper.

**(a)** $y = 2x + 1$ **(b)** $y = 2x - 5$

**(c)** $y = 5$ **(d)** $x = 0$

State the name of the geometrical figure formed by these four lines. Write down the vertices of this figure.

*In this chapter, you will learn how to:*

- *identify various types of angles*
- *calculate unknown angles involving adjacent angles on a straight line, vertically opposite angles, angles at a point, alternate angles, corresponding angles and interior angles between parallel lines.*

CHAPTER

# 14

# Basic Geometrical Concepts and Properties

## Introduction

Many sports involve the effective use of angles. This billard player needs to have a good judgement of angles when hitting the balls to score points.

# Points

The basic geometric figure is a **point**. All other geometric figures are made up of a collection of points. The smallest dot you can mark on your paper with a sharp pencil will give you an idea of what is meant by a geometric point. A point is only an idea in our mind; it is not a physical object and we regard it as having a position but not size or shape. We use a dot or sometimes a cross to mark the position of a point. We normally use capital letters to name points. Thus we speak of point *A*, point *B*, point *C*, etc.

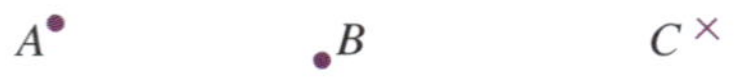


**For Your Information**

*The word "geometry" is derived from the greek words" "ge" (earth) and" "metrein" (to measure). Euclid's masterpiece, "The Elements", survived as the basic textbook for over 2000 years. The geometry that we are to study in this book is sometimes referred to as Euclidean geometry.*

# Lines, Rays, Line Segments

A **line** is the path described by a moving point. A straight **line segment** is formed when we use a ruler to join two points, say, *A* and *B*.

We call the line segment *AB* or *BA*. *A* and *B* are called the end-points. If we extend the line segment *AB* in each of the two directions indefinitely, we get a **line**. This is represented by the following diagram.

The diagrams below show parts of lines with only one end-point and extending in only one direction. We call them **rays**.

Thus we have ray *XY*, ray *HK* and ray *PQ* as in Fig. 14.1.

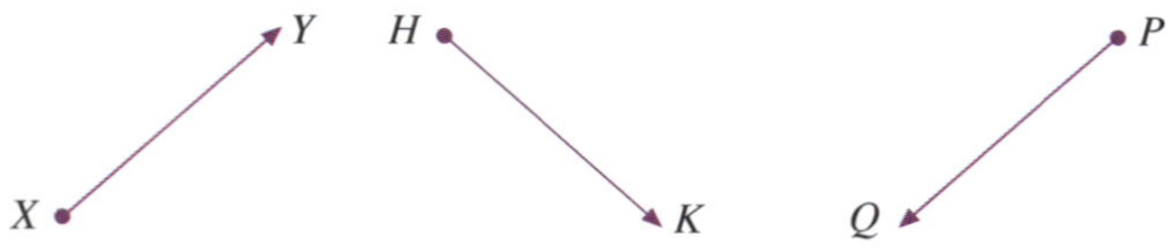


*Fig. 14.1*

In this book, when we speak of a line, it refers to a straight line whereas a curve refers to a curved line. Also, we shall simply use $AB$ to denote **line $AB$, line segment $AB$, ray $AB$** and the **length of $AB$.**

## Planes

A **plane** is a flat surface in which any two points are joined by a straight line lying entirely on the surface. The floor of a classroom is an example of a **horizontal plane** and the wall of a classroom is an example of a **vertical plane**. Can you name other examples of planes?

## Intersecting Lines

Fig. 14.2 shows two lines, $AB$ and $PQ$, on the same plane having a common point $X$. We say that the two lines **intersect** at $X$. Point $X$ is called the **point of intersection**.

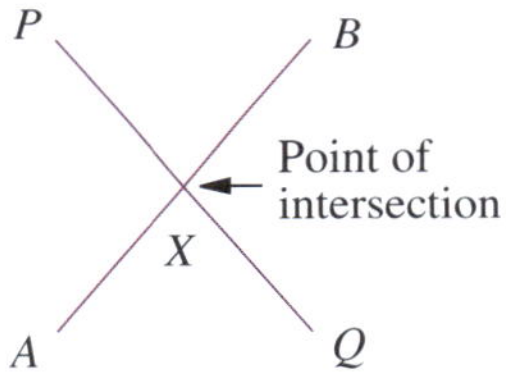


*Fig. 14.2*

## Angles

When two rays $OA$ and $OB$ meet at a point $O$, an **angle** is formed. $O$ is known as the vertex of the angle and $OA$ and $OB$ are the sides or arms of the angle.

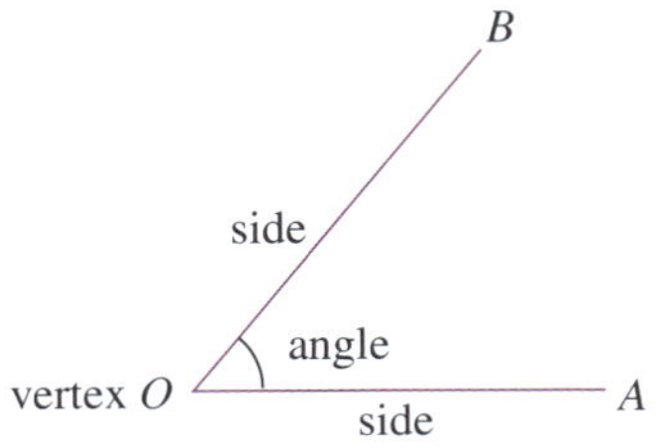


> **For Your Information**
>
> *The system of naming angles was first used by the Babylonians (3000–2000 BC). The idea of 360 parts was suggested to be a consequence of the Babylonians estimation of a year having 360 days.*

The angle is called angle $AOB$ or angle $BOA$ and is written as $A\hat{O}B$ *or* $B\hat{O}A$. Another way of writing this angle is $\angle AOB$ or $\angle BOA$. We may also call it angle $O$ and write $\hat{O}$ or $\angle O$ when it is clear which angle we are referring to.

How many different angles are there in the figure?

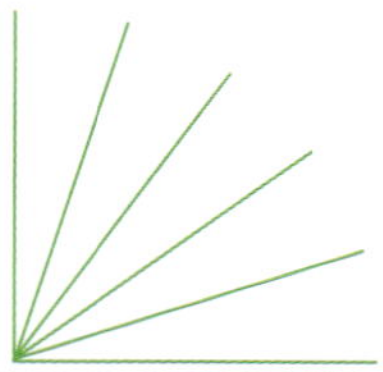

## The Protractor and Angle Measure

The standard unit for measuring angles is one degree (written as 1°). It is defined as $\frac{1}{360}$ of a complete revolution. By definition, one complete rotation about a point has an angle of 360°.

This is also referred to as angles at a point.

Fig. 14.3 shows a protractor which is used to measure angles. To measure an angle, place the protractor so that its centre *A* is at the vertex of the angle and its base *AB* along one side of the angle. Note under which graduation mark the other side passes. Thus, in Fig. 14.4(a), the angle *x* is 60°. Fig. 14.4(b) shows another way of measuring the angle.

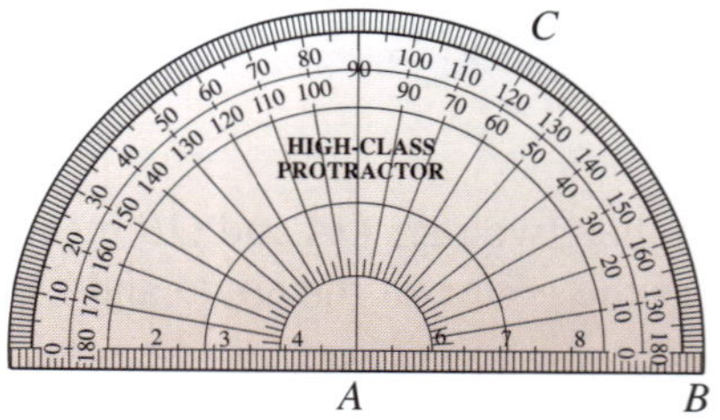


*Fig. 14.3*

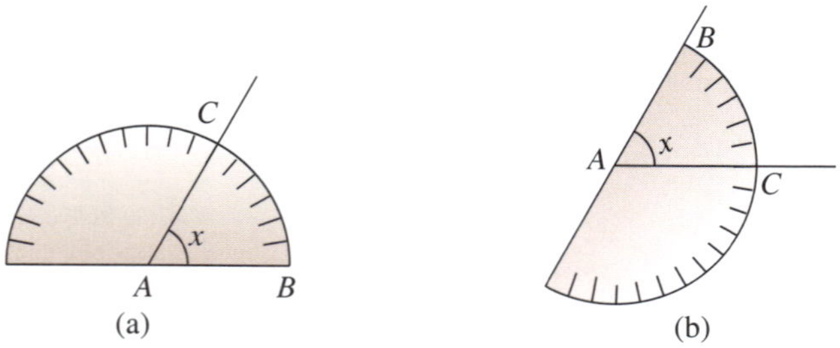


*Fig. 14.4*

Notice that the graduation marks on the protractor are marked with two sets of numbers, one greater than 90° and the other less than 90°. Hence, when using the protractor, you will need to choose the correct set of numbers. For example, if one arm of the angle to be measured lies along $AB$, the set of numbers to be used is the one in which the numbers increase as you read the graduations from $AB$ towards $AC$.

**Thinking Time**

An old man uses a magnifying glass which can enlarge things exactly three times. How many degrees will an angle of 2° appear to him?

## Different Kinds of Angles

Look at the following angles formed by everyday objects and instruments.

They are different in size.

An **acute angle** is less than 90°

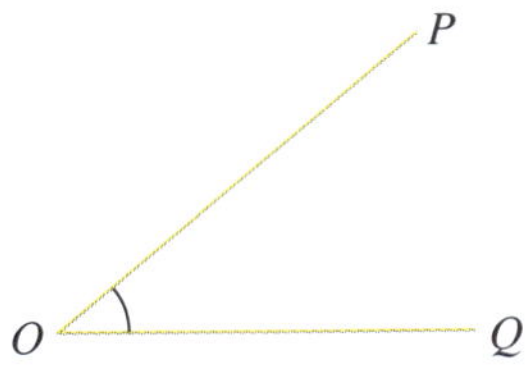


$P\hat{O}Q$ is an acute angle.

A **right angle** is equal to 90°

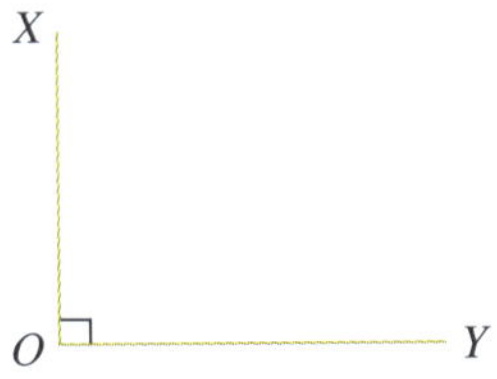


$X\hat{O}Y$ is a right angle.

An **obtuse angle** is larger than 90° but less than 180°.

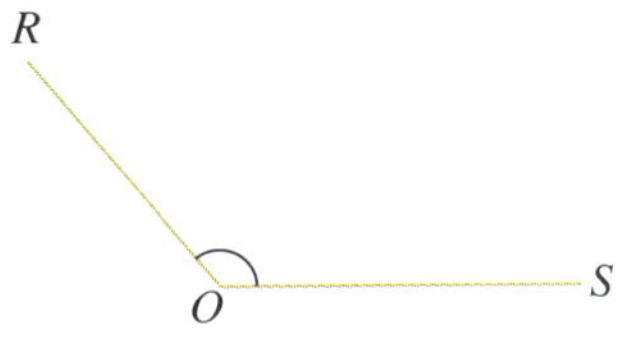


$R\hat{O}S$ is an obtuse angle.

A **reflex angle** is larger than 180° but less than 360°. Both $A\hat{O}B$ and $C\hat{O}D$ are reflex angles.

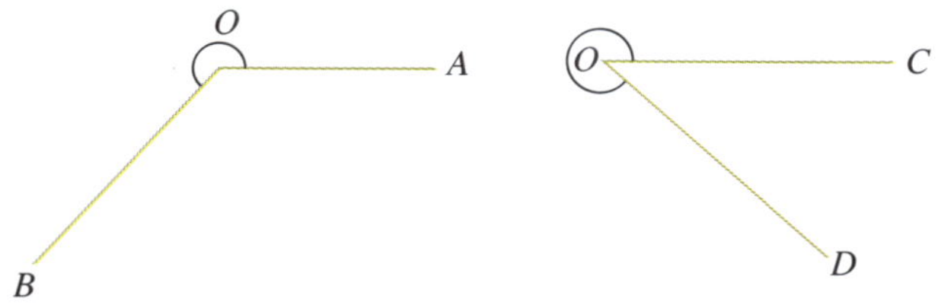


## Explorations

1. The hour hand of a clock is exactly covered by the minute hand at 12 00. What will be the next two times when the minute hand is exactly on top of the hour hand again?
   **(a)** How many degrees will the hour hand move when it moves through 1 hour?
   **(b)** How many degrees will the hour hand move when it moves through $x$ hours?
   **(c)** How many degrees will the minute hand move when it moves through 1 hour?
   **(d)** How many degrees will the minute hand move when it moves through $x$ hours?
   **(e)** Approximately between which two hours will the minute hand be on top of the hour hand again after 12 00?
   **(f)** Form an equation in $x$ and find the time at the first instance when the minute hand is on top of the hour hand.
   **(g)** Form another equation in $x$ and find the time at the second instance when the minute hand is on top of the hour hand.

2. The hands of a clock are at right angles at 09 00. What will be the next two times when the hands are at right angles again?
   **(a)** How many degrees will the hour hand move when it moves through 1 hour?
   **(b)** How many degrees will the hour hand move when it moves through $x$ hours?
   **(c)** How many degrees will the minute hand move when it moves through 1 hour?
   **(d)** How many degrees will the minute hand move when it moves through $x$ hours?
   **(e)** Approximately between which two hours will the hands be at right angle again after 09 00?
   **(f)** Form an equation in $x$ and find the time at the first instance when the two hands are at right angle again.
   **(g)** Form another equation in $x$ and find the time at the second instance when the two hands are at right angle again.

## Complementary Angles

Two angles are called **complementary angles** if their **sum** is **90°**. The angles 28° and 62° are said to be complementary and so are the angles 40° and 50°, 30° and 60°, etc.

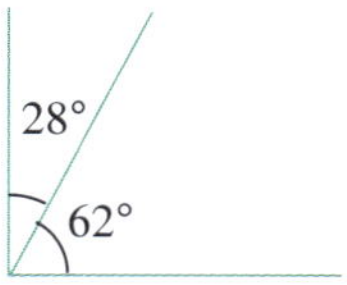


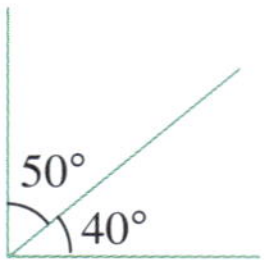


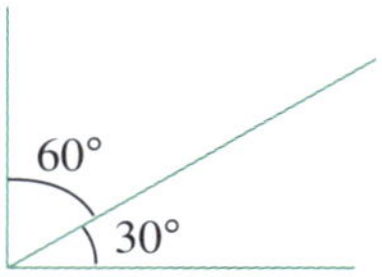


## Supplementary Angles

Two angles are called **supplementary angles** if their **sum** is **180°**. The angles 45° and 135° are supplementary angles and so are 53° and 127°, 105° and 75°, etc.

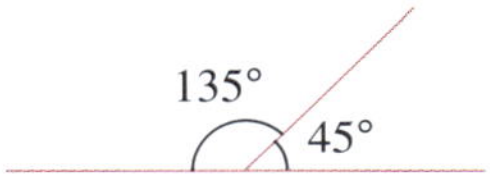


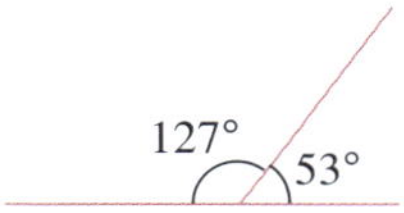


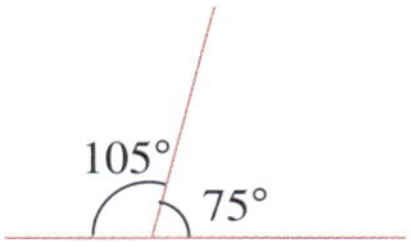


## Adjacent Angles on a Line

In Fig. 14.5, $X\hat{O}Z$ and $Y\hat{O}Z$ are called **adjacent angles** because
**(a)** they have a common vertex $O$,
**(b)** they have a common side $OZ$ and
**(c)** they lie on opposite sides of the common arm.

**The sum of adjacent angles on a line is equal to 180°.**
(Abbreviation for reference: **adj. ∠s on a line.**)

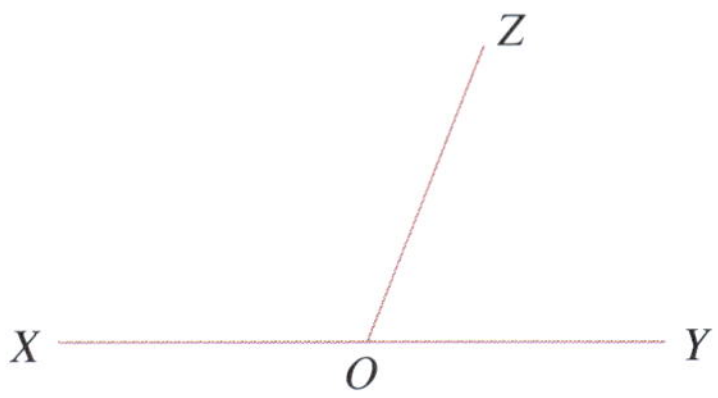


*Fig. 14.5*

**Thinking Time**

In Fig. 14.6, are
**(a)** $\angle a$ and $\angle b$ adjacent angles?
**(b)** $\angle a$ and $\angle c$ adjacent angles?
**(c)** $\angle b$ and $\angle c$ adjacent angles?

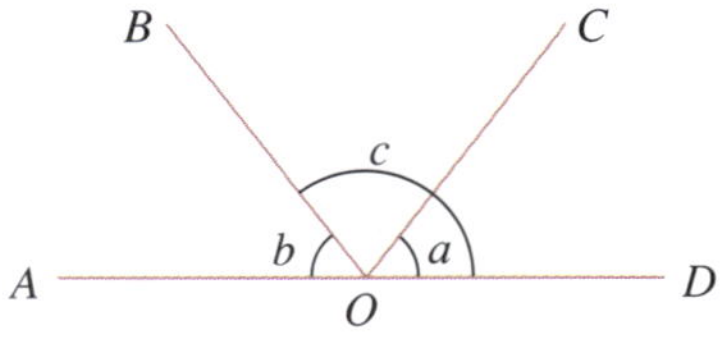

*Fig. 14.6*

## Example 1

*Find the values of a, b and c in the following diagrams:*

*(a)*

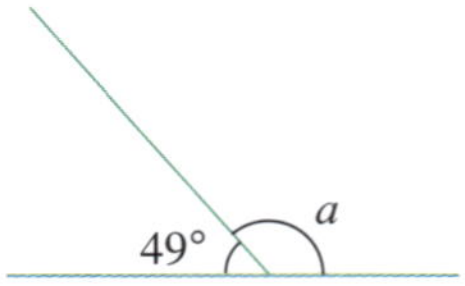


*(b)*

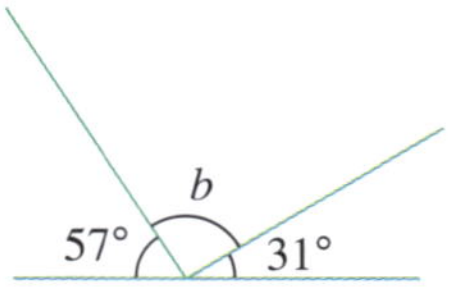


*(c)*

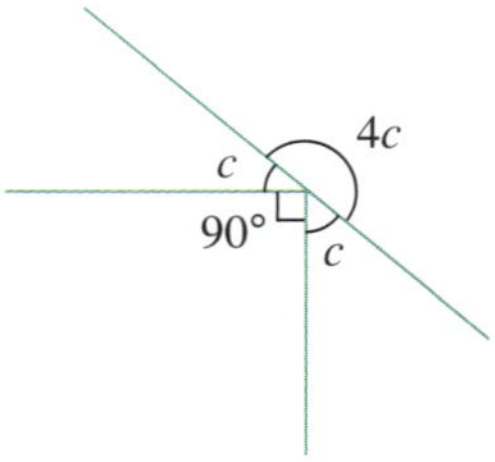


**Solution**

(a) $49° + a = 180°$ (adj. $\angle$s on a line)
$\therefore \quad a = 180° - 49° = 131°$

(b) $57° + b + 31° = 180°$ (adj. $\angle$s on a line)
$\therefore \quad b = 180° - 57° - 31° = 92°$

(c) $90° + c + c + 4c = 360°$ (angles at a point)
$6c = 270°$
$\therefore \quad c = 45°$

# Vertically Opposite Angles

Vertically opposite angles are formed when two straight lines intersect each other. The angles *BOC* and *AOD* in Fig. 14.7 are said to be vertically opposite. The angles *AOC* and *BOD* are also vertically opposite each other. Vertically opposite angles are equal.

The following is a simple proof. Referring to Fig. 14.7, we have

$a + b = 180°$ (adj. $\angle$s on a line)
$b + c = 180°$ (adj. $\angle$s on a line)
$\therefore \quad a + b = b + c$
$\therefore \quad a = c$

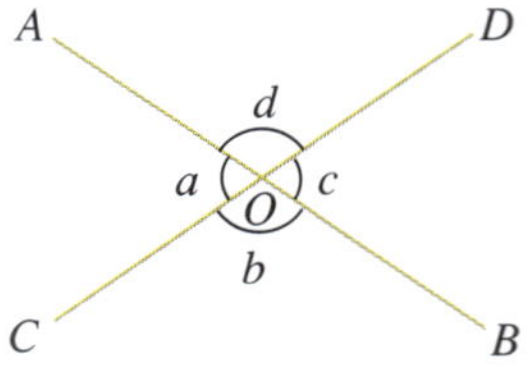

*Fig. 14.7*

Similarly, we can also show that $b = d$. Hence, **vertically opposite angles are equal**. (Abbreviation for reference: **vert. opp. $\angle$s.**)

## Example 2

*Find the values of x, y and z in the given diagram, where AOB and COD are straight lines.*

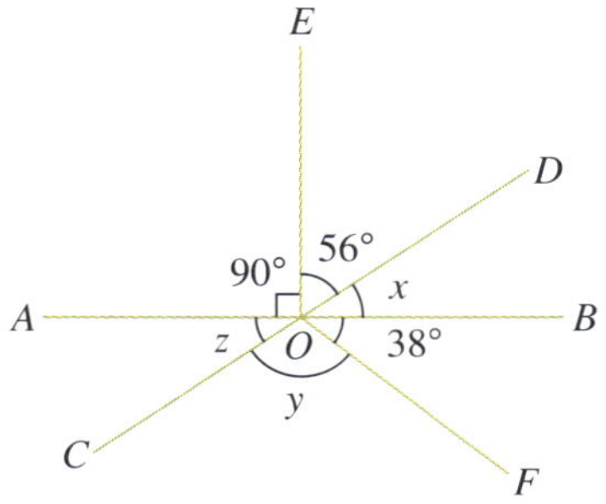


**Solution**

$90° + 56° + x = 180°$ (adj. ∠s on a line)

$\therefore \quad x = 180° - 90° - 56° = 34°$

$z = x = 34°$ (vert. opp. ∠s)

$\therefore \quad z = 34°$

$z + y + 38° = 180°$ (adj. ∠s on a line)

$34° + y + 38° = 180°$

$\therefore \quad y = 180° - 38° - 34°$

$= 108°$

## Exercise 14a

**1.** Classify each of the following either as acute, obtuse or reflex angle.

**(a)** $\frac{2}{7}$ of a right angle

**(b)** $\frac{7}{15}$ of a complete rotation

**(c)** $\frac{5}{8}$ of a complete rotation

**(d)** $1\frac{5}{8}$ of a right angle

**(e)** $\frac{3}{23}$ of a complete rotation

**2.** Measure each of the marked angles.

**(a)**

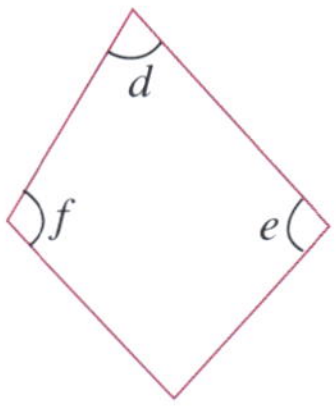


**(b)**

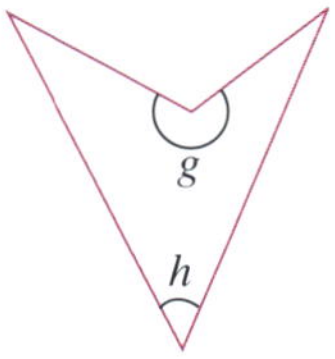

(c)

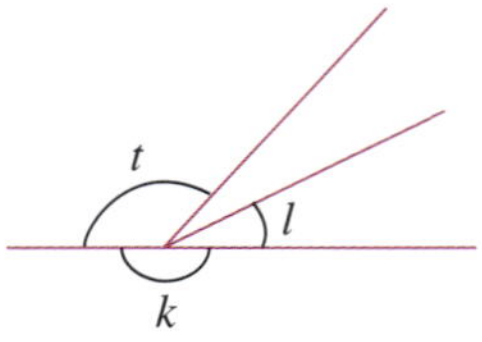


(d)

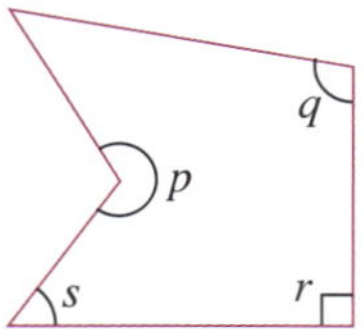


3. Find the measure of the complementary angle of each of the following angles:

(a) 18° (b) 46°
(c) 53° (d) 64°
(e) 7°

4. Find the measure of the supplementary angle of each of the following angles:

(a) 36° (b) 12°
(c) 102° (d) 171°
(e) 88°

5. Use a protractor to draw the following angles:

(a) 20° (b) 157°
(c) 197° (d) 242°
(e) 320° (f) 285°

6. Refer to the figure on the right in which $XOY$ is a straight line, and answer the following questions:

(a) Find $a$, if $b = 45°$ and $c = 86°$.
(b) Find $a$, if $b = 2a$ and $c = 3a$.
(c) Find $b$, given that $a + c = b$.
(d) Find $c$, given that $a = b = c$.

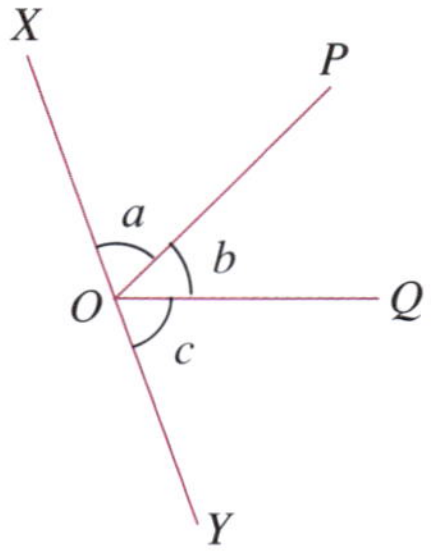


7. In the diagram below, $A\hat{O}B = p°$. If $B\hat{O}C$ is two times $A\hat{O}B$, $C\hat{O}D$ is four times $A\hat{O}B$ and $D\hat{O}A$ is five times $A\hat{O}B$, find the values of all the four angles.

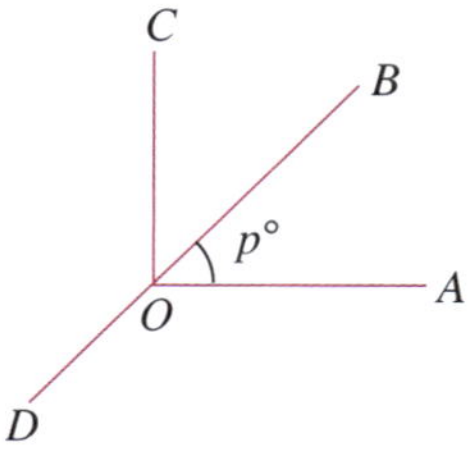


8. Calculate the unknown(s) in each of the following:

(a)

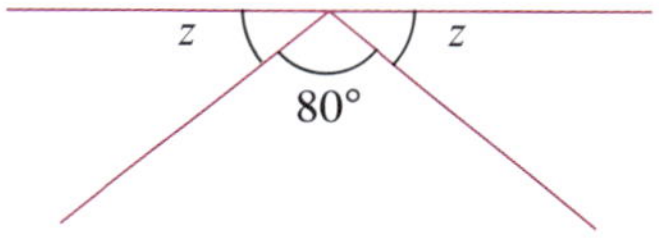


(b)

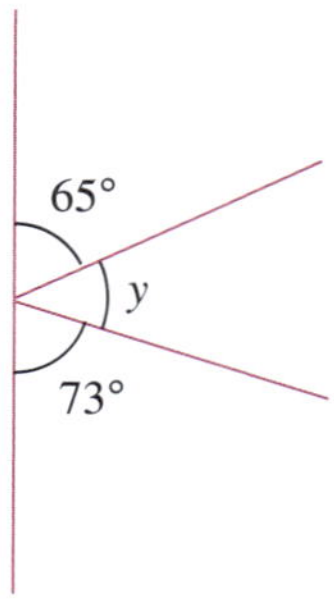

**(c)**

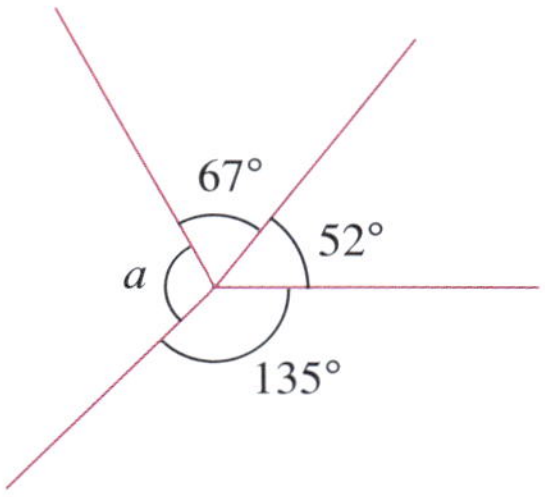


**(d)**

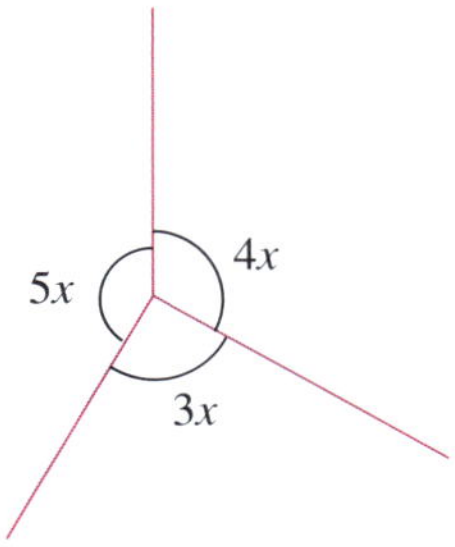


**(e)**

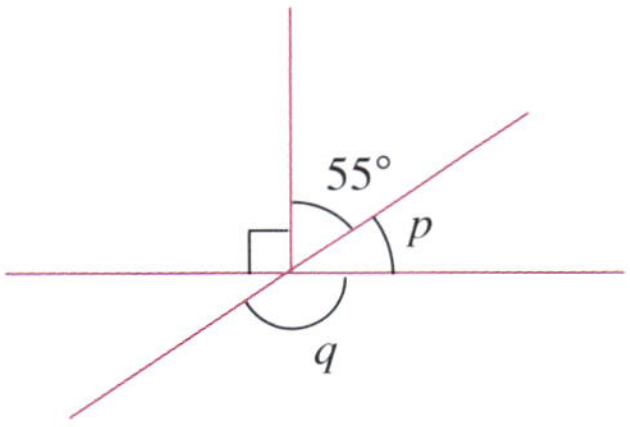


**(f)**

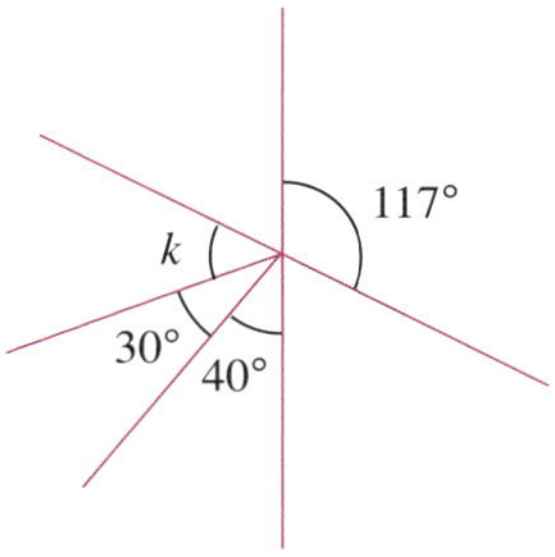


**(g)**

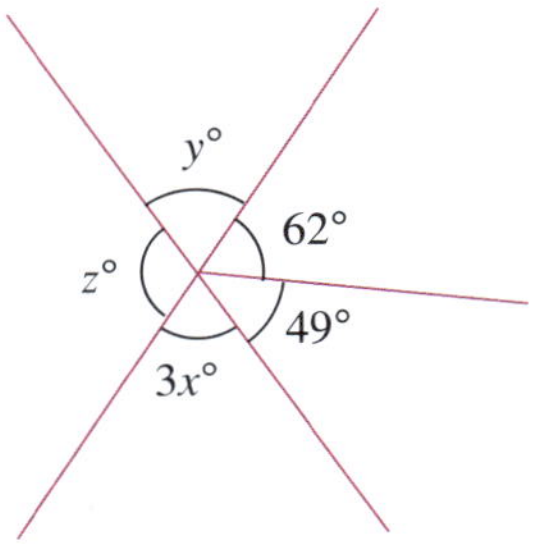


**(h)**

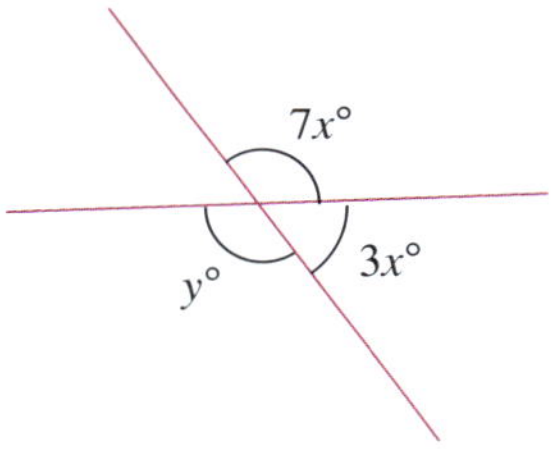

# Parallel Lines, Alternate Angles, Corresponding Angles, Interior Angles

Parallel lines are lines which extend in the same direction and remain the same distance apart. We can take parallel lines as two points moving in the same direction. The ruled lines in your exercise books are examples of parallel lines and so are railway lines. Can you identify other examples of parallel lines around you?

In geometry, a pair of parallel lines is represented by either a pair of single or double arrows (Fig. 14.8).

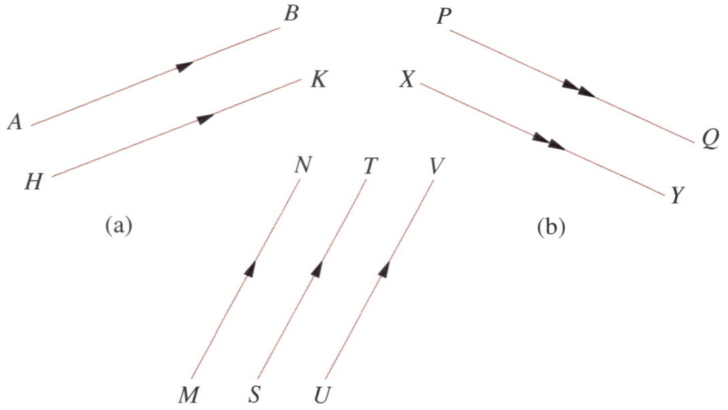


*Fig. 14.8*

We use the symbol "//" to represent "is parallel to". Thus, in Fig. 14.8(a), $AB$ // $HK$ means that $AB$ is parallel to $HK$ and in Fig. 14.8(b), $XY$ // $PQ$.

**Just for fun**

*Which of the following pairs of lines are parallel?*

*(a)*

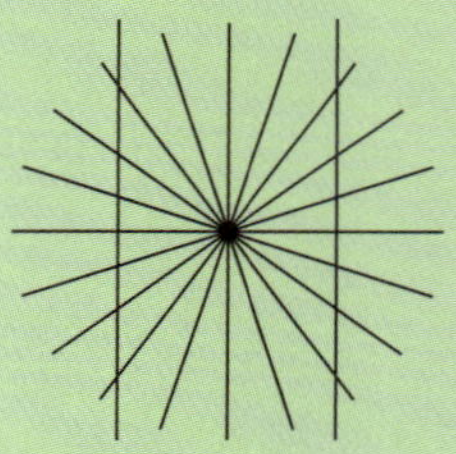

*(b)*

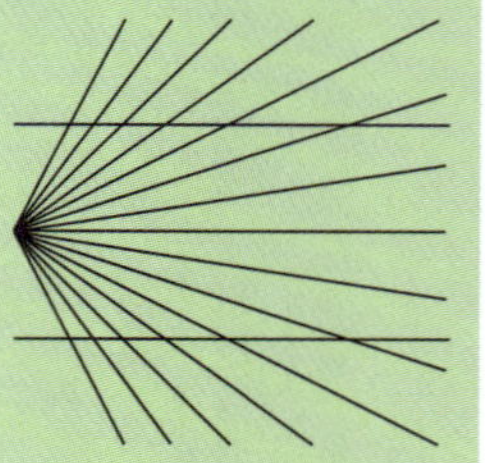

*(c)*

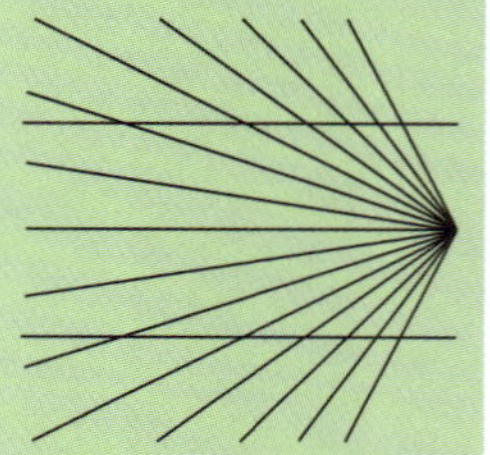

**Thinking Time**

In Fig. 14.8(c) MN // ST and ST // UV. How about MN and UV? Can you explain why?

In Fig. 14.9, the line *PQ* intersects a pair of parallel lines, *KL* and *MN*. It is called a **transversal**. The angles $k_1$ and $k_2$ are called **corresponding angles (corr. ∠s)**. Similarly, $l_1$ and $l_2$ are also called corresponding angles. What do you call the angles $n_1$ and $n_2$? How about the angles $m_1$ and $m_2$?

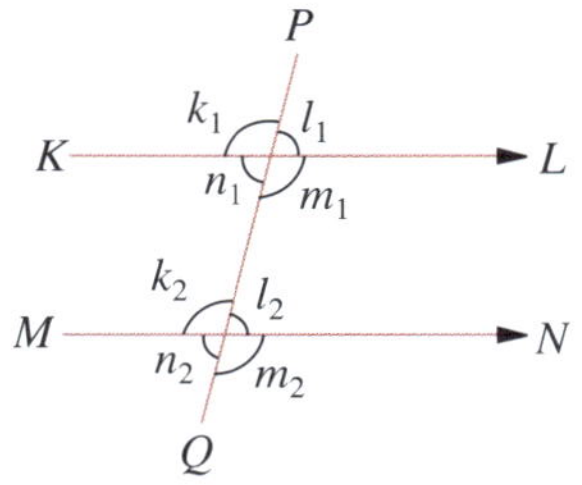


*Fig. 14.9*

Fig. 14.10 shows two lines *AB* and *CD* cut by a transversal *PQ*. The angles $a_1$ and $a_2$ are called **alternate angles (alt. ∠s)**. Similarly, $b_1$ and $b_2$ are also called alternate angles.

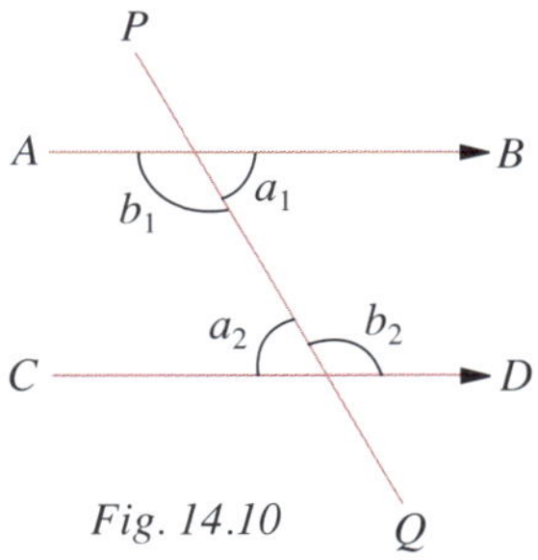


*Fig. 14.10*

Fig. 14.11 shows two lines *AB* and *XY* cut by a transversal *PQ*. The angles *a* and *b* are called **interior angles**. The angles *x* and *y* are also interior angles.

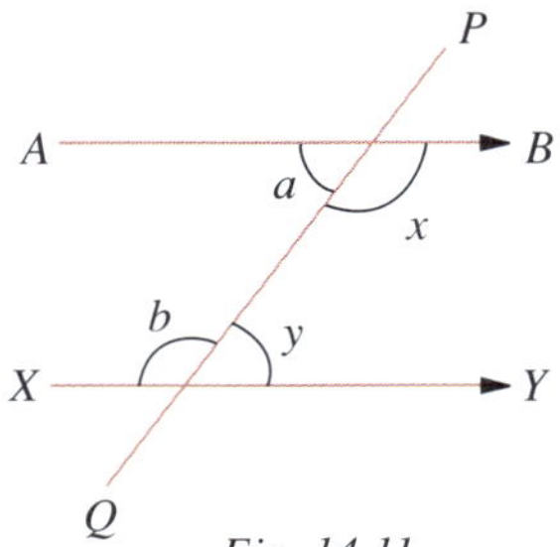


*Fig. 14.11*

The Geometer's Sketch Pad (GSP) is a powerful tool to do geometrical constructions. Here we use GSP to investigate some properties of parallel lines.

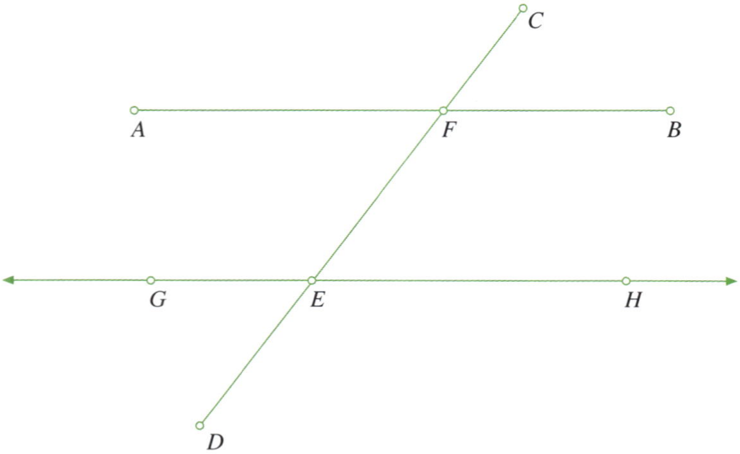


1. Choose Preferences from Display Menu and select Autoshow Labels for Points, Straight objects and Circles. Use  to mark 2 points *A* and *B*.

2. Join *AB* using .

3. Use  to mark another two points *C* and *D*, one on each side of line *AB*.

4. Join *CD* using .

5. Mark another point *E* on line *CD* using .

6. Use  to select point *E*. Press the shift key and click on line *AB*. Then select Parallel line from Construct in Menu bar. A line through *E*, parallel to *AB* will be drawn.

7. Use  to drag point *A* and see how the figure changes. Select points *B*, *C*, *D* and *E* each at a time and drag to see the changes.

8. Use  to mark point *F* on the point of intersection of the lines *AB* and *CD*. Mark two more points *H* and *G* on the parallel line *HEG*.

9. In the figure, line *CD* is a **transversal**. Angles *BFC* and *HEF* are corresponding angles. Can you name two more pairs of corresponding angles?

10. The angles *HEF* and *AFE* are alternate angles. Can you name another pair of alternate angles?

11. The angles *HEF* and *BFE* are interior angles. Can you name another pair of interior angles?

12. Use and click on *H*. Press the shift key and click on *E* and *F*. Choose Angle from Measure in Menu bar to measure the angle *HEF*.

13. Repeat the above to find the angles *AFE*, *BFE*, *BFC* and other angles that you want to measure.

14. What do you notice about the size of the corresponding angles *BFC* and *HEF*?
    What conclusion can you draw from this? When two parallel lines are cut by a transversal, then their corresponding angles are _______ .

    Use to drag point *A* and see how the figure changes. Select points *B*, *C*, *D*, *E* and *F* each at a time and drag to see the changes.
    Are the above conclusions still valid?

15. What do you notice about the size of the alternate angles *AFE* and *HEF*?
    What conclusion can you draw from this? When two parallel lines are cut by a transversal, then their alternate angles are _______ .

16. What do you notice about the sum of the interior angles *BFE* and *HEF*?
    What conclusion can you draw from this? When two parallel lines are cut by a transveral then the sum of their interior angles are _______ .

17. (a) If the corresponding angles *BFC* and *HEF* are equal, what can you conclude about the lines *AB* and *GH*?
    (b) If the alternate angles *AFE* and *HEF* are equal, what can you conclude about the lines *AB* and *GH*?
    (c) If the sum of the interior angles *BFE* and *HEF* is equal to 180°, what can you conclude about the lines *AB* and *GH*?

---

We can conclude from the above discussion that when two parallel lines are cut by a transversal,

> **1. the corresponding angles are equal;**
> **2. the alternate angles are equal;**
> **3. the interior angles are supplementary.**

The converse statements for the above are also true. That is, when two straight lines are cut by a transversal, and

> **1. if the corresponding angles are equal, then the two lines are parallel;**
> **2. if the alternate angles are equal, then the two lines are parallel;**
> **3. if the interior angles are supplementary, then the two lines are parallel.**

## Example 3

*Calculate the unknowns in the given figure.*

**Solution**

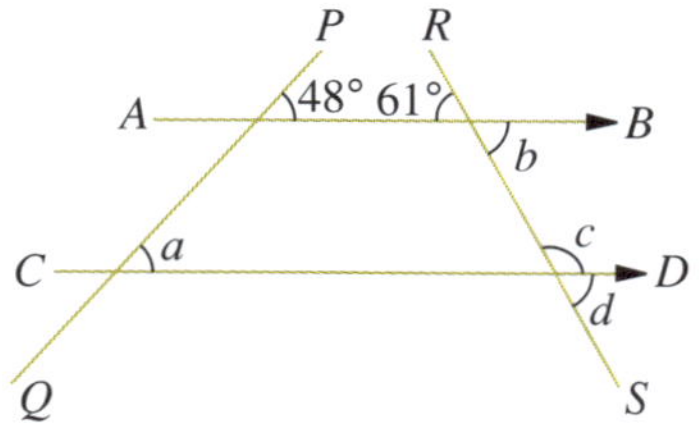


$a = 48°$ (corr. $\angle$s, $AB \parallel CD$)
$b = 61°$ (vert. opp. $\angle$s)
$d = b = 61°$ (corr. $\angle$s, $AB \parallel CD$)
$b + c = 180°$ (interior $\angle$s)
$c = 180° - 61° = 119°$

## Example 4

*Find the unknowns in the figure.*

**Solution**

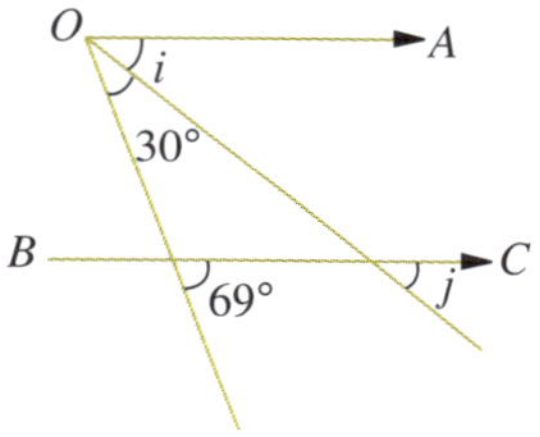


$i + 30° = 69°$ (corr. $\angle$s, $OA \parallel BC$)
$i = 69° - 30° = 39°$
$j = i$ (corr. $\angle$s. $OA \parallel BC$)
$j = 39°$

## Example 5

*Find the values of f, g and h in the figure.*

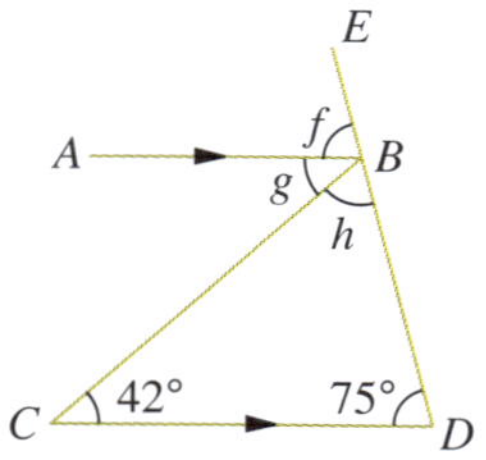


**Solution**

$g = 42°$ (alt. $\angle$s, $AB \parallel CD$)
$f = 75°$ (corr. $\angle$s, $AB \parallel CD$)
$f + g + h = 180°$ (adj. $\angle$s on a line)
$h = 180° - 42° - 75°$
$= 63°$

## Example 6

*Find the value of x in the figure.*

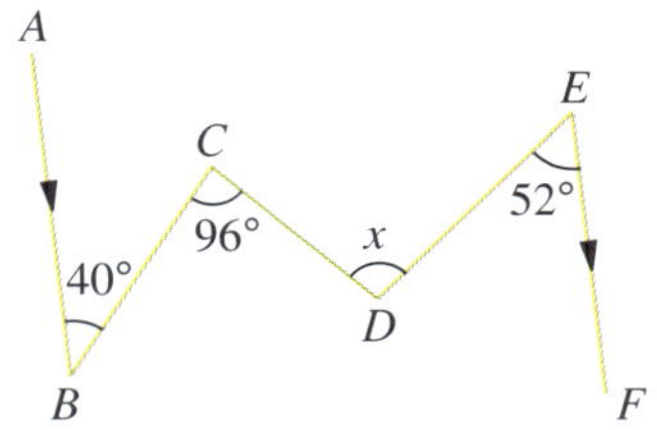


## Solution

At the points $C$ and $D$, two lines both parallel to $AB$ and $EF$ are drawn as shown in Fig. 14.12.

$y_1 = 40°$ (alt. ∠s, $AB$ // $PQ$)

$\therefore\ y_2 = 96° - 40°$
$= 56°$

$y_2 = y_3 = 56°$ (alt. ∠s, $PQ$ // $RS$)

$y_4 = 52°$ (alt. ∠s, $RS$ // $EF$)

$\therefore\ x = y_3 + y_4$
$= 56° + 52°$
$= 108°$

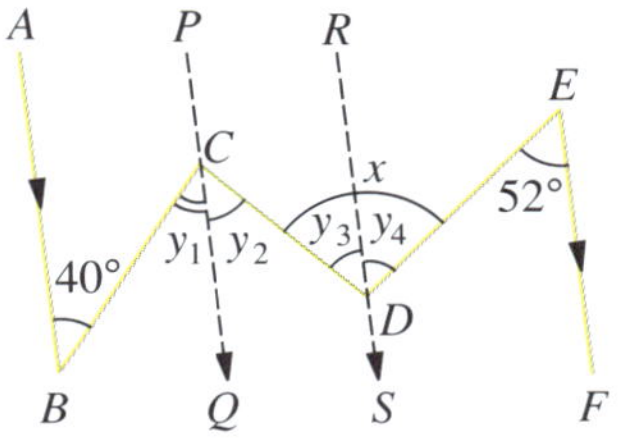


*Fig. 14.12*

## Exercise 14b

1. From the diagram below, list
   (a) three pairs of corresponding angles;
   (b) three pairs of alternate angles.

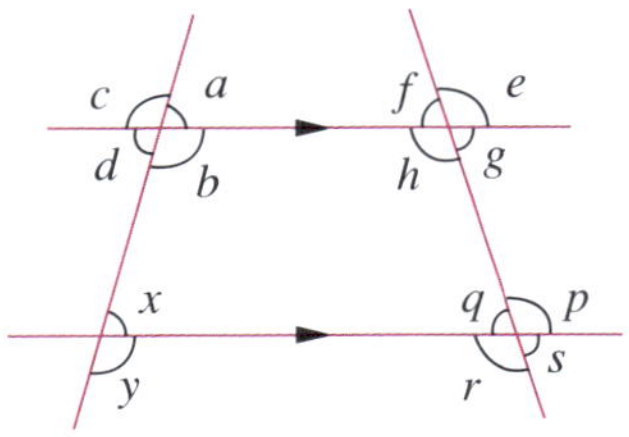


2. Refer to the diagram in Question 1. If $a = 73°$ and $e = 106°$, find the value of $p$, $q$, $r$, $s$, $x$ and $y$.

3. Calculate the unknown(s) in the following diagrams:

   (a)

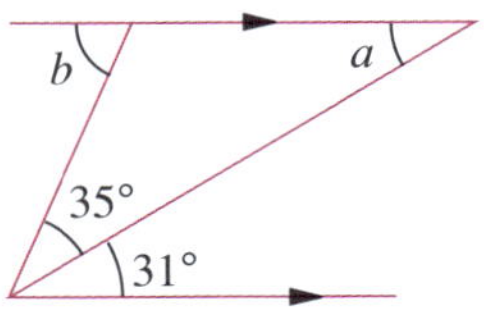

**(b)**

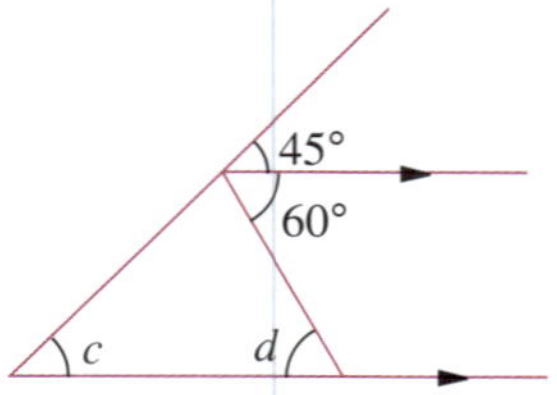


**(c)**

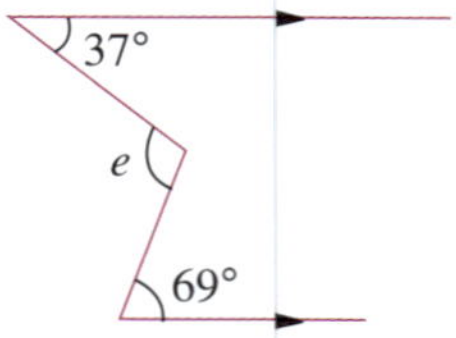


**(d)**

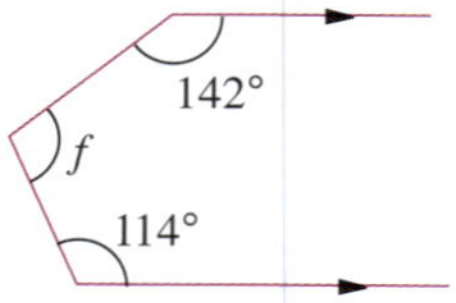


**(e)**

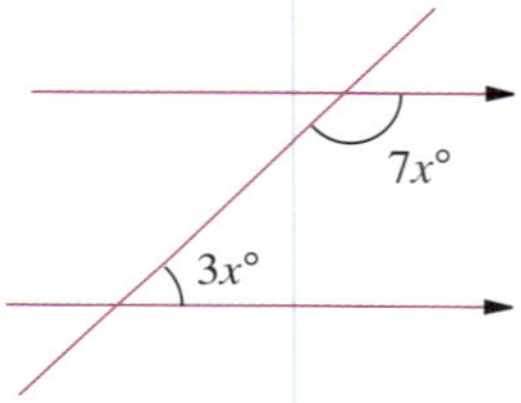


**(f)**

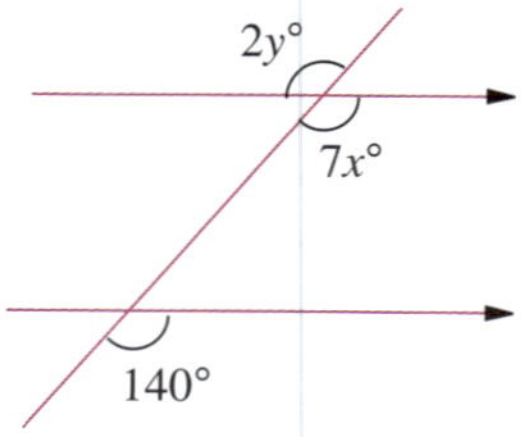


**(g)**

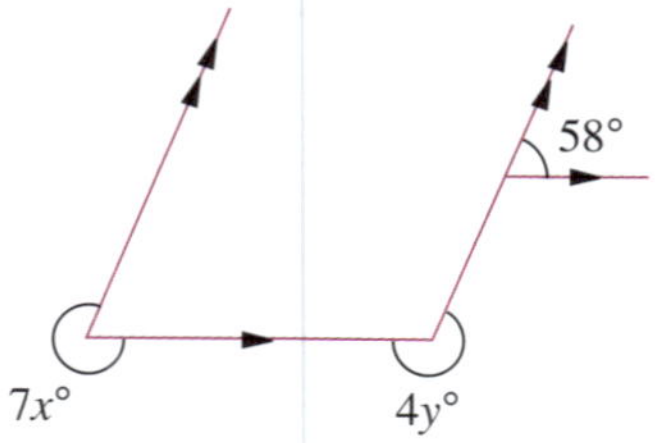


**(h)**

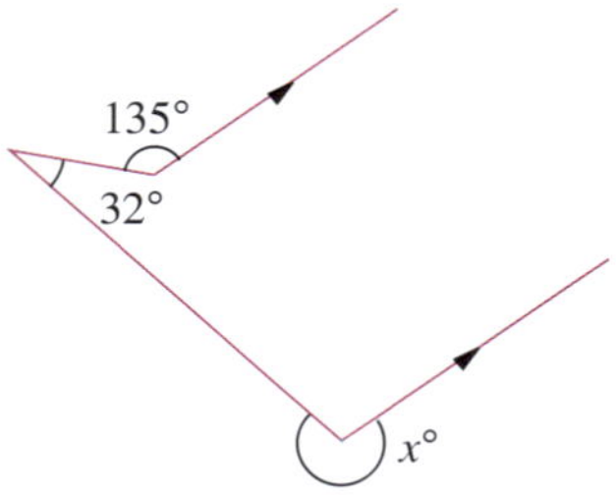


**(i)**

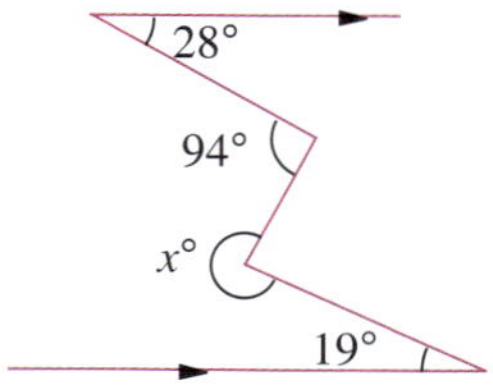


**(j)**

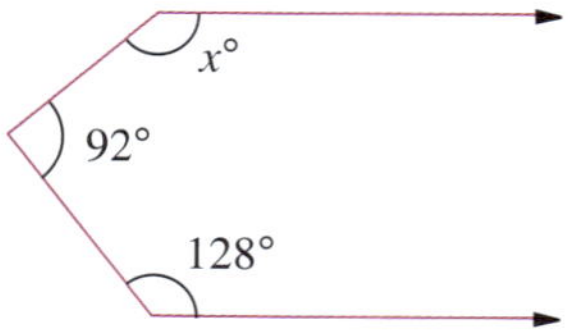


**4.** In the diagram, $RS // PQ // CD$, $A\hat{B}R = 47°$ and $B\hat{C}D = 86°$. Calculate

**(a)** $P\hat{B}C$
**(b)** $B\hat{R}S$

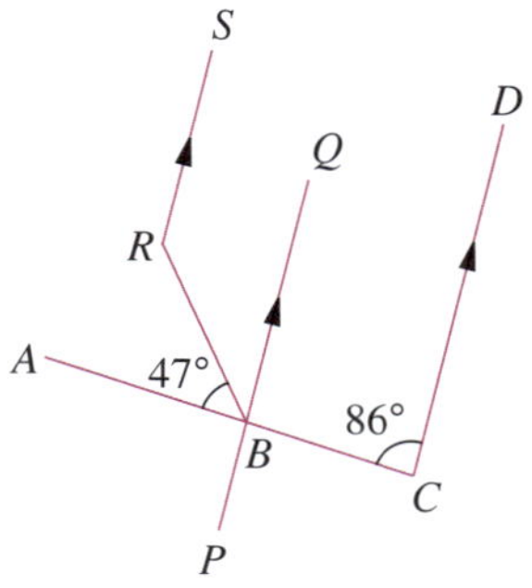

5. In the diagram, $AB // CD$, $BT // AD$, $X\hat{B}T = 68°$ and $X\hat{D}C = 58°$. Calculate

(a) $A\hat{X}B$
(b) $A\hat{B}X$

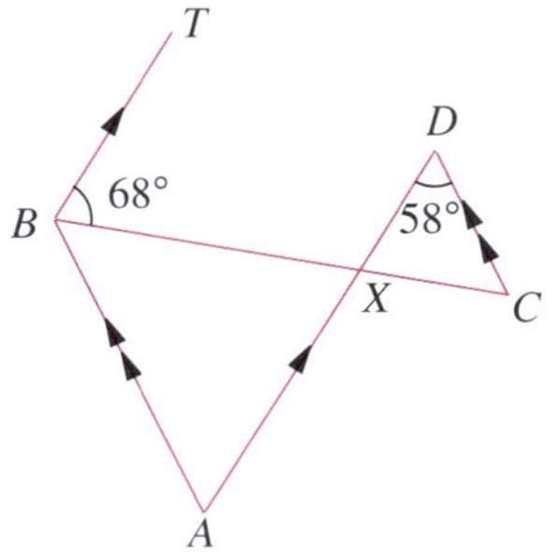


6. In the figure, $B\hat{A}C = a°$, $A\hat{C}D = b°$, $C\hat{D}F = c°$ and $D\hat{F}E = d°$. Form an equation connecting $a$, $b$, $c$ and $d$.

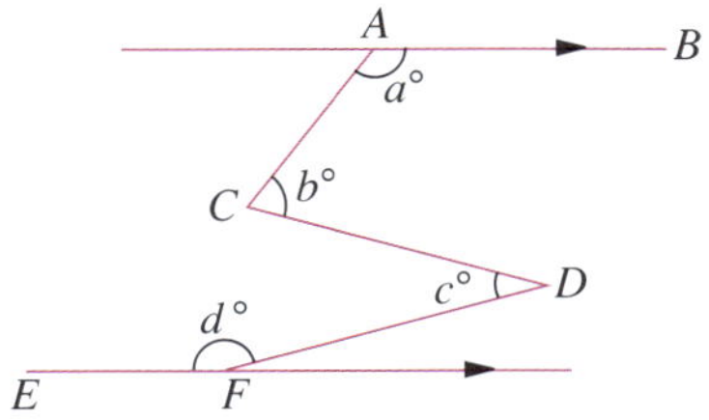


## SUMMARY

1. The **sum of all the angles** at a point is 360°.

2. The **sum of adjacent angles** on a straight line is 180°.

3. Fig. 14.13 shows two parallel lines, $AB$ and $CD$, with a transversal $PQ$.

(a) $x$ and $y$ are **vertically opposite angles** and are equal in size.
(b) $x$ and $q$ are **corresponding angles** and are equal in size.
(c) $e$ and $b$ are **alternate angles** and are equal in size.
(d) $e$ and $y$ are **interior angles**. Their sum is equal to 180°.

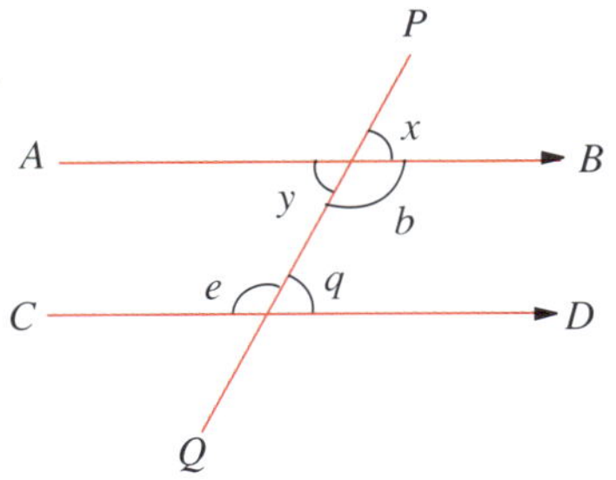

# Review Examples 14

## Example 1

*In the diagram, AB is parallel to DE,* $A\hat{B}C = 23°$, $B\hat{C}D = 3x°$ *and the reflex angle* $C\hat{D}E = 229°$, *calculate the value of x.*

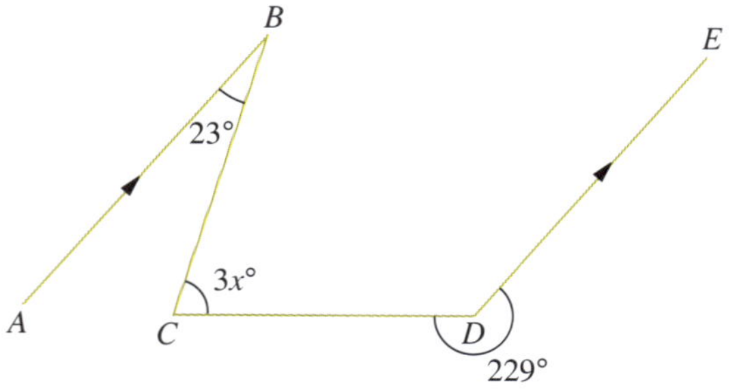


### Solution

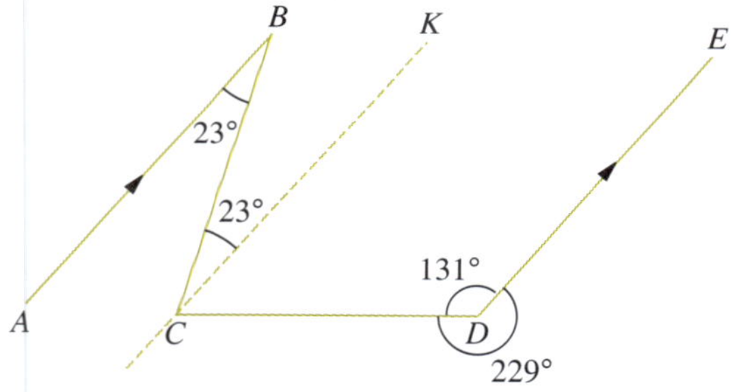


Construct a line through $C$ and parallel to $AB$ and $DE$.

Now $\quad B\hat{C}K = 23°$ (alt ∠s, $AB \,//\, CK$)

$C\hat{D}E = 360° - 229° = 131°$ (∠s at a point)

$\therefore \quad K\hat{C}D = 180° - 131° = 49°$ (interior ∠s, $CK \,//\, DE$)

thus $\quad 3x° = 49° + 23°$

$3x = 72$

$x = \dfrac{72}{3} = 24$

## Example 2

*In the figure, AB is parallel to XY,* $A\hat{B}P = b°$, $B\hat{P}Q = p°$, $P\hat{Q}Y = q°$ *and* $X\hat{Y}Q = y°$. *Form an equation connecting y, b, p and q.*

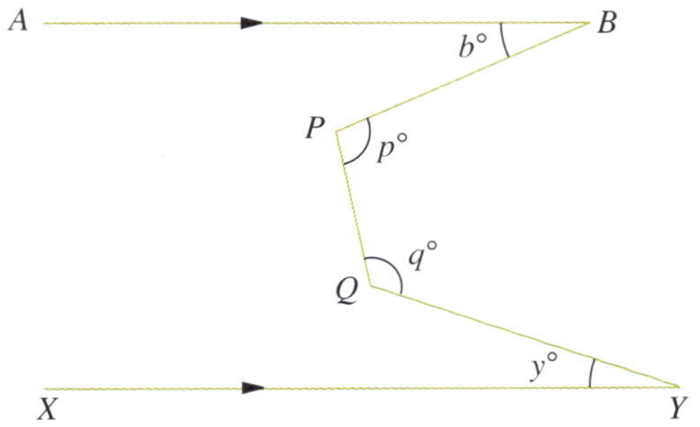


### Solution

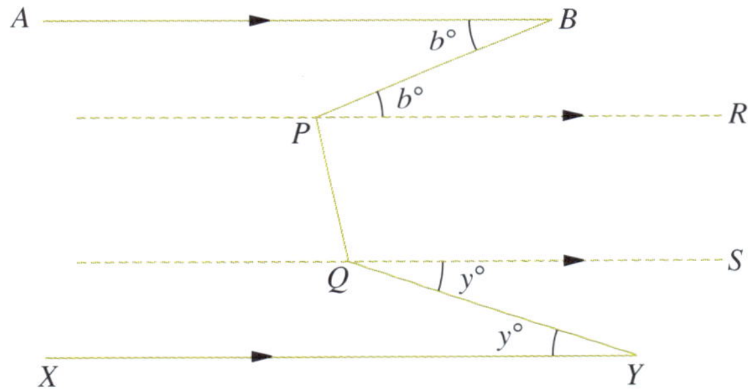


Construct $PR$ and $QR$ to be parallel to $AB$ and $XY$.

We have $B\hat{P}R = b°$ (alt ∠s, $AB$ // $PR$)

$S\hat{Q}Y = y°$ (alt ∠s, $QS$ // $XY$)

$\therefore$ $R\hat{P}Q = p° - b°$ and

$P\hat{Q}S = q° - y°$

Thus $p° - b° + q° - y° = 180°$ (sum of interior ∠s = 180°)

$\therefore$ $y = p + q - b - 180$

# Review Questions 14

Find the values of the unknowns in questions 1 to 4.

**1. (a)**

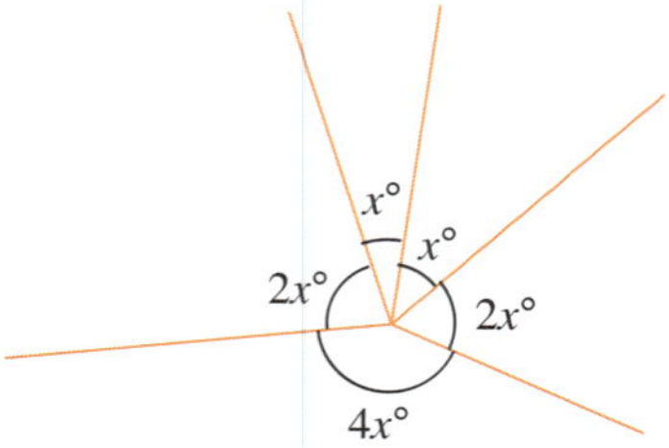


**(b)**

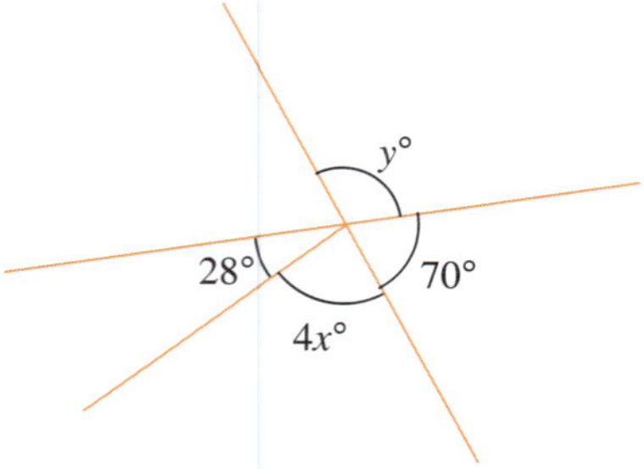


**(c)**

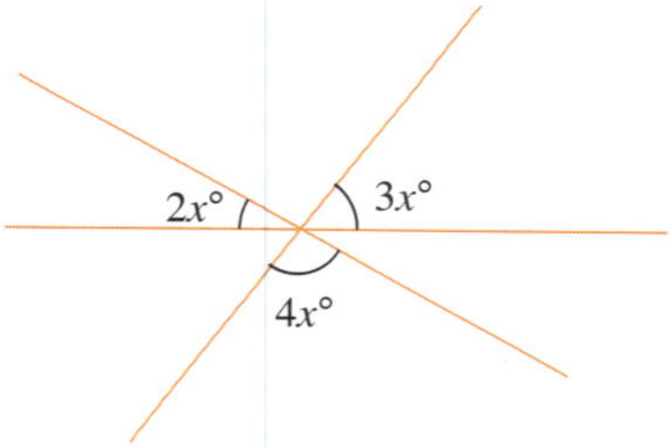


**(d)**

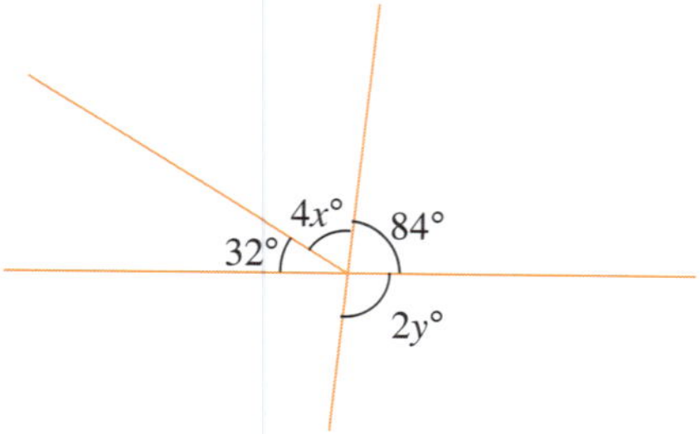


**2. (a)**

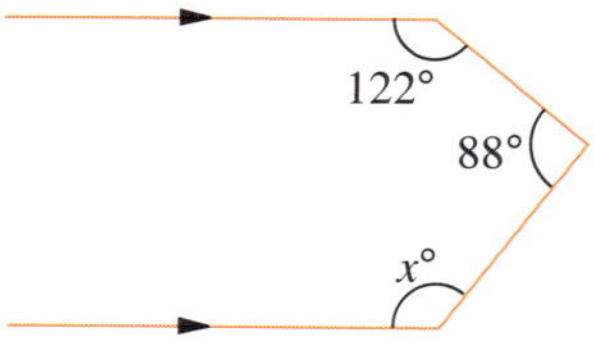


**(b)**

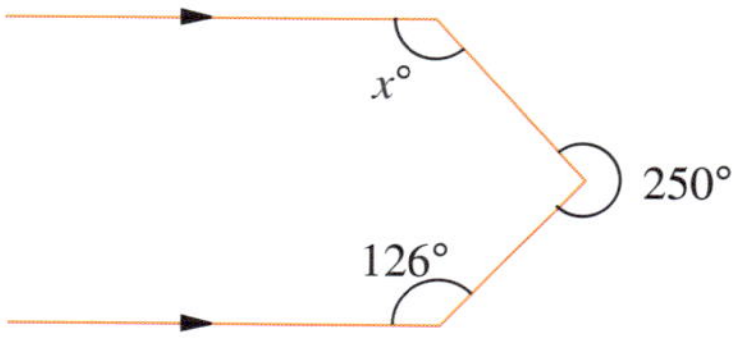


**3. (a)**

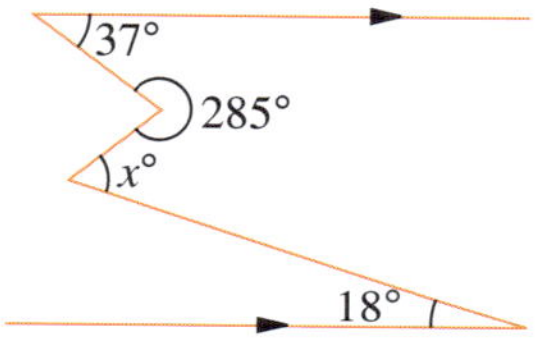


**(b)**

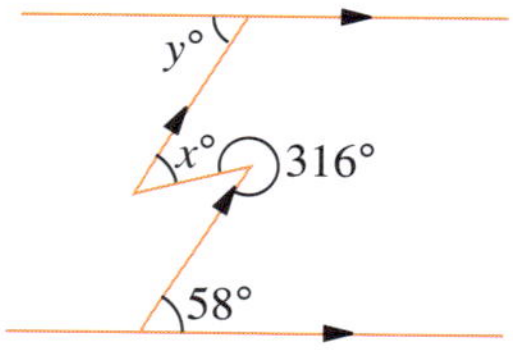


**(c)**

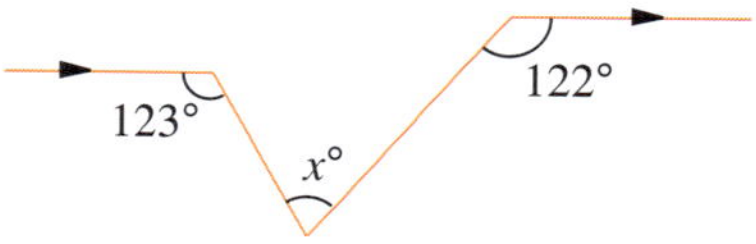


**(d)**

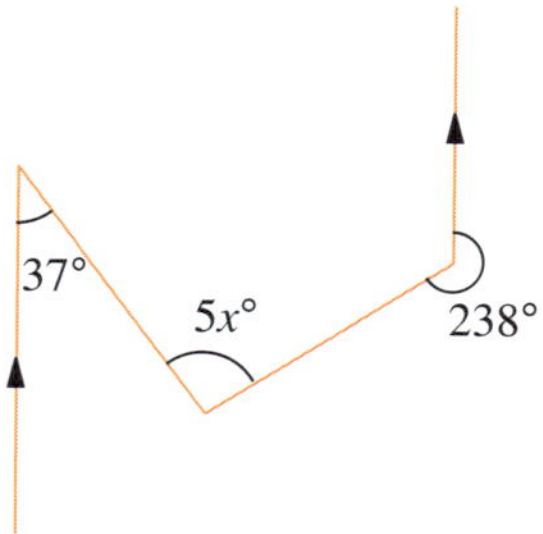

**4.** In the diagram, $RS \parallel PAQ$, $CD \parallel ASB$, $A\hat{D}C = 62°$ and $Q\hat{A}D = 26°$. Calculate

**(a)** $x$

**(b)** $y$

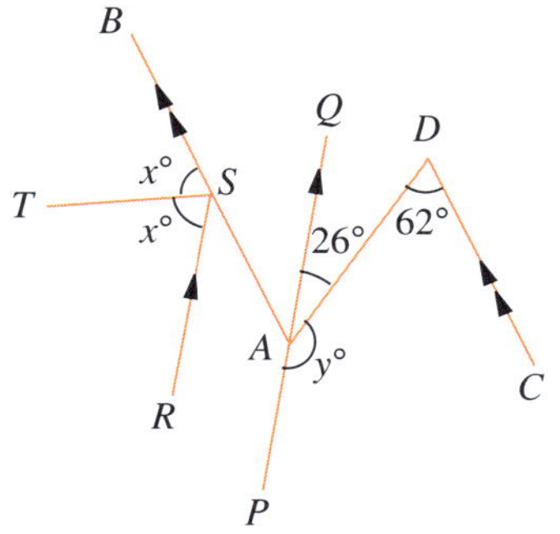


**5.** In the diagram, $PC \parallel DF$, $AB \parallel DE$, $E\hat{D}F = 84°$, $Q\hat{B}D = 74°$ and $P\hat{C}D = 148°$. Calculate

**(a)** $A\hat{B}Q$

**(b)** $C\hat{D}E$

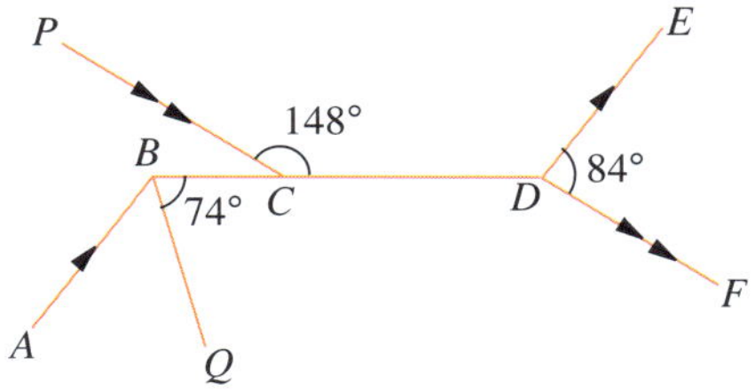

*In this chapter, you will learn how to*

- *calculate the unknown angles involving triangles, quadrilaterals and polygons using the angle properties of these figures*

# Geometrical Shapes

CHAPTER

# 15

# Angle Properties of Polygons

## Introduction

Look at the various stamps and the road signs on Singapore roads. Can you name the geometrical shapes of these stamps and road signs?

# Polygons

A plane figure with three or more straight edges as its sides is called a **polygon**. Each polygon is named after the number of sides it contains. The following names are given to some common polygons.

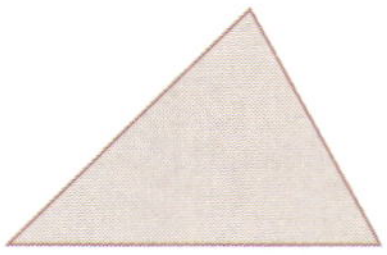

Triangle
(3-sided)

Quadrilateral
(4-sided)

Pentagon
(5-sided)

Notes

***Types of Polygons***
*Polygons have different names, depending on the number of corners or sides they have.*

Hexagon
(6-sided)

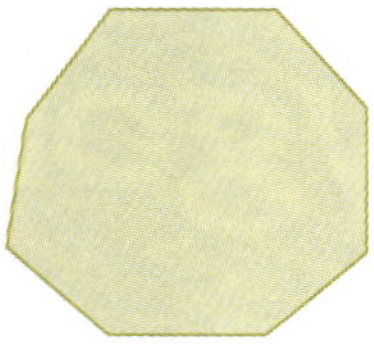

Octagon
(8-sided)

Decagon
(10-sided)

Notes

*In general, a polygon with n sides is called an n-gon. Thus a polygon with 12 sides is called a 12-gon and a polygon with 25 sides is called a 25-gon.*

A **regular polygon** is one in which all its sides and all its angles are equal. The following figures are some examples of regular polygons.

In this chapter, you will learn about properties of triangles, quadrilaterals and polygons.

# Triangles

A plane figure formed by having three straight edges as its sides is called a **triangle**.

Figure 15.1 shows a triangle *ABC*, which can be denoted by $\triangle ABC$, formed by the three sides *AB*, *BC* and *CA*. The points *A*, *B* and *C* are called the vertices (singular: vertex) of the triangle. $\angle ABC$, $\angle BAC$ and $\angle BCA$ are called the interior angles of the triangle, or simply the angles of $\triangle ABC$.

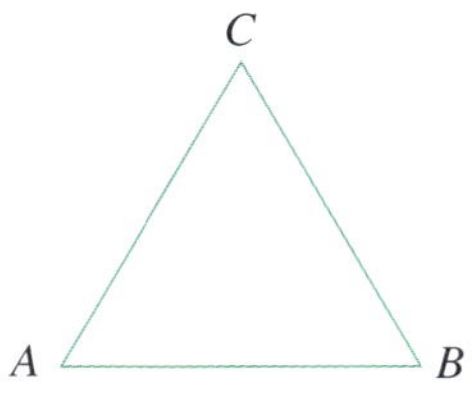


*Fig. 15.1*

Triangles can be classified according to:

(a) The number of equal sides

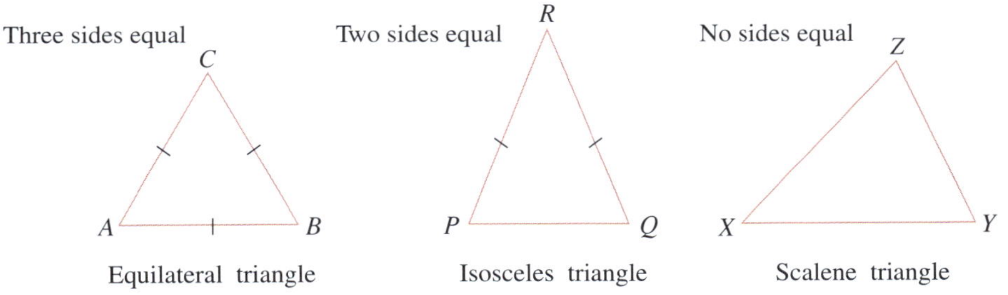


*Fig. 15.2*

(b) The types of angles

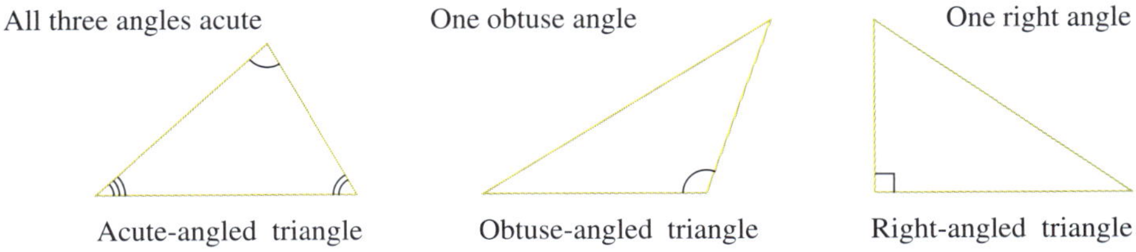


*Fig. 15.3*

**Notes**

*(1) All the three angles in an equilateral triangle are equal in size. Each angle is 60°.*

*(2) The two base angles of an isosceles triangle are equal, i.e., $\angle RPQ = \angle RQP$.*

*(3) All the three angles in a scalene triangle are different in size.*

## Angle Properties of Triangles

In primary school, we learned that the sum of the interior angles of a triangle is always equal to 180°, we shall now use the geometrical properties we learnt in the previous chapter to prove this statement.

Draw $PQ$ parallel to $AC$ and passing through $B$.

We have $\quad B\hat{A}C = P\hat{B}A \quad$ (alt. $\angle$s, $PQ//AC$)

and $\quad B\hat{C}A = Q\hat{B}C \quad$ (alt. $\angle$s, $PQ//AC$)

Now, $P\hat{B}A + A\hat{B}C + Q\hat{B}C = 180°$ (adj. $\angle$s on a str. line)

$\therefore \quad B\hat{A}C + A\hat{B}C + B\hat{C}A = 180°$

(**Abbreviation:** $\angle$ sum of $\triangle$)

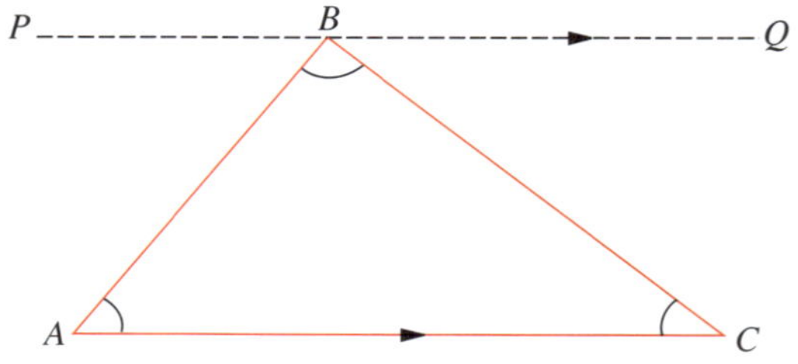


*Fig. 15.4*

### Explorations

Draw a triangle $ABC$. Use a protractor to measure the size of $B\hat{A}C$, $A\hat{B}C$ and $A\hat{C}B$. Use a ruler to measure the length of $AB$, $BC$ and $AC$. What do you notice about the size of the angle and the length opposite that angle, i.e. compare the size of $B\hat{A}C$ and $BC$, with $A\hat{B}C$ and $AC$. How about $A\hat{C}B$ and $AB$?
Draw another triangle and check your conclusion.

## Exterior and Interior Opposite Angles

Fig. 15.5 shows $\triangle ABC$ with $BA$ produced to $P$, $BC$ produced to $Q$ and $CB$ produced to $R$. $p$, $q$ and $r$ are called the exterior angles of $\triangle ABC$. In particular, $a$ and $b$ are referred to as the interior opposite angles with reference to $q$. Similarly, $a$ and $c$ are the interior opposite angles with reference to $r$. What are the interior opposite angles with reference to $p$?

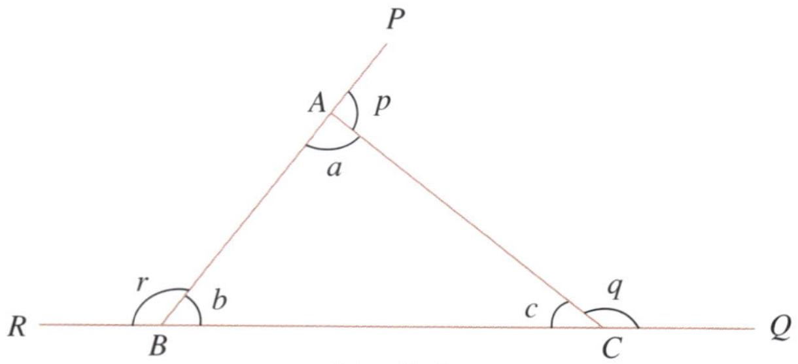


*Fig. 15.5*

# Explorations

In this exploration, we will use Geometer's Sketch Pad to explore the relationship between the sum of the interior opposite angles and their corresponding exterior angles. Step by step guides are provided.

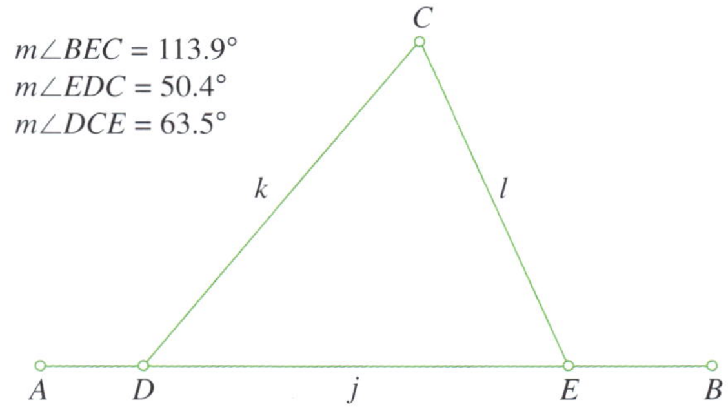


1. Choose Preferences from Edit on the Menu bar and select Autoshow Labels for Points, Straight objects and Circles. Choose Units for precision for angle measure. Use  to draw a long line *AB*. Draw *CD* and *CE* where *D* and *E* are on *AB*.

2. Angle *BEC* is the exterior angle and angles *EDC* and *DCE* are the interior angles opposite the exterior angle *BEC*.

3. Measure angle *BEC* by using ; click on *B* and press the shift key and click on *E* and then on *C*. Choose Angle from Measure on the Menu bar to measure angle *BEC*. What is the value of the angle *BEC*?

4. Use  and follow the procedure in step 4 to measure angles *EDC* and *DCE*.

5. Calculate mentally the sum of the angles *EDC* and *DCE*.
   What is the value of the angle *BEC*?

6. Use  to drag the point *C* to another position. Do you notice the change in the value of the angles *EDC* and *DCE*? What is the sum of the two angles? What is the value of the angle *BEC*?

7. Use  to drag the point *E* to another position. Do you notice the change in the value of the angles *EDC* and *DCE*? What is the sum of the two angles? What is the value of the angle *BEC*?

8. Use  to drag the point *D* to another position. Do you notice the change in the value of the angles *EDC* and *DCE*? What is the sum of the two angles? What is the value of the angle *BEC*?

9. What can you conclude?

We can also establish the fact that the sum of the interior opposite angles of a triangle is equal to its exterior angle by the following proof.

Refer to triangle $ABC$ with notations as given in Fig. 15.6.

We have $a + b + c = 180°$ ($\angle$ sum of $\triangle$)

and $q + c = 180°$ (adj. $\angle$s on a str. line)

$\therefore\quad a + b = q$

(**Abbreviation:** ext. $\angle$ = sum of int. opp. $\angle$s)

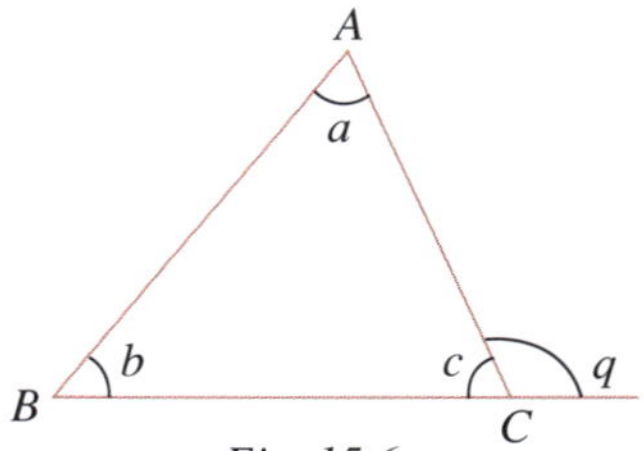


*Fig. 15.6*

## Example 1

*Find the unknown angles marked in each of the following diagrams:*

*(a)*

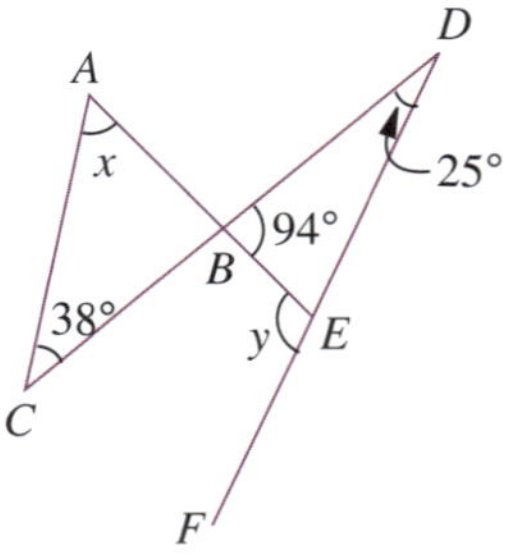


*(b)*

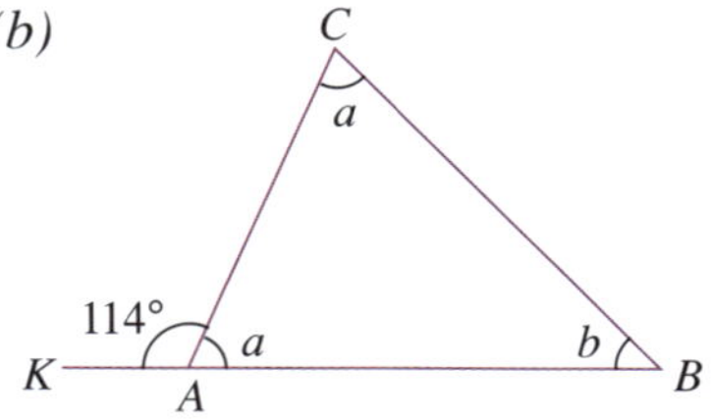


**Solution**

(a) $A\hat{B}C = 94°$ (vert. opp. $\angle$s)

$x + 94° + 38° = 180°$ ($\angle$ sum of $\triangle$)

$\therefore\quad x = 180° - 94° - 38° = 48°$

$y = 94° + 25°$ (ext. $\angle$ = sum of int. opp. $\angle$s)

$\therefore\quad y = 119°$

(b) $114° + a = 180°$ (adj. $\angle$s on a str. line)

$\therefore\quad a = 180° - 114° = 66°$

$114° = 66° + b$ (ext. $\angle$ = sum of int. opp. $\angle$s)

$\therefore\quad b = 114° - 66°$

$= 48°$

# Exercise 15a

**1.** Sketch, in each case, a triangle $ABC$ with the size of $\hat{A}$ and $\hat{B}$ given below. In each case, find $\hat{C}$ and classify each triangle (i) by its sides (ii) by its angles:

**(a)** $\hat{A} = 20°$, $\hat{B} = 60°$ **(b)** $\hat{A} = 70°$, $\hat{B} = 40°$ **(c)** $\hat{A} = 60°$, $\hat{B} = 60°$
**(d)** $\hat{A} = 42°$, $\hat{B} = 48°$ **(e)** $\hat{A} = 65°$, $\hat{B} = 50°$ **(f)** $\hat{A} = 25°$, $\hat{B} = 112°$

**2.** The following are base angles of isosceles triangles. In each case, find the third angle of the isosceles triangle:

**(a)** 42° **(b)** 82° **(c)** 18° **(d)** 64°

**3.** Calculate the values of the unknown in each of the following diagrams:

**(a)**

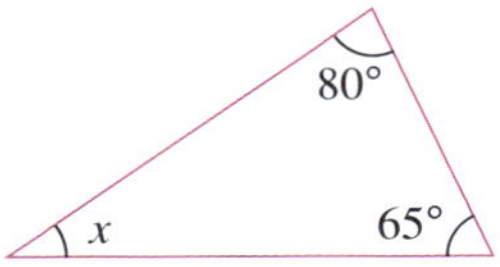


**(b)**

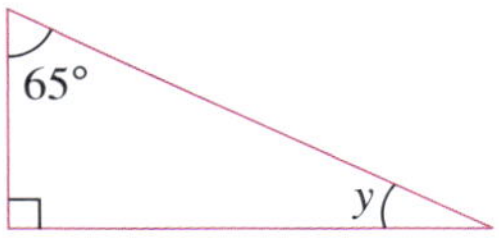


**(c)**

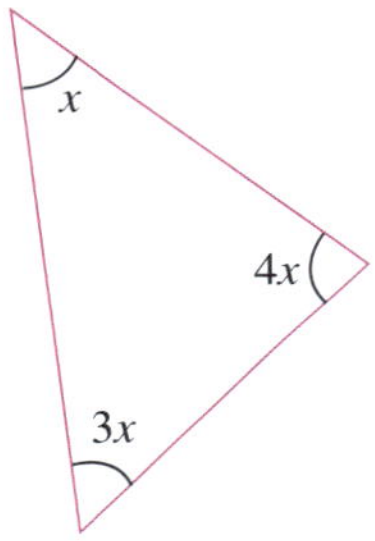


**(d)**

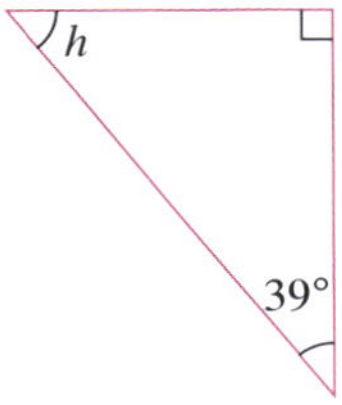


**(e)**

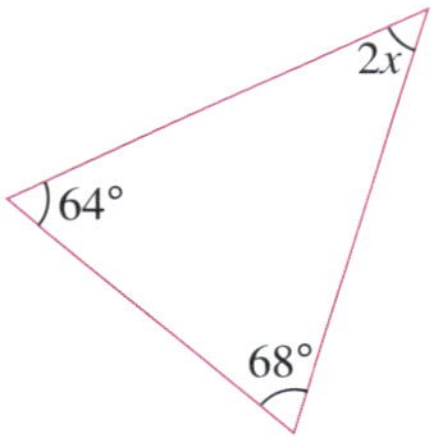


**(f)**

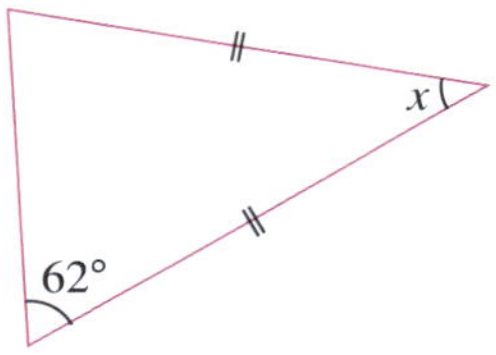


**(g)**

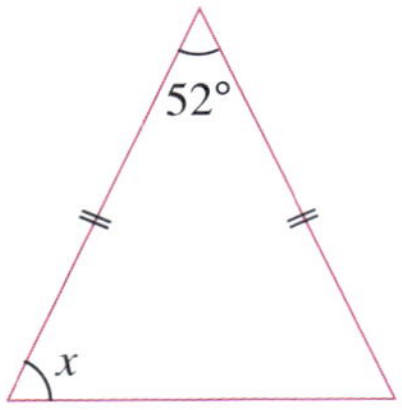


**(h)**

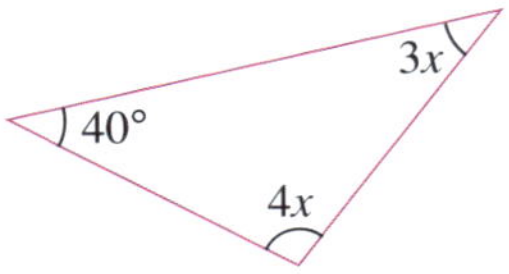


**4.** Calculate the values of the unknowns in the following diagrams.

**(a)**

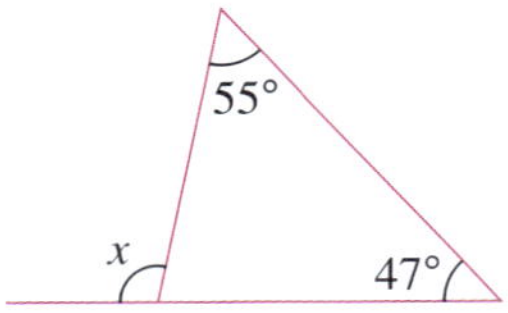


**(b)**

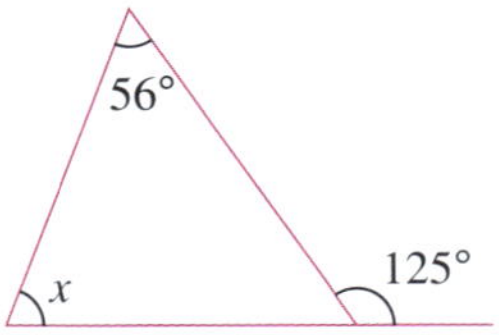


**(c)**

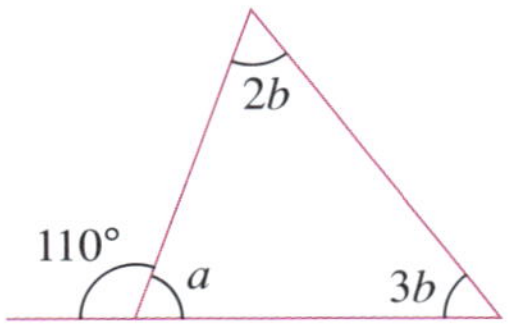

**(d)** 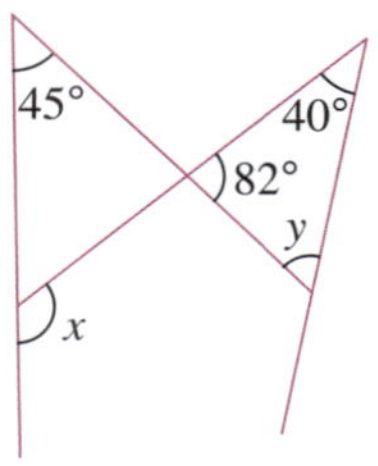


**(e)** 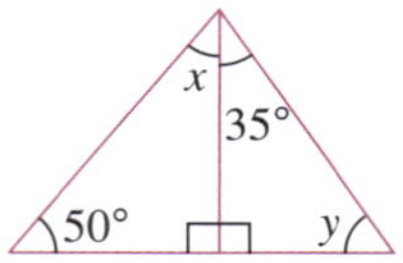


**(f)** 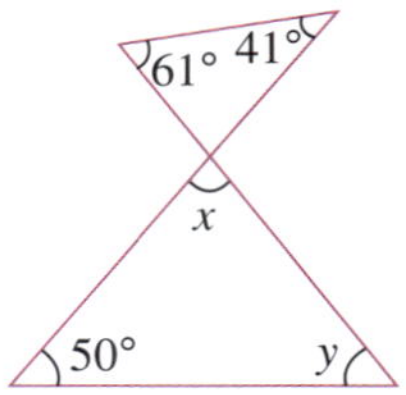


**(g)** 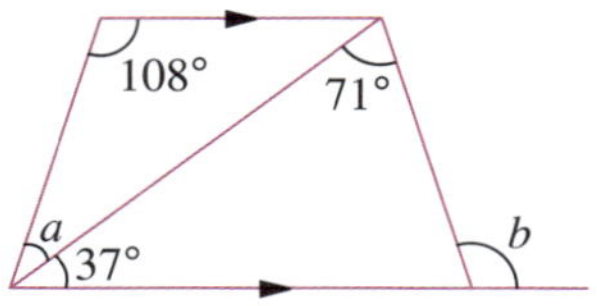


**(h)** 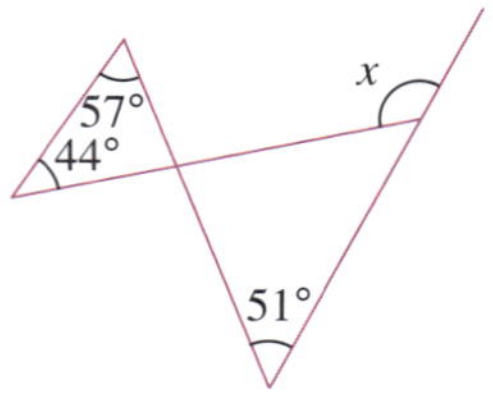


**5.** Calculate the values of the unknowns in the following diagrams.

**(a)** 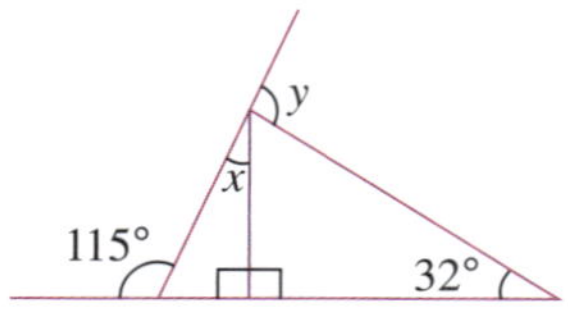


**(b)** 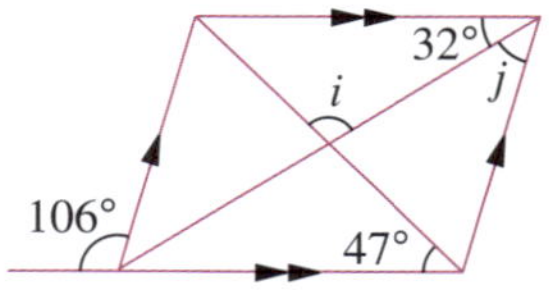


**(c)** 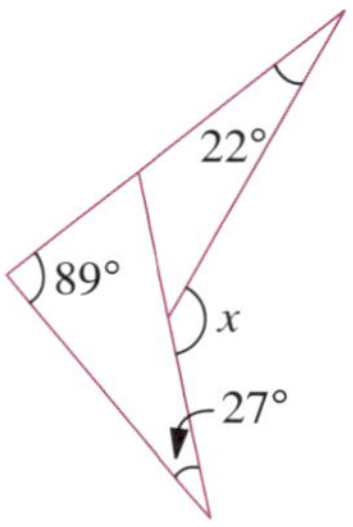


**(d)** 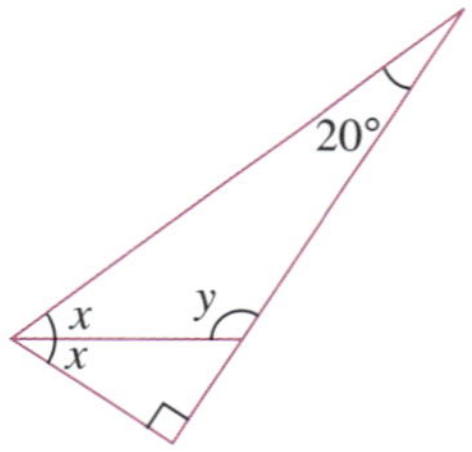


**(e)** 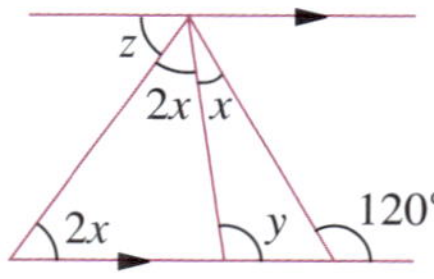


**(f)** 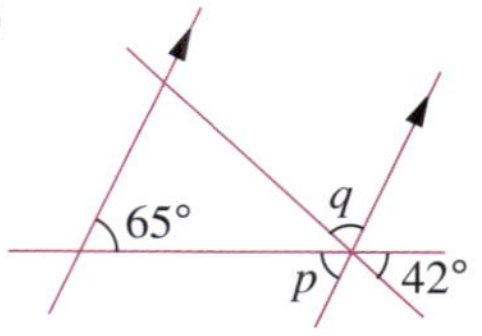


**(g)** 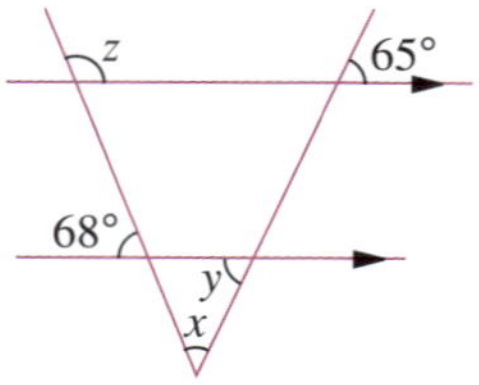


**(h)** 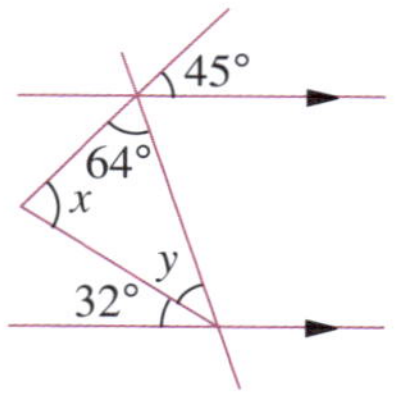


**6.** If the sizes of the angles of a triangle are $3x°$, $5x°$ and $4x°$, find the smallest angle of the triangle.

**7.** The angles of a triangle are $(x - 35)°$, $(x - 25)°$ and $\left(\frac{1}{2}x - 10\right)^{\circ}$. Form an equation in $x$ and hence find $x$.

**8.** In the triangle $ABC$, $\hat{A} = 50°$, $\hat{C} = 26°$ and $AB$ is produced to $D$. Find $A\hat{B}C$ and $C\hat{B}D$.

# Quadrilaterals

A quadrilateral is a plane figure with four sides and four angles. Fig. 15.7 shows a quadrilateral *ABCD*. It is named by taking the vertices in order either in a clockwise or anticlockwise direction. Thus, *ABCD*, *BCDA*, *DCBA* and *CBAD* are correct ways of naming the quadrilateral but *ABDC*, *ACBD* and *CDBA* are not. The line segments *AC* and *BD* that join the opposite vertices are called **diagonals**.

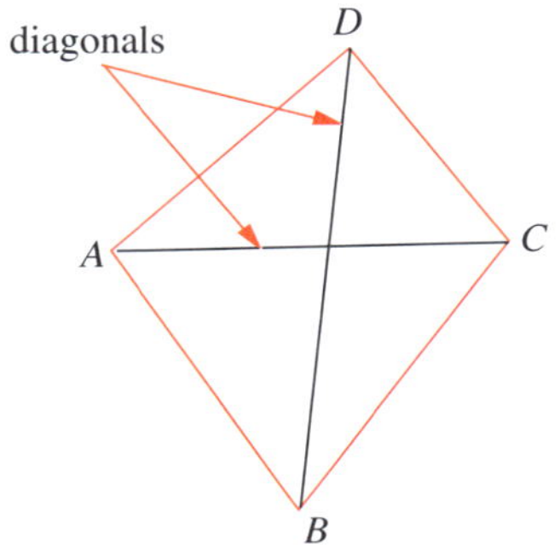


*Fig. 15.7*

Now let's find the sum of the four angles of a quadrilateral.

In Fig. 15.7, consider $\triangle ABD$. Then we have

$D\hat{A}B + A\hat{B}D + B\hat{D}A = 180°$ — (1) ($\angle$ sum of $\triangle$)

Consider $\triangle DBC$. Then we have

$D\hat{C}B + C\hat{B}D + B\hat{D}C = 180°$ — (2) ($\angle$ sum of $\triangle$)

Combining (1) and (2),

$D\hat{A}B + A\hat{B}C + B\hat{C}D + C\hat{D}A = 360°$.

**The sum of the four angles of a quadrilateral equals 360°.**

## Explorations

In this exploration, we will explore the properties of some special quadrilaterals with the help of the Geometer's Sketch Pad.

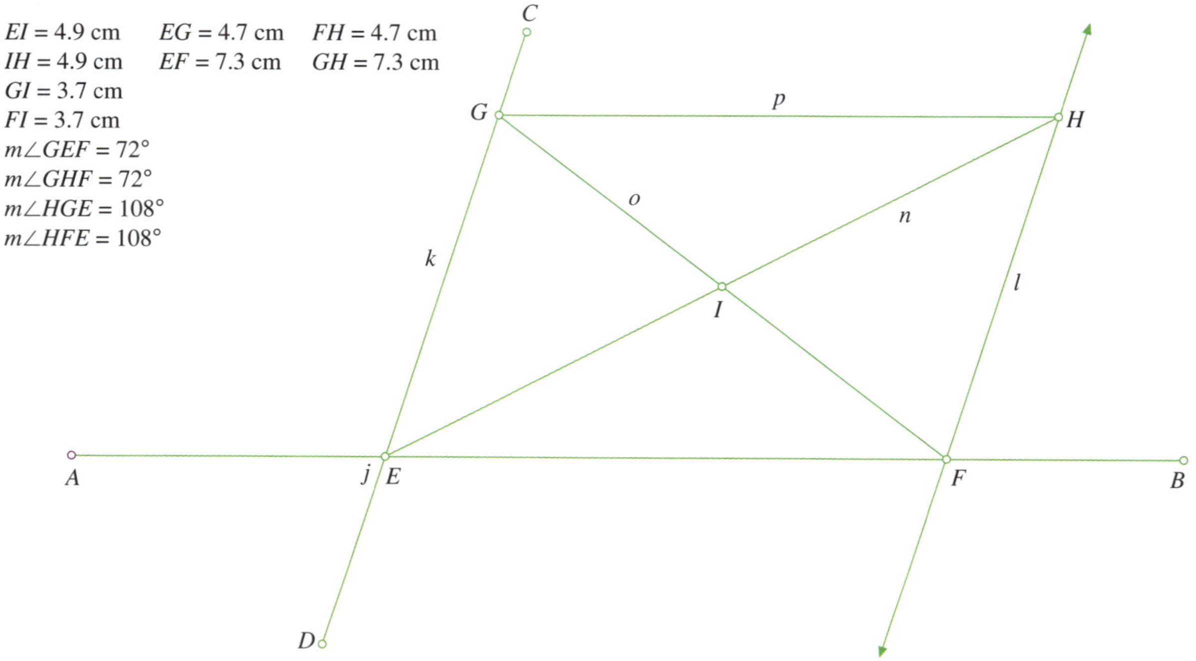


1. Choose Preferences from Edit on the Menu bar and select Autoshow Labels for Points, Straight objects and Circles. Use  to draw a long line $AB$. Also draw a long line $CD$ to cut the line $AB$.

2. Use  to mark the point $E$ at the intersection of the lines $AB$ and $CD$. Mark another point $F$ on $AB$.

3. Use  to select $F$. Then press the shift key and click to select line $CD$; choose Parallel Line from Construct on the Menu bar to draw the line through $F$ and parallel to $CD$.

4. Use  to select a point on $CD$ and label it $G$.

5. Use  to select point $G$. Then press the shift key and click to select line $AB$; choose Parallel Line from Construct on the Menu bar to construct a line parallel to $AB$.

6. Use  to mark the point of intersection of the lines parallel to $AB$ and $CD$ as point $H$. Now you get a parallelogram $EFHG$.

   [You can get rid of the excess length of the parallel by first using  to select the line $GH$, then choose hide line from Display on the Menu bar. Then we use  to join the line $GH$]

7. Use  to join the points *F* and *G* and points *E* and *H*.

8. Use  to mark the point of intersection and label it as point *I*.

9. Use  to select point *E*. Then press the shift key and click to select point *I*; choose distance from Measure on the Menu bar to measure the length of *EI*.

10. Repeat the above procedure to measure the lengths of *IH*, *GI*, *FI*, *EG*, *EF*, *FH* and *GH*.

11. What do you notice about the diagonals of the parallelogram?
The diagonals of a parallelogram ______ each other.

12. What do you notice about the lengths of the opposite sides of the parallelogram?
The opposite sides of a parallelogram are ______ in length.

13. Use  to select the point *G*. Then press the shift key and click to select point *E* and then *F*; choose Angle from Measure on the Menu bar to measure the angle *GEF*.

14. Repeat the above procedure to measure the angles *GHF*, *HGE* and *HFE*.

15. What do you notice about the opposite angles of the parallelogram i.e. (a) angles *GEF* & *GHF* and (b) angles *HFE* & *GEF*?
The opposite angles of a parallelogram are ______.

16. Use  to drag the point *C* until the interior angles of the figure becomes 90 degrees and you get a rectangle *EFHG*.

17. As in Step 9, measure the diagonals *GF* and *EH*. What do you notice about the diagonals of the rectangle?
The diagonals of a rectangle are ______ in length.

18. Use  to drag the point *H* until the lengths of *EF*, *EG*, *GH* and *HF* are all equal. You will now have a rhombus.

19. As in step 13, use  to measure the angle *FIH*.

20. What do you notice about the angle *FIH*? What can you conclude about the diagonals of a rhombus?
The two diagonals of a rhombus are _____.

21. Use  again to measure the angle pairs: $G\hat{E}H$ and $F\hat{E}H$, $G\hat{H}E$ and $E\hat{H}F$, $F\hat{G}H$ and $F\hat{G}E$; $E\hat{F}G$ and $H\hat{F}G$. What do you notice?
The diagonals of a rhombus _____ the interior angles.

22. Use  to drag the point *C* and then *H* until the angles of the quadrilateral are each equal to 90 degrees and the length of the sides are all equal. You will now have a square.

23. Can you name all the properties of a square?

## Comparison of the properties of some special quadrilaterals

| | Parallel sides | Equal sides | Interior angles | Diagonals |
|---|---|---|---|---|
| 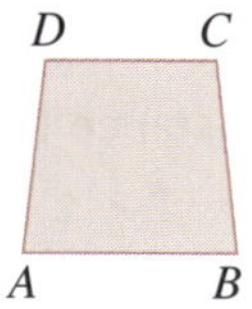  Trapezium | There is one pair of parallel opposite sides | | | |
| 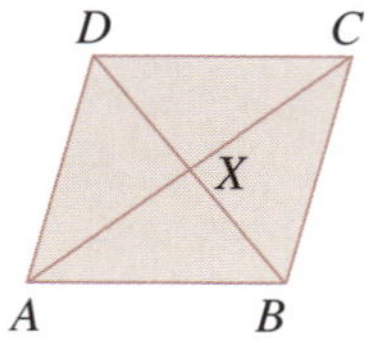  Parallelogram | There are two pairs of parallel opposite sides | The opposite sides are equal in length | Opposite angles are equal | Diagonals bisect each other i.e. $AX = XC$ and $BX = XD$. |
| 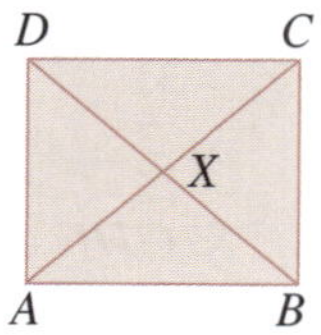  Rectangle | There are two pairs of parallel opposite sides | The opposite sides are equal in length | All four angles are right angles | (1) Two diagonals equal in length<br>(2) Diagonals bisect each other |
| 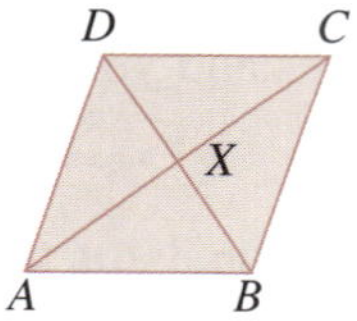  Rhombus | There are two pairs of parallel opposite sides | All sides are equal | Opposite angles are equal | (1) Diagonals bisect each other at right angles i.e. $B\hat{X}C = A\hat{X}B = 90°$<br>(2) Diagonals bisect the interior angles i.e. $A\hat{B}D = C\hat{B}D$ and $B\hat{C}A = D\hat{C}A$ |
| 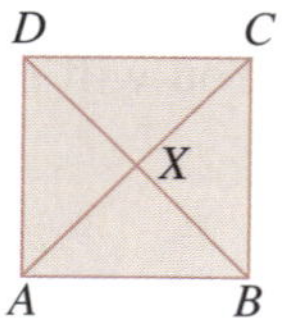  Square | There are two pairs of parallel opposite sides | All sides are equal | All four angles are right angles | (1) Two diagonals are equal in length<br>(2) Diagonals bisect each other at right angles<br>(3) Diagonals bisect the interior angles |
| 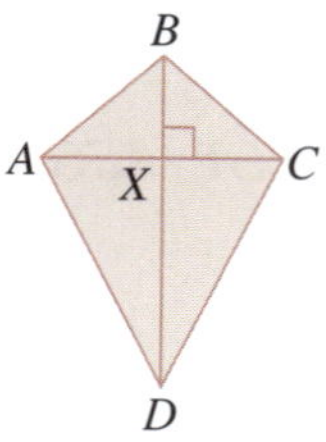  Kite | There are no parallel sides | There are two pairs of equal adjacent sides. | | (1) Diagonals intersect at right angles<br>(2) One diagonal bisects the interior angles i.e. $A\hat{B}D = C\hat{B}D$ |

## Example 2

*ABCD is a rectangle in which $A\hat{K}D = 85°$ and $A\hat{D}K = 45°$. Calculate (a) $B\hat{A}K$; (b) $B\hat{K}A$.*

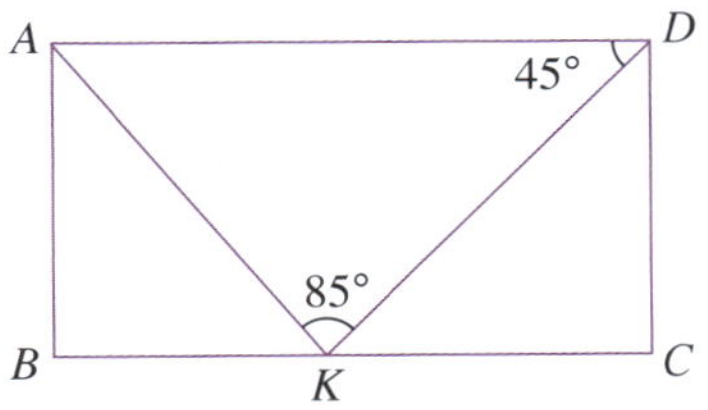


**Solution**

(a) $K\hat{A}D + 85° + 45° = 180°$ ($\angle$ sum of $\triangle$)

$K\hat{A}D = 180° - 85° - 45° = 50°$

$B\hat{A}K + 50° = 90°$ ($B\hat{A}D$ is a right angle)

$\therefore$ $B\hat{A}K = 90° - 50° = 40°$

(b) $B\hat{K}A + 40° + 90° = 180°$ ($\angle$ sum of $\triangle$, $A\hat{B}K = 90°$)

$\therefore$ $B\hat{K}A = 180° - 40° - 90° = 50°$

## Exercise 15b

**1.** For each of the rectangles given, calculate the unknown angles marked $x$ and $y$:

**(a)**

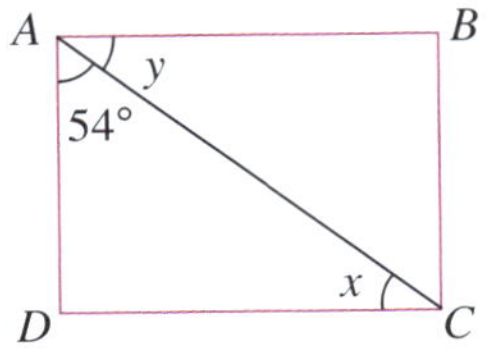


**(b)**

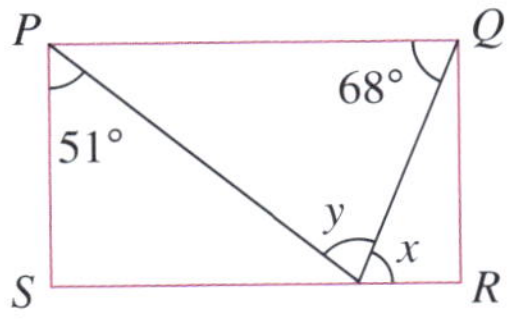


**(c)**

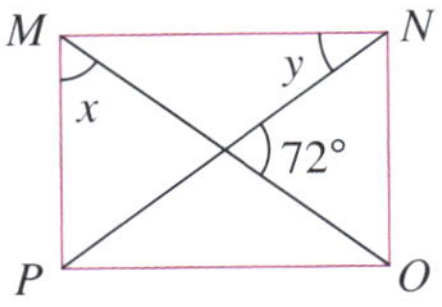


**2.** For each of the parallelograms given, calculate the values of $x$ and $y$:

**(a)**

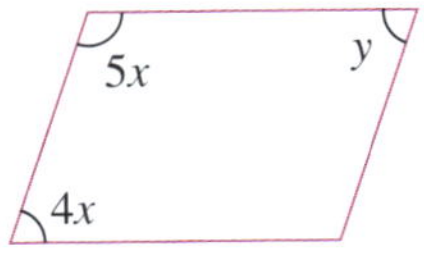


**(b)**

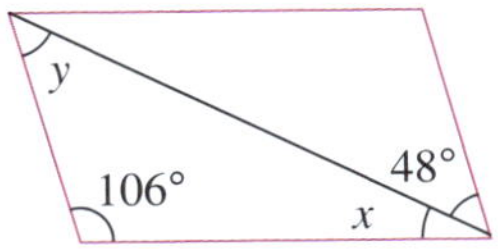


**(c)**

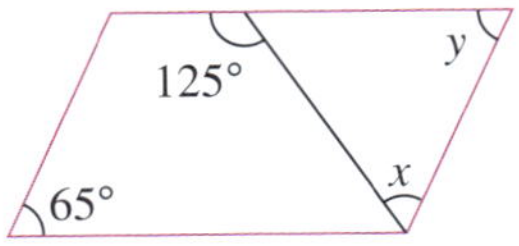

3. For each of the kites given, calculate the values of $x$ and $y$:

(a)

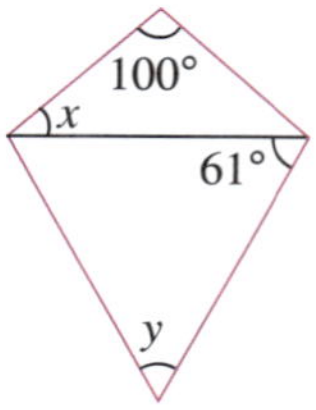


(b)

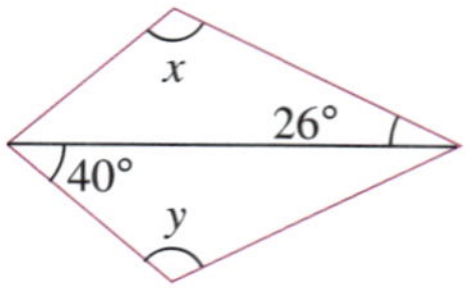


(c)

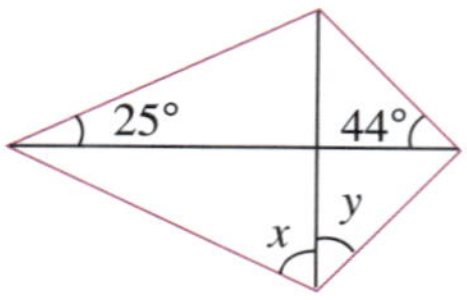


4. For each of the rhombuses given, calculate the unknown angles marked $x$, $y$ and $z$:

(a)

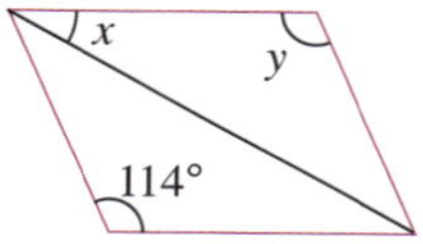


(b)

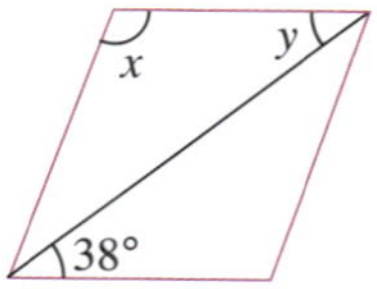


(c)

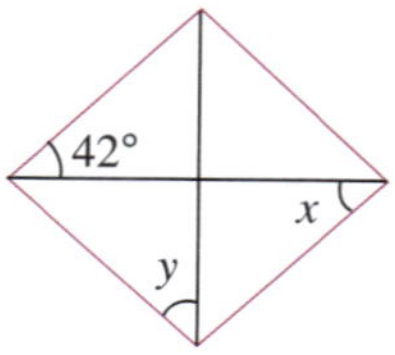


★5. For each of the figures given, calculate the unknown angles marked $x$ and $y$:

(a)

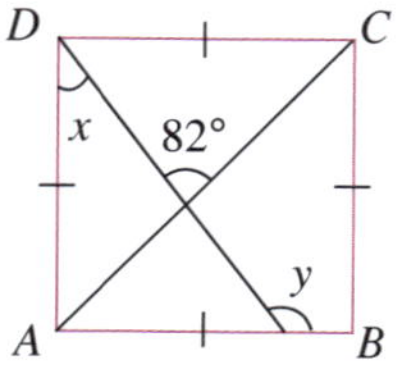


*ABCD* is a square.

(b)

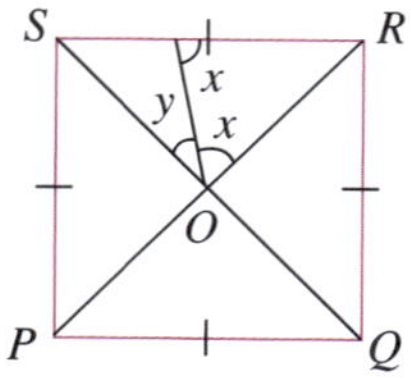


*PQRS* is a square.

(c)

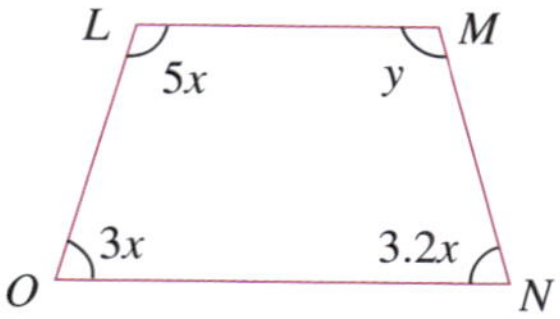


*LMNO* is a trapezium.

6. In a rectangle $ABCD$, $X$ is the mid-point of $AB$ and $C\hat{X}D = 118°$, calculate

(a) $A\hat{D}X$ (b) $X\hat{C}D$

7. In a parallelogram $PQRS$, $Q\hat{P}R = 42°$ and $Q\hat{R}S = 70°$, calculate

(a) $P\hat{Q}R$ (b) $P\hat{R}Q$

8. In a rhombus $PQRS$, $P\hat{Q}R = 108°$, calculate

(a) $Q\hat{R}S$ (b) $Q\hat{P}R$

(c) $Q\hat{S}R$

9. In a kite $ABCD$, $AB = BC$, $AD = CD$, $A\hat{D}C = 64°$ and $B\hat{A}C = 42°$, calculate

(a) $A\hat{C}D$ (b) $A\hat{B}C$

10. In a trapezium $PQRS$, $PQ$ is parallel to $SR$, $PQ = PS$, $P\hat{S}R = 62°$ and $Q\hat{R}S = 52°$, calculate

(a) $P\hat{Q}S$ (b) $S\hat{Q}R$

# Convex Polygons

A polygon in which none of its interior angles is a **reflex** angle is called a **convex** polygon.

The following are some convex polygons.

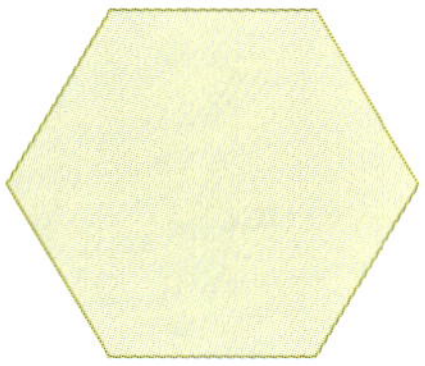

a hexagon

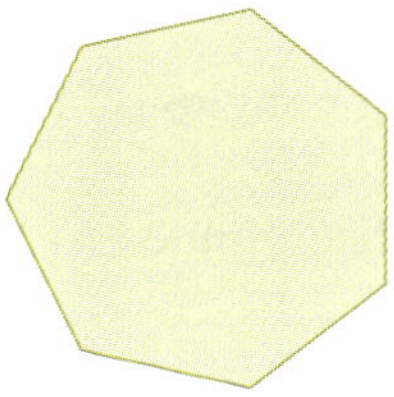

a heptagon

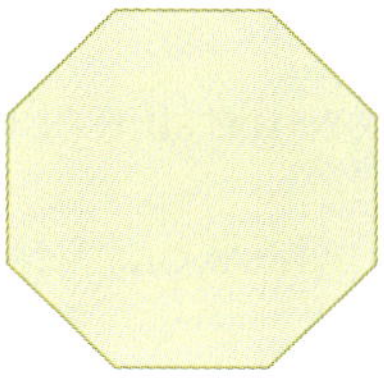

an octagon

*Notes*

*A reflex angle is more than 180° but less than 360°.*

**Thinking Time**

Are all regular polygons convex polygons?

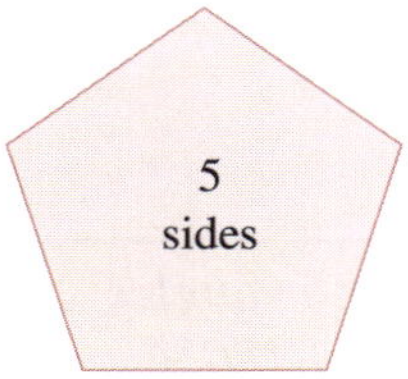


Regular Pentagon

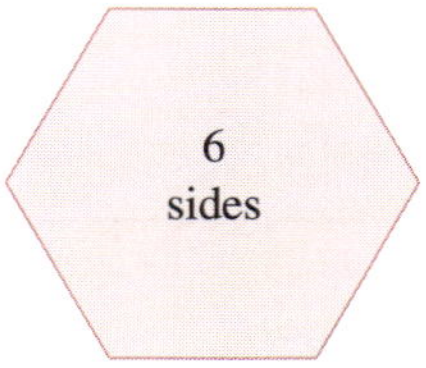


Regular Hexagon

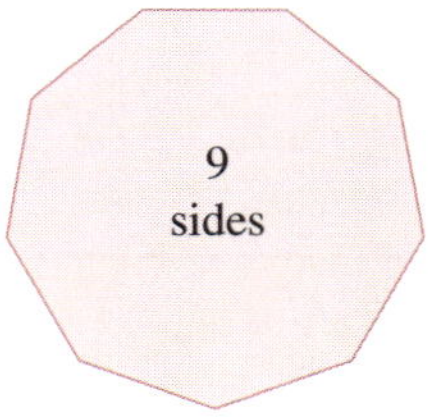


Regular Nonagon

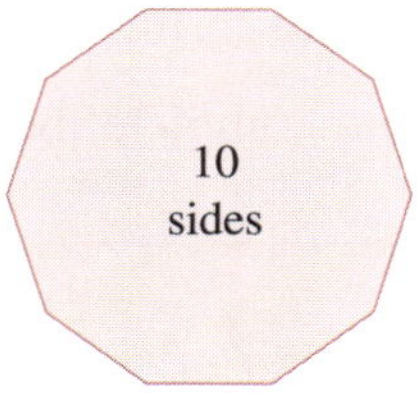


Regular Decagon

The figure below is not a convex polygon.

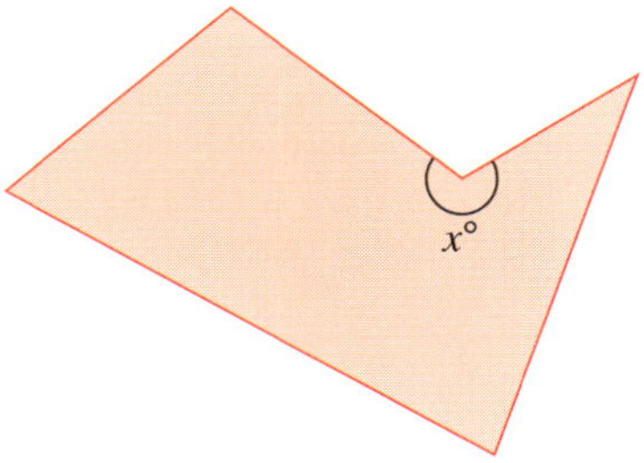


$180 < x < 360$

**Just For Fun**

*What types of polygons and how many of each kind can you find in this star?*

# Sum of the Interior Angles of a Convex Polygon

In this investigation, we will explore the sum of the interior angles of a polygon.

1. Draw the following polygons where one vertex has been joined to the other vertices.

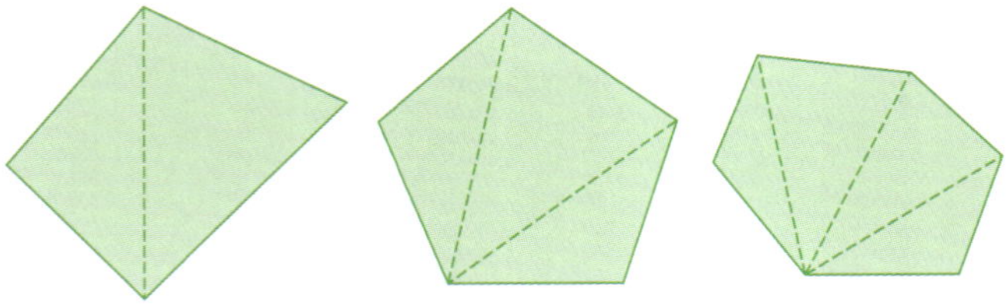

Quadrilateral Pentagon Hexagon

2. How many triangles does each of the above have? What is the sum of the interior angles of the above polygons?
3. Repeat the above process for polygons with 7, 8, 9 and 10 sides and find the number of triangles each polygon can form.
4. Copy and complete the following table:

| *Name of Polygon* | *No. of sides* | *No. of triangles formed* | *Sum of interior angles (in terms of 180°)* |
|---|---|---|---|
| Quadrilateral | 4 | 2 | $2 \times 180° = (4 - 2) \times 180°$ |
| Pentagon | 5 | 3 | $3 \times 180° = (5 - 2) \times 180°$ |
| Hexagon | 6 | 4 | $4 \times 180° = (6 - 2) \times 180°$ |
| Heptagon | 7 | | |
| Octagon | 8 | | |
| Nonagon | 9 | | |
| Decagon | 10 | | |
| $n$-gon | $n$ | | |

5. From the above table, what can you say about the number of triangles formed by a polygon in relation to its number of sides?

In general, we have the following conclusion:

**the sum of the interior angles of an *n*-sided polygon is $(n - 2) \times 180°$ or $(n - 2) \times 2$ rt ∠s or $(2n - 4)$ rt ∠s.**

## Sum of the Exterior Angles of a Convex Polygon

Let us consider a 6-sided convex polygon in which each side is produced. (Fig. 15.8). The interior angles of the polygon are named $a$, $b$, $c$, $d$, $e$ and $f$ while the exterior angles are named $p$, $q$, $r$, $s$, $t$ and $u$.

Now, $a + p = 180°$, $b + q = 180°$
$c + r = 180°$, $d + s = 180°$
$e + t = 180°$, $f + u = 180°$

i.e., $a + p + b + q + c + r + d + s + e + t + f + u = 6 \times 180°$
$(a + b + c + d + e + f) + (p + q + r + s + t + u) = 1080°$

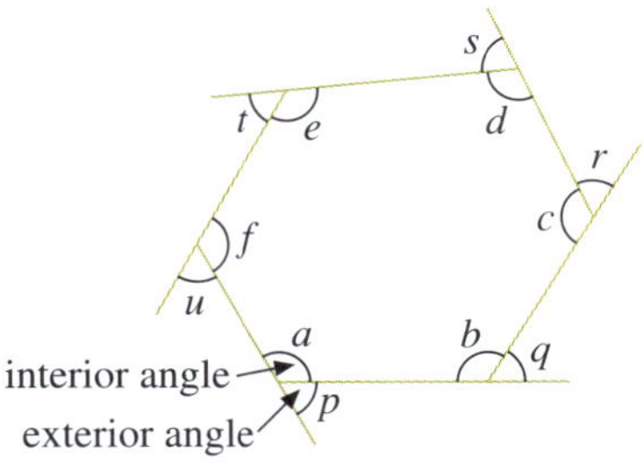

*Fig. 15.8*

But we know that the sum of the interior angles of a 6-sided polygon is 8 right angles, i.e., 720°.

Thus, $720° + (p + q + r + s + t + u) = 1080°$
$p + q + r + s + t + u = 360°$

By using this method, show that the sum of the exterior angles of a polygon with 7 sides and 8 sides is equal to 360°.

The diagram shows part of a polygon with $n$-sides in which each side is produced. The interior angles are named $a_1, a_2, a_3, a_4, \dots a_n$ and the exterior angles are named $x_1, x_2, x_3, x_4, \dots x_n$.

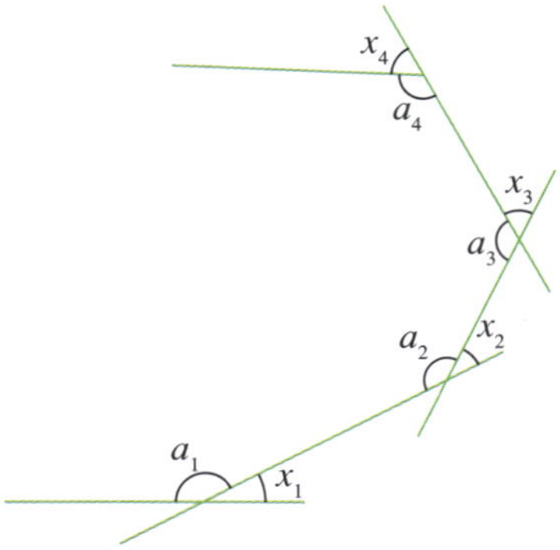


Now $\quad a_1 + a_2 + a_3 + \dots + a_n = (2n - 4) \times 90°$ —— (1)

and $\quad a_1 + x_1 = 180°, a_2 + x_2 = 180°, \dots, a_n + x_n = 180°$

Can you proceed to show that

$$x_1 + x_2 + x_3 + x_4 + \dots + x_n = 360°?$$

From the above exploration, we have

**The sum of the exterior angles of any polygon is 360°.**

## Example 3

*Calculate the size of the interior angle of a regular decagon.*

**Solution**

**Method I**

A regular decagon has 10 sides. Since the sum of all the exterior angles is 360°, each exterior angle $= \frac{360°}{10} = 36°$.

Since the sum of the interior and exterior angles at any vertex is 180°, an interior angle $= 180° - 36° = 144°$.

**Method II**

Sum of the interior angles of a 10-sided polygon = $(10 - 2) \times 180° = 1440°$,

$\therefore$ each interior angle = $\frac{1440°}{10} = 144°$

## Example 4

*Find the number of sides of a regular polygon if*

*(a) each interior angle of the polygon is 162°,*
*(b) each exterior angle of the polygon is 24°.*

**Solution**

(a) Let $n$ be the number of sides of the polygon.

Each interior angle of a regular polygon = $\frac{(n-2) \times 180°}{n}$

$$\therefore \quad \frac{(n-2) \times 180°}{n} = 162° \text{ i.e. } (n-2) \times 180° = 162 \times n$$

$$180n - 360 = 162n$$
$$180n - 162n = 360$$
$$18n = 360$$
$$\therefore \quad n = \frac{360}{18} = 20$$

Hence the polygon has 20 sides.

(b) Let the number of sides of the regular polygon be $n$.
There are $n$ equal exterior angles and we have
$24° \times n = 360°$

$$\therefore \quad n = \frac{360}{24} = 15$$

Hence the polygon has 15 sides.

## Example 5

*ABCDE is a regular polygon. If AB and DC are produced to meet at X, find the value of* $B\hat{X}C$.

**Solution**

The figure shows a sketch of the regular pentagon with $AB$ and $DC$ produced to meet at $X$.

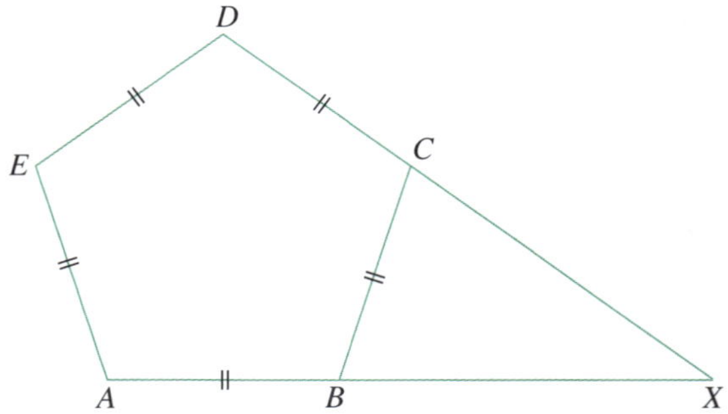


The size of each exterior angle of the regular pentagon $= \frac{360^\circ}{5} = 72^\circ$

$\therefore \quad X\hat{B}C = X\hat{C}B = 72^\circ$

and $\quad B\hat{X}C = 180^\circ - 72^\circ - 72^\circ$ ($\angle$sum of $\triangle$)

i.e. $\quad B\hat{X}C = 36^\circ$

## Exercise 15c

1. Calculate the sum of the interior angles of the following polygons:
   (a) a 10-gon, (b) a 12-gon,
   (c) a 15-gon, (d) an 18-gon.

2. Find the size of each interior angle of a regular polygon with
   (a) 5 sides, (b) 20 sides,
   (c) 24 sides, (d) 36 sides.

3. Find the number of sides of a regular polygon with an interior angle of
   (a) 140°, (b) 160°,
   (c) 162°, (d) 170°.

4. Find the number of sides of a regular polygon with an exterior angle of
   (a) 45°, (b) 36°,
   (c) 30°, (d) 12°.

5. For each of the following figures (not drawn to scale), find the value of $x$:

   **(a)**

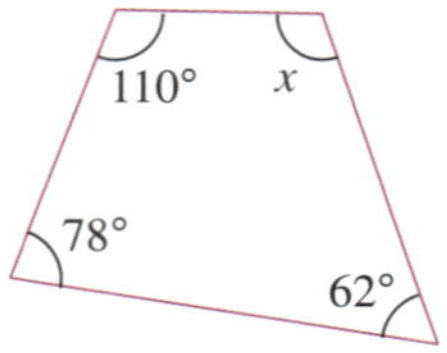


   **(b)**

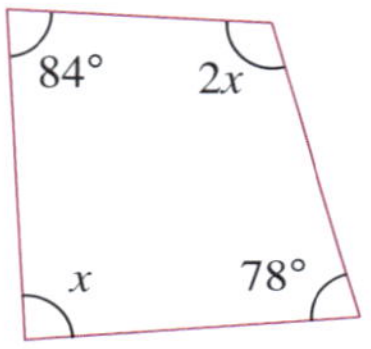


   **(c)**

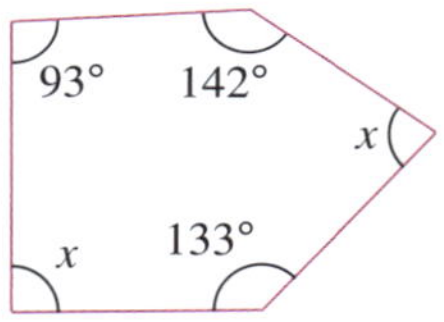


   **(d)**

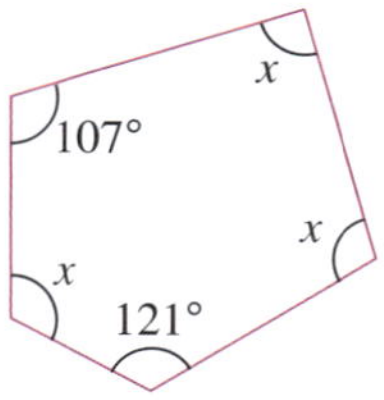


   **(e)**

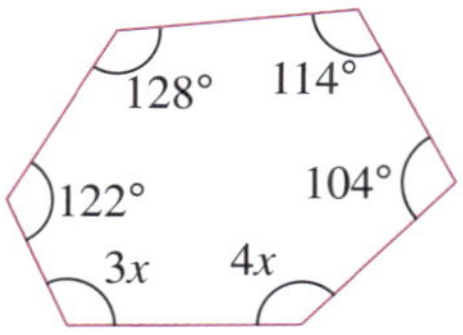

(f)

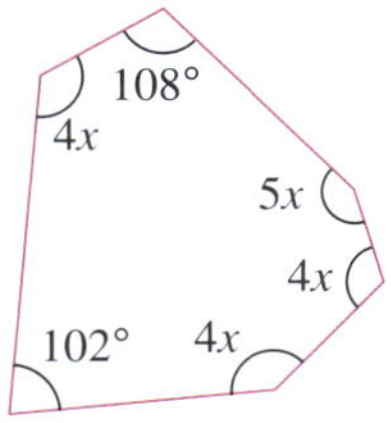


6. The angles of a quadrilateral are $3x$, $4x$, $5x$ and $6x$. Find $x$. What are the angles in degrees?

7. The exterior angles of a triangle are $3y°$, $4y°$ and $5y°$. Find $y$. What are the interior angles of the triangle?

8. In a quadrilateral $PQRS$, the angles $P$, $Q$, $R$ and $S$ are $3x$, $4x$, $5x$ and $8x$ respectively. Find **(a)** $\hat{P}$ and **(b)** $\hat{S}$.

9. A polygon has $n$ sides. Three of its exterior angles are 70°, 25° and 15° while the remaining $(n - 3)$ exterior angles are each equal to 50°. Find $n$.

10. Given that $ABCDE$ is a regular pentagon, calculate
**(a)** $A\hat{B}C$, **(b)** $B\hat{C}A$,
**(c)** $B\hat{D}E$.

★11. Given that $ABCDEFGH$ is a regular octagon, calculate
**(a)** $A\hat{B}C$, **(b)** $A\hat{C}D$,
**(c)** $A\hat{F}G$.

★12. The points $A$, $B$, $C$ and $D$ are consecutive vertices of a regular polygon with 20 sides. Calculate
**(a)** $A\hat{B}C$, **(b)** $A\hat{C}D$,
**(c)** $A\hat{B}D$.

## SUMMARY

1. The **sum of the angles of a triangle** is 180°.
The **exterior angle** of a triangle is equal to the **sum of the interior opposite angles**.

2. A quadrilateral is a 4-sided plane figure. The sum of the angles of a quadrilateral is equal to 360°.
   - The diagonals of a parallelogram bisect each other.
   - The diagonals of a rectangle are equal in length and they bisect each other.
   - The diagonals of a rhombus bisect each other at right angles.
   - The diagonals of a square are equal in length and they bisect each other at right angles.
   - The longer diagonal of a kite bisect the shorter diagonal at right angles.

3. The sum of the interior angles of an $n$-sided polygon is given by $(n - 2) \times 180°$ or $(2n - 4)$ right angles.

4. The sum of the exterior angles of an $n$-sided polygon is equal to 360°.

# Review Examples 15

## Example 1

*Two of the exterior angles of an n-sided polygon are 25° and 26°, three of its interior angles are each equal to 161° while the remaining interior angles are each equal to 159°, calculate the value of n.*

### Solution

The size of the different exterior angles are 25°, 26°, (180 – 161) ° and (180 – 159) °

$$\therefore \quad 25° + 26° + 3(180 - 161)° + (n - 5)(180 - 159)° = 360°$$

$$51 + 57 + 21n - 105 = 360$$

$$21n = 360 + 105 - 51 - 57$$

$$n = \frac{351}{21} = 17$$

# Review Questions 15

**1.** For each of the following figures, calculate the values of the unknowns:

**(a)**

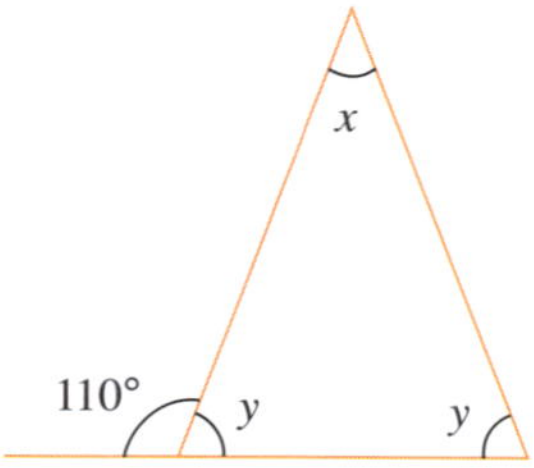


**(b)**

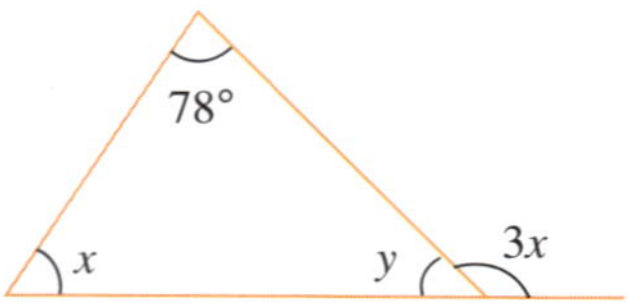


**(c)**

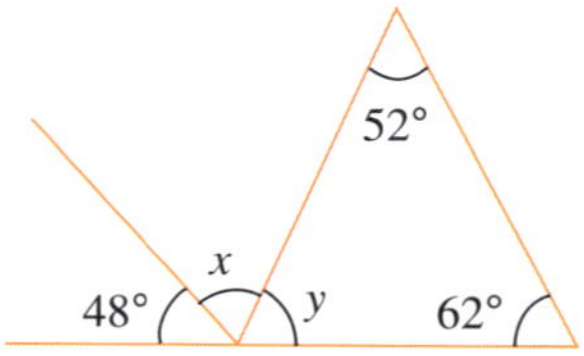


**(d)**

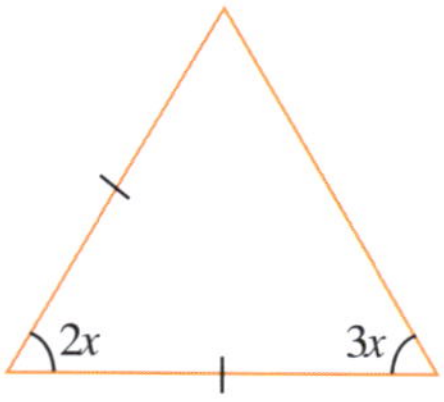

(e)

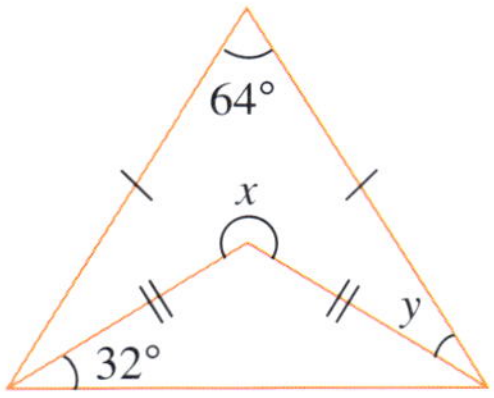


(f)

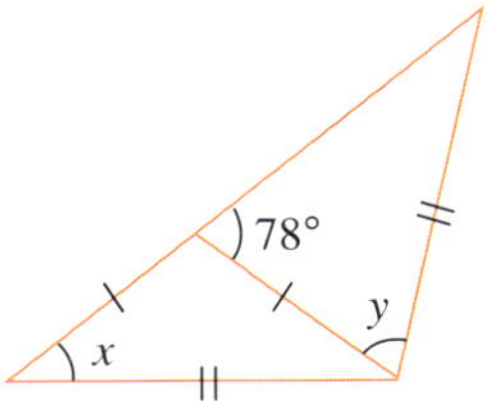


(g)

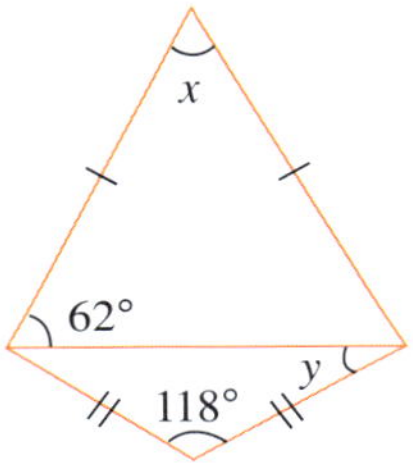


(h)

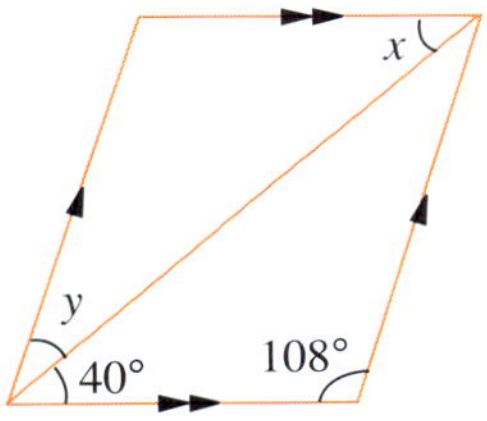


(i)

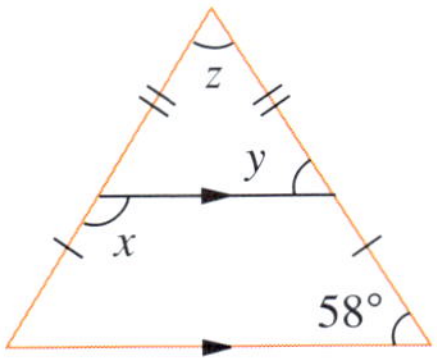


(j)

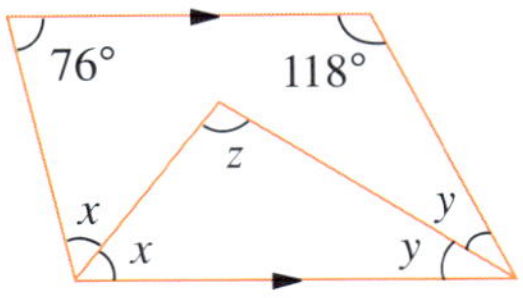


**2.** Find the unknown angles marked $x$ and $y$ in the following figures:

(a)

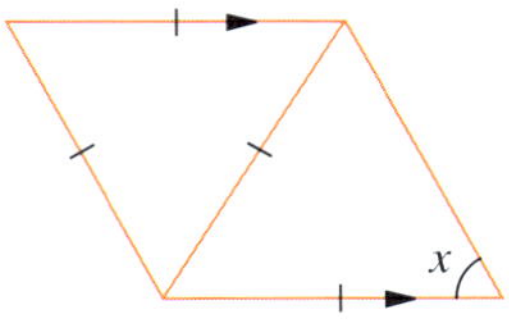


(b)

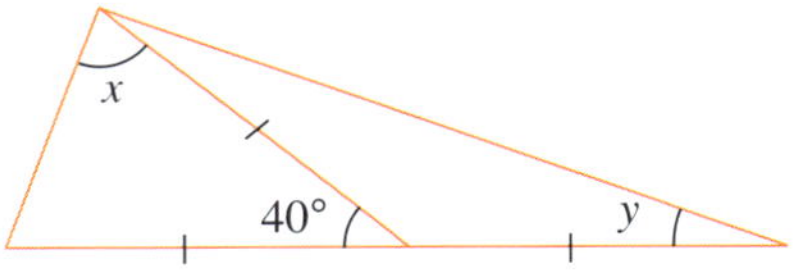


(c)

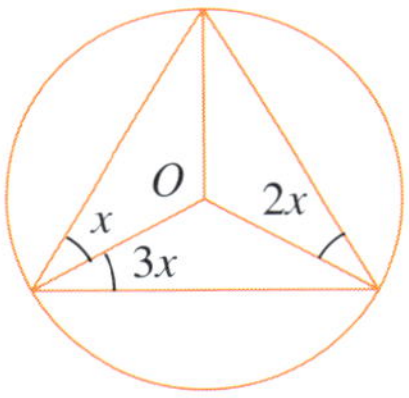


(d)

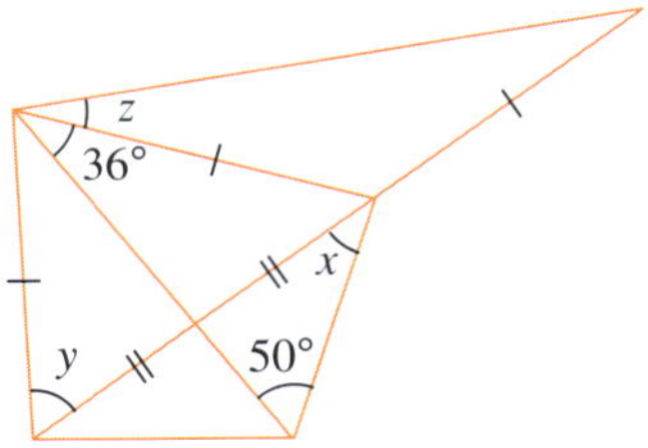

**3.** Find the values of the unknown marked in the following figures:

**(a)**

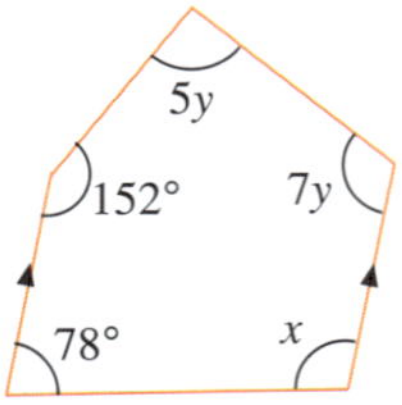


**(b)**

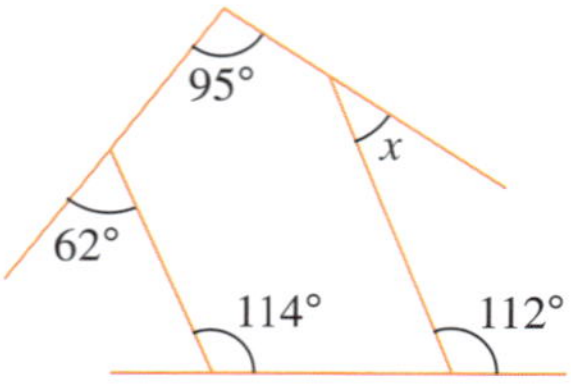


**(c)**

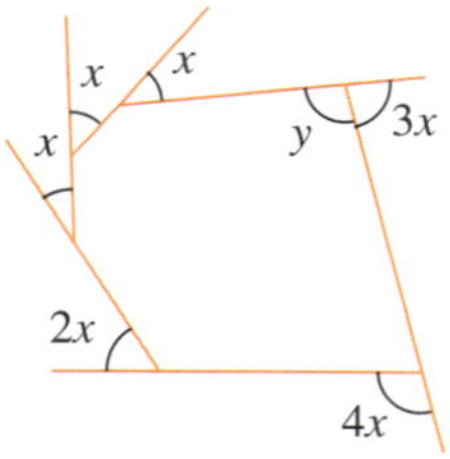


**(d)**

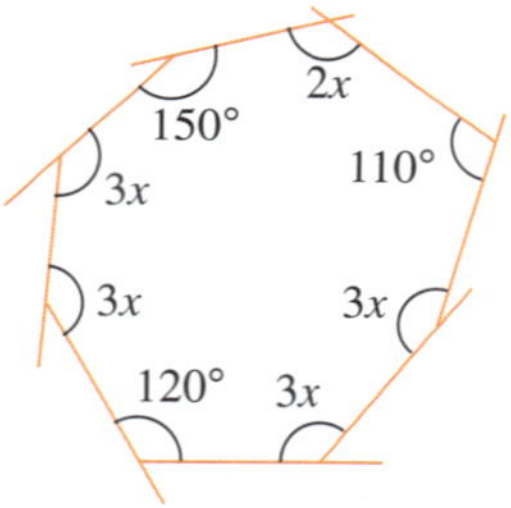


**4.** The sum of a polygon with $(2n - 3)$ sides is 62 right angles. Find the value of $n$.

**5.** The ratio of the interior angles of a pentagon is 4 : 4 : 5 : 5 : 6, find the largest interior and exterior angles.

**6.** A polygon has $n$ sides. One of its interior angles is 126° while the other angles are each equal to 162°, find the value of $n$.

**7.** The ratio of an interior angle to an exterior angle of a regular polygon is 13 : 2. Find the number of sides of the polygon.

**8.** If the sum of the interior angles of a polygon is six times the sum of its exterior angles, find the number of sides of the polygon.

**9.** A polygon has $n$ sides. Two of its exterior angles are 72° and 35° while the remaining exterior angles are each equal to 23°. Find the value of $n$.

***10.** Find the sum of the marked angles in each of the following diagrams:

**(a)**

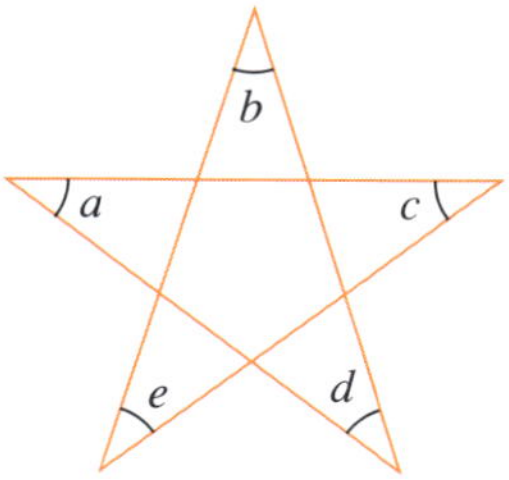


**(b)**

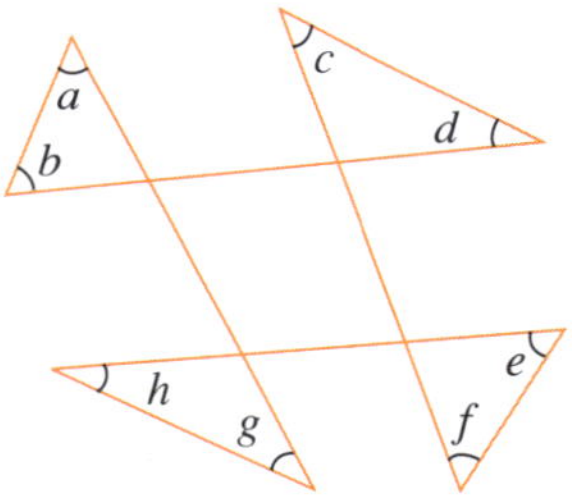


**11.** $ABCD$ is a square and $\triangle DPC$ is an equilateral triangle. Find $A\hat{P}B$.

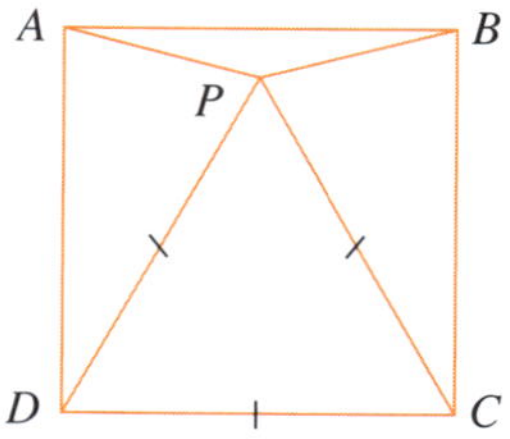

**12.** In the figure, *ABCDE* is a regular pentagon and *ABPQRS* is a regular hexagon. Calculate

**(a)** $A\hat{B}C$

**(b)** $A\hat{P}B$

**(c)** $A\hat{P}Q$

**(d)** $A\hat{C}D$

**(e)** $A\hat{S}E$.

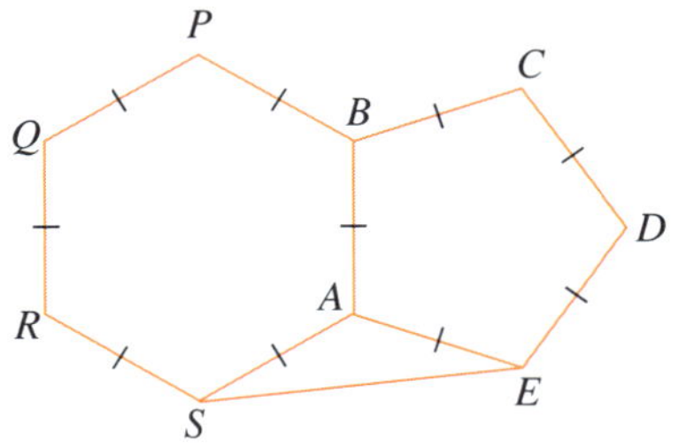

*In this chapter, you will learn how to*

- *use a ruler and a pair of compasses to construct angle bisectors, perpendicular bisectors, copy an angle and drop a perpendicular line from a point to a line;*
- *construct triangles and quadrilaterals.*

CHAPTER

# 16

# Geometrical Constructions

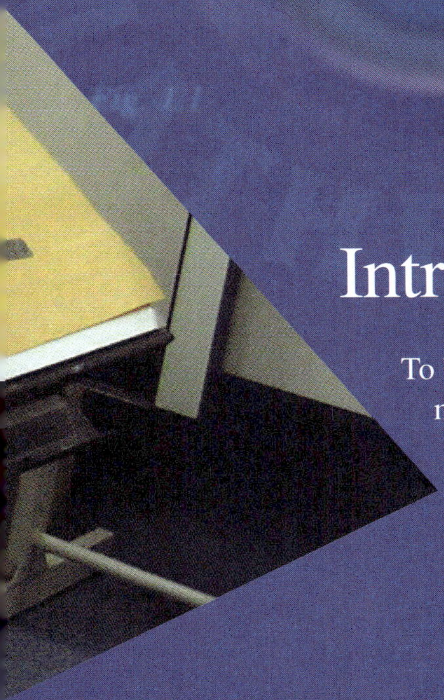

## Introduction

To be a good interior designer one must be able to use many of the mathematical tools such as compasses, set squares, protractors etc efficiently.

# Geometrical Constructions

In primary school, you learned how to draw triangles and quadrilaterals with the use of rulers, protractors and set square. We shall learn how to use a pair of compasses to construct geometrical figures.

# Use of Compasses

A pair of **compasses** is a mathematical instrument often used for drawing circles, marking off lengths and measuring the length of a line segment.

drawing a circle — marking off lengths — measuring a line segment

The following explains how to use a pair of compasses to mark off a length equal to $AB$ on the line $PQ$, i.e., $HK$:

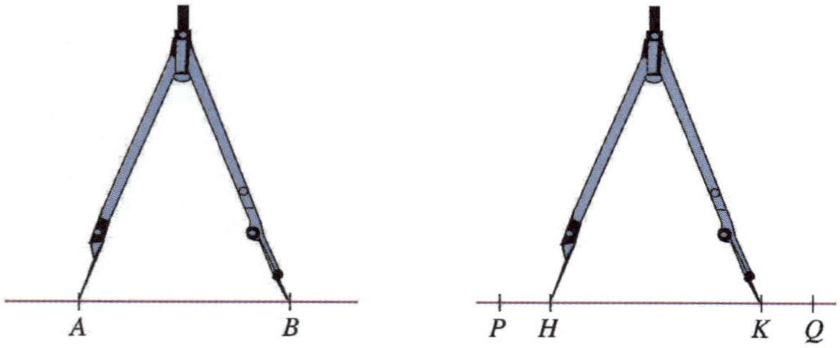


**Construction steps:**

(1) Adjust the arms of the compasses until they touch $A$ and $B$.
(2) Mark a point $H$ on another line, $PQ$.
(3) Without adjusting the arms of the pair the compasses, fix one arm at $H$, and move the other arm to draw an arc cutting $PQ$ at $K$. Hence, $AB = HK$.

## Explorations

You can create various designs with your pair of compasses. Some of these designs are shown below. See if you can create these designs. Try to create other designs of your own.

The following are useful points to note when doing constructions.

1. Use a hard pencil with a sharp point so that lines and points can be drawn finely and clearly.
2. Take great care to draw a line accurately through a point.
3. When making intersections with lines and arcs, make sure that the angles of intersection are as close to 90° as possible, e.g.

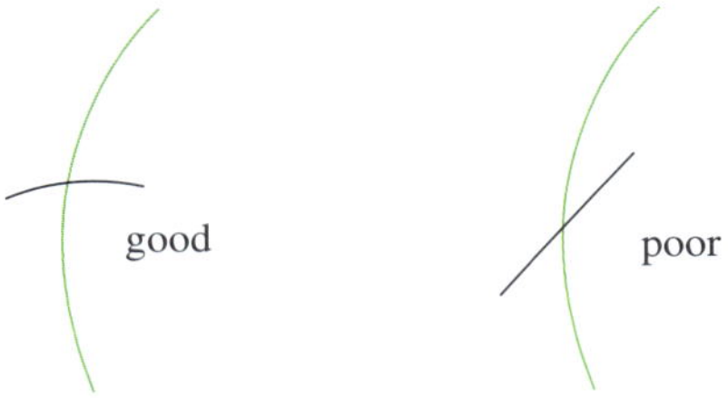


4. All construction lines must be clearly shown. Do not erase the construction lines you have already done.

## Bisecting an Angle

To bisect an angle is to divide an angle into two equal parts. The following shows two methods of doing this.

**Given:** $P\hat{Q}R$.

**To construct:** The angle bisector of $P\hat{Q}R$.

**Method 1:** Using ruler and Compasses

**Construction steps:**

1. Place the compass point at $Q$ and with a suitable fixed radius, draw an arc to cut $QP$ at $X$ and $QR$ at $Y$.
2. Using the same radius and with $X$ as centre, draw an arc 1 as shown.
3. With the same radius and with $Y$ as centre, draw an arc 2 to cut arc 1 at $Z$.
4. Join $QZ$.

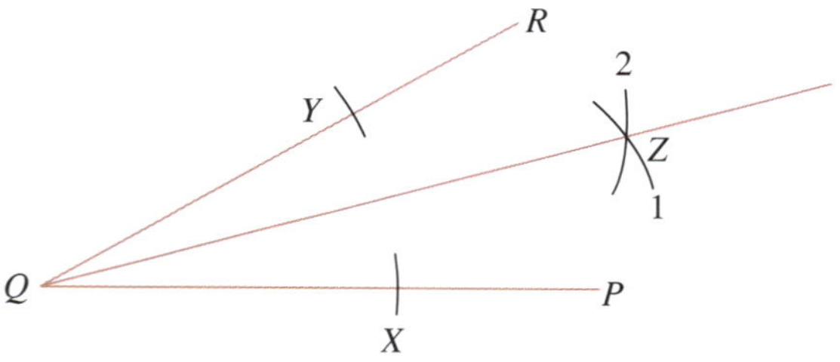


Use a protractor to measure the angles $P\hat{Q}Z$ and $R\hat{Q}Z$. You will find that $P\hat{Q}Z = R\hat{Q}Z$ and hence $QZ$ is the angle bisector of $P\hat{Q}R$.

Explain why line $QZ$ constructed as above is the bisector of $P\hat{Q}R$.

**Method 2:** Using ruler protractor

**Construction Steps:**

1. Use a protractor to measure $P\hat{Q}R$ as shown.
2. Keeping the protractor in the same position, mark a point $K$ such that $P\hat{Q}K = \frac{1}{2} P\hat{Q}R$.
3. Remove the protractor and join $QK$. $QK$ is the angle bisector of $P\hat{Q}R$.

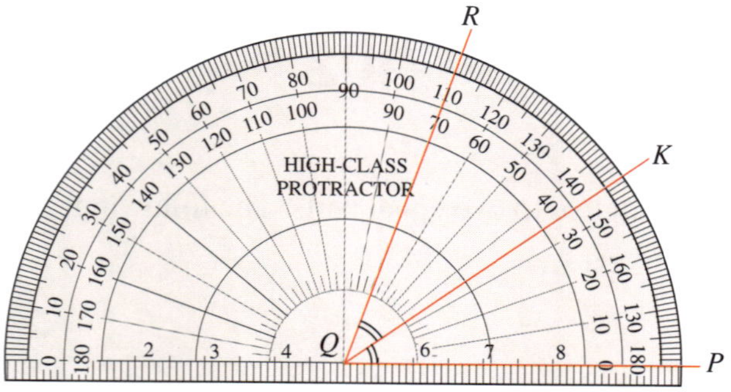


The angle bisector has the following property:

**Any point on the angle bisector is equidistant from the two sides of the angle.**

# Bisecting a Line Segment

**Given:** A straight line segment $PQ$.

**To construct:** A perpendicular bisector of $PQ$.

**Method 1:** Using ruler and compasses

**Construction steps:**

1. Place the compass point at $P$ and make arc 1 above the line $PQ$ and make an arc 2 below the line $PQ$ as shown.
2. Using the same radius as (1) and with the compass point at $Q$, make arc 3 to cut arc 1 at $X$ and arc 4 to cut arc 2 at $Y$.
3. Join $XY$ to cut $PQ$ at $S$. $XY$ is called the **perpendicular bisector** of $PQ$. It divides $PQ$ into two equal parts $PS$ and $SQ$. $S$ is called the midpoint of $PQ$.

**Notes**

*Any point on the perpendicular bisector of PQ is equidistant from P and from Q.*

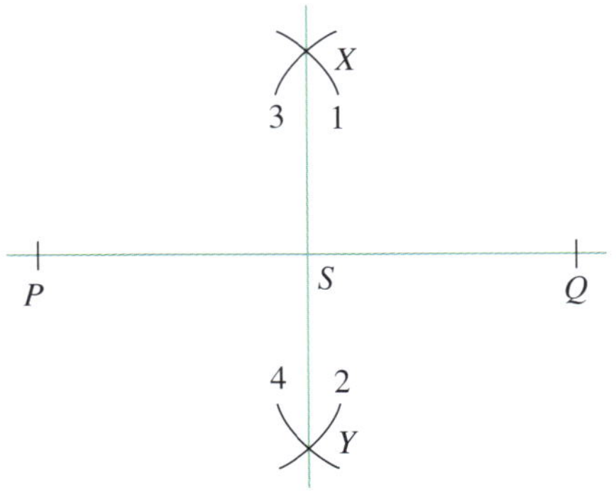


**Method 2:** Using ruler and protractor

**Construction steps:**

1. Measure $PQ$ accurately.
2. Mark a point $S$ on $PQ$ so that $PS = \frac{1}{2}PQ$.
3. Place the base of the protractor on $PQ$ with the centre at the point $S$ and mark a point $K$ at the 90° mark.
4. Remove the protractor and join $KS$, and produce to $H$. $KSH$ is the perpendicular bisector of $PQ$.

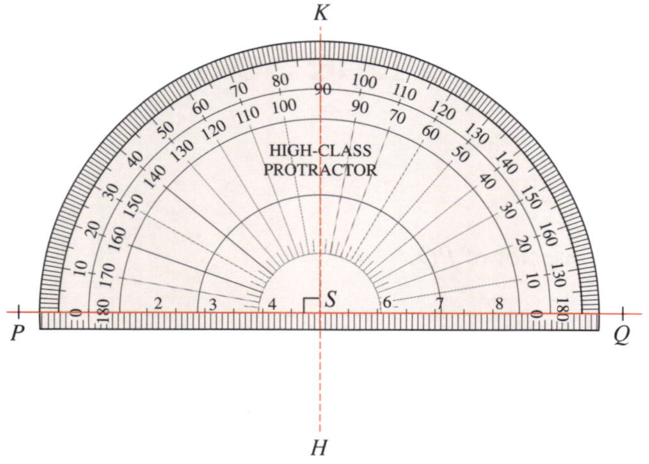


We can also draw the perpendicular bisector by using a ruler and a set-square. Find out how this can be done.

The property of the perpendicular bisector is stated as follows:

**Any point on the perpendicular bisector of *KSH* is equidistant from *P* and *Q*.**

Explorations

**Work in groups**

Construct a line parallel to *BC* and passing through *A*.

Explain why the line you have constructed is indeed parallel to *BC*.

What tools did you use? Can your group think of other ways to construct the parallel line?

Compare your results with other groups.

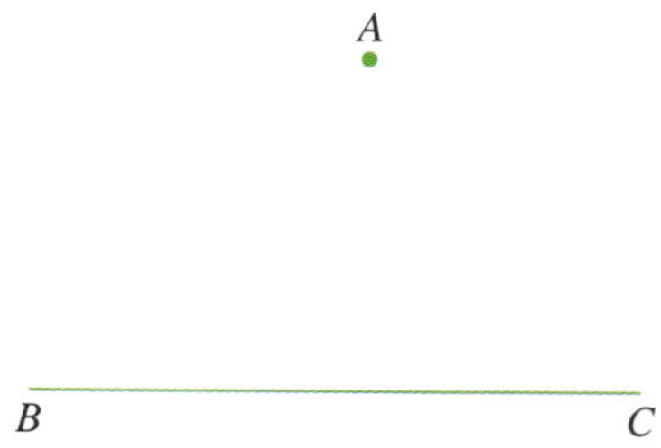

## Example 1

*Construct triangle PQR such that PQ = 10.5 cm, QR = 9.5 cm and PR = 8.5 cm. Construct the angle bisector of $Q\hat{P}R$. If the angle bisector of $Q\hat{P}R$ meets QR at X, measure and write down the length of QX.*

### Solution

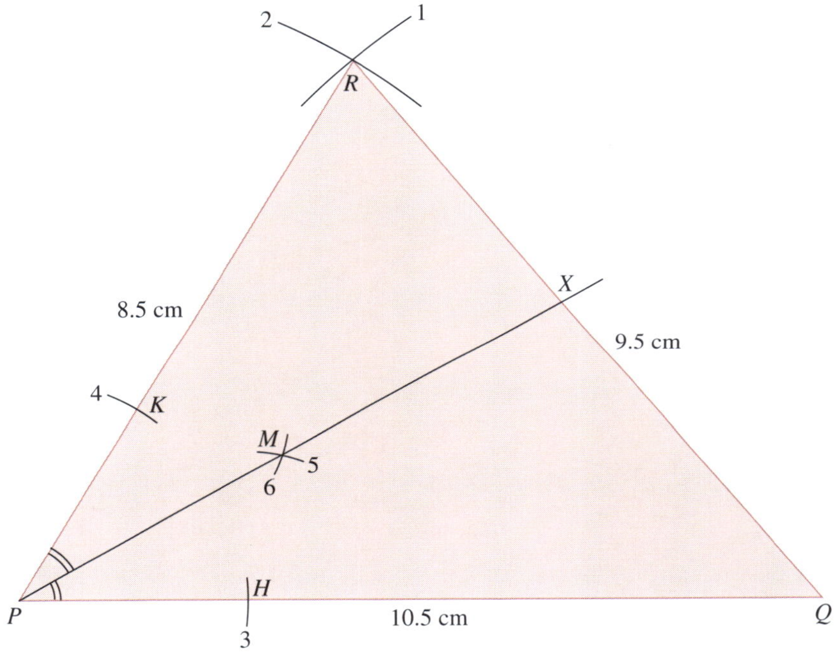


**Construction steps:**

1. Draw a line $PQ$ = 10.5 cm.
2. With centre at $Q$ and radius 9.5 cm, draw an arc 1.
3. With centre at $P$ and radius 8.5 cm, draw arc 2 to cut arc 1 at $R$.
4. Join $PR$ and $QR$.
5. With centre at $P$ and radius approximately 3 cm, draw arc 3 to cut $PQ$ at $H$ and arc 4 to cut $PR$ at $K$.
6. With centre at $H$ and radius approximately 2 cm, draw arc 5. With the same radius and centre at $K$ draw arc 6 to cut arc 5 at $M$.
7. Join $PM$ and produce to cut $QR$ at $X$. $PM$ is the angle bisector of $Q\hat{P}R$.
8. By measurement, $QX \approx 5.3$ cm.

## Example 2

*Construct $\triangle ABC$ such that $AB = 10.5$ cm, $A\hat{B}C = 60°$ and $BC = 7.5$ cm. Measure and write down the length of AC.*

Solution

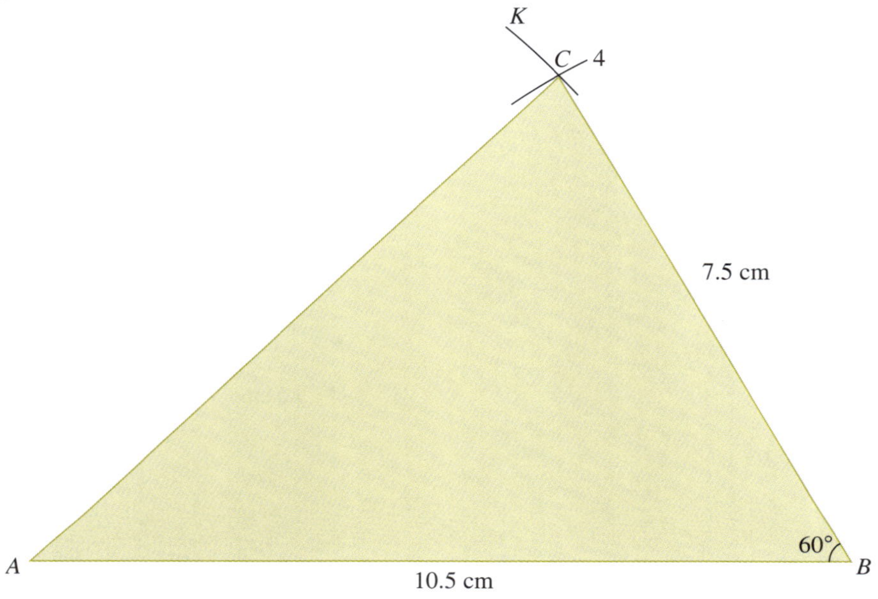


**Construction steps:**

1. Use ruler to draw a line $AB = 10.5$ cm.
2. Use a protractor to draw an angle $A\hat{B}K = 60°$.
3. Join $BK$ and produce. Now $A\hat{B}K = 60°$. Can you explain why this is so?
4. With $B$ as the centre and radius 7.5 cm, make an arc 4 to cut $BK$ at $C$.
5. Join $AC$ and measure its length, $AC \approx 9.4$ cm.

## xample 3

*Compasses only, construct a parallelogram ABCD such that AB = 8.5 cm, BC = 6.5 cm and ABC = 45°. Measure and write down the length of the diagonal AC.*

Solution

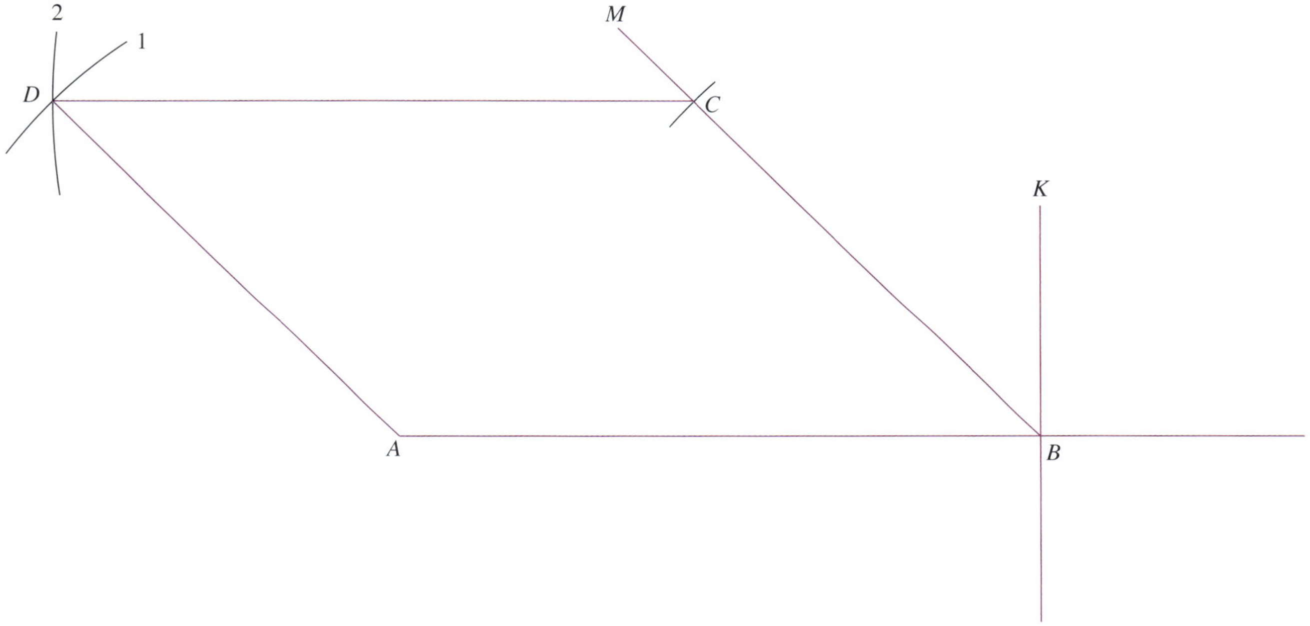


**Construction steps:**

1. Draw a line $AB$ = 8.5 cm.
2. Use a protractor to measure $A\hat{B}M = 45°$.
3. With $B$ as centre and radius 6.5 cm, cut $BM$ at $C$.
4. With $A$ as centre and radius 6.5 cm, draw an arc 1.
5. With $C$ as centre and radius 8.5 cm, draw an arc 2 to cut arc 1 at $D$.
6. Join $CD$ and $AD$. $ABCD$ is the required parallelogram.
7. By measurement, $AC \approx 6.0$ cm.

## Exercise 16a

1. Construct $\triangle ABC$ where $AB = AC = 10$ cm and $BC = 9$ cm. Measure and write down the size of $A\hat{B}C$.

2. Construct $\triangle PQR$ where $PQ = 9.5$ cm, $PR = 8.5$ cm and $QR = 9.8$ cm. Measure and write down the size of $P\hat{R}Q$. Construct the perpendicular bisector of $PR$.

3. Construct $\triangle XYZ$ where $XY = 8.8$ cm, $XZ = 9.2$ cm and $YZ = 10.4$ cm. Measure and write down the size of $Y\hat{X}Z$. Construct the angle bisector of $X\hat{Y}Z$.

4. Construct $\triangle ABC$ where $AB = 10$ cm, $AC = 8$ cm and $BC = 6$ cm. Construct the perpendicular from $C$ to $AB$. If this perpendicular cuts $AB$ at $X$, measure and write down the length of $AX$.

5. Construct $\triangle ABC$ where $AB = 5$ cm, $B\hat{A}C = 45°$ and $A\hat{B}C = 60°$. Construct the perpendicular bisector of $AB$. If the perpendicular bisector of $AB$ meets $AC$ at $X$, measure and write down the length of $BX$.

6. Construct $\triangle XYZ$ with $XY = 9.4$ cm, $XZ = 8.8$ cm and $X\hat{Y}Z = 60°$. Measure and write down the length of $YZ$.

7. Construct $\triangle PQR$ with $PQ = 9.8$ cm, $QR = 6.5$ cm and $P\hat{Q}R = 90°$. Measure and write down the size of $Q\hat{P}R$ giving your answer correct to the nearest degree.

8. Construct a parallelogram $PQRS$ where $PQ = 8.5$ cm, $QR = 5.5$ cm and $P\hat{Q}R = 120°$. Measure and write down the length of $PR$.

9. Construct a rectangle $ABCD$ where $AB = 10.5$ cm and $BC = 6.5$ cm. Draw the diagonal $AC$. Measure $AC$ and state its length.

10. Construct the rhombus $PQRS$ where $PQ = QR = RS = PS = 6$ cm and $PR = 9$ cm. Measure $QS$ and state its length. Also measure $P\hat{Q}R$ and state its size.

11. Construct the trapezium $ABCD$ with $AD$ parallel to $BC$, $AB = 4.5$ cm, $BC = 8$ cm, $A\hat{B}C = 60°$ and $AD = 6$ cm. Measure and write down the lengths of $AC$ and $CD$.

## Example 4

*Construct $\triangle PQR$ in which $PQ = 9$ cm, $\angle PQR = 38°$ and $\angle QPR = 67°$. Measure the angle PRQ and the length of the sides PR and RQ. What is the length of the side facing the smallest angle? And what is the length of the side facing the largest angle?*

*Just for fun*

*Are the sides of the triangle straight?*

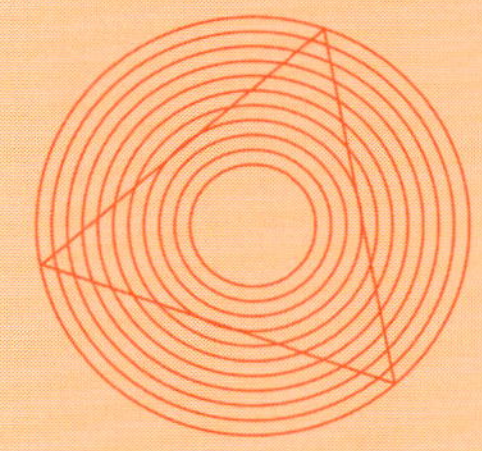

**Solution**

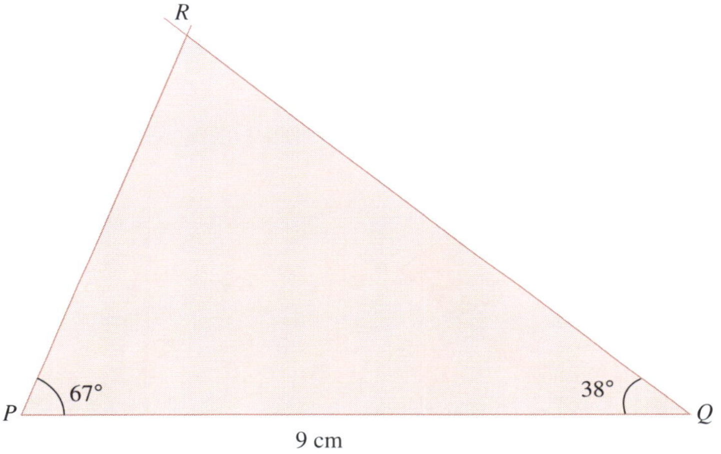


**Construction steps:**

1. Draw a line segment $PQ$ of length 9 cm long.
2. Using a protractor at points $P$ and $Q$, draw an angle of 67° and one of 38° respectively.
3. Produce the other arms of angles $P$ and $Q$ to meet at $R$. $\triangle PQR$ is the required triangle.

By measurement, $\angle PRQ = 75°$, $PR = 5.6$ cm and $QR = 8.5$ cm. The length of the side facing the smallest angle is 5.6 cm and that of the side facing the largest angle is 9 cm.

Do you notice that the smallest angle faces the shortest length and the largest angle faces the longest side?

## Example 5

*Construct a quadrilateral PQRS where PQ = 4.5 cm, QR = 5.6 cm. RS = 6.1 cm, PS = 4.3 cm and diagonal PR = 6.9 cm. Measure* $\hat{Q}$, $\hat{R}$ *and* $\hat{S}$.

### Solution

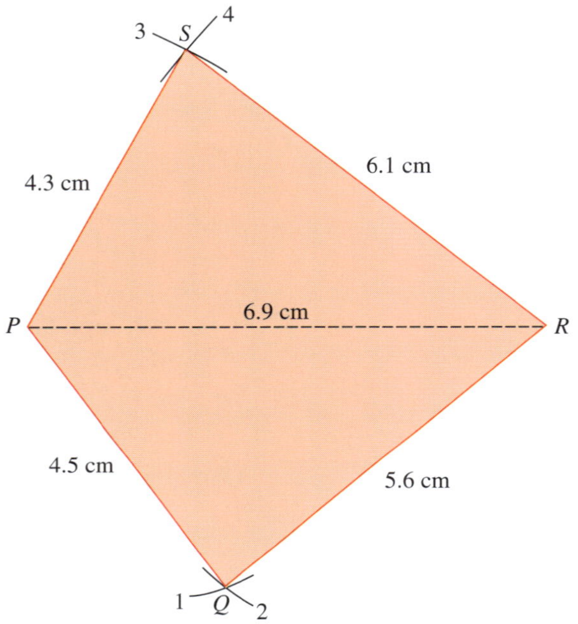


**Construction steps:**

1. Draw a line segment *PR* of length 6.9 cm.
2. With *P* and *R* as centres and radii 4.5 cm and 5.6 cm respectively, draw two arcs 1 and 2 to cut at *Q*.
3. With *P* and *R* as centres and radii 4.3 cm and 6.1 cm respectively, draw two arcs 3 and 4 to cut at *S*.
4. Join *PQ*, *QR*, *RS* and *PS*. *PQRS* is the required quadrilateral.

By measurement, $\hat{Q} = 86°$, $\hat{R} = 78°$ and $\hat{S} = 81°$.

## Example 6

*Construct a quadrilateral PQRS where PQ = 6.5 cm, QR = 4.8 cm, RS = 8.5 cm, $\hat{Q} = 75°$ and $\hat{R} = 98°$. Measure the length of PS and the angles P and S.*

**Solution**

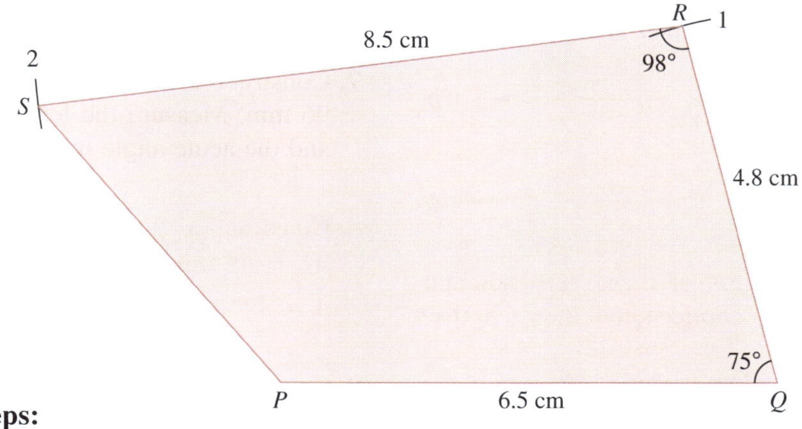


**Construction steps:**

1. Draw a line segment *PQ* of length 6.5 cm.
2. Using a protractor, construct $\hat{Q} = 75°$ and produce the arm of *Q*.
3. With *Q* as centre and radius 4.8 cm, draw an arc 1 to cut the produced arm of $\hat{Q}$ at *R*.
4. Using a protractor, construct $\hat{R} = 98°$ and produce the arm of $\hat{R}$.
5. With *R* as centre and radius 8.5 cm, draw an arc 2 to cut the produced arm of $\hat{R}$ at *S*.
6. Join *PS* and *PQRS* is the required quadrilateral.

By measurement, *PS* = 4.8 cm, $\hat{P} = 131°$ and $\hat{S} = 56°$.

## Explorations

Draw a rectangle of any size. Use your ruler to locate the midpoints of the sides carefully. Join these midpoints to form a new quadrilateral. What is the name of the quadrilateral you have obtained? Repeat the above by drawing

(a) a trapezium;
(b) a parallelogram;
(c) a kite;
(d) a rhombus;
(e) a quadrilateral of 4 unequal lengths.

What conclusion can you draw?

# Exercise 16b

1. In the figure, $AB$ and $PQ$ are parallel lines. Draw two perpendicular lines from points $X$ and $Y$, on $AB$, to $PQ$. Similarly, draw two perpendicular lines from $H$ and $K$ to $AB$.

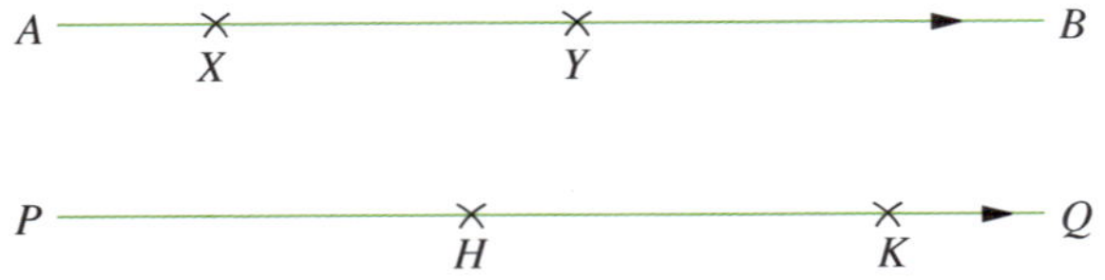


Measure the lengths of these perpendicular lines. What do you notice about their lengths?

2. Draw the following triangle $ABC$ with the given dimensions accurately and measure the angles marked $a$ and $b$.

Using a set square, draw a line from $C$ touching $AB$ and perpendicular to $AB$. Also, draw a line from $B$ touching $AC$ and perpendicular to $AC$. Measure the length of these perpendicular lines. Using a set square, draw the line through $A$ touching $BC$ and perpendicular to $BC$. Do these three perpendicular lines meet at the same point?

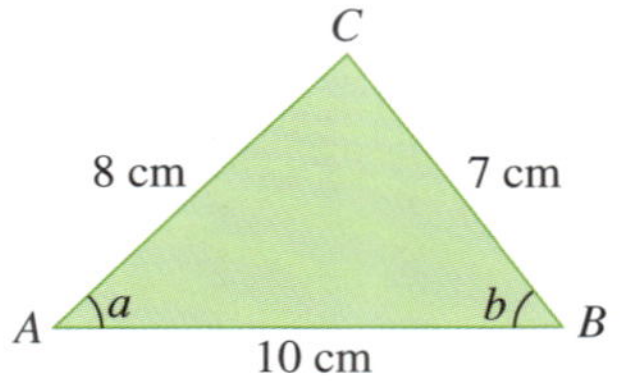


3. Construct a parallelogram $ABCD$ with $AB$ = 10 cm, $BC$ = 12 cm and $\hat{B} = 80°$. Measure $BD$.

4. Construct a trapezium $ABCD$ where $AB$ = 5.6 cm, $BC$ = 11.2 cm, $\hat{B} = 80°$ and $\hat{C} = 70°$. Measure $AC$.

5. Construct a rhombus $ABCD$ where $AB$ = 7.5 cm and $AC$ = 12 cm. Measure the length of the other diagonal.

6. Construct a rhombus $PQRS$ where $PQ$ = 6 cm and $\hat{Q} = 115°$. Measure the length of the diagonals.

7. Construct a rectangle of sides 84 mm and 96 mm. Measure the length of the diagonals and the acute angle made by these diagonals.

8. Construct a quadrilateral $ABCD$ given that $AB$ = 65 mm, $BC$ = 46 mm, $AD$ = 58 mm, $\hat{A} = 105°$ and $\hat{B} = 120°$. Measure $AC$ and $BD$.

9. Construct a quadrilateral $ABCD$ given that $AB$ = 5.3 cm, $BC$ = 6.3 cm. $CD$ = 6.7 cm. $\hat{B} = 75°$ and $\hat{C} = 60°$. Measure $AD$.

10. Construct a quadrilateral $PQRS$ where $PQ$ = 5.6 cm, $\hat{Q} = 80°$, $\hat{R} = 95°$, $QR$ = 6.2 cm and $RS$ = 9.2 cm. Measure $PS$.

11. Construct a quadrilateral $PQRS$ given that $PQ = PR = PS$ = 90 mm, $RS$ = 75 mm and $QR$ = 120 mm. Measure $Q\hat{P}S$.

12. Construct a quadrilateral $PQRS$ where $PS$ = 6 cm, $RS = QR$ = 9 cm and $P\hat{S}R = Q\hat{R}S = 110°$. Measure $PQ$.

13. The figure, not drawn to scale, shows a square $OPQR$ where $OQ$ = 10 cm and $OP = PQ = QR = OR$.
    - **(a)** Draw the diagram $OQ$ = 10 cm.
    - **(b)** Construct the perpendicular bisector of $OQ$.
    - **(c)** Locate the points $P$ and $R$ by using a pair of compasses.

(d) Measure and write down the length of $OP$.

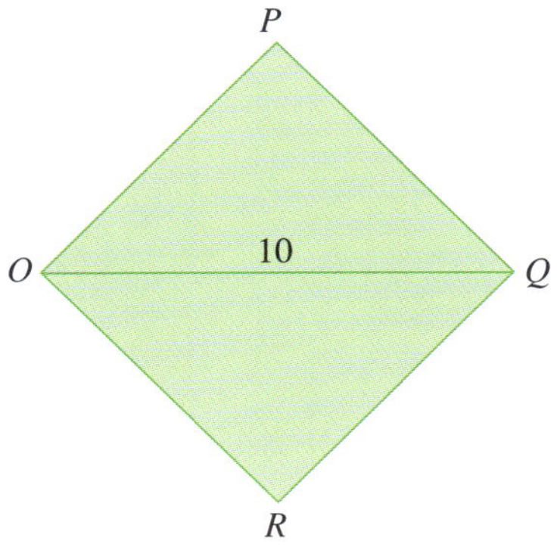


**14.** (a) Construct the trapezium $PQRS$ with $PQ = 5$ cm, $Q\hat{P}S = 120°$, $PS = 6$ cm, $SR = 9.8$ cm and $PQ$ parallel to $SR$.
(b) Measure and write down the length of $QR$.

**15.** (a) Construct the quadrilateral $ABCD$ in which the base $AB = 10$ cm, $BC = 6$ cm, $AD = 3.5$ cm, $A\hat{B}C = 45°$ and $B\hat{A}D = 60°$.
(b) Construct the perpendicular bisector of $BD$.
(c) Measure and write down the length of $BD$ and $CD$.

**16.** (a) Construct the quadrilateral $PQRS$ in which $PQ = 4$ cm, $SP = 3.6$ cm, $QR = RS = 4.8$ cm and $Q\hat{P}S = 90°$.
(b) Construct the perpendicular from $S$ to $QR$ and let it cut $QR$ at $X$.
(c) Measure and write down the length of $SX$.

**17.** (a) Construct the quadrilateral $ABCD$ such that $AB = 8$ cm, $AD = 6$ cm, $BC = 2$ cm, $B\hat{A}D = 60°$ and $A\hat{B}C = 90°$.
(b) Construct a line parallel to $AC$ and passing through $D$ to meet $BC$ produced at $P$.
(c) Measure and state the length of $BP$.

**18.** (a) Construct $\triangle ABC$ in which $AB = 4.5$ cm, $BC = 6$ cm and $A\hat{B}C = 60°$.
(b) Construct the angle bisector of $B\hat{A}C$ to meet $BC$ at the point $P$.

**19.** (a) Construct $\triangle ABC$ with $AB = 9$ cm, $BC = 7$ cm and $A\hat{B}C = 50°$.
(b) Measure and write down the length of $AC$ and the size of $B\hat{A}C$.
(c) Using ruler and compasses only, construct the angle bisector of $B\hat{A}C$ to meet $BC$ at $X$.

**20.** (a) Construct a parallelogram $PQRS$ in which $PQ = 9$ cm, $QR = 6$ cm and $P\hat{Q}R = 115°$.
(b) Measure and write down the length of $PR$.

**21.** (a) Construct $\triangle ABC$ in which $AB = 11$ cm, $BC = 8$ cm and $A\hat{B}C = 102°$.
(b) Measure and write down the length of $AC$.
(c) Measure and write down the size of $B\hat{A}C$.
(d) Construct the perpendicular bisector of $AC$ and let it cut $AC$ at $X$.

## SUMMARY

- When constructing geometrical figures, use a sharp hard pencil to draw points and lines finely and clearly.
- All construction lines must be clearly shown and do not erase the construction lines that you have drawn.

# Review Examples 16

## Example 1

*Construct, in a single diagram,*

*(a) the quadrilateral ABCD such that AB = 11 cm, AD = 7 cm, BC = 3.1 cm,* $B\hat{A}D = 60°$ *and* $A\hat{B}C = 90°$

*(b) a line parallel to AC and passing through D to meet BC produced at P.*

*Measure and write down the length of CP.*

### Solution

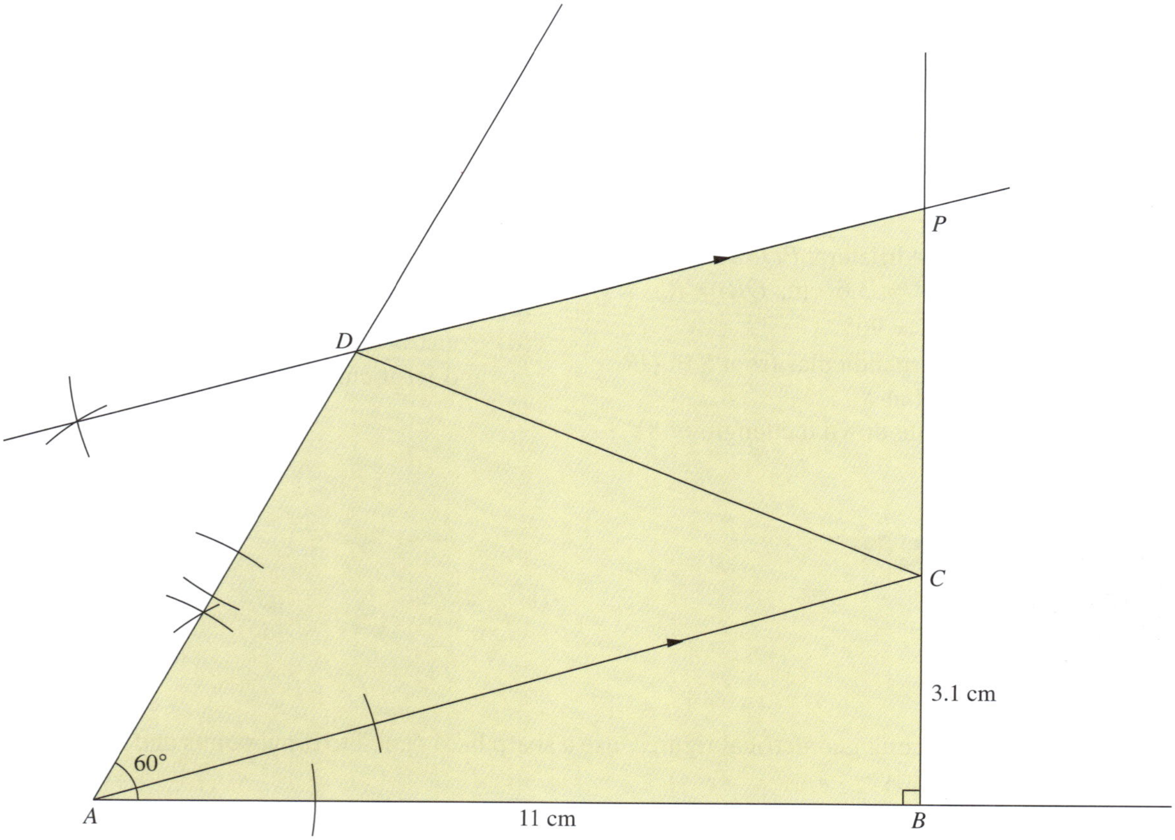


From the diagram, *CP* = 5 cm.

## Example 2

*Construct the following in a single diagram.*

(a) *Construct* $\triangle XYZ$ *in which* $XY = 10.2$ *cm,* $XZ = 11$ *cm and* $Z\hat{X}Y = 62°$.

(b) *Measure and write down the length of YZ.*

(c) *Construct a circle of radius 5 cm with the centre at Z.*

(d) *Construct the angle bisector of* $X\hat{Z}Y$ *and let this angle bisector cut the circle at point P inside the triangle XYZ and the point Q on XY.*

(e) *Measure and write down the length of PQ.*

### Solution

(b) From the diagram, $YZ = 10.9$ cm.

(e) From the diagram, $PQ = 4.7$ cm

# Review Questions 16

1. (a) Construct $\triangle ABC$ with $B\hat{A}C = 60°$, $AB = 7.5$ cm and $AC = 10$ cm.
   (b) Construct the perpendicular bisector of $AB$ and let it cut $AB$ at $X$ and $AC$ at $Y$.
   (c) Measure and write down the length of $XY$.
   (d) Measure and write down the size of $A\hat{C}B$.

2. Construct parallelogram $ABCD$ where $AB = 8$ cm, $A\hat{B}C = 120°$ and $BC = 5.5$ cm. Measure and write down the lengths of $AC$ and $BD$.

3. (a) Construct $\triangle PQR$ with $PQ = 9.8$ cm, $QR = 6.5$ cm and $P\hat{Q}R = 60°$;
   (b) Construct the perpendicular bisector of $PQ$ and let it cut $PQ$ at $X$.
   (c) Measure and write down the length of $XR$.

4. (a) Construct $\triangle PQS$ such that $QS = 8$ cm, $P\hat{Q}S = P\hat{S}Q = 60°$;
   (b) Construct the perpendicular bisector of $QS$ which passes through $P$ and let $X$ be the point where it cuts $QS$;
   (c) Construct the point $R$ on $PX$ produced such that $PX = XR$.
   (d) Join $QR$ and $SR$ to obtain the quadrilateral.
   (e) State the special name of the quadrilateral $PQRS$.

5. (a) Construct $\triangle ABC$ where $AB = 8$ cm, $AC = 5$ cm and $BC = 6$ cm;
   (b) Construct the perpendicular bisector of $AB$;
   (c) Construct the perpendicular bisector of $BC$;
   (d) Construct a circle which will pass through $A$, $B$ and $C$.
   (e) Measure and write down the radius of this circle.

6. (a) Construct $\triangle PQR$ in which $PQ = 7$ cm, $QR = 8$ cm and $PR = 6$ cm;
   (b) Construct the angle bisector of $P\hat{Q}R$;
   (c) Construct the angle bisector of $P\hat{R}Q$;
   (d) Construct a circle which will touch the sides of $PQ$, $QR$ and $PR$.
   (e) Measure and write down the radius of this circle.

7. (a) Construct a circle with diameter $AC = 10$ cm;
   (b) Construct a point $B$ on the circumference of the circle such that $AB = BC$;
   *(c) Construct the point $D$ on the circumference of the circle, and on the side of $AC$ opposite to $B$ such that $C\hat{A}D = 60°$.
   (d) Measure and write the length of $CD$.
   (e) Construct the angle bisector of $B\hat{A}C$ and let this angle bisector cut the circle at $P$.
   (f) Measure and write down the length of $PD$.

## Revision Exercise IV No. 1

1. (a) A triangle has sides 6 cm, 8 cm and 9 cm. Find the length of the shortest altitude of the triangle by means of an accurate drawing.

   (b) Draw a triangle with sides 12 cm, 8 cm and 10.5 cm. Bisect any two of the angles and let the bisectors meet at $X$. Construct the perpendicular from $X$ to the longest side of the triangle. Measure this perpendicular.

2. (a) A bus is supposed to start at 12 25 and to reach its destination at 13 50. It starts 3 minutes late and arrives 10 minutes late. How long did it take to reach its destination?

   (b) In the given diagram, find the values of $x$, $y$ and $z$.

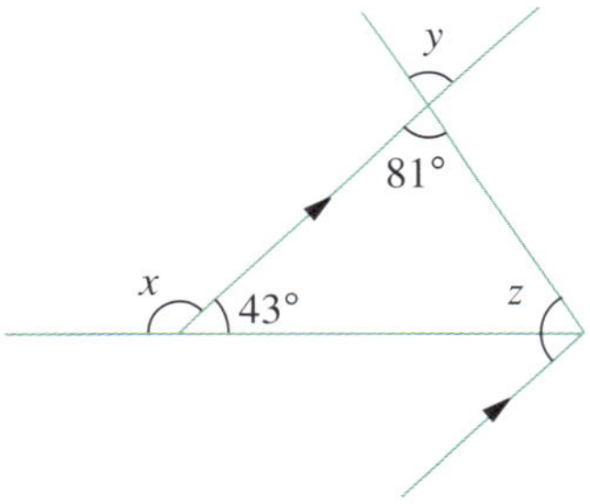


3. (a) A square piece of metal with side 14 cm costs \$4.90. How much will a similar piece of metal in the shape of a rectangle with sides 20 cm by 12 cm cost? (Assume that the cost of the metal is proportional to the area.)

   (b) The driver of a train averaging 32 km/h takes 3 hours for a journey. What is the length of the journey?
   How long would the driver of an express train averaging 48 km/h take for the same journey?

4. In the figure, $d = 90°$, show that
   (a) $a - c = 90°$, (b) $a + b = 180°$.

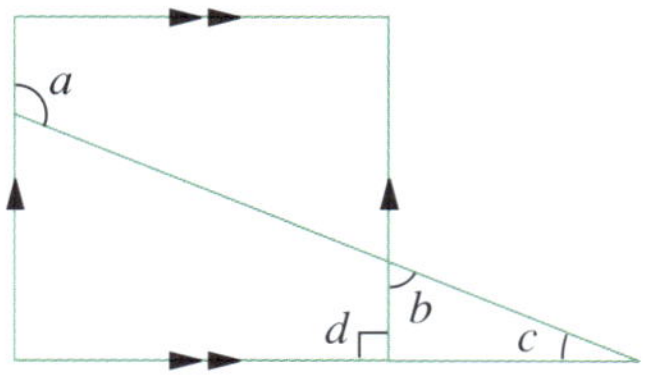


5. (a) Draw $\triangle PQR$ such that $PQ = 90$ mm, $QR = 100$ mm and $RP = 68$ mm. A point $T$ lies on $PQ$ such that $QT = 30$ mm. Draw a line through $T$ parallel to $QR$ to cut $RP$ at $S$. Measure the length of $RS$.

   (b) Construct $\triangle ABC$ such that $BC = 7.5$ cm, $AC = 6$ cm and $A\hat{C}B = 60°$. Also construct the angle bisector of $A\hat{B}C$. Measure the length of $AB$.

6. (a) A school librarian bought 56 books, some at \$3.50 each and others at \$4.50 each. If the total cost of these books was \$240, how many books that cost \$3.50 each were bought?

   (b) Four people contributed sums of money to the Community Chest of Singapore in the ratio 2 : 3 : 5 : 8. If the largest amount contributed is \$24, calculate the total amount contributed by the four people.

7. Solve the following equations:

   (a) $4 = \frac{1}{5}(2x - 3)$

   (b) $3x - 4 = 7 - 4(x - 5)$

8. Simplify each of the following:

   (a) $12x - 2[2x - 3(x - 5)]$

   (b) $\frac{x-2}{5} - \frac{2x-3}{4} + \frac{5-x}{10}$

**9.** In the figure, $ABC$ is a right-angled triangle with $A\hat{B}C = 90°$, $A\hat{C}B = 32°$ and $D\hat{A}B = D\hat{A}C$. Calculate

**(a)** $D\hat{A}B$,

**(b)** $A\hat{D}C$.

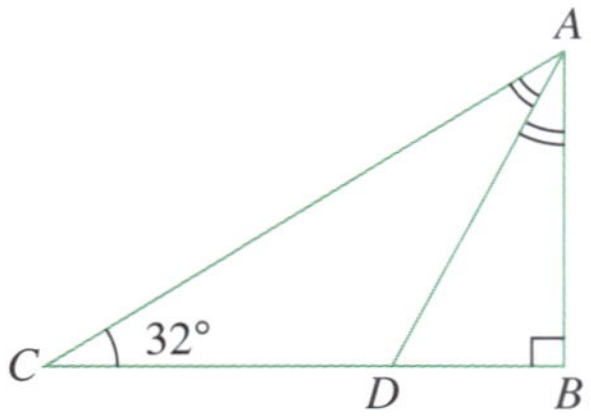


**10. (a)** Calculate the values of $x$ and $y$ in the diagram.

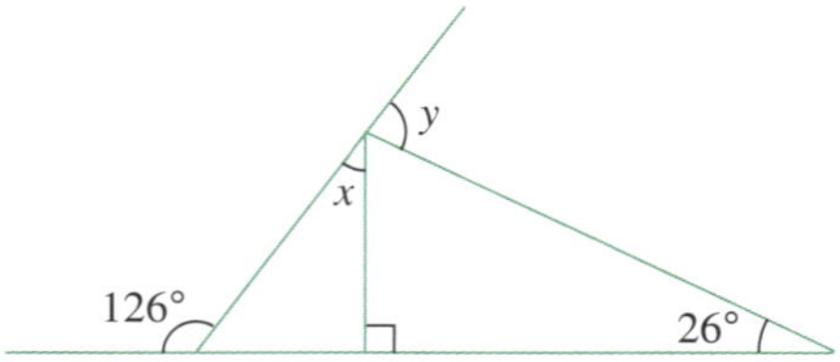


**(b)** The amount of pocket money Alvin received is just enough for him to buy 10 plates of chicken rice or 15 plates of fried noodles. If he wishes to buy equal number of plates of chicken rice and fried noodles, how many of each type can he buy with the money?

## Revision Exercise IV No. 2

**1. (a)** Draw $\triangle ABC$ with sides each of length 10.8 cm. Mark a point $P$ on $AB$ such that $PA = 2.4$ cm. On $BC$, mark a point $Q$ such that $QC = 8.6$ cm. Measure $PQ$.

**(b)** Construct a rhombus of side 6 cm and one of its diagonals 8 cm. Measure the length of the other diagonal.

**2. (a)** Two trains 245 m and 315 m long are travelling towards each other at 90 km/h and 54 km/h respectively on parallel lines. How long do the trains take to pass one another from the time they meet each other?

$\left(\textbf{Note: } 1 \text{ km/h} = \frac{5}{18} \text{ m/s}\right)$

**(b)** A father is four times as old as his son. Five years ago, the sum of their age was 70. Find their present ages.

**3.** In $\triangle ABC$, $A\hat{B}C = 64°$, $AB = AC$ and $BC$ is produced to $D$. Calculate

**(a)** $B\hat{A}C$; **(b)** $A\hat{C}D$.

**4.** In the given figure, $AX = XC$, $A\hat{X}C = 130°$ and $A\hat{B}C = 56°$. Calculate the value of

**(a)** $C\hat{A}X$; **(b)** $B\hat{A}X$.

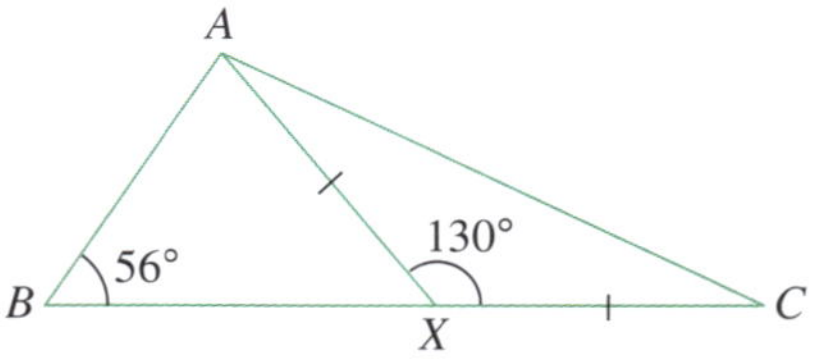


**5.** Find the size of the unknown angles marked $x$ and $y$ on the following diagrams:

**(a)**

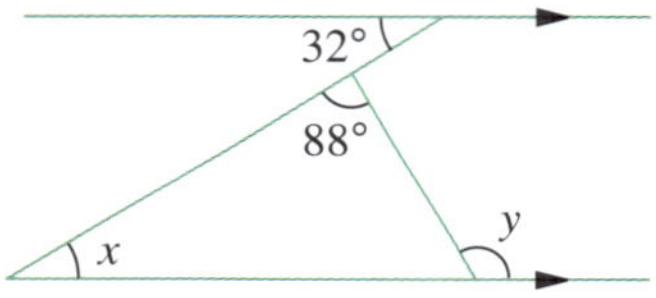


**(b)**

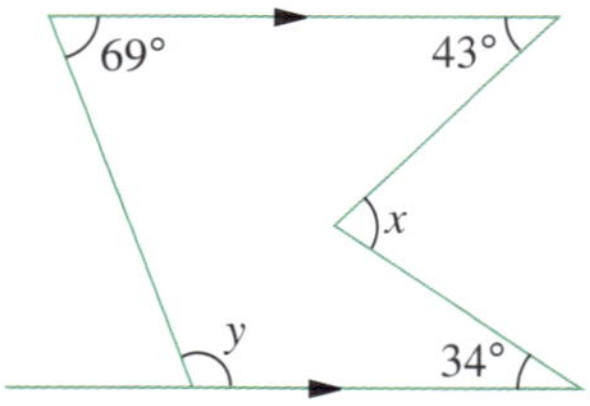


**6. (a)** Construct a parallelogram with diagonals 7.5 cm and 10.2 cm, and with the shorter sides 3.6 cm long. Measure the length of the longer sides.

**(b)** Construct $\triangle ABC$ where $AB = 8$ cm, $BC = 7$ cm and $CA = 10$ cm. Construct a point $X$ inside the triangle such that it is lying on the bisector of $A\hat{B}C$ and is 3 cm from $AC$.

**7.** **(a)** If $x\%$ of 300 is equal to 15% of 220, find $x$.

**(b)** Express 5.4 km as a percentage of 6.4 km
**(i)** exactly;
**(ii)** in decimal form correct to 3 significant figures.

**8.** Find the angles marked $p$, $q$, $r$ and $s$ in the figure.

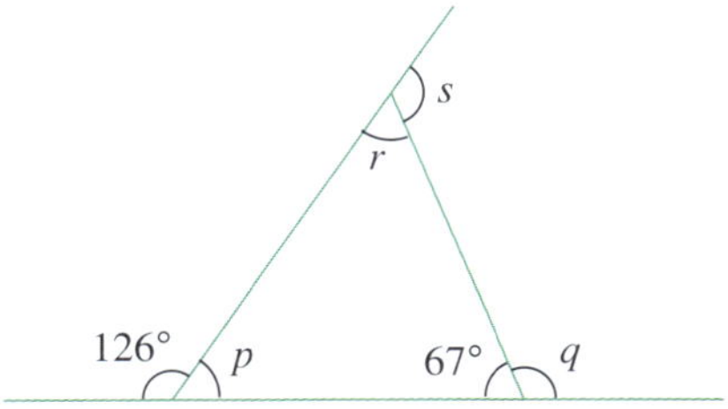


**9.** **(a)** The figure shows a pattern made up of straight lines, semicircles and arcs of quadrants of circles. Calculate the total area of the shaded regions.
(Take $\pi = 3.142$)

**(b)** A motorcycle wheel has a diameter of 42 cm. Find the number of complete revolutions it makes when it moves a distance of 1.25 km.

**10.** In a school, the pupils are divided into three groups, $A$, $B$ and $C$. $\frac{1}{3}$ of the pupils are in group $A$, $\frac{5}{12}$ in group $B$ and the rest in group $C$. If there are 150 pupils in group $C$, find
**(a)** the total number of pupils in the school;
**(b)** the ratio of the number of pupils in group $A$ to the number of pupils in group $B$.

## Revision Exercise IV No. 3

**1.** Construct a parallelogram with one of its diagonals 100 mm and two of its sides 48 mm and 109 mm. Measure the length of the other diagonal.

**2.** Construct a trapezium $PQRS$ in which $PQ$ is parallel to $SR$, $P\hat{S}R = 90°$, $QP = 3.6$ cm, $PS = 7.2$ cm and $SR = 12.6$ cm. Measure $QR$ and $P\hat{Q}R$.

**3.** Find the angles marked $p$, $q$ and $r$ in the following figures.

**(a)**

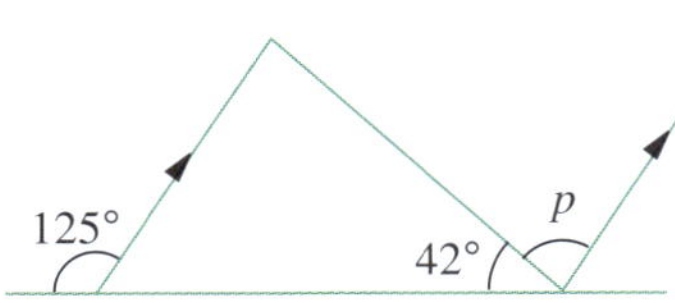


**(b)**

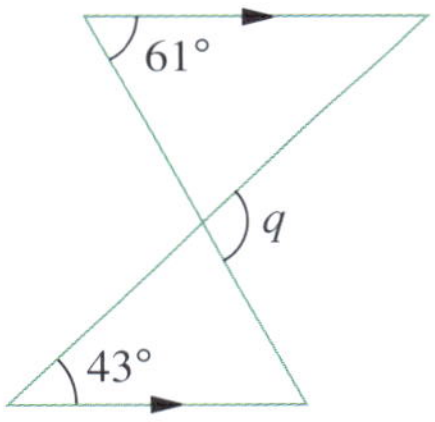


**(c)**

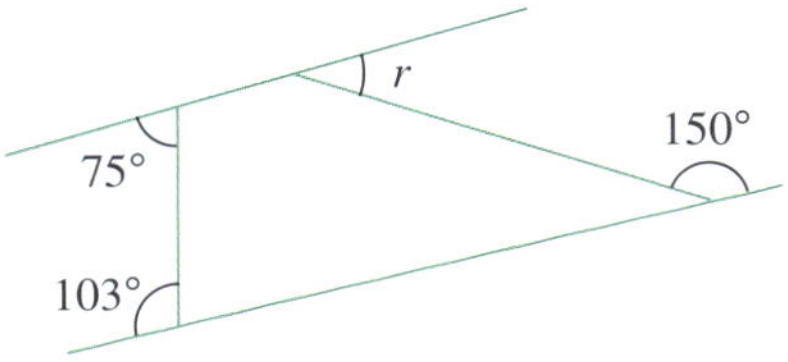


**4.** **(a)** The capacity of a tank is 60 000 litres. It is being filled at a rate of 1250 litres per hour. How long will it take to fill $\frac{5}{8}$ of the tank?

**(b)** A clock gains 12 min 15 s in one week. How many days will it take to gain 3 h 30 min?

**5.** Construct a quadrilateral $ABCD$ where $AB = CD = 8$ cm, $BC = 6$ cm, $A\hat{B}C = 90°$ and $B\hat{C}D = 130°$. Measure the length of $AD$.

6. (a) Find the unknown angles $x$, $y$ and $z$ in the figure below.

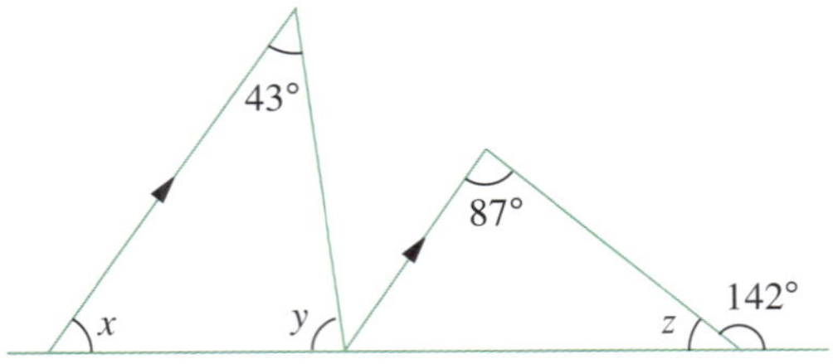


(b) Given that $7x = 2y$, find the ratio $x : y$.

7. Solve the following equations:
(a) $10 - 3y = 21 - 5y$
(b) $11x + 10 - 6x = 23 + 4x - 9$
(c) $\frac{4x - 2}{3} + \frac{2x + 17}{6} = \frac{7x - 3}{2}$

8. (a) A man cycles at $v$ km/h. Find the distance he travels in $t$ hours. How long will it take him to travel $s$ km?

(b) A car uses 1 litre of petrol to travel 9 km. If 1 litre of petrol costs \$1.14, find the cost needed for the car to travel a distance of 135 km.

9. (a) Find the angles marked $x$, $y$ and $z$ on the figure.

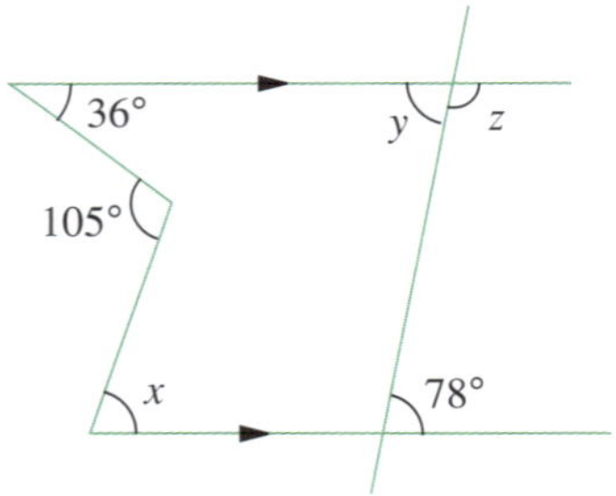


(b) Construct a parallelogram of sides 6.9 cm and 11.1 cm, and whose longer diagonal is 14.4 cm. Measure the length of the other diagonal.

(c) $ABCD$ is a parallelogram in which $A\hat{D}C = 108°$. Given that $KB$ is perpendicular to $BC$ and $K\hat{C}D = 23°$, find the values of $x$, $y$ and $z$.

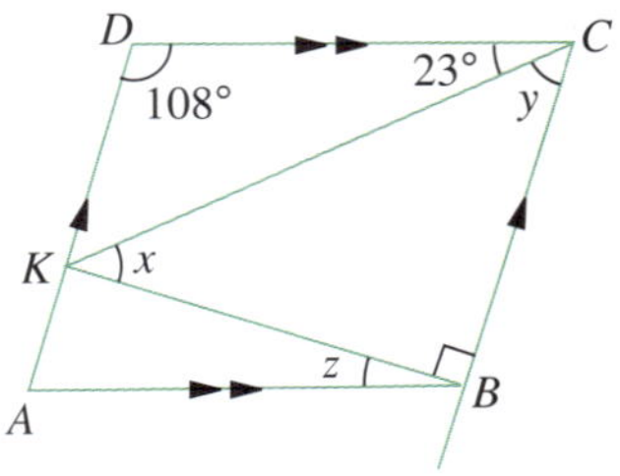


10. (a) The sum of three consecutive odd numbers is 63. Find the numbers.

(b) A woman is 7 times as old as her daughter. If in 5 years' time, she is 4 times as old as her daughter, what are their present ages?

(c) A housewife buys 11 m of cloth at \$8.20 per metre. How much change will be given to her if she pays with 2 fifty-dollar notes?

(d) Find the cost of painting the outer surface of a closed rectangular tank 3.6 m long, 2.5 m wide and 2 m deep with paint at 75 cents per square metre.

## Revision Exercise IV No. 4

1. (a) Construct a parallelogram with sides 7.5 cm and 9 cm and with one of its interior angles as 62°. Measure its diagonals.

(b) Construct $\triangle PQR$ such that $PQ = 10$ cm, $QR = 9$ cm and $RP = 8$ cm. A point $S$ on $PR$ is 2 cm from $R$. Draw a line through $S$ parallel to $RQ$ to cut $PQ$ at $T$. Measure $PT$.

2. Find, in the figure, the angles marked $x$ and $y$.

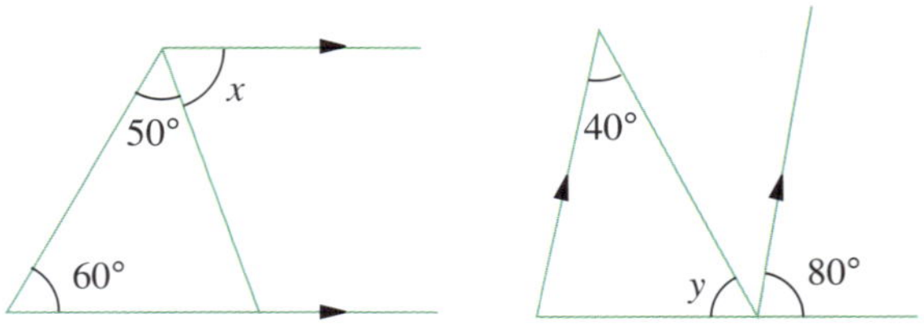

3. (a) 12 men takes 5 days to build a road 200 m long. How many days will 20 men take to build a road 400 m long?

(b) To ensure the long term availability of fresh water for its citizens, the Singapore Government decided to build a desalination plant scheduled for completion in the year 2004. Due to the Asian economic crisis in 1998, the cost of building the plant has dropped from an initial projected cost of \$1.2 billion to about \$912 million. Find the percentage savings for building the plant.

4. In the figure, $ABCD$ is a straight line, $BE = BC$, $A\hat{D}E = 35°$ and $A\hat{B}E = 140°$. If EC = 5 cm, find $CD$.

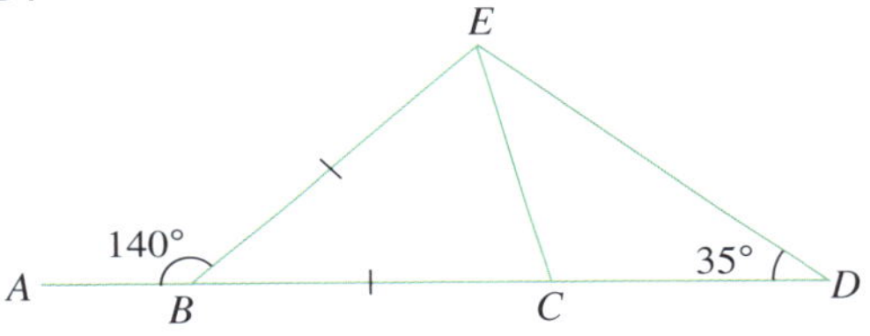


5. (a) In an isosceles triangle the base angle is 12° greater than the vertical angle. Find the size of the vertical angle.

(b) Construct a rhombus of side 8 cm with one of its interior angles measuring 70°. Measure the length of the diagonals.

6. Simplify

(a) $2(5x - y) - 3(6x - 5y) + (x - 7y)$;

(b) $\dfrac{x + 2}{3} - \dfrac{x - 5}{5} + \dfrac{2 - x}{4}$.

7. In the diagram, $AC$ is parallel to $PQ$, $CD = CB$, $B\hat{A}C = 44°$ and $C\hat{B}Q = 68°$. Calculate

(a) $A\hat{B}C$, (b) $A\hat{C}D$.

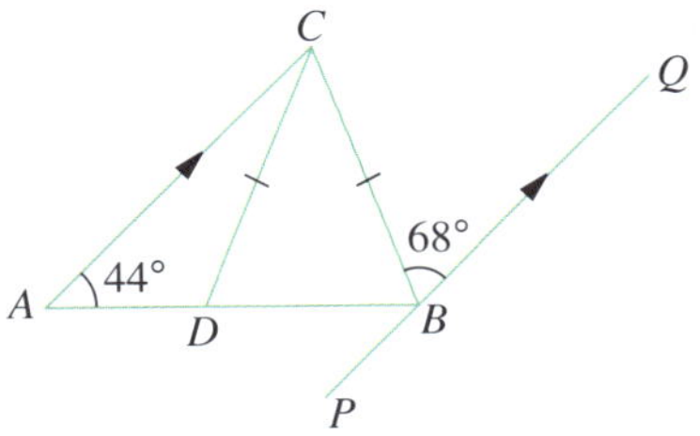


8. A motorist travels $x$ km in $t$ hours. If he travels at the same speed, how far will he travel in

(a) 3 hours;

(b) $y$ hours?

9. Find the value of $x$ in each of the following diagrams:

(a)

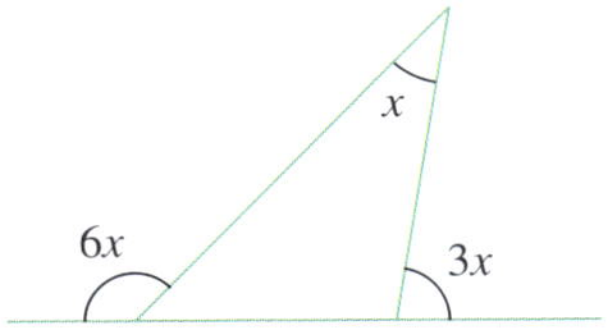


(b)

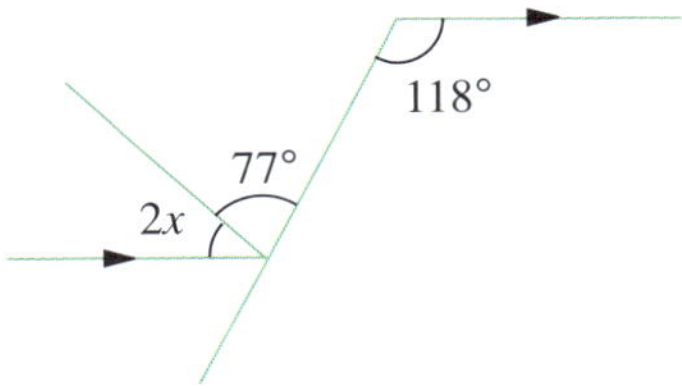


10. Construct the following in a single diagram:

(a) $\triangle ABC$ in which $AB = 8$ cm, $A\hat{B}C = 65°$ and $B\hat{A}C = 42°$.

(b) The perpendicular from $C$ to $AB$.

(c) The point $E$ such that $CE$ is parallel to $AB$ and $C\hat{B}E = 58°$.

## Revision Exercise IV No. 5

1. In the figure, $AQ = BQ$ and $PQ$ is parallel to $BC$. If $Q\hat{B}C = 28°$ and $Q\hat{C}B = 52°$, find $A\hat{P}Q$.

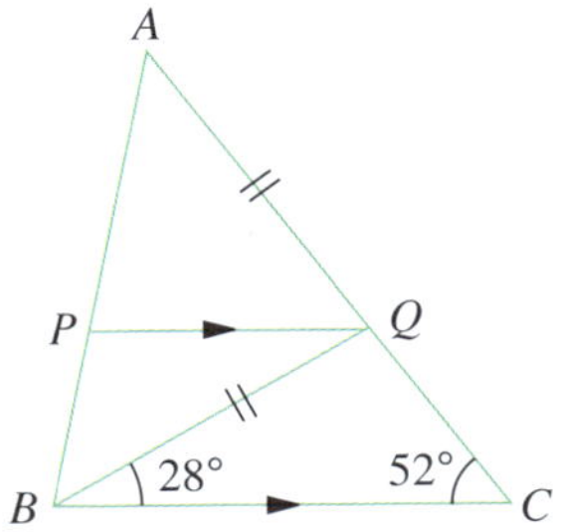


2. Construct, in a single diagram,

(a) $\triangle ABC$ where $AB = 9$ cm, $BC = 8$ cm and $AC = 7$ cm;

(b) a point $D$ such that $CD$ is parallel to $AB$ and $CD = 5$ cm;

(c) the perpendicular from $C$ to $AB$.

3. From a square wooden disc of side 10 cm, a circular disc of radius 5 cm is cut out. What percentage of wood remains?
(Take $\pi = 3.14$)

4. (a) In the figure, $BCD$ is a straight line and $\triangle ABC$ is an equilateral triangle. Given that $A\hat{D}C = 35°$ and $D\hat{A}C = x°$, find $x$.

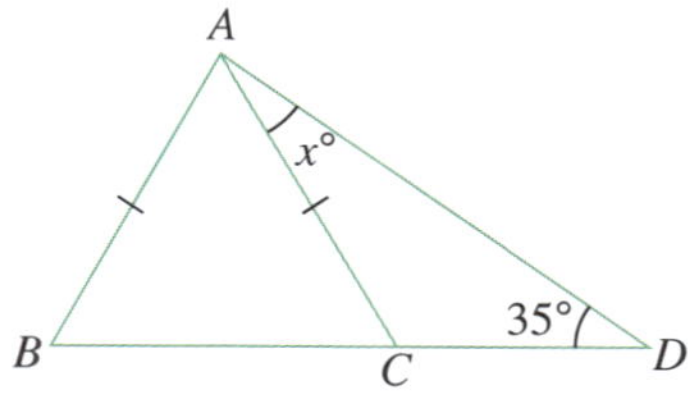


(b) The diagram shows a trapezium $ABCD$ with $AB$ parallel to $DC$. Given that $AB = 6$ cm, $DC = 12$ cm and the height between the parallel sides equals 9 cm, calculate the area of $ABCD$.

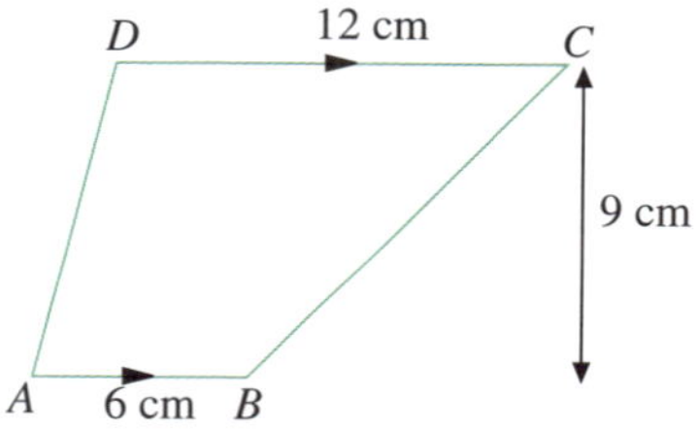


5. (a) Find the angles marked $x$ and $y$ in the figure.

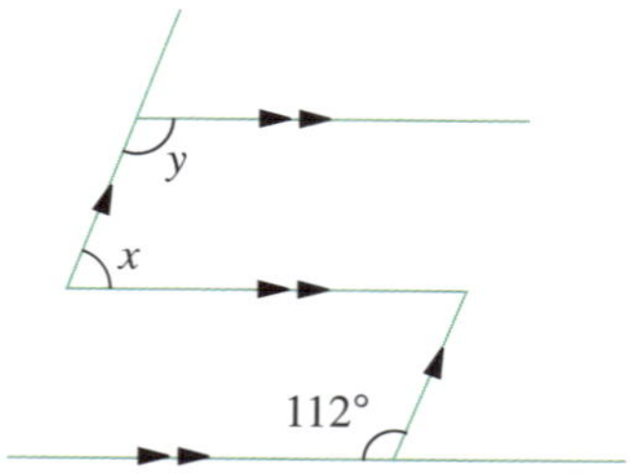


(b) Find $x$, $y$ and $z$ in the given figure.

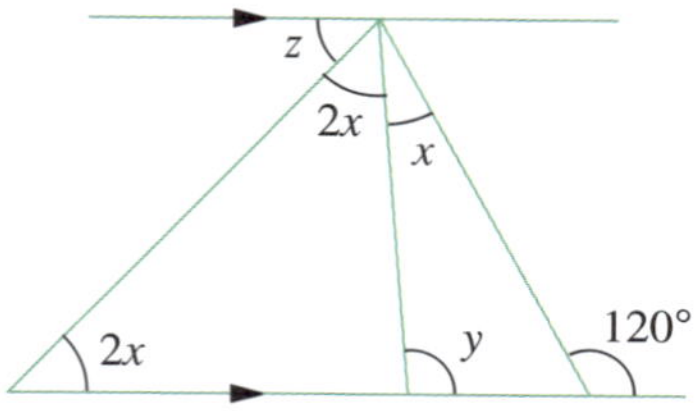


6. (a) A man uses a ruler to measure a rectangle 45 cm by 36 cm. His result is 44.6 cm by 35.8 cm. Find his error per cent in the value he obtains for the perimeter.

(b) A cylinder $P$ has radius $r$ cm and height $2h$ cm. A second cylinder $Q$ has radius $3r$ cm and height $h$ cm. Find the ratio of $\dfrac{\text{volume of } P}{\text{volume of } Q}$.

7. A man cycles at an average speed of $v$ km/h.
(a) Find the distance he travels in 5 hours.
(b) Find the time needed to travel $s$ km.

8. Express
(a) $c$ cents in dollars,
(b) $m$ millimetres in metres,
(c) $k$ kilograms in grams.

9. (a) $ABC$ is a triangle in which $\hat{B} = \hat{C} = 70°$. $AC$ is produced to $P$ and the bisector of $B\hat{C}P$ cuts $AB$ produced at $Q$. Find $A\hat{Q}C$.

(b) Construct a rhombus of side 7 cm with one of the diagonals 11 cm. Measure the other diagonal. Measure also one of the acute angles.

10. (a) Solve the equation
$$\frac{2}{3}(4x - 1) - \frac{5}{6}(2x + 1) = \frac{1}{2}.$$

(b) If water is running into a tank at the rate of 3 metres per second along a trough of rectangular cross-section 18 cm wide and 15 cm deep, find the amount of water that runs into the tank in 1 minute.

(c) If $3(x - 1) - 5(x - 3) = 3$, find the value of $7x - 5$.

# Answers

## Exercise 1a

1. (a) 1, 2, 4, 8, 16
   (b) 1, 2, 4, 7, 14, 28
   (c) 1, 2, 3, 4, 6, 8, 12, 16, 24, 32, 48, 96
   (d) 1, 2, 4, 5, 10, 20, 25, 50, 100
   (e) 1, 2, 3, 4, 5, 6, 8, 10, 12, 15, 20, 24, 30, 40, 60, 120
   (f) 1, 2, 3, 5, 6, 7, 10, 14, 15, 21, 30, 35, 42, 70, 105, 210
2. (a) 4, 8, 12, 16, 20, 24
   (b) 7, 14, 21, 28, 35, 42
   (c) 9, 18, 27, 36, 45, 54
   (d) 12, 24, 36, 48, 60, 72
   (e) 17, 34, 51, 68, 85, 102
   (f) 21, 42, 63, 84, 105, 126
3. 54, 126, 198
4. 1, 2, 3, 4, 8, 9, 12, 16, 48, 144
5. 24, 32, 56, 72, 64, 40, 96, 120
6. 4, 14, 28, 32, 56
7. (a) 1, 2, 3, 4, 5, 6, 8, 10, 12, 15, 16, 20, 24, 30, 32, 40, 48, 60, 80, 96, 120, 160, 240, 480
   (b) 1, 2, 3, 4, 5, 6, 8, 10, 12, 15, 20, 24, 25, 30, 40, 50, 60, 75, 100, 120, 150, 200, 300, 600
   (c) 1, 2, 3, 4, 5, 6, 8, 10, 12, 15, 16, 20, 24, 30, 32, 40, 48, 60, 64, 80, 96, 120, 160, 192, 240, 320, 480, 960
   (d) 1, 2, 3, 4, 6, 8, 9, 12, 18, 24, 26, 36, 39, 52, 78, 104, 117, 156, 234, 312, 468, 936
8. (a) 2 bags of 24 gums, 3 bags of 16 gums, 4 bags of 12 gums, 6 bags of 8 gums, 8 bags of 6 gums, 12 bags of 4 gums, 16 bags of 3 gums, 24 bags of 2 gums.
   (b) 3 bags of 15 gums, 5 bags of 9 gums, 9 bags of 5 gums, 15 bags of 3 gums
   (c) 3 bags each contains 16 orange-flavoured gums and 15 lime-flavoured gums
9. Prime numbers: (a), (c), (f);
   Composite numbers: (b), (d), (e)
10. 31, 37, 41, 43, 47
11. 2, 3 or 2, 5 or 7, 9 etc...

## Exercise 1b

1. (a) $7^2$ (b) $2^2 \times 5^2$
   (c) $3 \times 7^3$ (d) $5^2 \times 11^3$
   (e) $2^3 \times 13^2 \times 31$
   (f) $5^3 \times 19^2 \times 23 \times 29^2$
2. (a) $2^2 \times 7$ (b) $2^4 \times 3$
   (c) $2 \times 3^3$ (d) $2^3 \times 11$
   (e) $2^2 \times 3^3$ (f) $2^4 \times 3^2$
   (g) $2^6 \times 3$ (h) $2^8$
3. (a) $2 \times 2 \times 2$ (b) $2 \times 2 \times 3 \times 3$
   (c) $2 \times 3 \times 5$
   (d) $2 \times 2 \times 2 \times 3 \times 3$
4. (a) $2^2 \times 5^2$ (b) $5^3$
   (c) $3 \times 7^2$ (d) $2^3 \times 3^3$
   (e) $3^2 \times 5^2$ (f) $2^3 \times 3^2 \times 5$
   (g) $3^4 \times 7$ (h) $2^3 \times 3^4$

## Exercise 1c

1. (a) 3 (b) 3
   (c) 3, 9 (d) 3, 5, 15
2. (a) 6 (b) 15
   (c) 8 (d) 8
   (e) 12 (f) 28
   (g) 9 (h) 4
3. (a) 9 (b) 21
   (c) 12 (d) 15
   (e) 12 (f) 180
   (g) 350
4. (a) 30 cm (b) 12
5. (a) 14 cm (b) 37

## Exercise 1d

1. (a) 21 (b) 18
   (c) 21 (d) 36
   (e) 72 (f) 1755
   (g) 300 (h) 144
   (i) 250 (j) 4410
2. (a) 90 (b) 126
   (c) 5236 (d) 61 425
   (e) 72 (f) 72
3. (a) 1 417 500 (b) 16 200
   (c) 138 600
4. (a) 6, 126 (b) 7, 84
   (c) 13, 78 (d) 70, 420
   (e) 15, 450 (f) 112, 672
5. 8.01 p.m.
6. 5 p.m.
7. (a) 420 (b) 8
8. (a) 4 (b) 1

## Exercise 1e

1. 64, 81, 100, 121, 144
2. 16, 25, 49, 81, 100, 1, 144, 169, 225, 400
3. (a) 8 (b) 12
   (c) 17 (d) 60
4. (a) 6 (b) 9
   (c) 14 (d) 16
   (e) 21
5. (a) 34 (b) 36
   (c) 42 (d) 99
   (e) 105 (f) 186
6. (a) 11 (b) 19
   (c) 13 (d) 12
7. (a) $3^3 \times 5^3$; 15
   (b) $2^{12}$, 16
   (c) $2^6 \times 7^3$; 28
   (d) $2^6 \times 3^6$; 36
   (e) $3^6 \times 5^3$ ; 45
   (f) $2^9 \times 3^6$; 72
8. 3136 $cm^2$
9. 48 cm
10. 1331 $cm^3$
11. 14 cm

## Exercise 1f

1. (a) 1600 (b) 3600
   (c) 12 100 (d) 27 000
   (e) 64 000 (f) 1 000 000
   (g) 40 000 (h) 27 000 000
2. (a) 6 (b) 3
   (c) 8 (d) 5
   (e) 9 (f) 12
   (g) 10 (h) 30
3. (a) 676 (b) 1369
   (c) 6084 (d) 9801
   (e) 15 129 (f) 2197
   (g) 24 389 (h) 39 304
   (i) 300 763 (j) 1 295 029
   (k) 31 (l) 47
   (m) 59 (n) 106
   (o) 263 (p) 16
   (q) 41 (r) 68
   (s) 91 (t) 124
   (u) 1650 (v) 18 717
   (w) 15 704 (x) 426
   (y) 103
4. (a) Yes (b) Yes
   (c) No (d) No

## Review Questions 1

1. (a) F (b) F (c) T (d) F (e) F (f) T (g) F
2. (a) 18 (b) 6 (c) 13 (d) 5
3. (a) 60 (b) 144 (c) 630 (d) 2640
4. (a) 6 (b) 4 (c) 8
5. (a) 45 (b) 120 (c) 7689 (d) 8165 (e) 1183
6. 30 m
7. 8 cm
8. (4, 120), (12, 40), (20, 24), (8, 60)
9. (a) 70 (b) 84
10. 532 $(19 \times 7 \times 2 \times 2)$

## Exercise 2a

1. (a) +20 (b) +90 (c) −9
2. (a) 30 m above sea-level
   (b) −35
   (c) rotating 30° anticlockwise
   (d) speed of 45 km/h of a car moving towards west
3. (a) −1 (b) +2 (c) −4
4. (a) < (b) > (c) < (d) < (e) > (f) <
5. (a) $-5° < 12°$
   (b) $\$200 > -\$120$
   (c) $-40$ m $< -25$ m
6. (a) −90, −60, −6, −2, 0
   (b) −1500, −1, 1, 2
8. (a) > (b) < (c) > (d) > (e) < (f) >

## Exercise 2b

1. (a) 8 (b) 8 (c) 7 (d) 11 (e) −15 (f) −26 (g) −41 (h) −15 (i) −99
2. (a) −9 (b) −14 (c) 25 (d) 13 (e) −35 (f) 56 (g) 26 (h) −153
3. 9°C
4. (a) $(6000 − 2000 − 5500)
   (b) $1500 loss

## Exercise 2c

1. (a) 0 (b) −1 (c) −4 (d) 8 (e) 1 (f) 8 (g) 17 (h) −13 (i) −50
2. (a) −24 (b) −41 (c) −21 (d) −368 (e) −7 (f) 96 (g) −29 (h) −152 (i) −50

## Exercise 2d

(a) −21 (b) 6 (c) 280 (d) 16 (e) 36 (f) −2 (g) 61 (h) −55 (i) 192 (j) 0

## Exercise 2e

1. (a) ×, × (b) +, + (c) +, + (d) ×, ×
2. (a) −28, 60 (b) 12, −120 (c) −14, −50 (d) 8, 9, 11 (e) 10, −20 (f) −15, −9, 90
3. (a) 13 (b) −80 (c) 4 (d) 125 (e) 40 (f) −18 (g) 28 (h) 0 (i) −4 (j) −9 (k) −17

## Review Questions 2

1. (a) 67 (b) −28 (c) −22 (d) −100 (e) −36 (f) 390 (g) 1130 (h) −279
2. (a) −120 (b) −12 (c) 56 (d) 3 (e) 10 (f) −24 (g) 6 (h) 54 (i) 87
3. (a) 28 (b) −88 (c) −26 (d) −4 (e) 3 (f) −102
4. (a)

```
   324
 ×  57
 18468
```

(b)

```
  10503
 − 9726
    777
```

(c)

```
       68
 28 ) 1904
      168
       224
       224
         0
```

5.

```
        231
 117 ) 27027
       234
        362
        351
         117
         117
           0
```

6.

```
      (1)
    (5) (6)
  (9)     (7)
(2) (4) (8) (3)
```

7. (b) $(3 + 3) \div 3 + 3 - 3 = 2$
   (c) $3 + 3 - 3 - 3 + 3 = 3$
   (d) $(3 + 3 + 3 + 3) \div 3 = 4$
   (e) $3 + 3 \div 3 + 3 \div 3 = 5$
   (f) $3 + 3 + (3 - 3) \times 3 = 6$
   (g) $3 \times 3 - (3 + 3) \div 3 = 7$
   (h) $3 + 3 + (3 + 3) \div 3 = 8$
   (i) $3 + 3 + 3 + 3 - 3 = 9$
   (j) $3 + 3 + 3 + 3 \div 3 = 10$

## Exercise 3a

1. (a) $\frac{6}{17}$ (b) $\frac{7}{8}$
2. Peter
3. Susan
4. (a) $\frac{5}{8}, \frac{3}{4}, \frac{11}{12}$
   (b) $\frac{14}{33}, \frac{7}{11}, \frac{15}{22}$
5. (a) $\frac{11}{12}, \frac{4}{5}, \frac{3}{4}, \frac{7}{10}$
   (b) $\frac{5}{6}, \frac{7}{9}, \frac{13}{18}, \frac{2}{3}$

## Exercise 3b

1 (a) $3\frac{4}{5}$ (b) $5\frac{43}{56}$ (c) $3\frac{29}{200}$ (d) $3\frac{9}{10}$ (e) $4\frac{1}{15}$
2. (a) $1\frac{1}{14}$ (b) $1\frac{7}{8}$ (c) $5\frac{1}{60}$ (d) $\frac{29}{45}$
3. $4\frac{23}{60}$ km
4. $4\frac{19}{48}$ litres

5. (a) $2\frac{1}{3}$ hours or 2 hours and 20 minutes
(b) first classroom
(c) $\frac{1}{6}$ hour or 10 minutes

6. (a) $5\frac{1}{8}$ cups (b) $1\frac{5}{8}$ cups

7. $4\frac{1}{2}$ hours

## Exercise 3c

1. (a) 6 (b) $1\frac{7}{128}$
(c) $-5\frac{1}{3}$

2. (a) $\frac{9}{14}$ (b) $10\frac{1}{2}$ kg
(c) $3\frac{3}{4}$ kg
(d) (i) $5\frac{5}{8}$ kg
(ii) $1\frac{1}{8}$ kg

3. (a) $4\frac{2}{3}$ hours or 4 hours and 40 minutes
(b) (i) $9\frac{4}{5}$ hours or 9 hours and 48 minutes
(ii) $1\frac{19}{30}$ hours or 1 hour and 38 minutes

4. $\frac{12}{35}$

## Exercise 3d

1. (a) $1\frac{7}{10}$ (b) $\frac{13}{90}$
(c) $\frac{2}{9}$ (d) 10
(e) 0

2. (a) $\frac{3}{4}$ (b) –1
(c) $-1\frac{1}{10}$ (d) $-\frac{11}{12}$
(e) $-\frac{13}{20}$ (f) $\frac{3}{20}$
(g) $\frac{7}{12}$

3. (a) $4\frac{1}{2}$ (b) $-\frac{2}{15}$
(c) $1\frac{11}{16}$ (d) $-\frac{1}{18}$
(e) $\frac{3}{13}$

## Exercise 3e

1. 216 hectares
2. $39
3. (a) 24 boys (b) 12 girls
4. 25
5. 85
6. 35

## Exercise 3f

1. (a) $0.\dot{6}$ (b) $0.\dot{4}\dot{5}$
(c) $0.\dot{8}\dot{7}$ (d) $0.2\dot{2}\dot{7}$
(e) $0.\dot{5}71\,42\dot{8}$ (f) $0.486\dot{1}$

2. (a) $0.\dot{4}$, R (b) $2.\dot{3}$, R
(c) $0.6\dot{7}$, R (d) $0.\dot{5}\dot{9}$, R
(e) 0.776, N.R. (f) $0.2\dot{9}5\dot{4}$, R

## Exercise 3g

1. (a) R (b) R (c) R
(d) No (e) R (f) R
(g) No (h) R (i) R
(j) No

2. (a) T (b) F (c) T

3. (a) 6095 (b) 367 770
(c) –423 (d) –33 973
(e) $-\frac{65}{903}$ (f) $-3\frac{122}{539}$
(g) $2\frac{159}{184}$ (h) $-\frac{2821}{34\,542}$
(i) $\frac{598}{1225}$

4. (a) 11.36 (b) 4.33
(c) –2.47 (d) 2.37
(e) 0.50 (f) 23.60
(g) 19 940.66 (h) 113 269.73
(i) –0.07 (j) 0.79

## Review Questions 3

1. (a) $3\frac{3}{4}$ (b) $2\frac{10}{21}$
(c) $4\frac{1}{9}$ (d) $3\frac{29}{72}$

2. (a) 2 (b) $\frac{1}{4}$
(c) 0 (d) $\frac{3}{5}$
(e) $-\frac{1}{12}$ (f) $\frac{1}{3}$

3. (a) $\frac{9}{16}$ (b) 3.12
(c) $\frac{2000}{16\,947}$ (d) 1051.48
(e) $\frac{361}{432}$

4. $\frac{1993}{1994} = 1 - \frac{1}{1994}$ and
$\frac{1995}{1996} = 1 - \frac{1}{1996}$
Since $\frac{1}{1996} < \frac{1}{1994}$,
$\therefore \frac{1995}{1996}$ is larger

5. $\frac{1}{6} = \frac{1}{9} + \frac{1}{18} = \frac{1}{8} + \frac{1}{24}$
$= \frac{1}{10} + \frac{1}{15}$

6. 1 7. $\frac{11}{56}$

## Exercise 4a

1. 4 kg 2. No 3. B
4. (a) (iii) (b) (iii) (c) (ii)
(d) (i) (e) (i)
5. (a) (iii) (b) (iii) (c) (ii)
(d) (iii) (e) (iv)

## Exercise 4b

1. (a) (i) 5 (ii) 5.42
(b) (i) 16 (ii) 15.82
(c) (i) 8 (ii) 7.86
(d) (i) 131 (ii) 130.83

2. (a) 712.893 (b) 0.003
(c) 0.827 (d) 7.024

3. 0.118

4. (a) 0.21 (b) 0.64
(c) 0.64 (d) 0.73

## Exercise 4c

1. (a) 460 g (b) 700 g
(c) 3 cm (d) 123.5 cm
(e) 18 (f) 30 m
(g) 12.4 cm (h) 4300 pupils
(i) 850 km (j) 22.6 mm

2. (a) 3 (b) 3
(c) 5 (d) 3
(e) 2 (f) 2
(g) 4 (h) 4
(i) 1, 2, 3, 4, 5, 6 or 7
(j) 3

3. (a) 3.1 (b) 1.484
(c) 0.003 (d) 20
(e) 3.1 (f) 0.57
(g) 0.0568 (h) 217.01
(i) 3.60 (j) 15.70

4. (a) 4 (b) 400 000 000
5. 0.05
6. 28, 0
7. 0.211, 3
8. 0.0083, 2
9. 117.96, 5

## Exercise 4d

1. 0.0105
2. 0.442
3. 31.2
4. 11.1
5. 32.1
6. 9.99
7. 998 000 000
8. 1790 litres, 0.704 $m^3$
9. 949.7 $m^2$, \$24 220
10. (a) 2330 $cm^3$ (b) 13.26 cm

## Review Questions 4

1. (a) (iii) (b) (ii) (c) (ii) (d) (ii)
2. (a) (i) (b) (ii) (c) (ii) (d) (i)
3. (a) 0.086 (b) 0.0857 (c) 5.10 (d) 730 000 (e) 0.06
4. 20
5. (a) 50 (b) 2000
6. 44
8. 0.2
9. 190
10. 6
11. 16 years
12. B

## Exercise 5a

1. (a) $2x + 14$ (b) $5a - 14$ (c) $8k$ (d) $\frac{x}{3y}$ (e) $2x + 6y$ (f) $\frac{y}{2} - 5x$
2. (a) $2x + y$ (b) $7k$ (c) $2t - 15$ (d) $3u - 4$ (e) $\frac{v}{2} + 8$ (f) $\$\left(\frac{h}{2} + 2k\right)$ (g) $\$(1.82x + 2.22y)$
3. (a) $1.70x$ (b) $\frac{3x}{3 + k}$ (c) $5(h + 3)$ (d) $\frac{1}{4}(m - 4)$ (e) $kn$ eggs (f) $xz$ km
4. (a) $3(x + 1)$ (b) $5(x - 6)$ (c) $4x + 8$ (d) $\frac{x - 14}{3}$

## Exercise 5b

1. THE OSTRICH
2. APRIL FOOL

## Exercise 5c

1. (a) $9a$ (b) $8b + 5$ (c) $9a + c$ (d) $7h + 4k$ (e) $5p + 5q$ (f) $5u + 8v$ (g) $2p + 3$ (h) $a - 6b$ (i) $2c - d$
2. (a) $60n$ (b) $-12a$ (c) $8k$ (d) $\frac{3}{2}m$ (e) $-8n$ (f) $2a$ (g) $3u$ (h) $-\frac{c}{11}$ (i) $9v$ (j) $-\frac{3}{d}$

## Exercise 5d *(Pg 137)*

1. (a) $3x + 6$ (b) $4k - 20$ (c) $24 - 6m$ (d) $6x + 14$ (e) $16y - 40$ (f) $6x + 15y$ (g) $5x - 20z$ (h) $24h - 32k$ (i) $-3x - 18$ (j) $-4y + 24$
2. (a) $5a + 7b$ (b) $15v - 2u$ (c) $3b - 5a$ (d) $11m - 8n$ (e) $7k - 18h$ (f) $27b - a$ (g) $9p - 14q$ (h) $7a - 9b + 10c$ (i) $9y - 5x - 6$ (j) $6q - 5p + 20$
3. (a) $2x - 3$ (b) $4p - 2$ (c) $a - b + c + d$ (d) $28 - 4x$ (e) $4p - 2q$ (f) $-6a - 16$ (g) $25c + 5d$ (h) $9w - 11r - 6t$ (i) $3y - 6x$ (j) $-24x - 28y$

## Exercise 5e

1. (a) $2a + c$ (b) $6p - 2q + 3r$ (c) $x + 2y - 4z$ (d) $3p + 5q + 3r - 5s$
2. (a) $2a + 5b - 10c$ (b) $p - 6q - 3r + s$ (c) $2a + 2b - 9c - 4d$ (d) $3p - 13q + 13r$
3. (a) $-3a + 3b$ (b) $13a - 8b - 2c$ (c) $3a + 6b - 10c$ (d) $-37a + 12b - 37c$ (e) $7a - 7b - 12c$ (f) $-12a - 53b + 54c$ (g) $61a - 68b + 7c$ (h) $18a + 60b - 52c$

## Exercise 5f

1. (a) $\frac{9x}{10}$ (b) $\frac{x}{12}$ (c) $\frac{17x + 7}{35}$ (d) $\frac{x - 4}{8}$ (e) $\frac{a}{12}$ (f) $\frac{5y + 3}{12}$ (g) $\frac{13x - 1}{12}$ (h) $\frac{8x - 42}{15}$ (i) $\frac{a + 5}{4}$ (j) $\frac{13x + 11}{9}$
2. (a) $\frac{17x - 20}{35}$ (b) $\frac{28x + 26}{15}$ (c) $\frac{23x - 3}{10}$ (d) $\frac{10x - 25}{28}$ (e) $\frac{18 - 17x}{20}$ (f) $\frac{23x - 6}{20}$

## Exercise 5g

1. (a) $4(x + 3)$ (b) $6(x - 3)$ (c) $5(-2x + 5)$ (d) $8(2 - 3x)$
2. (a) $18a(1 - 3y)$ (b) $28(x + 2y)$ (c) $p(2x - 3y)$ (d) $3a(1 + 2x - 3y)$ (e) $4(x + 2y + 4z)$ (f) $2m(2 - 3y - 9z)$

## Exercise 5h

1. (a) $(3x + 2)(a + 2)$ (b) $(2y + 3)(3a + h)$ (c) $(5x + y)(a - c)$ (d) $(3x + 5y)(6h + 1)$ (e) $(z - 3)(3k - h)$
2. (a) $(x + y)(a + 3)$

(b) $(x + 3y)(2h - k)$
(c) $(x - 2y)(6a + 5)$
(d) $(5k + h)(3x - y)$
(e) $(2a + b)(x + 3y)$

3. (a) $5(m + 3n)(y + 2z)$
(b) $3(x - 3)(a - 2k)$
(c) $(a + b)(x - z)$
(d) $(p + q)(3y - 4z)$
(e) $(2a + 3b - 4c)(3x - 2y)$

### Exercise 5i

1. $(a + 4)(x - 5)$
2. $(4h - 3)(3y + 5)$
3. $(2a - 5)(3z - 1)$
4. $(a + b)(x + y)$
5. $(a + b)(y - z)$
6. $(h - k)(y - x)$
7. $(k + 4h)(3y - x)$
8. $(2h + 3k)(3x + 4y)$
9. $(4v + 3u)(6x - 5y)$
10. $(2c - 5d)(7x - 2y)$

### Review Questions 5

1. (a) 27 (b) 8 (c) –34
(d) –18 (e) 3 (f) 81

2. $-\frac{1}{2}$

3. 6

4. (a) $60f - 68e$
(b) $2m - 9$
(c) $4 - k$
(d) $7 - 7x$
(e) 17
(f) $-10x - 2$
(g) $6x + 4$
(h) $a - b + c - d - e$
(i) $126 - 45a$ (j) $3x + 7y$

5. (a) $\frac{7x}{12}$
(b) $\frac{5x + 15}{12}$
(c) $\frac{13x - 2}{15}$
(d) $\frac{17y - 43}{30}$
(e) $\frac{a + 13}{12}$
(f) $\frac{113a - 4}{30}$
(g) $\frac{34a - 5b - 21c}{30}$
(h) $\frac{58a - 14b}{15}$
(i) $\frac{10x - 9y + 25}{15}$
(j) $\frac{x + 11y - 40}{14}$

6. (a) $4(x - 2y + 4z)$
(b) $5(1 + 2b + 2c)x$
(c) $(a + b)(x - z)$
(d) $(5a - 3)(2x + y)$
(e) $(3a - 4b)(x + y)$
(f) $(a - b)(x - y)$
(g) $(4a - 1)(2x - 1)$
(h) $(3c - 4d)(2x - y)$
(i) $(2a + 7d)(x - 3y)$
(j) $(5m - 3)(x - 2)$

7. (a) $2\frac{1}{4}$
(b) $2\frac{1}{10}$
(c) $\frac{3}{8}$
(d) $\frac{6}{275}$

8. (a) $2\frac{1}{2}a$
(b) $\frac{2}{5}b$
(c) $\frac{2}{7}c$

9. $\frac{(m + f)A - mB}{f} = A - \frac{m}{f}(A - B)$

10. (a) $5x$¢
(b) $30x$¢
(c) $11x$¢

### Revision Exercise I No. 1

1. (a) $404x$
(b) $20y - 8x$
(c) 3, 180

2. (a) 900
(b) 30
(c) 2

3. (a) $3a$
(b) $28c^2$
(c) $5ab$
(d) $10a + c$
(e) $14a^2$
(f) 0
(g) $20a$
(h) $8x$

4. (a) $89
(b) $11

5. (a) 5, 85
(b) 3, 5, 7; 51, 85, 119

6. (a) False
(b) True
(c) False
(d) True

7. $6k - h + 18m$

8. (a) 30
(b) (i) 0.1508
(ii) 0.002 32

9. (a) $\frac{1}{2}$
(b) 251
(c) $(20x + 12y)$¢

10. (a) > (b) <
(c) > (d) =

### Revision Exercise I No. 2

1. (a) (i) 504
(ii) 280
(iii) 3465
(b) (i) 8
(ii) 6

2. (a) $-\frac{45}{368}$
(b) 26, 10 cm

3. (a) (–2, 6)
(b) (–1, 5)
(c) (1, 3)
(b) (–2, –1), (1, 2)
(c) (–1, 5, 6), (1, 3, 6), (1, 4, 5), (2, 3, 5)

4. (a) 38°C
(b) –3°C

5. (a) (i) 40
(ii) 20
(b) (i) 49.735
(ii) 14.442
(iii) 111.595

6. (a) 180 (b) 52

7. (a) $155y$ sec
(b) 500 g per pack, 10¢ per kg.

8. (a) –1 (b) 3
(c) 14 (d) –18

9. (a) 0.3125 (b) $\frac{9}{40}$
(c) (i) 9.91 (ii) 9.9

## Revision Exercise I No. 3

1. (a) > (b) <
(c) < (d) >
(e) > (f) <

2. (a) $\frac{11}{13}$ (b) $-5\frac{1}{18}$
(c) $\frac{3}{4}$ (d) $\frac{57}{68}$
3. (a) (i) $3 \times 2^4$ (ii) $2^2 \times 3^4$
(b) (i) 7 (ii) 3
(c) (i) 504 (ii) 210
4. (a) $100
(b) $\frac{3}{4}, \frac{5}{6}, \frac{8}{9}, \frac{11}{12}$
(c) $\frac{2}{5}, \frac{3}{8}, \frac{1}{3}, \frac{1}{4}$
(d) $x = 4, y = 1, z = 2$
5. (a) (i) $12x - 7y$
(ii) $7z - 11x$
(b) (i) 8 (ii) 21
6. (a) $(12a + 7b)$¢
(b) (i) $5 < 7 < 10$
(ii) $c < a < b$ or $b < a < c$
7. $\frac{5}{24}$
8. (a) 294
(b) 15, 1800
(c) $2^6 \times 7^2$, 56
9. (a) (i) 0.016 26
(ii) 37 560
(b) $x - 6$, 42
10. (a) 20 (b) –15
(c) 32 (d) 36

## Revision Exercise I No. 4

1. (a) 5.34 (b) 0.090
(c) 5000 (d) 12 100
2. (a) $6z - 2y$
(b) $-\frac{1}{30}x - 1\frac{1}{6}y$
3. (a) 10 (b) –18
(c) $1\frac{3}{7}$ (d) $-\frac{4}{13}$
4. (a) $\frac{xy}{15}$ (b) 5 h 36 min
5. 9
6. (a) (i) 96, 4 (ii) 840, 4
(b) $-\frac{2}{5}$ (c) $-\frac{5}{6}$
7. (a) $70x$ km (b) $\frac{50x}{11}$
(c) $1000k$
8. (a) $2^3 \times 3^6$, 18 (b) 30
(c) $\frac{4x + 7}{15}$
9. $2\frac{3}{5}$
10. $48.09; $1.91

## Revision Exercise I No. 5

1. (a) $\frac{4x + 6}{3}$ (b) 720 m
2. (a) $15 580
(b) (i) 8.840 (ii) 212.3
(iii) 2094 (iv) 1.895
3. (a) $3\frac{3}{20}, \frac{22}{7}, 3.142, \frac{47}{15}$
(b) (i) 16, 192 (ii) 3, 756
4. (a) 13 (b) –280
(c) $-1\frac{11}{20}$ (d) $\frac{1}{4}$
5. (a) 5, 3780 (b) 5, 6, 7, 8
(c) –9
6. (a) (i) 2 (ii) –1
(iii) $\frac{1}{4}$ (iv) 7
(b) (i) $a - 4b$ (ii) $-5m - 4n$
7. (a) $2^8 \times 3^2$, 48
(b) $-\frac{4}{9}, -\frac{2}{3}, -\frac{5}{7}, -\frac{8}{11}$
8. $502
9. $5\frac{7}{24}$
10. (a) 80¢ (b) $565
(c) –6°C

## Exercise 6a

1. (a) 14, 17, 20
(b) 40, 50, 60
(c) 80, 87, 94
(d) 48, 40, 32
(e) 46, 73, 82
(f) 55, 43, 39
2. (a) multiply the preceding term by 3; 81, 243, 729
(b) multiply the preceding term by 2; 96, 192, 384
(c) divide the preceding term by 2; 100, 50, 25
(d) multiply the preceding term by 3; 324, 972, 2916
3. (a) add 5 to preceding term; 34, 39
(b) add 11 to preceding term; 72, 83
(c) subtract 6 from preceding term; 49, 43
(d) subtract 9 from preceding term; 63, 54
(e) multiply preceding term by 2; 240, 480
(f) divide preceding term by 3; 27, 9

4. (a) 15, 21, 28
(b) 37, 42, 47
(c) 33, 32, 27
(d) 21, 34, 55

## Exercise 6b

1. (a) $1 + 3 + 5 + 7 + 9 + 11 = 36 = 6^2$
$= (5 + 1)^2$
$1 + 3 + 5 + 7 + 9 + 11 + 13 = 49$
$= 7^2 = (6 + 1)^2$
(b) $1 + 3 + 5 + 7 + 9 + 11 + 13 + 15$
$+ 17 + 19 + 21 + 23 = 144$
$= 12^2 = (11 + 1)^2$
(c) $a = 25, c = 13, d = 12$
2. (a) $2 + 6^2 = 38$
(b) 8
3. (a) $10^2 - 9^2 = 19 = 10 + 9$
(b) 1195
(c) $m = 41, n = 40$
4. (a) $\frac{11 \times 12}{2} + (11 - 1)^2 = 166$
(b) $p = 10, q = 136$
5. (a) 9, 12 (b) $t = 3n$
(c) (i) 150 (ii) 29
6. (a) 16, 25
(b) $T = (N + 1)^2$
(c) (i) 100 (ii) 10

## Exercise 6c

1. (a) $3 + 1 = 4, 4 + 1 = 5,$
$5 + 1 = 6, 6 + 1 = 7$
(b) 50 (c) 100
2. (a) (i) $\frac{8-2}{2} = 3, \frac{10-2}{2} = 4,$
$\frac{12-2}{2} = 5, \frac{14-2}{2} = 6$
(ii) $2(3) + 2 = 8,$
$2(4) + 2 = 10,$
$2(5) + 2 = 12,$
$2(6) + 2 = 14$
(b) (i) 9 (ii) 14
(c) (i) 46 (ii) 74
3. (a) $\frac{4 \times (4-1)}{2} = 6,$
$\frac{5 \times (5-1)}{2} = 10,$
$\frac{6 \times (6-1)}{2} = 15,$
$\frac{7 \times (7-1)}{2} = 21$
(b) 190
4. (b) 19 (c) 26
(d) $2n + 1$ (e) $2n + 8$

5. 37 min

6. (a) $\frac{2}{8 \times 9 \times 10} = \frac{1}{8} - \frac{2}{9} + \frac{1}{10}$

(b) $\frac{2}{10 \times 11 \times 12} = \frac{1}{660}$

(c) $p = 19$

7. (a) (i) 11, 13

(ii) 24, 28

(iii) 84, 112

(iv) 85, 113

(b) $13^2 + 84^2 = 85^2$

$15^2 + 112^2 = 113^2$

## Review Questions 6

1. (a) 25, 31, 37
(b) 42, 57, 75
(c) 16, 26, 42
(d) 85, 81, 77
(e) 21, 25, 29
(f) 27, 38, 51

2. (a) 9 (b) 6 (c) 16
(d) 28 (e) 26

3. (a) 1 111 111 111 – 22 222 $= 33\ 333^2$

(b) $x = 1\ 111\ 111\ 111\ 111\ 111$

$y = 22\ 222\ 222$

4. (a) (i) $1 + 2 + 3 + 4 + 5 + 6 + 7 = 28 = \frac{7 \times (7 + 1)}{2}$

(ii) $k = 9$

(b) (i) $1^3 + 2^3 + 3^3 + 4^3 + 5^3 + 6^3 = 1 + 8 + 27 + 64 + 125 + 216 = 441 = 21^2$

(ii) $x = 8^3$, $y = 512$, $z = 1296$

5. (a) 108 (b) 630, 2460

6. (a) 28 (b) $\frac{n(n + 1)}{2}$

## Exercise 7a

1. (a) 2 (b) 8
(c) 8 (d) 12
(e) 12 (f) 9
(g) 2 (h) 10

2. (a) 2 or –2 (b) 3 or –3
(c) not possible
(d) 25 (e) 27
(f) –8 (g) 60
(h) 24

## Exercise 7b

1. (a) 12 (b) –7
(c) 7 (d) 2
(e) 4 (f) 1.2
(g) –5 (h) –3
(i) 12 (j) –12

2. (a) 2 (b) 2
(c) 1 (d) 4
(e) –3 (f) $\frac{3}{5}$
(g) $1\frac{1}{8}$ (h) $-\frac{3}{8}$
(i) 0.75 (j) 2

3. (a) 2 (b) 9
(c) $2\frac{1}{2}$ (d) $\frac{1}{2}$
(e) $-\frac{1}{2}$ (f) $4\frac{1}{3}$
(g) 2.1 (h) 0.6
(i) $1\frac{1}{30}$

## Exercise 7c

1. (a) 4 (b) –5
(c) 3 (d) 3
(e) $\frac{1}{2}$ (f) 1
(g) 3 (h) –4
(i) 4 (j) 8

2. (a) –2 (b) –3
(c) –7 (d) 38
(e) $-13\frac{1}{2}$ (f) –18
(g) 13 (h) $-1\frac{6}{7}$

3. (a) 3 (b) –40
(c) 25 (d) 24
(e) 4 (f) –18
(g) 9 (h) –10
(i) $-4\frac{1}{2}$ (j) $-3\frac{1}{2}$

4. (a) 10 (b) $-2\frac{2}{5}$
(c) $1\frac{19}{21}$ (d) $2\frac{2}{5}$
(e) –60 (f) $-2\frac{1}{2}$
(g) 15 (h) $2\frac{2}{3}$
(i) $\frac{-1}{3}$ (j) $-4\frac{3}{4}$

5. (a) $7\frac{1}{2}$ (b) 3
(c) 3 (d) $-1\frac{1}{3}$
(e) 4 (f) $-13\frac{1}{2}$
(g) 22 (h) $-\frac{1}{3}$

## Exercise 7d

1. 86 2. 86
3. 1386 4. 22
5. 5 6. 32
7. $1\frac{5}{6}$ 8. $2\frac{1}{2}$
9. 15 10. 3
11. $4\frac{1}{8}$ 12. 2
13. –123 14. –30
15. $6\frac{1}{3}$ 16. $5\frac{3}{10}$

## Exercise 7e

(a) $S = a + b + c$

(b) $P = xy$

(c) $D = a - e$ or $e - a$

(d) $A = \frac{1}{2}\pi r^2$

(e) $C = \frac{3m}{25}$

(f) $A = \frac{1}{4}(m + n + p + q)$ years old

(g) $x = 180 - 2y$

(h) $T = 60a + b$

(i) $T = dp + cq$

(j) (i) $r = p + q$

(ii) $\frac{xy(p + q)}{py + qx}$ kmh$^{-1}$

## Exercise 7f

1. 5000 g; $1000x$ g
2. \$24; \$$4p$; \$$pq$
3. $\frac{1}{4}m$ kg 4. $2b$ km
5. $60m$ minutes 6. $\frac{y}{7}$ weeks
7. $20x$ 8. $\frac{1}{2}b$
9. $\frac{21}{v}$ hours 10. \$$(a + b)$
11. $6b$ years; $(7b + 2y)$ years
12. $(2x + 4)$ years
13. $n - 1, n, n + 1$
14. $(5u + 3v)$ km

## Exercise 7g

1. (a) $x + 7 = 18$

(b) $(x - 2) \times 3 = 24$

(c) $(x - 5) \times 7 = 63$
(d) $(24 - x) \div 5 = 4$
(e) $x + (x + 1) + (x + 2) = 63$
(f) $x + (x + 3) = 43$
(g) $6x = 2x + 16$
(h) $2[x + (x + 5)] = 32$
(i) $2(x + 2x) = 54$

**2.** \$70 **3.** 15
**4.** $3x + x + 2x = 4800$
**5.** $3x + 2(2x) = 450$

## Exercise 7h

**1.** 11, 12, 13, 14 **2.** 14
**3.** 4 **4.** 16, 64
**5.** 19, 15 and 13 **6.** \$20, \$10
**7.** 8700 kg **8.** 27
**9.** $\frac{9}{14}$ **10.** \$3.50
**11.** \$32, \$128, \$96
**12.** \$152 **13.** 12

## Exercise 7i

**1.** (a) > (b) <
(c) < (d) =
(e) < (f) =
(g) < (h) >
(i) > (j) <
(k) < (l) =
(m) = (n) >
(o) < (p) <

## Exercise 7j

**1.** (a) $x > 7$ (b) $y < 35$
(c) $h > 1\frac{1}{6}$ (d) $a > 2.8$
(e) $z < -6.1$
**2.** 14

## Review Questions 7

**1.** (a) $2\frac{9}{11}$ (b) $-4\frac{1}{4}$
(c) 30 (d) $\frac{1}{4}$
(e) –6 (f) 4
(g) $\frac{17}{19}$ (h) $6\frac{3}{5}$
**2.** (a) $x < 9$ (b) $k < 21$
**3.** 45
**4.** 120 kg, 240 kg, 120 kg
**5.** 120 **6.** 16 hours
**7.** 9 **8.** 24 days

**9.** 8 **10.** $3\frac{3}{5}$ km
**11.** 8, 40 **12.** 21, 24
**13.** 36 km/h **14.** 25, 13
**15.** 25 cents **16.** 5 min
**17.** 22 years old **18.** 48 km
**19.** $A = 3, B = 5, C = 6, D = 8$
**20.** $A = 2, B = 4, C = 7, D = 9$

## Exercise 8a

**1.** (a) 0.000 85 (b) 0.025
(c) 63 000 (d) 4060
(e) 4440 (f) 31 000
(g) 0.000 053 7
(h) 280 000 (i) 0.0532
(j) 0.000 694 5
**2.** (a) 20 m, $62\frac{6}{7}$ m, $314\frac{2}{7}$ m$^2$
(b) 28 mm, 56 mm, 2464 mm$^2$
(c) 14 m, 28 m, 88 cm
(d) 1.8 m, $11\frac{11}{35}$ m, $10\frac{32}{175}$ m$^2$
**3.** (a) 72 cm, 308 cm$^2$
(b) 56 cm, 217 cm$^2$
(c) 96 cm, $661\frac{1}{2}$ cm$^2$
(d) 22.4 cm, 35 cm$^2$
(e) $28\frac{4}{7}$ cm, $50\frac{3}{7}$ cm$^2$
(f) 396 cm, 10 234 cm$^2$
**4.** 2460 cm$^2$
**5.** 0.38 m$^2$
**6.** 810 cm$^2$
**7.** \$2152.50
**8.** 275 m$^2$
**9.** 11.73 cm/min
**10.** 55.2 cm$^2$
**11.** 5.65 ha
**12.** 18 cm, 180 cm$^2$
**13.** 157.5 cm$^2$
**14.** (a) 102 cm$^2$
(b) 17.5 cm$^2$
(c) 21 cm
**15.** (a) 762 cm$^2$
(b) 63 cm$^2$
(c) 659 cm$^2$
(d) 220 cm$^2$
**16.** 972
**17.** 3684 cm$^2$

## Exercise 8b

**1.** (a) 84 (b) 7
(c) 5.5
**2.** (a) 60 cm$^2$ (b) 168 cm$^2$
(c) 1040 m$^2$ (d) 105 cm$^2$
**3.** (a) 54 (b) 10
(c) 13
**4.** \$14 080
**5.** (a) 8 (b) 23
(c) 12 (d) 21

## Exercise 8c

**1.** (a) 76 cm$^2$ (b) 810 cm$^2$
(c) 671.91 cm$^2$ (d) 363.89 cm$^2$
(e) 247.63 cm$^2$ (f) 161.96 cm$^2$
**2.** (a) 80 cm$^2$ (b) 20 cm$^2$
(c) 40 cm$^2$
**3.** 45.4 cm **4.** 35.2 m/s
**5.** 301 **6.** $8.89r^2$

## Review Questions 8

**1.** 42 cm$^2$ **2.** 18 cm$^2$
**3.** 128 cm$^2$
**4.** 13 cm, 117 cm$^2$ **5.** 864
**6.** 12 cm; 8 cm, 4 cm
**7.** 14 **8.** 44 m
**9.** 88 cm, 154 cm$^2$ **10.** 60 m
**11.** 576 cm$^2$

## Exercise 9a

**1.** (a) 480 cm$^3$, 376 cm$^2$
(b) 420 cm$^3$, 358 cm$^2$
(c) 115 200 mm$^3$, 27 360 mm$^2$
(d) $7\frac{1}{2}$ cm$^3$, $41\frac{1}{2}$ cm$^2$
(e) $\frac{21}{64}$ cm$^3$, $3\frac{43}{160}$ cm$^2$
(f) 4.095 cm$^3$, 19.26 cm$^2$
**2.** (a) 2160 mm$^3$, 1284 mm$^2$
(b) 8 cm, 158 cm$^2$
(c) 2.5 cm, 89.5 cm$^2$
(d) 8 m, 432 m$^2$
(e) $3\frac{1}{4}$ cm, $102\frac{5}{8}$ cm$^2$
(f) 5 cm, 540 cm$^3$
**3.** (a) 69 300 $l$
(b) 33 345 $l$
(c) 6.84 $l$
(d) 35.568 $l$
**4.** 96 cm$^2$
**5.** 16 cm, 1420 cm$^2$
**6.** 2.9 m

**7.** 31.35 cm
**8.** **(a)** 456 000
**(b)** \$25.08 million, \$836
**9.** 6703.2 $m^3$

## Exercise 9b

**1.** **(a)** 369 840 $cm^3$
**(b)** 16 644 $cm^3$
**(c)** 960 $cm^3$
**(d)** 1332 $cm^3$
**(e)** 770 $cm^3$
**(f)** 4725 $cm^3$
**2.** **(a)** 6 $cm^2$, 42 $cm^3$
**(b)** 14 cm, 693 $cm^3$
**(c)** 32 cm, 240 $cm^2$
**(d)** 400 cm, 95.94 $cm^2$
**3.** 102 480 $m^3$
**4.** 153 $cm^3$, 250 $cm^2$
**5.** 5625 $m^3$, 1937.5 $m^2$
**6.** 3.66 $m^2$
**7.** **(a)** 180 $cm^3$
**(b)** 264 $cm^2$
**8.** 1728 *l*, 0.872 $m^3$
**9.** 49 500 $cm^3$
**10.** 25 cm
**11.** 138.6 *l*

## Exercise 9c

**1.** **(a)** 1848 $cm^3$, 836 $cm^2$
**(b)** 4.526 $m^3$, 17.35 $m^2$
**(c)** 44 550 $mm^3$, 7354.3 $mm^2$
**2.** **(a)** 8 cm **(b)** 28 cm
**3.** **(a)** 42 cm **(b)** 21 cm
**4.** 7.425 *l*
**5.** 1056 $cm^3$
**6.** 19 250 $cm^2$
**7.** 20.1 $m^2$
**8.** 11 550 $m^3$, 154
**9.** 123.75 *l*, 206
**10.** 0.9856 $cm^3$, 5.32 g, 10 348.8 $m^3$, 55 884 tonnes

## Review Questions 9

**1.** **(a)** **(i)** 108 $cm^3$, 168 $cm^2$
**(ii)** 88 $cm^3$, 152 $cm^2$
**(b)** 1217.86 $cm^3$, 1233.57 $cm^2$
**2.** **(a)** **(i)** 14 $cm^3$, 50 $cm^2$
**(ii)** 16 $cm^3$, 58 $cm^2$
**(b)** 539 $cm^3$, 438 $cm^2$
**3.** 3000
**4.** 64
**5.** 1800
**6.** 1350 *l*
**7.** **(a)** 7250 $cm^3$
**(b)** 8808 cm
**(c)** 56.55 kg
**8.** 1487.2 kg
**9.** 51.47 $cm^3$, \$1.11
**10.** 92 000 $cm^3$, 18 400 $cm^2$
**11.** 1.216 kg
**12.** 1 h 55 min 30 sec
**13.** 291 000 $m^3$, 366 660 kg

## Revision Exercise II No. 1

**1.** **(a)** 6 **(b)** $5\frac{2}{11}$
**(c)** $12\frac{3}{7}$ **(d)** 11
**2.** **(a)** 138 **(b)** 10
**3.** **(a)** 24 yrs **(b)** 64 kg
**4.** **(a)** **(i)** 21, 25 **(ii)** $\frac{5}{6}, \frac{6}{7}$
**(iii)** $5 \times 7, 6 \times 8$
**(b)** **(i)** $\frac{1}{8 \times 9} = \frac{1}{8} \times \frac{1}{9}$
**(ii)** $\frac{1}{2550}$
**(iii)** 18, 19
**5.** **(a)** $6\frac{1}{5}$
**(b)** $-\frac{3}{11}$
**6.** **(a)** $\frac{4x + 27y}{15}$
**(b)** $\frac{-38x - 31}{15}$
**7.** **(a)** 91 $cm^2$
**(b)** **(i)** 135 000 $m^3$
**(ii)** 9
**8.** $\frac{60}{77}$ cm
**9.** **(a)** $3(x + 3y + 4z)$
**(b)** $2(5h - k)(x - 2y)$
**(c)** $(2c + 5d)(x - 3y)$
**(d)** $(2a - 3b)(x - 2y)$

## Revision Exercise II No. 2

**1.** **(a)** $\frac{1}{6}$
**(b)** $-\frac{4}{7}$
**2.** **(a)** 42
**(b)** 9 mths
**3.** **(a)** **(i)** $6^2 - 2 \times 6 = 24$
**(ii)** 9
**(b)** **(i)** 69, 61, 52
**(ii)** 0, 12, 24
**4.** **(a)** 24 $cm^2$, 3 cm
**(b)** 2288 $cm^2$, 7392 $cm^3$
**5.** **(a)** 690 $cm^2$, 5175 $cm^3$
**(b)** 18 cm
**6.** **(a)** 9 yrs
**(b)** \$11.25
**7.** **(i)** $1 + 3 + 5 + ... + 17 + 19 = 100 = 10^2$
**(ii)** 225 **(iii)** 13
**8.** 5
**9.** **(a)** 14, 21
**(b)** $\frac{14a - 19}{20}$
**(c)** \$11
**10.** **(a)** $(3x - y)(3a + 4b)$
**(b)** $(3a - 4b)(2x - 3y)$
**(c)** $(5x + y)(3h - k)$

## Revision Exercise II No.3

**1.** **(a)** 9
**(b)** 0.15
**(c)** 15
**(d)** $3\frac{14}{15}$
**2.** **(a)** 135
**(b)** 45 yrs
**3.** **(a)** **(i)** 3, 6
**(ii)** 77, 88
**(iii)** 65, 78
**(b)** **(i)** 30, 38
**(ii)** $7\frac{2}{3}$, 10
**4.** **(a)** $2\frac{1}{2}$ **(b)** 7
**5.** **(a)** $<$ **(b)** $<$
**(c)** $>$
**6.** **(a)** 85 **(b)** –3
**7.** **(a)** \$23.50
**(b)** 9 yrs, 36 yrs
**8.** **(a)** 42.9 cm
**(b)** 1106 $cm^2$, 2814 $cm^3$, 10.69 kg
**9.** **(a)** 48 cm, 161 $cm^2$
**(b)** 156 $cm^3$ **(c)** 32 $cm^2$
**(d)** 2250 $m^3$
**10.** **(a)** **(i)** $x < 8$
**(ii)** $x > 12$
**(iii)** $x > -8$

(b) (i) $3x(1 + 6h - 4k)$
(ii) $(5x - y)(2h + k)$
(iii) $(3x - y)(h - 5k)$

## Revision Exercise II No. 4

1. (a) 6 (b) $7\frac{1}{5}xy$ km
2. (a) $1\frac{2}{3}$ (b) 8
(c) 14 (d) 4
(e) 2 (f) 5
3. (a) 7 yrs (b) 56 yrs
4. (a) $-\frac{1}{2}$ (b) 15 yrs, 45 yrs
(c) $5\frac{2}{45}$
5. 81
6. (a) $29\frac{1}{3}$ cm
(b) 76 $cm^3$, 114 $cm^3$
7. 3000 litres, 19
8. (a) (i) $10^2 - 9^2 = 19$
(ii) 197
(iii) $x = 79$, $y = 78$
(b) $4x^2y^3 - 1 - 3x$
9. (a) $-24\frac{1}{2}$
(b) (i) 441 (ii) 105
10. (a) (i) $x < 33$ (ii) $y \geq 11$
(iii) $x > -\frac{2}{5}$
(b) (i) $7x(1 - 2h + 3hk)$
(ii) $(2a - 3b)(3x - y)$
(iii) $(4x - 5y)(3a - 2b)$

## Revision Exercise II No. 5

1. (a) 5 (b) 3.5
2. (a) 6 (b) $2\frac{2}{3}$
(c) 11 (d) 5
(e) 8 (f) $-1\frac{1}{4}$
3. (a) $28a$ (b) $8.5a$
4. (a) 26, 37, 50
(b) 2, 0, –2
(c) 23, 37, 60
(d) $\frac{1}{16}, \frac{1}{32}, \frac{1}{64}$
5. $k = 5$
(a) 9 (b) 1200
6. (a) 49 (b) 49
7. 1

8. (a) 2 cm (b) 32 cm
9. (a) 18 (b) 25 yrs
(c) $\frac{6}{17}$
10. (a) (i) $x < 5$
(ii) $x \geq 5$
(b) (i) $(2h - k)(x - 3y)$
(ii) $(x - y)(2a - 3b)$

## Exercise 10a

1. (a) 25 : 21 (b) 13 : 15
(c) 4 : 5 (d) 1 : 50
(e) 3 : 4 (f) 4 : 5 : 3
(g) 8 : 4 : 1 (h) 3 : 5 : 7
(i) 21 : 6 : 5 (j) 2 : 3 : 5
2. (a) (i) 5 : 16 (ii) $\frac{5}{16}$
(b) (i) 7 : 12 (ii) $\frac{7}{12}$
(c) (i) 5 : 2 (ii) $\frac{5}{2}$
(d) (i) 2 : 3 (ii) $\frac{2}{3}$
(e) (i) 5 : 12 (ii) $\frac{5}{12}$
(f) (i) 7 : 120 (ii) $\frac{7}{120}$
3. (a) 14 : 39
(b) 14 : 18 : 39
(c) $\frac{28}{195}$
4. 5 : 6 : 20
5. 4 : 21 : 2
6. (a) 8 : 3 (b) 5 : 8
7. 13 : 12 : 7
8. 2 : 3 : 6 : 7
9. (a) 4 : 9 (b) 2 : 3
10. 3 : 7 : 24 : 5

## Exercise 10b

1. (a) 25 kg (b) 64 m
(c) 87.5 ha (d) 1 $cm^2$
2. (a) 7 : 5
3. 3 : 4
4. 7.7 cm by 12.6 cm
5. $7.20
6. $35 200

## Exercise 10c

1. 25 g, 40 g, 45 g
2. $3920
3. $70, $154
4. 120 kg, 240 kg, 120 kg
5. (a) 10 kg, 30 kg
(b) $3
6. (a) 9 kg, 15 kg, 21 kg
(b) $10.80
7. (a) 150 g, 650 g
(b) 40 g, 480 g
(c) 19 : 93 : 48
(d) 95 g

## Exercise 10d

1. (a) 15 (b) 200
(c) 19 (d) 5
2. $150
3. 272 km
4. 240
5. 221 g
6. $172.50
7. $5000
8. $35.10
9. 500 g
10. (a) 25 g (b) 350 g
(c) 18 $m^2$
11. 15

## Exercise 10e

1. (a) 0.15 shot/s
(b) 9 shots/min
2. (a) 0.13 sandwich/g
(b) $23.87
3. (a) $14.25 per m
(b) $39.90
4. 684
5. 40 m
6. 40 min

## Exercise 10f

1. (a) 08 00 (b) 21 42
(c) 00 00 (d) 02 42
2. (a) 3.30 a.m. (b) 11.12 p.m.
(c) 7.15 p.m. (d) 12.00 a.m.
3. (a) 20 45 (b) 03 35
(b) 06 35 the next day
(c) 2 h 5 min
(d) 12 h 28 min
(e) 10 h 40 min
(f) 22 35
4. 6 h 37 min
5. 07 13 the next day
6. 10 51
7. 7 h 10 min, 05 30
8. (a) 4 h 55 min
(b) 7 h 20 min
(c) 6 h 50 min
(d) 6 h 25 min
(e) 13 h 15 min

## Exercise 10g

1. (b) 8 m/s (c) $\frac{2}{3}$ m/s
   (d) 440 km (e) 500 m
   (f) 5 s
2. (a) 5 m/s (b) 20 m/s
   (c) 25 m/s
3. (a) 36 km/h (b) 126 km/h
   (c) 1800 km/h
4. 22 s
5. 10 53
6. (a) 13 42 (b) 112 km/h
7. 80 m
8. (a) 19 km (b) 110 km/h
9. 10 km/h
10. 48 km/h
11. (a) 5 m/s (b) 4 s
    (c) 7.5 m/s
12. (a) 60 m (b) 4 s
    (c) 16 m/s
13. 280 m
14. 37 800 m

## Review Questions 10

1. 3 : 4
2. 240 cm by 288 cm
3. 333 000
4. 9 : 10 : 12
5. (a) 45 (b) 15
6. (a) 06 47 (b) 1.2 km
7. (a) 09 51 (b) 88 km/h
8. (a) $18.54 (b) 5 kg
9. (a) (i) 2 h 40 min
       (ii) 6 h 25 min
   (b) (i) 53 min
       (ii) 1219 words
10. (a) 63 m (b) 7 m
11. $53\frac{1}{3}$ km/h
12. (a) $1050, $750, $300
    (b) $1225, $875, $350, $175, $175
13. (a) 12 30
    (b) 3 h 20 min; 58.5 km/h
14. (a) 7 h 15 min
    (b) $11.90
15. 14 km
16. $3\frac{3}{4}$ h
17. 13 : 36
18. 3 h
19. 10 days

## Exercise 11a

1. 6%
2. 20%, 30%, 45%, 5%
3. (a) $99.20 (b) 325
   (c) 62.5 cm (d) 91.8 litres
4. 7225 people
5. 1190 tyres
6. (a) 47.4 kg (b) 237 000 kg

## Exercise 11b

1. $32 000
2. Mulu 13.3%, Koka 11.2%
3. Ben—Gold, James & Roy—Silver
4. 45%
5. 52¢
6. 8.1%
7. $2700
8. 2070 g
9. 560 m*l*
10. the same
11. first choice; Alan will have more money to spend in the first year though the 2nd year will end up with the same amount as in the 2nd choice.

## Exercise 11c

1. 63
2. 135
3. 140
4. 240
5. $122 400
6. $65
7. 355 houses
8. 5.4 m
9. (a) $55.80
   (b) (i) $9.18 (ii) $367.20
10. 15%
11. $86 400
12. 20%
13. 16.64%
14. (a) $11\frac{1}{9}$%
    (b) 40 minutes
    (c) 180 seconds
    (d) 192 seconds
    (e) James

## Exercise 11d

1. (a) 12% (b) 20%
2. (a) (i) $4.50 (ii) $40.50
   (b) (i) $11 (ii) $99
3. (a) $700 (b) $810
4. (a) $87.50 (b) $164
5. $500
6. (a) $72 (b) $72.20
7. $28
8. $700
9. $1600, $1120. same.
10. (b) 38 cents (c) $15\frac{5}{9}$%
11. $1310
12. $18 000
13. $90
14. $36, $48, $60
15. $96.90
16. $252
17. $250

## Review Questions 11

1. $935
2. (a) $291.20 (b) $1164.80
3. $1\frac{3}{4}$, 1.74, 173%, $1\frac{2}{3}$, 1.56
4. $\frac{7}{80}$
5. $\frac{3}{40}$
6. 144, 200
7. 5%
8. (a) $3
   (b) (i) $33 (ii) $9.00
9. Yes, $15
10. $17\frac{11}{17}$%
11. Increase by 5.8%
12. 2 : 1
13. 300

## Exercise 12a

1. $A(-4, -3)$, $B(-2, 3)$, $C(2, 4)$, $D(4, 2)$, $E(1, 1)$, $F(3, -1)$
2. $A(-4, -6)$, $B(-2, 5)$, $C(1, 0)$, $D(6, 5)$, $E(2, -3)$, $O(0, 0)$
5. (a) rectangle
   (b) rhombus
   (c) isosceles triangle
   (d) quadrilateral
   (e) isosceles triangle
   (f) trapezium
   (g) trapezium
6. 24 units$^2$
7. The points lie on a straight line.
8. (c) A straight line.

## Exercise 12b

1. (a) –3 (b) 17
   (c) 33
2. (a) 40 (b) 25
   (c) –20
3. (a) $-1\frac{2}{3}$ (b) $1\frac{1}{3}$
   (c) $2\frac{1}{6}$
4. (a) 50 (b) 14
   (c) –34
6. (a) 1 (b) 1
   (c) –2 (d) –3
7. (a) $\frac{3}{2}$ (b) $-\frac{3}{2}$
   (c) $-\frac{4}{3}$ (d) $-\frac{4}{5}$

## Review Exercise 12

1. (a) rectangle (b) square
   (c) trapezium (d) kite

2. (a) 150 (b) 450
(c) –1150

3. (a) $-3\frac{1}{2}$ (b) $8\frac{1}{2}$
(c) $46\frac{1}{2}$

4. $2\frac{1}{2}$

5. (a) 5 (b) –7
(c) $-\frac{1}{2}$ (d) $\frac{2}{3}$

6. (a) $40\frac{5}{9}\,l$ (b) $22\frac{17}{54}$;
$y = 45 - \frac{x}{10.8}$;
$y = 47\frac{17}{54}$; $-6\frac{2}{3}$
There is no meaning for $x$ to be negative as reversing the car will also use up fuel. The fuel tank can only take up 45 litres of petrol. When $x = 558$, $y$ is negative which implies the car cannot possibly be driven with no petrol.

## Exerice 13a

1. (a) 2 : 1 (b) 66.7%
2. (a) 53 (b) 50%
3. (a) 1000 (b) 900
(c) $\frac{4}{7}$
4. (a) Product 1 (b) April
(c) \$37 500 (d) \$45 000
(e) \$140 000 (f) \$125 000
(g) $\frac{28}{53}$
5. (a) 2005 — 260 000
(b) 2001 — 140 000
2002 — 160 000
2003 — 180 000
2004 — 220 000
(c) \$90 million
(d) 18.2%
8. (a) 4 : 5 (b) $33\frac{1}{3}\%$

## Exercise 13b

3. (c) A pie chart
(d) A bar chart
4. Hospital (a) \$324 m
(b) 45%
Research (a) \$25 m
(b) 3.5%
Public Health (a) \$25 m
(b) 3.5%
Dental Clinics (a) \$36 m
(b) 5%
Drugs (a) \$40 m
(b) 5.6%
Doctors (a) \$130 m
(b) 18.1%
Nursing Homes (a) \$80 m
(b) 11.1%
Others (a) \$60 m
(b) 8.3%
5. (a) 20
(b) (i) 540 (ii) 225
6. (a) Red
7. (a) 90° (b) 100°
(c) 36 (d) 27.8%
8. 252° 9. 9
10. (a) (i) 50% (ii) 20%
(b) 63

## Exercise 13c

1. (a) 14 tonnes (b) 4 tonnes
(c) 55 tonnes
2. (b) 39°, 38°
3. (a) 1999 and 2000
(b) $56\frac{1}{4}\%$

## Exercise 13d

1. (c) 3 (d) 6
2. (b) 26 (c) 6 (d) 2
3. (b) Country *A*: 9; Country *B*: 8
(c) Country *A*: 349; Country *B*: 99
4. (c) 30% 6. (b) $26\frac{2}{3}\%$

## Exercise 13e

1. 4, 4, 6, 8, 10, 5, 4, 4, 3
5. (a) 314 (b) 110 (c) 54

## Review Questions 13

2. (a) \$55 million
(b) (i) 22% (ii) 64.8
(iii) $\$24\frac{3}{4}$ million
3. (a) 8 4. (b) 0
5. (b) $73\frac{1}{3}\%$
7. $x = 70$, $y = 400$

## Revision Exercise III No. 1

1. 50 min 2. 7 : 12
3. \$24 4. \$8.40
5. (a) 48 (b) 64
(c) $\frac{2}{5}$
6. (a) 1 + 2 + 3 + 4 + 5 + 4 + 3 + 2 + 1 = 25
(b) 81 (c) 13
7. (a) 32 m (b) 42.24 $m^2$
8. 18.48 $l$, 3256 $cm^2$
9. (a) 25 (b) 10

## Revision Exercise III No. 2

1. (a) 23, \$6.80 (b) 85 $l$
2. (a) $\frac{7}{200}$ (b) 1.225
(c) 2.00 (d) 0.025
3. (a) 1200
(b) $18y - 6x - 21$
4. $\frac{3}{11}$
5. (a) 60 cm (b) 144 000 $l$
6. (a) 11 × 11 = 10 × 12 + 1
(b) 14 (c) 39 601
7. (a) 35 (b) $\frac{1}{5}$
(c) $\frac{2}{7}$
8. (a) 0.462 $m^3$ (b) \$4760
(c) $\$\left(\frac{1}{3}xz - xy\right)$
9. (a) 220 cm (b) 880 m

## Revision Exercise III No. 3

1. 24 days
2. 147, 245, 392
3. (a) 88 (b) 30% (c) 2024
4. 126 $cm^2$
5. (a) 112.5 (b) \$1175
6. 288 $cm^2$
7. 320 $cm^3$, 224 g
8. (a) (i) 20% (ii) \$30 200
(b) 9
9. $2^4 \times 3$; $x = 3$
10. –1, 0, 1, 2, 3, 4

## Revision Exercise III No. 4

1. (a) \$66.25 (b) \$40
2. (a) $A = 12, B = 10, C = 9$
(b) 33.84 km/h
3. 108°

4. (a) (i) 0.09 (ii) 0.086
(b) $3^4 \times 5^2 \times 7^2$, 315

5. (a) $2\frac{12}{19}$ (b) 12

6. $Q$, 1545

7. (a) 50
(b) There are school holidays in the months of March and June.

8. (a) $3\frac{3}{5}$ (b) 43.75%
(c) 21, 23

9. (a) 6252 $cm^3$ (b) 9.378 kg

10. 25 $units^2$

## Revision Exercise III No. 5

1. (a) 87.5% (b) 510
(c) 5400 m
2. (a) \$80.75 (b) \$25.20
3. 4 yrs, 32 yrs
4. 5 h
5. (a) 52.5 (b) 34, 36, 38
6. (a) 84 (b) 154
(c) 29.8%
7. 220
8. (a) 81 $cm^2$ (b) 4 cm
9. 184 ha
10. trapezium, (0, 1), (0, –5), (5, 5), (2, 5)

## Exercise 14a

1. acute: (a), (e)
obtuse: (b), (d)
reflex: (c)
2. (a) $d = 74°, e = 97°, f = 107°$
(b) $g = 245°, h = 50°$
(c) $k = 180°, t = 132°, l = 28°$
(d) $p = 248°, q = 99°, r = 90°, s = 55°$
3. (a) 72° (b) 44°
(c) 37° (d) 26°
(e) 83°
4. (a) 144° (b) 168°
(c) 78° (d) 9°
(e) 92°
6. (a) 49° (b) 30°
(c) 90° (d) 60°
7. 30°, 60°, 120°, 150°
8. (a) 50° (b) 42°
(c) 106° (d) 30°
(e) $p = 35°, q = 145°$
(f) 47°
(g) $x = 23, y = 69, z = 111$
(h) $x = 18, y = 126$

## Exercise 14b

1. (a) $b = y, h = r, e = p$
(b) $x = d, q = g, p = h$
2. $p = 106°, q = 74°, r = 106°, s = 74°, x = 73°, y = 107°$
3. (a) $a = 31°, b = 66°$
(b) $c = 45°, d = 60°$
(c) $e = 106°$
(d) $f = 104°$
(e) $x = 18$
(f) $x = 20, y = 70$
(g) $x = 34, y = 59.5$
(h) $x = 257$
(i) $x = 275$
(j) $x = 140$
4. (a) 86° (b) 141°
5. (a) 68° (b) 54°
6. $a° + b° = c° + d°$

## Review Questions 14

1. (a) $x = 36$
(b) $x = 205, y = 110$
(c) $x = 20$
(d) $x = 16, y = 48$
2. (a) $x = 24$
(b) $x = 30$
(c) 150
(d) 124
3. (a) $x = 56$
(b) $x = 44, y = 58$
(c) $x = 65$
(d) $x = 19$
4. (a) 72 (b) 154
5. (a) 54° (b) 128°

## Exercise 15a

1. (a) 100°
(i) scalene △
(ii) obtuse-angled △
(b) 70°
(i) isosceles
(ii) acute-angled △
(c) 60°
(i) equilateral △
(ii) acute-angled △
(d) 90°
(i) scalene △
(ii) right-angled △
(e) 65°
(i) isosceles △
(ii) acute-angled △
(f) 43°
(i) scalene △
(ii) obtuse-angled △
2. (a) 96° (b) 16°
(c) 144° (d) 52°
3. (a) 35° (b) 25°
(c) 22.5° (d) 51°
(e) 24° (f) 56°
(g) 64° (h) 20°
4. (a) 102° (b) 69°
(c) $a = 70°, b = 22°$
(d) $x = 127°, y = 58°$
(e) $x = 40°, y = 55°$
(f) $x = 78°, y = 52°$
(g) $a = 35°, b = 108°$
(h) $x = 130°$
5. (a) $x = 25°, y = 97°$
(b) $i = 101°, j = 42°$
(c) 138°
(d) $x = 35°, y = 125°$
(e) $x = 24°, y = 96°, z = 48°$
(f) $p = 65°, q = 73°$
(g) $x = 47°, y = 65°, z = 112°$
(h) $x = 77°, y = 39°$
6. 45°
7. $(x - 35) + (x - 25) + \left(\frac{1}{2}x - 10\right) = 180; x = 100$
8. 104°, 76°

## Exercise 15b

1. (a) $x = 36°, y = 36°$
(b) $x = 68°, y = 73°$
(c) $x = 54°, y = 36°$
2. (a) $x = 20°, y = 80°$
(b) $x = 26°, y = 48°$
(c) $x = 60°, y = 65°$
3. (a) $x = 40°, y = 58°$
(b) $x = 114°, y = 114°$
(c) $x = 65°, y = 46°$
4. (a) $x = 33°, y = 114°$
(b) $x = 104°, y = 38°$
(c) $x = 42°, y = 48°$
5. (a) $x = 37°, y = 127°$
(b) $x = 67.5°, y = 22.5°$
(c) $x = 22.5°, y = 108°$
6. (a) 59° (b) 31°
7. (a) 110° (b) 28°
8. (a) 72° (b) 36°
(c) 54°
9. (a) 58° (b) 96°
10. (a) 31° (b) 97°

### Exercise 15c

**1.** **(a)** 1440° **(b)** 1800° **(c)** 2340° **(d)** 2880°
**2.** **(a)** 108° **(b)** 162° **(c)** 165° **(d)** 170°
**3.** **(a)** 9 **(b)** 18 **(c)** 20 **(d)** 36
**4.** **(a)** 8 **(b)** 10 **(c)** 12 **(d)** 30
**5.** **(a)** 110° **(b)** 66° **(c)** 86° **(d)** 104° **(e)** 36° **(f)** 30°
**6.** 20°; 60°, 80°, 100°, 120°
**7.** $y = 30$; 90°, 60°, 30°
**8.** **(a)** 54° **(b)** 144°
**9.** 8
**10.** **(a)** 108° **(b)** 36° **(c)** 72°
**11.** **(a)** 135° **(b)** 112.5° **(c)** 45°
**12.** **(a)** 162° **(b)** 153° **(c)** 153°

### Review Questions 15

**1.** **(a)** $x = 40°, y = 70°$
**(b)** $x = 39°, y = 63°$
**(c)** $x = 66°, y = 66°$
**(d)** 22.5°
**(e)** $x = 244°, y = 26°$
**(f)** $x = 39°, y = 63°$
**(g)** $x = 56°, y = 31°$
**(h)** $x = 40°, y = 32°$
**(i)** $x = 122°, y = 58°, z = 64°$
**(j)** $x = 52°, y = 31°, z = 97°$
**2.** **(a)** 60°
**(b)** $x = 70°, y = 20°$
**(c)** 15°
**(d)** $x = 40°, y = 54°, z = 27°$
**3.** **(a)** $x = 102°, y = 17\frac{1}{3}°$
**(b)** $x = 35°$
**(c)** $x = 30°, y = 90°$
**(d)** $x = 50°$
**4.** $n = 18$ **5.** 135°, 90°
**6.** 64 **7.** 15
**8.** 14 **9.** 13
**10.** **(a)** 180° **(b)** 360°
**11.** 150°
**12.** **(a)** 108° **(b)** 30° **(c)** 90° **(d)** 72° **(e)** 24°

### Exercise 16a

**1.** 63° **2.** 62°
**3.** 71° **4.** 6.4 cm
**5.** 3.5 cm **6.** 8.0 cm
**7.** 34° **8.** 12.2 cm
**9.** 12.3 cm **10.** 7.9 cm, 41°
**11.** 6.9 cm, 3.9 cm

### Exercise 16b

**2.** $a = 44°, b = 53°$, 5.6 cm, 7.0 cm
**3.** 16.9 cm **4.** 11.6 cm
**5.** 9 cm
**6.** 6.4 cm, 10.1 cm
**7.** 128 mm, 82°
**8.** 97 mm, 98 mm
**9.** 1.7 cm **10.** 7.0 cm
**11.** 133° **12.** 14.4 cm
**13.** **(d)** 7.1 cm
**14.** **(b)** 5.5 cm
**15.** **(c)** 8.8 cm, 4.2 cm
**16.** **(c)** 4.5 cm
**17.** **(c)** 6.4 cm
**19.** **(b)** 7 cm, 50°
**20.** **(b)** 12.8 cm
**21.** **(a)** 14.9 cm **(c)** 32°

### Review Questions 16

**1.** **(c)** 6.5 cm **(d)** 46°
**2.** 11.8 cm, 7.1 cm
**3.** **(c)** 5.9 cm
**4.** **(e)** rhombus
**5.** **(e)** 4.1 cm
**6.** **(e)** 1.9 cm
**7.** **(d)** 8.7 cm **(e)** 10 cm

### Revision Exercise IV No. 1

**1.** **(a)** 5.2 cm **(b)** 2.7 cm
**2.** **(a)** 1 h 32 min
**(b)** $x = 137°, y = 81°, z = 99°$
**3.** **(a)** $6 **(b)** 96 km, 2 h
**5.** **(a)** 22.5 m **(b)** 6.9 m
**6.** **(a)** 12 **(b)** $54
**7.** **(a)** $11\frac{1}{2}$ **(b)** $4\frac{3}{7}$
**8.** **(a)** $14x - 30$ **(b)** $\frac{17 - 8x}{20}$
**9.** **(a)** 29° **(b)** 119°
**10.** **(a)** $x = 36°, y = 80°$
**(b)** 6 of each

### Revision Exercise IV No. 2

**1.** **(a)** 7.5 cm **(b)** 8.9 cm
**2.** **(a)** 14 sec **(b)** 64 yrs, 16 yrs
**3.** **(a)** 52° **(b)** 116°
**4.** **(a)** 25° **(b)** 74°
**5.** **(a)** $x = 32°, y = 120°$
**(b)** $x = 77°, y = 111°$
**6.** **(a)** 8.2 cm
**7.** **(a)** 11
**(b)** **(i)** 84.375%
**(ii)** 84.4%
**8.** $p = 54°, q = 113°, r = 59°, s = 121°$
**9.** **(a)** 308 $cm^2$ **(b)** 947
**10.** **(a)** 600 **(b)** 4 : 5

### Revision Exercise IV No. 3

**1.** 136 mm
**2.** 11.5 cm, 141°
**3.** **(a)** $p = 83°$ **(b)** $q = 104°$
**(c)** $r = 32°$
**4.** **(a)** 30 h **(b)** 120 days
**5.** 11.3 cm
**6.** **(a)** $x = 55°, y = 82°, z = 38°$
**(b)** 2 : 7
**7.** **(a)** $5\frac{1}{2}$ **(b)** 4 **(c)** 2
**8.** **(a)** $tv$ km, $\frac{s}{v}$ h
**(b)** $17.10
**9.** **(a)** $x = 69°, y = 78°, z = 102°$
**(b)** 11.6 cm
**(c)** $x = 41°, y = 49°, z = 18°$
**10.** **(a)** 19, 21, 23
**(b)** 5 yrs, 35 yrs
**(c)** $9.80
**(b)** $31.80

### Revision Exercise IV No. 4

**1.** **(a)** 8.6 cm, 14.2 cm
**(b)** 7.5 cm
**2.** $x = 70°, y = 60°$
**3.** **(a)** 6 days **(b)** 24%
**4.** 5 cm
**5.** **(a)** 52°
**(b)** 13.1 cm, 9.2 cm
**6.** **(a)** $6y - 7x$ **(b)** $\frac{130 - 7x}{60}$
**7.** **(a)** 68° **(b)** 24°
**8.** **(a)** $\frac{3x}{t}$ km **(b)** $\frac{xy}{t}$ km
**9.** **(a)** 22.5° **(b)** 20.5°

### Revision Exercise IV No. 5

1. 78°
3. 21.5%
4. (a) 25 (b) 81 $cm^2$
5. (a) $x = 68°$, $y = 112°$
   (b) $x = 24°$, $y = 96°$, $z = 48°$
6. (a) 0.74% (b) $\frac{2}{9}$
7. (a) $5v$ km (b) $\frac{s}{v}$ h
8. (a) \$$\frac{c}{100}$ (b) $\frac{m}{1000}$
   (c) $1000k$
9. (a) 15° (b) 8.7 cm, 76°
10. (a) 2 (b) 4860 $l$